The sweet potato is at present grown in more than 100 of the independent countries of the world. Most of the producer nations are situated in the tropical developing world where a high proportion of the poorest people live. Increasing recognition of the exciting potential which sweet potato holds for combating food shortages and malnutrition has resulted in intensified research efforts to enhance production and consumption. This book reviews the current knowledge about the varied aspects of the sweet potato as a human food and animal feedstuff.

Sweet potato: an untapped food resource

SWEET POTATO

an untapped food resource

JENNIFER A. WOOLFE

Published in collaboration with the International Potato Center, Peru

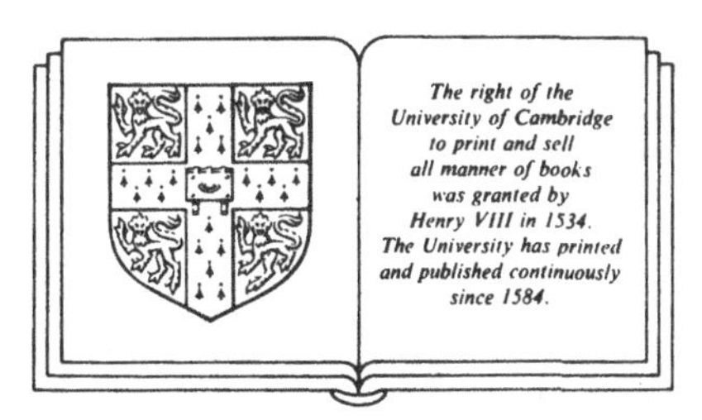

CAMBRIDGE UNIVERSITY PRESS
Cambridge
New York Port Chester Melbourne Sydney

CAMBRIDGE UNIVERSITY PRESS
Cambridge, New York, Melbourne, Madrid, Cape Town, Singapore, São Paulo

Cambridge University Press
The Edinburgh Building, Cambridge CB2 8RU, UK

Published in the United States of America by Cambridge University Press, New York

www.cambridge.org
Information on this title: www.cambridge.org/9780521402958

First published 1992
This digitally printed version 2008

A catalogue record for this publication is available from the British Library

Library of Congress Cataloguing in Publication data
Woolfe, Jennifer A.
Sweet potato : an untapped food resource / Jennifer A. Woolfe.
p. cm.
Includes bibliographical references and index.
ISBN 0 521 40295 6 (hardcover)
1. Sweet potato products. I. Title.
TP444.S94W66 1991
664′.80522 – dc20 91-10460 CIP

ISBN 978-0-521-40295-8 hardback
ISBN 978-0-521-05053-1 paperback

'When you have rice to eat, don't forget sweet potatoes.'
(Chinese saying)

CONTENTS

To the one
who always believes in me –
my husband and best friend

ACKNOWLEDGEMENTS

I owe a particular debt of gratitude to those who have read my manuscript so painstakingly and have contributed with immensely helpful comments and suggestions. My thanks are especially due to Professor Stanley Kays, who reviewed almost the entire manuscript. I am also extremely grateful, for reading of individual chapters, to the following: Dr J. Howard Bradbury, Dr Douglas Horton, Mr Dennis Morgan, Professor Ikuzo Uritani and Ms Erica Wheeler.

My thanks go to the many scientists all over the world with whom I have corresponded, and who supplied me with information, translations and encouragement. Outstanding among these is Professor Ikuzo Uritani, who has given so generously of his time and knowledge to me through correspondence and also during my visit to Japan. I would also especially like to thank the following people for the help they have so kindly given in various ways towards the preparation of this book: Ms Marisela Benavides, Dr Wanda Collins, Professor Barry Duell, Ms Jane Earland, Ms Paulette Foss, Professor Machiko Ono, Ms Linda Peterson, Dr Satoshi Sakamoto, Dr Truong Van Den and Dr Yoshiki Umemura.

This work was carried out through the auspices of the International Potato Center. I would finally like to thank the Center's Director, Dr Richard Sawyer, for giving me the chance to write this book and Dr Douglas Horton for his constant support and encouragement throughout its preparation.

ABBREVIATIONS FOR ORGANIZATIONS

ACIAR: Australian Centre for International Agricultural Research
AFRC: Agriculture and Food Research Council (UK)
AVRDC: Asian Vegetable Research and Development Center
CARICOM: Caribbean Community
CGIAR: Consultative Group for International Agricultural Research
CIAT: Centro Internacional de Agricultura Tropical (Colombia)
CIP: Centro Internacional de la Papa (International Potato Center)
CSIRO: Centre for Scientific and Industrial Research (Australia)
CSPWG: Caribbean Sweet Potato Working Group
DSIR: Department of Scientific and Industrial Research (New Zealand)
EEC: European Economic Community
EMBRAPA: Empresa Brasileira de Pesquisa Agropecuaria
ENCA: Encuesta Nacional de Consumo de Alimentos
FAO: Food and Agriculture Organization of the United Nations
FONAGRO: Fondo de Fomento Agropecuaria (Chincha, Peru)
HITAHR: Hawaii Institute of Tropical Agriculture and Human Resources
ICAR: Indian Council of Agricultural Research
IDRC: International Development Research Centre
IITA: International Institute of Tropical Agriculture (Nigeria)
ILRAD: International Laboratory for Research on Animal Diseases
INCAP: Institute of Nutrition of Central America and Panama
INIAA: Instituto Nacional de Investigacion Agraria y Agroindustrial

INTA: Instituto Nacional de Tecnologia Agropecuaria (Argentina)

PCARRD: Philippine Council for Agriculture, Forestry and Natural Resources Research and Development

UNDP: United Nations Development Programme

UNICEF: United Nations Children's Fund

UNU: United Nations University

USDA: United States Department of Agriculture

ViSCA: Visayas State College of Agriculture

WHO: World Health Organization

Introduction

The sweet potato, *Ipomoea batatas* L. (Lam.), is a dicotyledonous plant which belongs to the family Convolvulaceae. Amongst the approximately 50 genera and more than 1000 species of this family, only *I. batatas* is of major economic importance as a food (Figures I.1 and I.2). However, *Ipomoea aquatica* is also grown as a food plant in Malaysia and China, where it is eaten as a raw salad or a cooked green vegetable or used as animal fodder. The number of wild *Ipomoea* species is estimated at more than 400, but *I. batatas*, a 'cultigen', is not found in the wild; nor, so far, has a direct ancestor of this species been positively identified. The intervention of humans by first domestication and then artificial selections of the sweet potato, as well as the occurrence of natural hybridization and mutations, has resulted in the existence of a very large number of cultivars; there is more diversity in the sweet potato than in, for example, cassava, yam or cocoyam. Cultivars differ from one another in the colour of the root skin (white, cream, brown, yellow, red or purple), or flesh (white, cream, yellow, orange or reddish-purple), in the size and shape of the roots and leaves, in the depth of rooting, the time to maturity, the resistance to disease and in the texture of the cooked roots.

I. batatas is an extremely important crop in many parts of the world, being cultivated in more than 100 countries. As a world crop, it ranks seventh from the viewpoint of total production (see Table I.1). In monetary terms, it ranks thirteenth globally in the production value of agricultural commodities, and is fifth on the list of the developing countries' most valuable food crops (see Table I.2). The sweet potato's adaptation to, and hence presence in, the tropical areas where a high proportion of the world's poorest people live, together with its nutritional advantages, make it an attractive focus for further increase in its production and consumption, both directly (as fresh or processed food)

Figure I.1. Sweet potato seller, Central Java (G.A. Watson).

Table I.1. *World production of leading food crops, 1984*

Crops	Production (million tonnes)	Dry matter (million tonnes)	Edible portion: Energy (trillion kJ)	Edible portion: Protein (million tonnes)
Wheat	530	463	5526	53.5
Rice, paddy	478	421	4785	21.4
Maize	456	393	5760	35.8
Potatoes	317	64	804	5.4
Barley	175	155	1754	10.1
Cassava	131	53	461	0.5
Sweet potatoes	119	35	452	1.6
Soybeans	91	82	1515	31.2
Sorghum	73	65	946	7.6
Bananas and plantains	62	21	222	0.5
Tomatoes	60	4	46	0.6

Note:
Adapted from Horton, 1987; derived from FAO data.

Table I.2. *Value of production of ten principal food crops in all developing countries*

	Number of producing countries	Production (in 10^6 tonnes)	Producer price ($/tonne)	Value (in 10^9 $)
Rice	97	383	170	65
Wheat	69	162	148	24
Maize	119	154	119	18
Potatoes	95	91	142	13
Sweet potatoes	100	137	89	12
Cassava	95	127	70	9
Bananas and plantains	119	62	107	7
Sorghum	69	44	123	5
Groundnuts	92	17	297	5
Millet	53	27	144	4

Note:
From: FAO, 1987.

Table I.3. *Comparative energy yields of sweet potato and other major crops*

Crop	Average tropical yield (tonnes/ha)	Edible energy value (MJ/kg)	Proportion of edible energy (%)	Edible energy per ha (in 10^3 MJ)	Average crop growth period (days)	Edible energy (MJ/ha per day)
Roots/tubers and bananas[a]						
Sweet potato	7	4.8	88	27.2	140	194
Cassava	9	6.3	83	45.6	330	138
Yam	7	4.4	85	26.2	280	94
Banana	13	5.4	59	41.4	365	113
Cereals[a]						
Rice[b]	2	14.8	70	20.8	140	149
Maize	1	15.2	100	18.8	130	145
Sorghum	<1	14.9	90	11.1	110	101
Millet	<1	15.0	100	8.2	100	82

Notes:
Based on de Vries et al., 1967.
[a] Cereals, air-dry; roots/tubers/bananas fresh.
[b] Paddy rice.

and indirectly (as animal feed). It also has potential as a raw material for the manufacture of a wide range of industrial products.

As a crop the sweet potato combines a number of advantages which give it an exciting potential role in combating the food shortages and malnutrition that may increasingly occur as a result of population growth and pressure on land utilization. In many parts of the world, as population grows, fertile arable land available per head diminishes. This tends to create a shift to the use of marginal land in those densely populated areas where low incomes allow only modest investment in land improvement and crop production. In addition, the increasing flow of people away from the land and into already densely populated urban areas puts a heavy burden on the diminishing number of rural food producers. Therefore, those crops which yield the greatest amount of food per unit area per unit of time, and which are capable of yielding even in marginal conditions, will have to be accorded their rightful place in the world food production system. On a world scale, sweet potato provides significant amounts of energy and protein (Table I.1). Its production efficiency of edible energy and protein are outstanding in the developing world. Table I.3 shows that sweet potato heads the list of eight important developing-world crops in terms of the quantity of energy per hectare per day which they can produce. Sweet potato also has a higher energy efficiency ratio than the cereals when produced in a non-mechanized situation. The average energy output/input ratios for rice and sweet potato on Fijian farms were 17:1 and 60:1, respectively (Norman et al., 1984). Furthermore, even though the percentage of protein on a fresh weight basis is low in sweet potato and other roots and tubers, Table I.4 demonstrates that the average protein production per hectare from sweet potatoes, yams and bananas in the tropics is of the same order as that from the cereals, beans and chickpeas.

As previously mentioned, the sweet potato has a broad genetic base with a tremendous variability (greater even than that of the potato), which is increasingly being tapped by plant breeders searching for means of potential crop improvement. Yields of most crops are higher in temperate than in tropical areas, but sweet potatoes yield relatively better in the tropics than do some other major crops (Bouwkamp, 1985; see also Table I.4). Although yields are still disappointingly low in many countries, it has been shown that there is a tremendous potential for increasing yields by the introduction of improved clones and more efficient cultivation practices (see Table I.5). High yields have generally been demonstrated by experimental stations under what might be considered optimum conditions. Nevertheless, some nations have attained increases in country-wide yields under farm conditions (Figure I.3). China, for example, increased its average yield from 8 tonnes/ha in

Figure I.2. A store in Kyushu, Japan, selling a large number of different cultivars of fresh sweet potato roots and, in the background, sweet potato products (J.A. Woolfe).

1961 to 18 tonnes/ha in 1985 (Horton, 1988). Individual counties in Shandong, one of the major sweet potato growing provinces of China claim yields of up to 60 tonnes/ha (International Potato Center, 1987, p. 15). The International Institute of Tropical Agriculture (IITA) in 1972 produced 18 to 37 tonnes/ha during the dry season without the benefit of irrigation using selected clones. Sweet potato production lends itself to both high and low technology input agricultural systems. High yields have been attained in labour intensive, low technology systems such as those of China, as well as with high technology input levels in countries such as the United States and Japan.

The sweet potato possesses the following additional advantages (Bouwkamp, 1985). Sweet potatoes do not normally require high levels of input. Sweet potato vines grow very rapidly and cover the ground within a few weeks of planting, so that the necessity for weeding by either cultivation or the use of herbicides may be kept to a minimum. Applications of insecticides and fungicides are often unnecessary as, with the exception of the sweet potato weevil, most insect damage is cosmetic and not yield-reducing, and fungal diseases are not usually a problem in the growing crop. Sweet potatoes grow well in soils at pH 4.5 to 6.5, where it is therefore unnecessary to apply lime, unless the soil has

Table I.4. *Comparative protein yield of crops grown in the tropics*

Crop	Average tropical yield (tonnes/ha)	Protein (%)	Average protein yield (kg/ha)
Roots, tubers etc.			
Cassava	9	1.0	90
Sweet potato	7	1.6	110
Yams	7	2.0	140
Bananas	13	1.1	143
Legumes			
Soybean	>1	38.0	505
Groundnut	<1	25.5	217
Beans	<1	22.0	132
Chickpea	<1	20.0	132
Cereals			
Rice	2	7.5	151
Maize	>1	9.5	118
Sorghum	1	10.5	87
Millet	<1	10.5	58

Note:
Adapted from: Norman et al., 1984.

Table I.5. *Yields (tonnes/ha) obtained on experimental stations compared with national averages*

	Experimental yield	Farmers' yield (1979)	Yield gap	Possible improvement (%)
Tropical				
India	37	7	30	428
Philippines	35	5	30	600
Nigeria	32	13	19	146
Temperate				
Japan	35	20	15	75
Korea	43	20	23	115
USA	45	13	32	246

Note:
From: Kay, 1987.

Figure I.3. A progressive farmer who is anticipating a yield of 30–40 tonnes/ha after irrigating his sweet potato field in Henan, China (M. Iwanaga).

a heavy aluminium concentration. Sweet potatoes are very sensitive to aluminium toxicity and will die within about 6 weeks after planting in high aluminium soil if lime is not applied at planting (Beaufort-Murphy, H., personal communication). Sweet potatoes can grow and produce under relatively dry conditions. Hence supplementary irrigation may not be required.

There have been indications that nitrogen-fixing bacteria of the *Azospirillum* genus are found in association with the roots of sweet potato (Hill et al., 1983). Nitrogenase activity varies among different cultivars. Evaluation and enhancement of microbial associations promoting nitrogen-fixation in sweet potato roots could be an alternative to the use of expensive chemical fertilizers for increasing plant production.

The method of vegetative propagation, used by farmers, has several advantages. The use of vine cuttings for propagation precludes the need to set aside and store part of each year's crop for seed, or the expense of buying seed in from outside. There is not a high risk of disease being carried over from one year's planting to the next. Flowering and pollination, affected by adverse weather conditions leading to a reduction in yield, are unnecessary. Thus adverse weather rarely causes a total crop failure and many farmers plant sweet potato as an 'insurance crop' to fall back on for family food in case of emergency. Moreover, being a perennial crop planted as an annual, the sweet potato crop does not ripen or mature, thus the grower is often able to use the whole of a variable length growing season at a particular location. A cereal crop may not utilize all of an unusually long season, or may not ripen in an unnaturally short season. This characteristic of the sweet potato also makes it possible to produce up to two crops a year in some locations.

Finally, sweet potato yields two useful food types from the same plant, namely fleshy storage roots and green tops (vines consisting of stems and leaves). Both can be used as a nutritious food for human and animal feeding. In some parts of the world, sweet potato is grown almost exclusively for one or the other product, but in many countries both are utilized in a variety of ways. The sweet potato is, or can be, a tropical spinach or salad green, a staple or vegetable food, a sweet dessert, a variety of convenience processed products, a fast food (fries (chips)), a snack (chips (crisps)), a multi-purpose flour, an alcoholic or non-alcoholic drink, a starch, an animal feed or a basic industrial raw material (Figures I.4 and I.5). It has a number of outstanding nutritional characteristics (described in Chapter 3) which could make it a valuable tool for combating certain severe and widespread nutritional problems in the developing world.

In spite of the combination of desirable traits which the sweet potato possesses, its use has declined in many countries. The low status accorded both roots and vines due to their image as a subsistence crop, a 'poor man's food' or something to be eaten only in times of dire need such as famine or war may have been a limiting factor in their exploitation as foods of high nutritional quality. Pre- and postharvest losses, resulting in excessive waste, have increased prices to levels unattractive to those searching for a low cost nutritious substitute for more expensive, but prestigious, foods. In many areas, the lack of cultivars with characteristics catering to consumer preferences for colour, texture, flavour and low fibre levels, combined with the difficulties of handling and storage of a highly perishable commodity under tropical conditions of elevated temperatures and humidities, has frequently resulted in the sale of inferior quality sweet potatoes. The high levels of sweetness and the strong flavour associated with many cultivars may have reduced its popularity as a staple food, and make it difficult to combine with other foods in a variety of dishes. Sweet potato leaves and tips are often considered to be tough and too strongly flavoured in comparison with other green leafy vegetables. These factors have helped to reduce the esteem of the sweet potato in the eyes of the consumer. Furthermore there is little available as yet in developing countries in the way of new, tasty, interesting and nutritious processed forms of sweet potato, appropriate to local dietary preferences, which

TROPICAL
DELIGHTS
SWEET POTATO
BEVERAGE
SWEET POTATO
BEVERAGE
SWEET POTATO
BEVERAGE
Sweet Potato
JAM
Sweet Potato
JAM
KETCHUP
ISLA
CATSUP
MAYON
CATSUP

a

新口味
高筋麵
萬隆食品有限公司

b

Figure I.5. Sweet potato tops, here being sold in China as animal fodder, can also be eaten by humans as a salad or vegetable green (M. Iwanaga).

would help to promote and raise its status in the eyes of those who would benefit most from increased intakes. There is thus a pressing need to draw the attention of policy makers and extension workers to the nutritional advantages of increasing local dietary intakes, and to encourage researchers to tackle the outstanding problems associated with this aim.

Research into the improvement of sweet potato (along with that of other root and tuber crops) has been neglected in favour of the more prestigious cereals, plantation crops or other export cash crops. This has been at least partly attributed to cultural–historical factors leading to an identification of European cultures with grain and its products, and the consequent assumption of the inferiority of cultures based on non-grain-crops (Coursey, 1977). The bulkiness of sweet potatoes, their relatively low cash value per unit of weight, and difficulties associated with their storage and transportation in tropical conditions have resulted in a very low level of importance in international trade and the bulk of the crop is still used or sold for domestic purposes.

Figure I.4. (a) A wide variety of products can be made from sweet potato: Left to right: a 'dried fruit-like' product, a non-alcoholic beverage, jam and catsup, Philippines (ViSCA). (b) Sweet potato and wheat flour noodles, Taiwan (Chia-yi Agric. Expt. Stn).

Economic policies have adversely affected domestic food production in many developing countries. Protection of the farm sector in industrial nations has resulted in expanded grain supplies and depressed international prices. Faced with food scarcities, governments of many developing countries have chosen to subsidize the importation of food grains rather than invest in research, extension and related programmes in support of domestic agricultural production. The results have been to depress food crop prices in developing areas and to discourage local production of crops such as sweet potatoes.

As indicated above, agricultural research and development efforts have been concentrated on export crops and on the major temperate zone food crops. Domestic food crops in the tropics and sub-tropics have received less attention (Technical Advisory Committee, 1987). Successful agricultural programmes have greatly improved yields and postharvest systems for cereals and export crops, reducing their per-unit cost and stimulating their use. By the same token, other domestic food crops, such as sweet potato, that have not benefited from crop improvement programmes have become more expensive to produce and consume.

Organized sweet potato programmes with published results have been confined to a very few countries, principally China, Japan, Korea, Taiwan and the United States. Recently, two international centres (AVRDC, Taiwan, and IITA, Nigeria) have made much progress with sweet potato. Elsewhere little has been done and, even in the five countries listed above, research programmes have tended to be fragmented and discontinuous. The small base for sweet potato research is clearly evident from the FAO's Agris documentation system, where for every ten references on potato there is only one for sweet potato. Recently several countries have reduced their research and development efforts with sweet potato. The result is a further constriction of new information and technologies for sweet potato improvement. The sources of training for sweet potato researchers are also limited. In contrast to the situation of most of the world's major food crops, extremely few universities or research institutes provide training related to the sweet potato. While professional associations and journals exist for rice, wheat, maize, potatoes and other crops, not a single professional association or journal is dedicated to the sweet potato.

Basic reference materials are needed to provide scientists and policy makers with authoritative, up-to-date information on compositional, nutritional, quality, consumption and utilization aspects of the sweet potato. Although a great many technical research papers have been published on the various aspects of the sweet potato as a food, they are often not readily accessible to researchers and policy makers in developing countries. This publication seeks to remedy this state of affairs by synthesizing the available information in a concise and accurate review, which it is hoped will be a valuable source of reference for a broad group of agricultural professionals such as policy makers, researchers, and extension workers, as well as nutritionists, food and home economics scientists, dieticians, nutritional extension workers and all students in the fields of agriculture, food and nutrition. Although it is hoped that this volume will be of particular value to those in developing countries with the most pressing problems of food supplies and nutritional deficiencies, it should also be of interest to similar professionals in the developed world.

References

Bouwkamp, J.C. 1985. Introduction-Part 1. In: Bouwkamp, J.C. (ed.), *Sweet potato products: a natural resource for the tropics.* CRC Press, Inc., Boca Raton, FL.

Coursey, D.G. 1977. The status of root crops: a cultural-historical perspective. Regional meeting on the production of root crops, 1975, Suva, Fiji. *S. Pacific Comm. Tech. Paper* No. 174, pp. 125–30.

FAO 1987. Dossier. Roots and tubers. Their role in food security. *The Courier* No. 101, pp. 62–5.

Hill, W.A., Bacon-Hill, P., Crossman, S.M. and Stevens, C. 1983. Characterization of N_2–fixing bacteria associated with sweet potato roots. *Can. J. Microbiol.* **29** (8): 860–2.

Horton, D. 1987. *Potatoes: production, marketing, and programs for developing countries.* Westview Press, Boulder, CO.

1988. *Underground crops. Long term trends in production of roots and tubers.* Winrock International, Morrilton, AR.

International Potato Center, 1987. *Sweet potato research in the People's Republic of China*, a CIP/AVRDC/IFPRI study. International Potato Center, Lima.

Kay, D.E. (Revised by Gooding, E.G.B.) 1987. *Root crops*, Crop and Product Digest No. 2, 2nd edn. Tropical Development and Research Institute, London (now Natural Resources Institute, Chatham Maritime).

Norman, M.J.T., Pearson, C.J. and Searle, P.G.E. 1984. *The ecology of tropical food crops.* Cambridge University Press, Cambridge.

Technical Advisory Committee. 1987. *CGIAR priorities and future strategies.* FAO, Rome.

Vries, C.A. de, Ferwerda, J.D. and Flack, M. 1967. Choice of food crops in relation to actual and potential production in the tropics. *Netherlands J. Agric. Sci.* **15**: 241–8.

CHAPTER 1

Sweet potato – past and present

Origin and history

The brief discussion which follows, dealing with the current knowledge of this fascinating topic, draws heavily on the review of O'Brien (1972), whose conclusions are in turn based on an assessment of the evidence and opinions of others; also included are some more recent findings summarized by Yen (1982). The exact centre of origin of the sweet potato, and the routes and times for its dispersal to some of its present locations, are still in dispute and are likely to remain so until further archaeological or ethnobotanical evidence is available.

All archaeological, linguistic and historical evidence so far establishes the origin of the sweet potato as the New World, in either the Central or South American lowlands. The oldest remains so far discovered of dried roots are those from the caves of the Chilca Canyon of Peru (Engel, 1970), which have been radiocarbon dated at 8000–10,000 years old. However, it is not certain whether these had been domesticated, or simply collected from wild plants. Additional archaeological evidence comes from the discovery of actual remains of cultivated sweet potato from the Casma valley of Peru, dated at approximately 2000 B.C. (Ugent, Pozorski and Pozorski, 1983), and linguistic evidence pointing to the presence of sweet potato in the Mayan area of Central America between 2600 and 1000 B.C. It has been suggested (Austin, 1977) that sweet potato originated in a region bounded by the Yucatan Peninsula to the north and the Orinoco river to the south, with secondary centres of high diversity in Guatemala and southern Peru. Other work on variation in *I. batatas* (Yen, 1982) indicated maximum diversity in the area comprising Colombia, Equador and northern Peru.

Whether it originated from Central or South America, the sweet potato was already widely established in the New World by the time

Europeans first arrived. From there it spread in two waves of dispersal. The best known and documented was the post-Columbian spread by Europeans. The other, and actually earlier of the two, details of which are still in dispute by scholars, is the prehistoric appearance of the crop in Polynesia.

Columbus apparently discovered the sweet potato in Hispaniola and Cuba during his first voyage in 1492 and, on his return, introduced it into Spain, whence it spread to the rest of Europe. There it was first known as *batata* or *padada*, later being called the Spanish potato or sweet potato to avoid confusion with the Irish potato (*Solanum tuberosum* L.), which reached Europe 80 years later than *I. batatas*. There is no evidence to suggest that the sweet potato was growing in North America before European settlement, a surprising finding in view of its pre-Columbian introduction to Polynesia.

Further spread of the sweet potato in historic times was by two lines of transmission: (a) the *batatas* line, which followed on from the Spanish introduction into Europe, continuing after 1500 A.D. by the transfer of European-grown clones to Africa, India and the East Indies through Portuguese exploration; and (b) the *kamote* line whereby Mexican clones were carried to the Philippines by Spanish trading galleons (Yen, 1982).

From the wide distribution of sweet potato names such as *batata*, *tata*, *mbatata* and so on throughout the African continent, it was concluded (Conklin, 1963) that the sweet potato was introduced into Africa by the Portuguese from the Atlantic coast regions of mid-latitude America. O'Brien, however, reviews evidence to suggest that it was introduced via the Portuguese to Mozambique and possibly Angola in the sixteenth century directly from Portugal; other common names such as *bombe*, *bambai*, *bambaira* etc., associated with the Indian trading city of Bombay, acquired by the British in 1662, may be linked to a later spread of the plant in the seventeenth to nineteenth centuries by British colonial influences.

Sweet potato was apparently brought to one part of China from the Philippines and appeared in Fukien by 1594. However, an observation (Ho, 1955) that it was found in a region close to the Burmese border by 1563 suggests the possibility, therefore, that the earliest introduction was overland from India or Burma.

An early English introduction into Japan in 1615 was unpopular and subsequently lost. The sweet potato was reintroduced from China in about 1674 and, as had been the case in China, was used to combat famine. The Spanish brought sweet potato, along with its Mexican Nahuatl name *camote* (*camotl*) to the Philippines after their arrival in 1521.

Apparently, the Portuguese spread the Carib Arawak name *batata* and the Spanish the Nahuatl name *camote* (O'Brien, 1972). Hence, in India and Southeast Asia it is possible to trace the source of sweet potato introduction from its local names. For example, in Malaysia it is called Spanish tuber, in the Philippines *camote*, in Guam both *camote* and *batat*, and in Ambon, Timor and the Northern Moluccas varieties of the word *batata*.

Although it was originally held by some that the Spanish also introduced the sweet potato into Polynesia, strong evidence was presented in 1932 by Dixon that established its presence there before the 1521 round-the-world voyage by Magellan. The first explorers to land on New Zealand, Hawaii and Easter Island, the most distantly separated points of the triangle which defines the limits of Polynesia, reported that sweet potato was the basic food plant grown, and extremely important economically in all three locations. The sweet potato's pre-Magellanic introduction into Polynesia now seems beyond doubt. There are still many conflicting opinions, however, as to exactly how the sweet potato arrived there and to what extent it spread to the west (into Melanesia and New Guinea) without European influences.

The linguistic links between the South American and Polynesian names for sweet potato, *kumara* and its variations, suggest human transfer, either by Peruvian or Polynesian voyagers. Heyerdahl's famous *Kon Tiki* theory of a Peruvian origin for the population of Polynesia, has not been confirmed by archaeological evidence. The latter in fact seems to suggest the opposite view that the islands were settled from the west (Asia) at various times in prehistory and that there was no Amerindian input into Polynesian culture. Thus it is possible that Polynesians made the return voyage to the New World taking the sweet potato back with them. O'Brien cites evidence, however, to suggest that the word *kumara* was not used in prehistoric times by the inhabitants of the coasts of Equador, Peru or Chile and that the Polynesian name could have been introduced into South America by the Spanish, who picked it up in the Pacific during their trading voyages in the colonial period.

O'Brien goes on to give two alternative possible explanations for the prehistoric presence of the sweet potato in Polynesia. The first concerns human agency, but as the result of accidental rather than purposeful introduction; for example a ship-wrecked vessel which had accidentally drifted from the New World, carrying sweet potato, was cast up somewhere in Polynesia. For some reason the crew did not survive but the sweet potato did, either as a result of direct adoption, or by reverting to a feral form which was later redomesticated. The other alternative might have been the carriage of sweet potato seeds by birds either in their intestines or in mud on their feet. At present neither alternative can be proved, but O'Brien feels that such accidental introductions explain the early arrival of the plant and also account for the absence of other cultural traits (certain animals, plants and types of pottery) which might have been expected to occur if any sustained exchange between living persons had taken place. Purseglove (1974) sees no reason to invoke the aid of man at all in the transference of the sweet potato from the New World to Polynesia. It has been shown that sweet potato capsules float in water and that the seeds, which have an almost impervious testa, can germinate after immersion in sea water. Purseglove therefore suggests that sweet potato capsules were carried by currents from South America to Polynesia or, alternatively, that a fruiting plant of the twining type could have floated across attached to a branch or log.

By whichever means the sweet potato arrived and was put into cultivation, it has been proposed that the plant spread with the dispersal of Polynesian peoples from the region of Samoa at about the time of the birth of Christ to the Marquesas Islands and thence to the furthest reaches of Polynesia. Both Yen and O'Brien feel that there is some evidence of a prehistoric penetration of sweet potato from Polynesia into Melanesia.

The arrival time and source of the sweet potato in New Guinea, where it is of great importance, is also subject to contradictory theories. O'Brien reviews these, summarizing them as follows: the sweet potato was introduced from Indonesia, from the Malayan Archipelago, from the Solomon Islands, from Polynesia; via Europeans, via Melanesians, via Polynesians; the plant is an ancient cultural element, is a recent cultural element; the sweet potato expanded populations, displaced populations, did not affect populations etc. A hypothesis propounded by Golson in 1975 (cited by Yen, 1982), based on evidence of ancient forest clearance, is that sweet potato had already reached the New Guinea Highlands as long as 1200 years ago. This early existence of sweet potato in New Guinea has yet to be proved and is treated with caution even by Golson (1976). Yen feels that it would be difficult to account for the early presence of the plant in New Guinea if there had not been a prehistoric introduction of it from Polynesia into Melanesia. O'Brien also favours a Melanesian agency for the transport of the sweet potato to New Guinea but refuses to date its arrival until much more archaeological evidence is available.

The sweet potato in developing countries

Knowledge of historical trends and of the current status of global sweet potato production and utilization is very limited. The best source of statistics on global root crop production and use is the Basic Data Unit of the FAO. Unfortunately, however, available statistics are not very reliable, particularly in developing countries where many sweet potatoes are grown for on-farm use in isolated areas on small, irreguarly shaped or intercropped parcels of land. A review of the available literature and on-going field studies conducted by the International Potato Center and by national researchers indicate that official, national and FAO statistics generally underestimate sweet potato production. Despite these limitations, the FAO statistics are useful for characterizing broad regional patterns and trends of sweet potato production.

Production trends

The sweet potato crop, which has its origin in the tropical Americas (see pp. 15–16), has spread to most of the world's tropical, sub-tropical and warmer temperate regions. According to the FAO, sweet potatoes are grown in 111 countries, of which 101 are classified as 'developing nations'. Among the world's root crops the sweet potato ranks second only to the potato in economic importance (Horton, 1988a, p.10).

Developing countries produce and consume nearly all of the world's sweet potatoes. Approximately 90% are grown in Asia, just under 5% in Africa and only about 5% in all the rest of the world (Table 1.1). Only about 2% of the world's sweet potatoes are grown in industrialized countries, mainly in the United States and Japan. With an annual harvest of nearly 100 million tonnes, China is the world's largest producer of sweet potatoes. In China, sweet potato yields are double those of the rest of the world, and production per head is 20 times that of other countries. Indonesia, Uganda and Vietnam, which follow China in production, each harvest about 2 million tonnes of sweet potato roots annually. Several small countries, including the Solomon Islands, Tonga, Rwanda, Burundi and some of the Caribbean islands, have high levels of per capita production, with sweet potato playing an important economic and dietary role (Table 1.2).

According to FAO estimates, world sweet potato production increased by 50% from 1961 to 1973 and then declined to a level about 15% higher than that of the early 1960s. Over the last quarter of a century, production has fallen sharply in Japan, the United States, and other industrialized countries. In Latin America, sweet potato production rose in the 1960s and then fell to about 80% of its initial level. Sweet potato production in Asia, dominated by China's figures, followed a similar but less pronounced trend; production now stands approximately 25% above the early 1960 level. The only world region in which sweet potato production increased throughout the period is Africa, where it is now more than 80% above its 1960 level (Figure 1.1).

Table 1.1. *World sweet potato production and changes since the early 1960s*

	1983/1985 average				
	Production (in 10^3 tonnes)	Yield (tonnes/ha)	Harvested area (in 10^3 ha)	Production per capita (kg)	Change in production (1961/63–1983/85) (%)
World	114,185	14	7,998	24	13
Asia	104,603	16	6,413	38	12
(China)	(93,550)	(18)	(5,067)	(91)	23
Africa	6,100	6	1,094	11	78
North and Central America	1,442	7	213	4	10
South America	1,371	9	153	5	−37
Oceania	560	5	116	23	52
Europe	108	11	10	0	−44
Developing countries	111,979	14	7,867	32	20
Developed countries	2,206	17	131	2	−70

Note:
From: CIP Root Crop Data Bank. Derived from FAO Basic Data Unit (unpublished).

Table 1.2. *Countries with highest sweet potato production and production per capita 1983/85 average*

Production		Production per capita	
Country	(in 10^3 tonnes)	Country	(kg)
China	93,550	Solomon Islands	193
Indonesia	2,142	Tonga	161
Uganda	1,867	Rwanda	150
Vietnam	1,867	Papua New Guinea	136
India	1,565	Uganda	125
Japan	1,435	Burundi	112
Korea Republic	898	Niue	100
Rwanda	879	China	91
Philippines	865	Equatorial Guinea	90
Brasil	732	Cook Islands	75

Note:
From: CIP Root Crop Data Bank. Derived from FAO Basic Data Unit (unpublished).

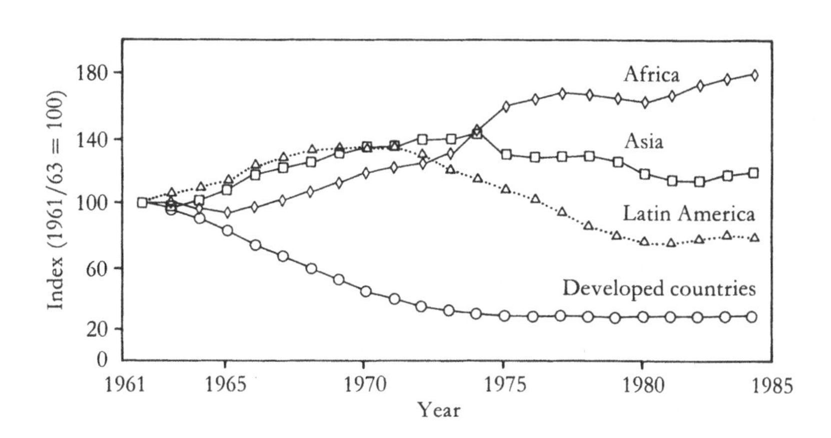

Figure 1.1. Sweet potato production trends in developing countries of Africa, Asia and Latin America and in developed countries (3-year moving averages: from Horton, 1988b; by permission of the publishers, Butterworth & Co. (Publishers) Ltd ©.

Figure 1.2. Sweet potato field in Uganda (M. Iwanaga).

Food systems

Available publications and early results of field studies being conducted by the International Potato Center and national researchers indicate that the sweet potato has a broad ecological and agroeconomic adaptability, which is related to the genetic diversity within the cultivated species (International Potato Center, 1989). The crop is grown from 35°N to 35°S and from sea level to almost 3000 m above sea level. In South America, it is grown in the Andes mountains, in the Amazon jungle, on the great sub-tropical and temperate plains of the southern cone, and, under irrigation, in the desert on the Pacific coast. In the Caribbean and the Pacific it is grown on small tropical islands, in Africa at mid-elevations and in parts of the tropical lowlands (Figure 1.2), and in Asia at a wide range of altitudes and from temperate to tropical zones. Many of the environments where sweet potato grows have poor, degraded soils that support few other crops.

Sweet potato plays many roles in diverse food systems across the world. Perhaps the most 'typical' system involves small-scale sweet potato production, primarily for household consumption and secondarily for livestock feed and sale. Intercropping is a common feature (Figure 1.3). While globally this is perhaps the most common system, in many areas sweet potato is important as a cash crop, livestock feed, or for industrial input.

Figure 1.3. Sweet potato intercropped with maize in Central Java, Indonesia (G.A. Watson).

Field studies in Asia, Africa, and Latin America show a diversity of sweet potato production practices and final uses in different kinds of food system. For example on the island of Java, sweet potato is intensively cultivated as a cash crop by farmers who are linked to a well-organized marketing chain supplying the major cities for fresh consumption or food processing into snacks. In the neighbouring island of Sumatra, sweet potato is grown as an off-season staple food between rice crops (Watson, 1988, 1989).

In Rwanda, where sweet potato is one of the major staples (see Chapter 8), the intensity of production and the importance of marketing vary according to topography and altitude. In Kenya, where there are very large urban populations, large-scale commercial sweet potato production occurs. One example is the coastal strip that supplies Mombasa, where both fresh roots and processed sweet potato snacks are consumed (Ewell, P., unpublished material).

On the coast of Peru, where desert conditions restrict agriculture to irrigated valleys with little or no pasture land, sweet potato foliage is the major source of animal feed (see Chapter 8). Cultivars are selected by farmers for their two qualities: good root production and abundant foliage.

On the southern plains of the South American continent, in Argentina and Uruguay, sweet potato is found both as a small kitchen garden crop and in extensions of 200 ha or more. Most production here is for household use or for sale to the large cities. Buenos Aires annually consumes about 75,000 tonnes of fresh sweet potato. Despite high levels of production, only a relatively small percentage of production in Argentina and Uruguay is processed for making a sweet cake eaten for dessert.

The most sophisticated and diversified system of postharvest utilization occurs in China (see Chapter 8). Sweet potato is grown for its fresh roots or greens for human consumption and animal feed, for many food products including snacks and noodles and for use as an industrial raw material which is the starting point for a variety of food and non-food items, derived from starch and alcohol.

General description of the plant and its cultivation

Major parts of the plant

The sweet potato plant can be divided into three basic parts, each of which has its own function. Above ground, the photosynthetic canopy absorbs light energy and converts it to a manageable chemical form (carbon compounds); the petioles and vines transport this energy, and the resources acquired by the root system, from one site to another within the plant. Below ground, the root system absorbs water and nutrients and acts as an anchor for the plant. It also stores excess energy, i.e. that not needed for maintenance or structural development, in the form of carbohydrate in large fleshy or storage roots. It should be noted from the outset that the storage organ of the sweet potato is a root and not a tuber. This fact seems to be ignored by many specialists who write about the sweet potato. Storage roots are true roots whereas tubers are modified stems, and as such, differ from roots in both anatomy and physiology. The sweet potato storage root therefore undergoes enlargement or development and not 'tuberization'. Each of the three major parts of the plant and its functions are closely related, changes in one often resulting in rapid and significant changes in the others.

Reproduction and propagation

The sweet potato can reproduce by three means. Asexually it reproduces and colonizes an area by production of storage roots which subsequently sprout to give new plants, or it can allocate its major resources to producing vines which may form roots at the nodes, producing daughter plants. Of least importance in the numerical sense is sexual reproduction in the form of seed, very little energy being allocated to this. Flowers are borne singly or on inflorescences that grow vertically upward from the leaf axils. They are trumpet-shaped and characteristic of the morning glory family. A mature flower opens before dawn, stays open only a few hours and closes and wilts before noon on the same day. Pollination is by insects. As flowers are open and receptive for a very short time, the chances of failure of pollination are quite high. Incompatibility complexes (for example self-incompatibility and occasional cross-incompatibility) exist, which restrict the chances that pollination will result in fertilization and seed production. Variations in stamen height with respect to the style also complicate the pollination mechanism. These features make seed production difficult. The sweet potato fruit is a capsule 5–8 mm in diameter. It usually contains only one or two black 3 mm long seeds in its chambers. The seed testa or outer coat is very hard and almost impervious to water and oxygen; hence seeds germinate only with difficulty. For all these reasons, the asexual or vegetative method of propagation is used by growers in the production of storage roots and/or foliage, the sexual method being used almost exclusively by plant breeders at present, for the development of new or improved cultivars. The possibility of the use of true seed for propagation purposes is only beginning to be explored. However, frequent germination of seed in the field has been noted in Papua New Guinea (Kimber, 1972), indicating that vegetative propagation of these chance seedlings by subsistence cultivators could lead to the appearance of new cultivars in a relatively short time.

Sweet potato is a perennial plant but is normally grown as an annual. It is usually propagated from vine cuttings in the tropics, but in temperate regions it may also be grown from rooted sprouts ('slips') pulled from bedded storage roots (Figures 1.4–1.6). Vine cuttings, not having roots and adhering soil have the advantage of not harbouring disease organisms which normally attack only below-ground parts.

Cuttings 30–45 cm long are taken from the apical growth of mature plants. The bottom leaves are removed and the lower half to two thirds of the cutting inserted into the soil at an angle. In some parts of East Africa, cuttings may be wilted for a few days before planting to encourage root initiation, but there is no scientific basis for this practice (Kay, 1987). In India, the central portion of a cutting is buried in the soil, leaving a node exposed at either end (Purseglove, 1974).

Sprouts are obtained by planting small or medium sized roots close together in nursery beds. When resulting sprouts reach 22–30 cm in length, they are removed by pulling them from the storage roots and planted in the field.

Cuttings and sprouts are planted on mounds or ridges (Figure 1.7), or on the flat. Growing on the flat should be discouraged unless soils are

Figure 1.4. Sweet potato vine beds for planting stock, next to rice terraces being ploughed and mounded for following sweet potato planting, Central Java, Indonesia (G.A. Watson).

Figure 1.5. Indonesian farmer harvesting sweet potato vines for planting (G.A. Watson).

Figure 1.6. Planting sweet potato vines on mounded ridges in West Java, Indonesia (G.A. Watson).

sufficiently friable and well drained to make mounds or ridges unnecessary, and deep cultivation is required so that roots have ample room for growth thus improving yields. Cultivation on mounds is extensively practised throughout the tropics, especially where the water table is high, as it increases drainage. It also increases the amount of top soil, but the formation of individual mounds is more suitable for manual than mechanized labour and is very time consuming. Mounds may be up to 60 cm high and are usually about 90–120 cm apart, with several cuttings planted on each mound. Ridging is suitable for mechanical preparation of the ground. Ridges are approximately 45 cm high and 90–120 cm apart with cuttings planted at 30 cm intervals.

Growth conditions

Sweet potatoes are grown from 40°N to 32°S of the equator. On the equator they are grown from sea level to 3000 m. The optimum conditions for growth and production have been described as follows (Onwueme, 1978): growth is best at or above 24°C; when temperatures fall below 10°C it is severely retarded. The crop is damaged by frost and this restricts its cultivation in temperate regions to areas with a minimum frost-free period of 4–6 months, with relatively high temperatures during this period. Sweet potato grows best where light intensity is

a

b

Figure 1.7. Sweet potato planted (a) on mounded ridges, Indonesia (G.A. Watson) and (b) on mounds/hills in Busia district, western Kenya to take advantage of loose soil and good drainage (P. Ewell).

relatively high, but at the same time both flowering and root formation are promoted by short day lengths such as those found in the tropics. Optimum rainfall is 75–100 cm per annum, with approximately 50 cm falling during the growing season. The rest of the rain, falling during the the non-growing season, enables propagation and maintenance of the vine growth to be used as planting material in the following season. Sweet potato prefers sandy-loam soils of a high organic matter content and with a permeable subsoil, doing poorly on clay soils. Good drainage is essential as the plants do not withstand water logging. Soils with a high bulk density or poor aeration retard storage root formation and result in poor yields. A soil pH of 5.6–6.6 is preferred, the plant being sensitive to alkaline or saline conditions.

The adaptability of the crop to a wide range of conditions, resulting from its broad genetic base, is illustrated by the case of Papua New Guinea (Bourke, 1985). Here the crop is grown from sea level to over 2700 m and in areas with a mean annual rainfall of 170 to more than 500 cm. It is cultivated on heavy clay and peat soils as well as sandy loams; on flatlands as well as slopes of up to 40°. Moreover, the sweet potato is very drought resistant. Vines remain green and healthy during severe droughts although root growth is negligible. In parts of East Africa, for example, where the dry season is very pronounced, the vines provide an important source of animal feed. Planting is also carried out in swampy areas or by rivers or seepage areas during the drought to ensure a supply of planting material for the following season (Jana, 1982).

Plant growth and development

After planting, the growth and development of the sweet potato plant comprises three more or less distinct though slightly overlapping phases. An initial phase is characterized by a slow growth of the vines, and a rapid growth of the adventitious roots which arise from the underground stem within a few days of planting. These roots may eventually penetrate the soil to a depth of 2 m or more, depending on the soil conditions. This relatively deep penetration enables the crop to survive drought conditions as it can absorb water from the deeper soil layers. However, yields are reduced if drought occurs within 6 weeks after planting or at the start of storage root formation. An intermediate phase consists of a rapid growth of the vines and hence a large increase in leaf area, accompanied by initiation of storage root development. In the final phase further growth of vines ceases and a rapid bulking of storage roots takes place. The respective durations of the three phases can vary with cultivar and environment. The first phase may last from planting until about 9.5 weeks later, the second from 9.5 to 16 weeks, and the

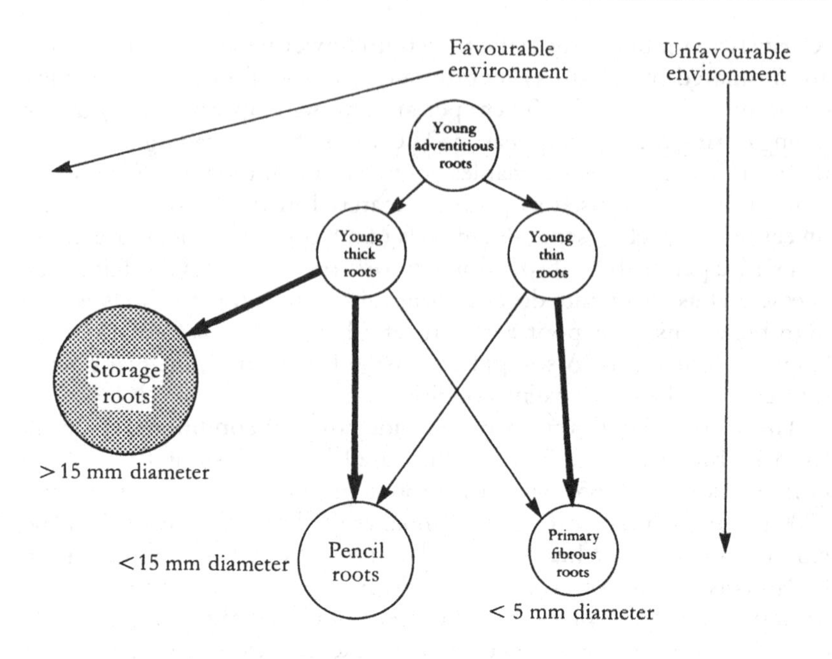

Figure 1.8. Development of adventitious roots into three main root types found in sweet potato (after Kays, 1985).

third phase for the rest of the season (Edmond and Ammerman, 1971). Under tropical conditions root initiation can start as early as 4 weeks after planting with most occurring 4–7 weeks after planting. The rest of the season is devoted mainly to root enlargement.

The following description of the root system and its development (Kays, 1985) is based on several studies. The adventitious roots arising from the underground stem portion of a vine cutting or transplant, or from a root piece are of three types: storage, primary fibrous and pencil (Figure 1.8). From any of these, lateral roots may arise. Young adventitious roots may be divided into two types: 'thin' and 'thick' roots. The former arise mainly from the internodal areas and are typically tetrarch in the arrangement of their primary vascular tissue, i.e. four xylem and phloem points are found within the vascular cylinder. The 'thick' roots, however, which arise from the nodes of the underground stem are pentarch or hexarch in structure, i.e. with five or six xylem points. The development of young adventitious roots into 'thin' or 'thick' roots is dependent on the environment both above and below ground. If the environment is favourable, young 'thick' roots subsequently develop into storage roots. Under unfavourable conditions of high N or low O_2 in the soil, these roots develop into primary fibrous

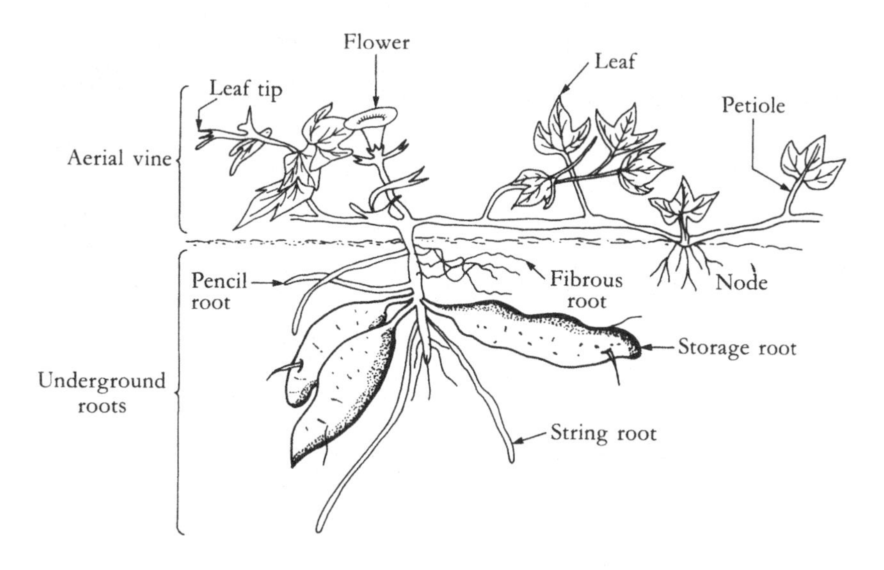

Figure 1.9. Gross morphology of the sweet potato plant. In practice, the proportion of foliage to root is somewhat greater than that shown here.

roots. Young 'thin' roots develop into either primary fibrous roots or, in some cases, pencil roots. Under dry compacted soil conditions, young 'thick' roots begin to enlarge; however, this is brought to a halt by rapid lignification which results in the formation of pencil roots, greater in diameter than fibrous roots but much smaller than storage roots. A stout stalk attaches the storage root to the rest of the plant and the root tapers thinly at the distal end. The gross morphology of the entire sweet potato plant is shown in Figure 1.9.

The storage root

The mature storage root ranges in shape from almost spherical to spindle-shaped, in length from a few centimetres to more than 30 cm and in weight from 0.1 kg to several kilograms. The external appearance of the harvested storage roots of various types of sweet potato is shown in Figures 1.10 and 1.11. The structure of the storage root in cross-section is illustrated in Figure 1.12. The principal tissues are the periderm or skin, the ring of secondary vascular bundles just under the periderm, and the tracheids, sieve tubes and laticifers (latex ducts) interspersed among a large quantity of storage parenchyma between the secondary xylem of the vascular ring and the centre of the root.

The surface of the root is covered by a thin layer of cork (part of the periderm) and may be smooth or irregularly ribbed. The skin, as well as

Figure 1.10. Storage roots of a sweet potato cultivar from the United States (W. Collins).

the flesh, contains carotenoid and anthocyanin pigments which determine its colour. The combinations and intensity of these two groups vary to produce skin and flesh of a whitish, yellow, orange, pink or purple colour, depending on the cultivar under consideration. Laticifers, which produce a sticky white latex when cut, are present throughout the flesh.

Figure 1.11. Many different cultivars for sale in Tawan Mangu, Central Java, Indonesia (G.A. Watson).

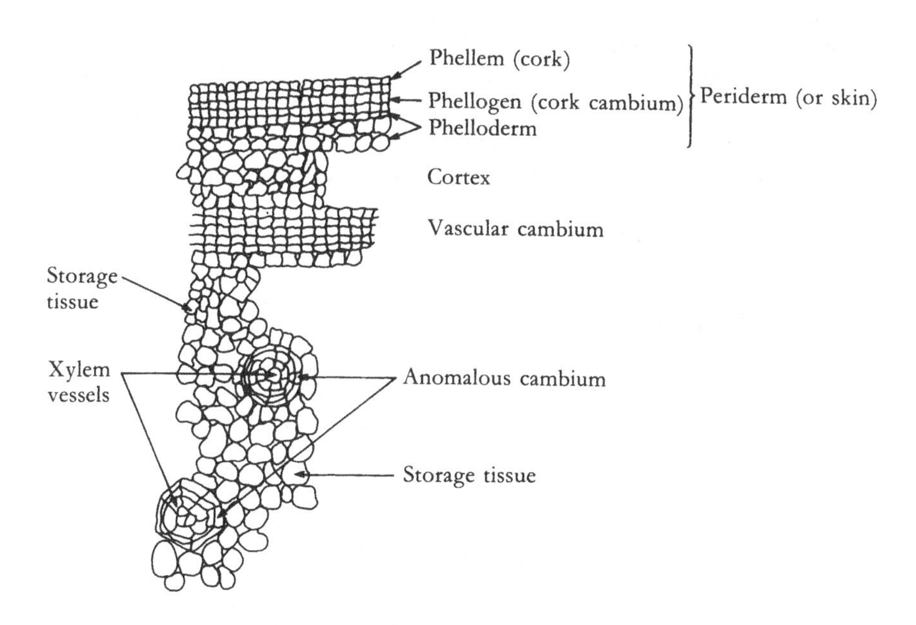

Figure 1.12. Cross-sectional structure of young sweet potato storage root (based on Edmond and Ammerman, 1971).

Storage roots may also be classified according to their texture when cooked, i.e. cultivars with firm or dry flesh, or cultivars with soft or moist flesh, although a complete gradient of textures exists between these two extremes. In the United States, the moist-fleshed types are frequently referred to as 'yams', an erroneous term, since true yams are botanically distinct from sweet potatoes and are not even closely related to them. The true yam is a member of the Dioscoreaceae whereas it will be remembered that the sweet potato is a member of the Convolvulaceae.

The vine

The above-ground parts of the plant, some of which may also be consumed by humans or animals, comprise: (a) the leaves, which absorb light energy converting it into carbohydrate through the fixation of atmospheric carbon; and (b) the leaf petioles and stems, which form the conduits for transport of this carbon throughout the plant. Sweet potato genotypes are predominantly prostrate vining plants (although they can be ascending and sometimes also twining) and in contrast to most agricultural plants they establish a relatively shallow and largely two-dimensional canopy. The sweet potato's long thin stems trail along the soil surface, sending roots into the soil at the nodes. The stem length varies with cultivar and may be approximately 0.5 m to more than 4 m. Some clones have stems, known as 'invaders', of up to 16 m in length. Such clones do not produce storage roots but may be employed for cattle food, or erosion control. The sweet potato stem is predominantly green but some lines also contain purple pigmentation.

The leaves are arranged spirally on the stem and have petioles varying from 5 to 30 cm in length. Leaves are variable in size and shape even within the same plant, but lines can be divided into two basic groups from the viewpoint of leaf shape: lines with deeply lobed leaves and lines with entire margins. Within these two basic groups exist a variety of leaf forms, some of which are illustrated in Figure 1.13. Leaves are usually green but may also contain a considerable amount of purple pigmentation, especially along the veins. In addition, leaves may be glabrous (smooth) or lightly, moderately or heavily hairy. Leaves and/or petioles or leaf tips about 10–15 cm in length are consumed in various parts of the world as a vegetable, the above-mentioned characteristics, as well as degree of tenderness, i.e. fibre content, influencing the cultivars grown for this purpose and their acceptability or otherwise.

Planting and harvesting

Planting of vine cuttings or sprouts is done by hand in much of the tropical developing world, although single- or multiple-row planters are

Figure 1.13. Sweet potato leaf shapes.

available. Many of these are designed to supply water or a nutrient solution to plants during placement in the field, thus providing the possibility of planting during a dry spell before the onset of the rains. The best time for planting is early in the rainy season, so that the whole of the rainy season can be utilized for growth. If the rainy season is very long, planting may be delayed or timed so that the approach of the harvest season coincides with decline in rainfall. Sweet potato is planted in the spring in temperate or subtropical regions as soon as the soil has warmed and any danger of frost is past.

The rapid establishment of a horizontal and relatively shallow canopy has the advantage of maximizing incoming radiation, although it has been claimed that when vines are tied up to a wire mesh, the light exposure of the canopy is increased and yields are improved. If left untied, which is usually the case, the vines' rapid growth completely covers the ground within about 6–8 weeks after planting, hence suppressing weed growth. For this reason traditional farmers may not weed the crop at all or carry out only a single weeding about 4 weeks after planting.

Sweet potato roots are ready for harvesting 3–8 months after planting, often much sooner than other root and tuber crops. In Hawaii, for example, growing periods were 150, 225 and 365 days for sweet potato, yam and cassava, respectively (Vander Zaag, 1979). In many parts of the humid tropics it is possible to grow two crops of sweet potato per year. The short growth cycle in some areas has other advantages; for example, in Fiji, it ensures rapid regeneration following disruptions of production occurring in the aftermath of hurricanes. Indications of increases in the extent of sweet potato cropping have been observed in some islands with a high incidence of hurricanes (Haynes, 1977). In drier areas, as well as in the temperate zones, only one crop per year can be produced. The exact time to harvest of the roots differs with environmental conditions and cultivar. If all roots are to be harvested at the same time the appropriate lifting time must be correctly judged. If lifting is carried out too early, yields are low. On the other hand, if roots are left too long in the ground they are increasingly prone to attack by the sweet potato weevil, and to various rots. In many traditional settings, sweet potato is in fact harvested as needed and there is no fixed harvest time. The fact that sweet potatoes can be left in the ground until needed is invaluable to subsistence farmers who use the crop only for home consumption. In parts of the world, for example Papua New Guinea, where temperature and rainfall are favourable to crop production all year round, it is common to grow the sweet potato as a perennial and harvest individual roots from the plant as necessary. In this case vines are earthed up to encourage production of storage roots at the nodes.

In most parts of the tropics harvesting is done by hand. In countries where sweet potato is grown on large farms or plantations, mechanical harvesters are used. These first remove the vines with cutters and then plough out the roots. The roots are then gathered up by hand, or mechanically sorted and loaded into trucks to be sent for further postharvest handling.

Curing and storage

In temperate climates, roots are commonly subjected, immediately after harvest, to a process known as curing. This treatment involves holding roots at 30–33°C and 85–95% relative humidity for 5–7 days and takes place in specially heated storage houses. One of the major purposes of curing is to promote wound healing through the formation of wound periderm. Since at the minimum wounds occur at the stem and root ends of the storage roots, wound healing is essential for holding or storage. In addition, curing reduces desiccation and invasion by pathogens causing storage rots. In those parts of the humid tropics where prevailing temperatures and humidities are very close to those recommended for curing, artificial curing is rarely carried out but can occur naturally. The roots should be allowed to cure for 4–20 days before storage depending upon environmental conditions. A traditional method of curing is typified by the Indian practice of spreading the roots in the sun for about one week, covering them with a waterproof protection during the night.

In countries such as the United States, where the sweet potato processing industry is well developed, curing is followed by storage in specially constructed cold stores at 13–16°C and 80–85% relative humidity. Storage in tropical countries is seldom practised, as roots are usually consumed immediately after harvest. However, traditional storage practices are in use in some areas (see Chapter 5); for example in Indonesia roots are stored in pits between layers of wood ash, earth and straw (Winaro, 1982).

Diseases and pests

Sweet potatoes are subject to a number of diseases including various fungal diseases which attack the root in the field and may be visible at harvest, or which remain quiescent in the field but develop during storage (see Chapter 5). Fungal diseases are not normally very serious in the tropics. Stem rot, *Fusarium oxysporum* Schlect. f. sp. *batatas* is destructive in the United States. A non-specialized form of *F. oxysporum*, which causes a disease called surface rot on storage roots often as a result of rough handling during harvest, is widespread in the tropics. Lesions on the periderm are circular and dark brown in colour. *F. solani*, also present throughout the tropics, produces similar symptoms on storage roots, except that dark brown rings are occasionally present on the surface of the lesion. Black rot, *Ceratocystis fimbriata* Ell. and Halst. has been reported to occur wherever sweet potatoes are grown. The organism causes a disease of the roots and underground stems of sweet potato plants, and can occur prior to harvest or during storage. A soft rot during storage is caused by *Rhizopus* spp. Infected root tissue becomes soft, moist and stringy.

Virus diseases may attack the root or the leaves. They include internal cork disease and mosaic virus. The identity of the viruses involved is not well established. Viruses are important in parts of East and West Africa, but not in the rest of the world.

Pest attack includes that of various nematodes and insects. *Meloidogyne* spp. (root-knot) and *Rotylenchulus reniformis* are the major known nematode pests of sweet potatoes in the tropics. They attack the fibrous roots and reduce yield. They also attack fleshy roots, reducing quality and causing wounds through which other pathogens can penetrate.

The chief insect pest of sweet potato is the sweet potato weevil, *Cylas* spp. *Cylas formicarius* Fab. is the major pest in most countries. Larvae and adults feed on the roots, causing extensive damage both in the field and in storage, in many parts of the world. The scarabee, *Euscepes postfasciatus* (Fairm.), is a serious pest in the drier parts of South America, the Caribbean and the Pacific. Larvae and adults feed on roots and stems of sweet potato. The larvae produce narrow tunnels in the roots and, like the sweet potato weevil, cause the roots to produce bitter, toxic terpenoid compounds making them inedible (see Chapter 4). The sweet potato stem borer *Omphisa anastomosalis* (Guernée) has been reported in India, Malaysia and China, where it is considered to be as destructive as the sweet potato weevil. Wild pigs, rats and other herbivores cause damage and loss in some countries.

The major diseases and insect pests to which sweet potato is susceptible have been fully reviewed (Edmond and Ammerman, 1971; Moyer, 1985; Schalk and Jones, 1985).

References

Austin, D.F. 1977. Another look at the origin of the sweet potato (*Ipomoea batatas*). Paper presented at the 18th Annual Meeting of the Society for Economic Botany, 11–15 June, University of Miami and Fairchild Tropical Garden.

Bourke, R.M. 1985. Sweet potato (*Ipomoea batatas*) production and research in Papua New Guinea. *Papua New Guinea J. Agric. Forest. Fish.* **33** (3–4): 89–108.

Conklin, H.C. 1963. The Oceanian-African hypotheses and the sweet potato. In:

Barrau, J. (ed.), *Plants and the migrations of Pacific peoples*. Bishop Museum Press, Honolulu, pp. 129–36.

Dixon, R.B. 1932. The problem of the sweet potato in Polynesia. *Am. Anthropol.* **34** (1): 40–66.

Edmond, J.B. and Ammerman, G.R. 1971. *Sweet potatoes: production, processing, marketing*. AVI Publ. Co. Inc., Westport, CT.

Engel, F. 1970. Exploration of the Chilca Canyon, Peru. *Curr. Anthropol.* **11** (1): 55–8.

Golson, J. 1976. The making of the New Guinea highlands. In: Winslow, J. (ed.), *The Melanesian environment*. Australian National University Press, Canberra, pp. 45–6.

Haynes, P.H. 1977. Root crop production and use in Fiji. Regional meeting on the production of root crops, 1975, Suva, Fiji. *S. Pacific Comm. Tech. Paper*, No. 174. pp. 44–50.

Ho, P-T. 1955. The introduction of American food plants into China. *Am. Anthropol.* **57** (2): 191–201.

Horton, D. 1988a. *Underground crops. Long-term trends in production of roots and tubers*. Winrock International Institute for Agricultural Development, Morrilton, AR.

1988b. World patterns and trends in sweet potato production. *Trop. Agric.* **65** (3): 268–70.

International Potato Center. 1989. Achievement, impact and constraints. Working paper, Lima.

Jana, R.K. 1982. Status of sweet potato cultivation in East Africa and its future. In: Villareal, R.L. and Griggs, T.D. (eds.), *Sweet potato*. Proceedings of the First International Symposium. AVRDC, Shanhua, T'ainan, pp. 63–72.

Kay, D.E. (Revised by Gooding, E.G.B.) 1987. *Root crops*, 2nd edn, Crop and Product Digest No. 2. Tropical Development and Research Institute, London (now Natural Resources Institute, Chatham Maritime).

Kays, S.J. 1985. The physiology of yield in the sweet potato. In: Bouwkamp, J.C. (ed.), *Sweet potato products: a natural resource for the tropics*. CRC Press, Inc., Boca Raton, FL, pp. 82–6.

Kimber, A.J. 1972. The sweet potato in subsistence agriculture. *Papua New Guinea Agric. J.* **23** (3/4): 80–95.

Moyer, J.W. 1985. Major disease pests. In: Bouwkamp, J.C. (ed.), *Sweet potato products: a natural resource for the tropics*. CRC Press, Inc., Boca Raton, FL, pp. 35–57.

O'Brien, P.J. 1972. The sweet potato: its origin and dispersal. *Am. Anthropol.* **74** (3): 342–65.

Onwueme, I.C. 1978. *The tropical tuber crops. Yam, cassava, sweet potato, cocoyams*. J. Wiley & Sons Ltd, London.

Purseglove, J.W. 1974. *Tropical crops, dicotyledons*. Longman Group Ltd, London, pp. 80–1.

Schalk, J.M. and Jones, A. 1985. Major insect pests. In: Bouwkamp, J.C. (ed.), *Sweet potato products: a natural resource for the tropics*. CRC Press, Inc., Boca Raton, FL, pp. 59–77.

Ugent, D., Pozorski, S. and Pozorski, T. 1983. [Archeological remains of potato and sweet potato tubers from the Casma valley in Peru] Spanish. *Bol. Lima* **5** (25): 28–44.

Vander Zaag, P. 1979. The phosphorus requirements of root crops. Ph.D. thesis, University of Hawaii, Honolulu.

Watson, G. 1988, 1989. Research progress reports on sweet potato in Indonesian food systems. International Potato Center, Lima, [Mimeo].

Winaro, F.G. 1982. Sweet potato processing and by-product utilization in the tropics. In: Villareal, R.L. and Griggs, T.D. (eds.), *Sweet potato*. Proceedings of the First International Symposium. AVRDC, Shanhua, T'ainan, pp. 373–84.

Yen, D.E. 1982. Sweet potato in historical perspective. In: Villareal, R.L. and Griggs, T.D. (eds.), *Sweet potato*, Proceedings of the First International Symposium. AVRDC, Shanhua, T'ainan, pp. 17–30.

CHAPTER 2

Chemical composition

The parts of the sweet potato used for food, namely the roots and leaves or tips, are complex organs possessing a variety of chemical compounds utilized in the life processes of the plant. Only those compounds which are relevant to the use of sweet potato as a food, from the viewpoint of nutrition, quality or food processing, will be discussed here. A brief description of the chemical composition of the raw roots and leaves, and the factors influencing compositional variations, is given in this chapter. Components with nutritional importance are compared, on a raw and cooked basis, with those of other plant foods and discussed in relation to their contribution to diets in the developing world in Chapter 3. The most recent analyses to date of the composition of sweet potato roots from the South Pacific region (Bradbury and Holloway, 1988) are summarized; further details of these and comparisons of them with contemporary analyses of other Pacific root and tuber crops can be found elsewhere (Bradbury and Holloway, 1988).

Roots

Dry matter

In common with other roots and tubers the sweet potato has a high moisture content, resulting in a relatively low dry matter content. The average dry matter content is approximately 30%, but varies very widely depending on such factors as cultivar, location, climate, day length, soil type, incidence of pests and diseases, and cultivation practices (Bradbury and Holloway, 1988). Dry matter content varies from 13.6% to 35.1% in a number of sweet potato lines grown in Taiwan (Anon., 1981) and from 22.9% to 48.2% in 18 cultivars grown in Brazil (Cereda et al., 1982). The average composition of sweet potato root dry matter is shown in

Table 2.1. *The approximate composition of raw sweet potato root dry matter*

	% in DM	
Constituent	Average value	Ranges[a]
Starch	70	30–85
Total sugars	10	5–38
Total protein (N × 6.25)	5	1.2–10
Lipid	1	1–2.5
Ash	3	0.6–4.5
Total fibre (NSP + lignin)[b]	10	?
Vitamins, organic acids and other components in low concentrations	<1	

Notes:
DM, dry matter.
[a] Published ranges.
[b] Non-starch polysaccharides (NSP) and lignin calculated by difference.

Table 2.1. In practice the composition is extremely variable and the concentration of each component depends on one or more of the same factors which influence dry matter content. Indications of the extent of such variations are given in the following sections.

Carbohydrates

Approximately 80–90% of sweet potato dry matter (24–27% fresh weight) is made up of carbohydrates, which consist mainly of starch and sugars, with lesser amounts of pectins, hemicelluloses and cellulose. The relative composition varies not only with cultivars and maturity of the root, but also with storage time and cooking or processing, and has considerable influence on quality factors such as texture, including firmness, dryness, mouthfeel, and taste. Such postharvest changes are detailed in Chapter 5, the composition of the cooked or processed sweet potato being of greatest importance to the consumer. It should also be noted that the starch in sweet potato undergoes ready enzymic transformations into sugars. Even preparing the roots for analysis by peeling, cutting or drying may bring about these changes if great care is not taken to avoid them. The description of carbohydrate composition given below is therefore intended merely as an introduction to the fuller details given in Chapter 5 concerning the role of carbohydrates in quality aspects of sweet potatoes.

Location of growth is apparently an important factor influencing total carbohydrate concentration. When three cultivars were grown at seven production sites using plants from a common source, the total carbohydrate content for any one cultivar varied among production sites more than between cultivars at a particular production site (Hammett, 1974).

Starch

Content

Compounds produced as a result of photosynthesis in the leaves are translocated to the fleshy roots and eventually transformed into starch. On average, starch constitutes 60–70% of dry matter, but the proportion of starch to other carbohydrates varies greatly.

Factors affecting starch content

There are notable genetic differences in starch content, for example 42.6–78.7% (dry weight basis) among 18 cultivars grown in a single Brazilian location (Cereda et al., 1982) and 33.2–72.9% (dwb) in Filipino and American cultivars (Truong, Bierman and Marlett, 1986). In the raw fresh root, starch averages about 18% of the total weight, but was found to vary, in cultivars grown under similar conditions, from 11.0% to 25.5% in 31 Indian cultivars (Shanmugan and Venugopal, 1975), from 7% to 22.2% in 292 Taiwanese cultivars (Li and Liao, 1983) and from 4.1% to 26.7% in 75 Thailand cultivars (Prabhuddham et al., 1987). Table 2.2 shows ranges of starch content in South Pacific clones. A highly significant positive correlation ($r = 0.926$) between percentage starch and dry matter was found for the Taiwanese cultivars, indicating that the percentage dry matter of roots can be used to evaluate their percentage starch content. There was no significant correlation between fresh weight of roots and percentage starch. The same study investigated the effect of environmental factors, including location, year, crop season and length of growing season on the percentage crude starch of roots. The average starch content of eight cultivars varied from 13.1% to 15.9% between four locations and of 15 cultivars from 17.1% to 18.5% between 2 years. The cultivar × location or cultivar × year interactions were highly significant, indicating the importance of testing in different locations and years. The average starch contents of four cultivars ranged from 9.8% to 14.9% in different crop seasons, those of the autumn and winter plantings being higher than those of the spring and summer. Starch content was also significantly higher at harvest 150 or 180 days after planting than at 120 days after planting in the same season. High doses of potassium fertilizer (124.4 and 186.7 kg/ha) significantly increased the percentage starch in dry matter (Sharfuddin and Voican, 1984), a finding attributed to greater uptake of potassium by the leaves leading to increased leaf photosynthesis and hence more accumulation of starch in the roots. There was also a small increase in starch brought

Table 2.2. *The composition of South Pacific sweet potato roots (fwb)*

Constituent	Mean[a]	Ranges[a]
Moisture (%)	71.1	61.2–89.0
Energy (kJ/100 g)	438[b]	125–635[b]
(kcal/100 g)	105	30–152
Protein (%)	1.43	0.46–2.93
Starch (%)	20.1	5.3–28.4
Sugar (%)	2.38	0.38–5.64
Dietary fibre (%)	1.64	0.49–4.71
Lipid (%)	0.17	0.06–0.48
Ash (%)	0.74	0.31–1.06
Minerals (mg/100 g)		
Ca	29	7.5–74.5
P	51	41.0, 70.0[c]
Mg	26	18.4–35.7[c]
Na	52	13.8–84.0[c]
K	260	129–382[c]
S	13	
Fe	0.49	0.16–0.94
Cu	0.17	0.08–0.28[c]
Zn	0.59	0.27–1.89[c]
Mn	0.11	0.05–0.26[c]
Al	0.82	0.24–1.14[c]
B	0.10	0.07–0.14[c]
Vitamins (mg/100 g)		
Vitamin A (β-Car./6)[e]	0.011	0.008–0.014[d]
Thiamin	0.086	0.073–0.099[d]
Riboflavin	0.031	0.025–0.041[d]
Niacin	0.60	0.38–0.77[d]
Pot. niacin[f]	0.32	
Ascorbic acid	24	

Notes:
fwb, fresh weight basis.
From: Bradbury and Holloway, 1988.
[a] Mean and ranges of 164 samples from five South Pacific countries: Solomon Is., Tonga, Papua New Guinea lowlands and highlands, Western Samoa, Fiji.
[b] Calculated.
[c] Solomon Is. and Tonga only.
[d] Solomon Is., Tonga and Papua New Guinea.
[e] White/cream-fleshed types. car., carotene.
[f] Potential niacin (tryptophan ÷ 60).

about by decreasing plant densities in the field from about 27,000 to 18,500 plants/ha, which presumably provided the space needed for optimum development and adequate plant nutrition.

Properties of sweet potato starch

Starch is quantitatively the most important component of sweet potato root dry matter. The extraction of this starch on a home, village and commercial scale, and its use in various food products is widely practised in some countries, for example China. It is therefore important to consider the properties of sweet potato starch and briefly how they compare with those of other starch sources.

Starch occurs in plant tissues in the form of discreet granules whose characteristics of size, shape and form are unique to each botanical species. Sweet potato granules are oval, round or polygonal, with a central hilum and, within an individual cultivar, vary greatly in size (Madamba, Bustrillos and San Pedro, 1975). For example granule size of one cultivar ranged from 7 to 43 μm. Between cultivars the mean granule size ranged from 12.3 to 21.5 μm. The average granule size of sweet potato starch was reported to be similar to that of cassava starch, about one third to one half that of yam starch and only about one fifth that of arrowroot starch (Lii and Chang, 1978). Similarly, Szylit et al. (1978) found the mean diameters of cassava, sweet potato and yam starch granules to measure 12, 25 and 75 μm, respectively. There is some evidence that the smallest starch granules (about 5% of the total) in a sweet potato starch sample are colloidal in nature and are lost when the starch is extracted with water (Upadhya, M., personal communication). The efficiency of extraction of starch from sweet potato roots can, perhaps for this reason, be less than that from cassava roots. Phosphorus content, which can affect enzymic degradation and pasting properties of starch, was found to vary from 9 to 22 mg/100 g amongst starches from different cultivars (Madamba et al., 1975) and was 20 mg/100 g in two Taiwanese samples (Lii and Chang, 1978).

Starches fall into two major groups of crystalline organization as characterized by their X-ray diffraction spectra: an A-type pattern corresponding to a double crystalline structure with three relatively intense spectrum lines, and a B-type with a single crystalline structure. Sweet potato in common with cassava starch has an A-type structure, as in the common cereals (Szylit et al., 1978). In contrast, yam and potato starches have a B-type structure.

As with most types of starch sweet potato granules are made up of amylopectin and amylose molecules, the sweet potato starch amylopectin:amylose ratio being variable but generally about 3:1 or 4:1. Amylopectin is a large, highly branched polymer of alpha-1,4-linked glucose chains branching through alpha-1,6-glucosidic links. The amylose molecule is a smaller, unbranched, straight-chained polymer with its glucose subunits being joined by alpha-1,4 links. The amylose contents of sweet potato starch granules have variously been determined as ranging from 17.5% to 38% in cultivars from the United States, Philippines, Korea and Puerto Rico (Doremus, Crenshaw and Thurber, 1951; Bertoniere, McLemore and Hasling, 1966; Madamba et al., 1975; Shin and Ahn, 1983; Martin and Deshpande, 1985). The amylopectin: amylose ratio does not change in sweet potatoes during curing and storage according to some researchers (Bertoniere et al., 1966). It increases on curing, but remains stable during subsequent storage according to others (Hammett and Barrantine, 1961). It has been suggested that a low amylose–high amylopectin content is responsible for the 'moist', sticky texture, when baked, of certain sweet potato cultivars popular in the United States (Hammett and Barrantine, 1961), but other work found no correlation between 'moistness' and amylopectin content (Swingle, 1966).

A range of intrinsic viscosities and reducing power was found amongst the starches of seven cultivars of varying sweetness in Puerto Rico (Martin and Deshpande, 1985). Some cultivars had relatively low intrinsic starch viscosities and high starch reducing power, suggesting low molecular weights and highly branched polymers; others appeared to have a high molecular weight and a highly branched polymer.

The temperature and type of gelatinization (the swelling which takes place when starch granules are heated in water), swelling ability, hot-paste viscosity and gel-forming properties of individual starches are important in determining their behaviour in food formulations. Gelatinization of sweet potato starch is of the single stage type (Madamba et al., 1975; Shin and Ahn, 1983) and has been variously reported to take place between 58°C and 69°C (Lii and Chang, 1978), 58°C and 75°C (Madamba et al., 1975) and 65°C and 80°C (Shin and Ahn, 1983). The characteristics on heating in water and then cooling of various root and tuber starches have been compared (Rasper 1969a,b). The swelling ability of sweet potato starch was much less than that of potato, similar to that of cassava and cocoyam (*Xanthosoma sagittifolium*) and greater than that of yam (*Dioscorea* spp.) or cocoyam (*Colocasia antiquorum*) (Rasper, 1969b). The degree of association of the molecules in sweet potato starch granules is therefore much greater than those in potato, similar to those in cassava and cocoyam (*Xanthosoma*) and lower than those of yam.

At its pasting temperature (above that of the gelatinization temperature) sweet potato starch attains a peak viscosity which then falls somewhat as the temperature rises further. This behaviour is similar to that of cereal starches, but contrasts with yam (*Dioscorea esculenta*) or legume starches which do not peak (Goshima et al., 1984). Legume starch viscosity is very much lower than that of root or cereal starches. Sweet potato starch has a fairly high peak and initial hot-paste viscosity, thinning rapidly on prolonged cooking at boiling temperature. Its maximum viscosity is lower than that of some yam starches, similar to that of cassava starch and higher than that of cocoyam or maize starch (Rasper, 1969a). The hot paste viscosity of sweet potato starch is lowered when the starch contains a high level of impurities as in commercial starch produced by traditional methods (Lii and Chang, 1978). The rapid thinning of sweet potato starch paste on prolonged heating is in contrast to some yam starch pastes which show great stability on heating. When cooled, sweet potato, like cassava and cocoyam, gives very poor gels of low consistency, in contrast to yam gels, which are firm and increase in consistency with time (Rasper, 1969a).

One group of workers is of the opinion that, for certain uses, sweet potato starch is superior to corn starch in its lower gelatinization temperature, higher maximum viscosity, and thin boiling character (Garcia, Querido and Cahanap, 1970). However, the amylose of sweet potato starch and other root starches was found to have a lower degree of polymerization than that from mung beans which is traditionally used for making starch noodles in Taiwan (Chang, 1983). This caused inferior quality in noodles prepared with sweet potato and other root starches, the noodles having a soft texture and high solids losses during cooking.

When starch solutions, pastes or gels age they undergo physicochemical changes known as retrogradation. Such changes include a reorientation of the starch molecules and subsequent hydrogen bonding to form a crystalline structure. Retrogradation, the degree of which varies with the starch source, is important in the staling of bread and changes in the texture of canned soups and of other foods with high starch concentrations. Sweet potato starch was found to retrograde more slowly and to a lesser extent than some other starch sources (del Rosario and Pontiveros, 1983), which apparently explained a slower rate of staling in sweet potato-substituted breads as compared with whole wheat bread.

In common with most other root and tuber starches, raw sweet potato starch is more resistant than the raw cereal starches to the action of digestive enzymes such as amylase. Sweet potato starch was only 2.4% digested by bacterial alpha-amylase as compared to 9.2% for maize and 17.6% for wheat (Rasper, 1969b). Other workers (Szylit et al., 1978), comparing the degradation by bacterial alpha-amylase of raw sweet potato starch and cassava and yam starches, found sweet potato to be 15% degraded in 6 hours compared to 20% for cassava and only 10% for yam. Susceptibility to amylase degradation was increased by rupture of the starch granules by pelleting (for animal feed purposes). Moreover, the susceptibility to breakdown of sweet potato starch by alpha-amylase greatly increases on cooking (Cerning-Beroard and Le Dividich, 1976).

After boiling sweet potato in water for 30 min, 54% of the starch was easily hydrolysable, and rats fed on a diet based on cooked sweet potato showed a growth performance similar to that of those fed on a diet based on maize starch.

Uses of starch

Sweet potato is one of the most important starch-producing crops in the world, although its production in temperate areas is limited to 4 to 6 months only of every year. Starch plants in the United States closed down in the 1950s due to high production costs, poor starch quality and low returns per unit of cultivated area, which made competition with foreign imports of starch difficult. Similarily, consumption of sweet potato starch declined in Japan following relaxation of import restrictions on less expensive supplies of corn starch. Unfortunately there are at present no known special characteristics of sweet potato starch which would justify a premium price. About 8%, 16% and 28% of production is now used as starch raw material in Korea, Taiwan and Japan, respectively (Sakamoto and Bouwkamp, 1985). Sweet potato starch is used in the food industry as an ingredient of bread, biscuits, cakes, juices, icecream and noodles. It is often, however, converted to glucose syrup, or isomerized glucose syrup in which some of the glucose has been converted to fructose for increased sweetness. Glucose syrup is utilized in candies, icecream, jams etc. and isomerized glucose syrup in lactic acid beverages, soft drinks, bread and many other foods (Kainuma, 1984). Two novel products which can be made from sweet potato starch have been developed in Japan, namely cyclodextrin, which has a variety of applications in the food and pharmaceutical industries, and oligosaccharides used as reagents for blood tests. These products are expensive; their commercialization thus raises the value of starch raw material.

The content of starch and other fermentable carbohydrates in sweet potatoes also makes them a suitable raw material for the production of alcoholic beverages, which takes place in countries such as Japan and Korea. About 5% of total production is used for making the distilled spirit *shochu* in Japan. Other commodities produced by fermentative processes using sweet potato include lactic acid, acetone, butanol, vinegar and yeast.

There is increasing interest in the use of starchy raw materials for the production of industrial alcohol. This can be employed as a solvent, beverages, food and animal feed (via single cell protein), for petrochemical synthesis and, especially since the world oil crisis, motor fuel. Sweet potato could be a viable alternative to cereal grains as a substrate for alcohol production. It has a higher starch yield per unit of cultivated land than cereal grains, which in any case are in urgent demand for use as food and feeds. The starch is first converted to sugar which is fermented by yeast to alcohol and carbon dioxide. The resulting dilute alcohol is purified by fractional distillation. Sweet potatoes for industrial use must be high yielding and high in starch content. Saccharification of starch is the key step in determining the economics and efficiency of alcohol production. Processes such as the use of acid and heat (Azhar and Hamdy, 1981a), intrinsic or crystalline sweet potato beta-amylase enzymes (Azhar and Hamdy, 1981b) or high pressure extrusion (Kim and Hamdy, 1987) have been investigated for the hydrolysis of starch to fermentable sugars. These investigations are continuing. They are coupled with the need to minimize side-reactions, such as the production of hydroxymethylfurfural, which can inhibit growth and alcoholic fermentation by yeast in the ensuing stages of the process. In addition, research into the alcohol production potential of different sweet potato genotypes and its economic feasibility compared with other biomass sources has been pursued (Hall and Smittle, 1983). In Indonesia it is planned to increase alcohol production substantially with preference given to sweet potato as a raw material, since it can be harvested three times a year (Winaro, 1982).

The establishment of starch industries in the tropics, utilizing sweet potato as a raw material requires consistent, year-round supplies of sweet potatoes with a high starch content. Renewed interest in breeding sweet potatoes for high starch concentrations has taken place as a result of the increasing production of animal feed, products such as flakes and chips, which require high dry matter sweet potatoes, and industrial products such as alcohol.

Sugars

Variability in total sugars between sweet potato samples is notable, ranging from 0.38% to 5.64% (fwb) among cultivars from various regions of the South Pacific (Bradbury and Holloway, 1988; and see Table 2.2) and from 2.9% to 5.5% (fwb) in American cultivars (Picha, 1985c). On a dry weight basis, total sugars varied from 5.6% in a popular local Filipino cultivar to 38.3% in a Louisiana sample (Truong et al., 1986), and from 6.3% to 23.6% in cultivars grown in Puerto Rico ranging from staple types, through lines intermediate in sweetness, to a dessert type (Martin and Deshpande, 1985; and see Figure 2.1 below). Considerable variability in total sugars exists even within different roots of the same cultivar (Tamate and Bradbury, 1985).

Time of harvest had a significant effect on total sugar content in six cultivars grown in one location in Brazil (Menezes et al., 1976), a higher concentration being recorded in roots harvested 6 months after planting than in those harvested 4 or 8 months after planting. Significant cultivar

Table 2.3. *Composition of sugars in raw and cooked sweet potato roots*

	% Sugar (fwb)							
	Glucose		Fructose		Sucrose		Maltose	
Cultivar	Raw	Baked	Raw	Baked	Raw	Baked	Raw	Baked
'Centennial'[a]	0.24	0.27	0.30	0.43	4.10	5.17	0	9.33
'Jasper'	0.44	0.42	0.43	0.41	3.63	5.14	0	7.75
'Travis'	1.50	2.73	1.15	1.99	2.87	3.26	0	4.02
'Jewel'	1.22	1.29	1.01	1.20	2.78	3.98	0	7.55
'White Star'	0.40	0.39	0.39	0.40	2.50	3.35	0	14.12
'Rojo Blanco'	0.95	1.22	0.65	0.97	1.30	1.59	0	10.77
Tongan cvs.[b]	0.45	0.37	0.33	0.26	2.03	2.43	0.64	7.09

Notes:

[a] Grown in the USA: high-performance liquid chromatography (HPLC) analysis. From: Picha, 1985c.

[b] Averages for five cultivars grown in Tonga. HPLC analysis. From: Bradbury et al., 1988.

differences were enhanced by interactions between cultivar and harvest time.

In recent years significant advances have taken place in the precision of quantitative measurements for individual sugars in sweet potatoes by the use of such techniques as high-performance liquid chromatography (HPLC) and nuclear magnetic resonance (NMR) spectroscopy. The major sugars occurring in raw roots are sucrose, glucose and fructose. Maltose has also been recorded in low concentrations in raw roots by some authors (Truong et al., 1986; Tamate and Bradbury, 1985; Bradbury et al., 1988), but was found by others to be absent (Picha, 1985c; Losh et al., 1981). The concentration of maltose increases significantly during cooking due to starch hydrolysis, which also produces polysaccharides of varying chain length known as dextrins. In all cases of cultivars analysed raw the concentration of sucrose exceeded that of the other sugars (Tamate and Bradbury, 1985; Truong et al., 1986; Martin and Deshpande, 1985; Picha, 1985c). In some cultivars the concentration of glucose is higher than that of fructose, in others they are present in approximately equal amounts. Some values determined for individual sugars in different cultivars, both raw and baked, are shown in Table 2.3. It was noted for sweet potatoes grown in Papua New Guinea that the sucrose content of all lines grown in one location was consistently higher than that of any lines from elsewhere (Tamate and Bradbury, 1985).

There has been some interest recently in the determination of oligosaccharides including raffinose, stachyose and verbascose in sweet potato roots in connection with the poorly documented occurrence of flatulence among sweet potato consumers. This will be dealt with in more detail in Chapter 4. No raffinose was observed in root samples from Papua New Guinea or the Solomon Islands (Tamate and Bradbury, 1985). The small amounts of raffinose apparently found in raw Filipino and American samples (Truong et al., 1986) by HPLC analysis were found to be largely due to the coelution of raffinose and maltotriose, most of the raffinose being accounted for as maltotriose by gas chromatography. No stachyose or verbascose was detected.

Non-starch polysaccharides

Role as dietary fibre

The compounds classed together as non-starch polysaccharides include the pectic substances, hemicelluloses and cellulose, which are found in the middle lamella or plant cell wall. As a group they are classed as dietary fibre, and play a role in the nutritional value of the sweet potato.

The pectic substances are polysaccharides found in the plant intercellular or middle lamella region and the cell wall. They consist of water-insoluble protopectin, soluble pectic and pectinic acids and pectins and may contain small amounts of the sugars D-galactose, L-arabinose and L-rhamnose. Protopectin, whose nature is still not fully understood, is the parent molecule from which pectin is formed by hydrolysis. Pectic acids consist of colloidal polyuronides composed of D-galacturonic acid units joined by alpha-1,4 linkages. Cold water-soluble pectinic acids and pectins have varying proportions of the carboxyl groups esterified with methyl groups and are capable of forming gels in the presence of sugar and acid. Pectins with low methoxyl contents (soluble in solutions of metallic sequestering agents such as Calgon or oxalate) form gels by cross-linkages between free carboxyl groups and metallic ions, especially calcium.

Hemicelluloses are cell wall polysaccharides classed according to the sugars present in their molecules, and are therefore known as mannans, xylans etc. Cellulose is a straight-chain polysaccharide made up of glucose units held together by beta-1,4 linkages. It is a highly ordered fibrillar structure occurring in the cell wall, and is water-insoluble and largely indigestible by humans.

In recent years substantial epidemiological research and physiological studies have shown that dietary fibre has important functions in the diet, with suggestions that it gives protection against diverticulosis, cardiovascular disease, colon cancer and diabetes, the various fibre components having differing roles in this respect.

Older food composition tables and even some recent publications still give values for the crude fibre content of sweet potato. Crude fibre measures only a partial and variable amount of the total dietary fibre and its determination should be discontinued in favour of that of dietary fibre. However, the determination of dietary fibre is by no means standardized throughout the world, and discussion as to the exact components to be included in its definition continue. Authors generally include lignin in calculations of total dietary fibre. Lignin, however, is not a carbohydrate, but an insoluble, high molecular weight polymer of coumeryl, coniferyl and sinapyl alcohols, acting as a hydrophobic filler and producing secondary thickening in plant cell walls. Therefore it has been argued (Englyst et al., 1989) that lignin should not be grouped with non-starch polysaccharides both because of its non-carbohydrate nature, and the difficulty of determining it accurately in methods for dietary fibre, which simply isolate a collection of material better referred to as 'substances measuring as lignin'. There have also been moves to include resistant starch (a retrograded fraction of the starch in starch-rich processed foods, resistant to enzymic hydrolysis) in fibre determinations. Resistant starch constitutes only a small fraction of the starch which in fact escapes digestion in the human small intestine and its physiological role has yet to be determined. In the meantime, dietary fibre may be defined for analytical purposes as non-starch polysaccharides and values for dietary fibre of sweet potato will be given using this definition. Values for 'lignin' will be given separately.

It is difficult to compare the values for dietary fibre of sweet potato recently determined by various groups of researchers as they have used different methods of analysis and have not all included the same components in their determinations. The range of dietary fibre contents found among South Pacific samples is shown in Table 2.2. A pooled sample of sweet potatoes grown in Tonga contained 12.6% total dietary fibre + 1.4% lignin (dwb) which could be calculated as approximately 4.0% and 0.4%, respectively, on a fresh weight basis (Holloway et al., 1985). Individual component percentages are shown in Table 2.4. The mean dietary fibre (fresh weight basis, fwb) of four cooked American cultivars (calculated as the sum of pectins, hemicellulose and cellulose) was comparable at 3.6% (0.97% + 0.93% + 1.7%) (Reddy and Sistrunk, 1980). The total dietary fibre (components unspecified) of raw sweet potato samples from the Solomon Islands and Papua New Guinea ranged from 1.20% to 2.62% (fwb) (Bradbury et al., 1985a). The total non-starch polysaccharide content of sweet potato flesh boiled for only 5 min was 2.4 g/100 g (fwb) or 8.1 g/100 g (dwb) according to British researchers (Englyst et al., 1988). These comprised 40% cellulosic and 60% non-cellulosic polysaccharides. A method measuring neutral deter-

Table 2.4. *Dietary fibre components of raw sweet potato roots from Tonga*

Component	% dwb	% fwb[b]
SNSP[a]	4.4	1.4
Pectin	2.5	0.8
Hemicellulose	3.8	1.2
Cellulose	1.9	0.6
'Lignin'[c]	1.4	0.4

Notes:
From: Holloway et al., 1985.
[a] SNSP, soluble non-starch polysaccharides (polyuronide + neutral sugars).
[b] Calculated from dry weight figures using moisture content of original samples; no allowance made for moisture remaining in dry samples.
[c] Residue left after determination of other dietary fibre constituents.

gent fibre (hemicellulose + cellulose) in American and Puerto Rican sweet potatoes (Lund and Smoot, 1982) showed that the peel (fwb) contains a higher percentage of these components than the flesh, though the situation was reversed on a dry weight basis. As might be expected the peel contained a much higher percentage of lignin than the flesh. The dietary fibre components of sweet potato are compared with those of other foods and their significance discussed in the following chapter.

Role in textural attributes

In addition to the contribution of the pectic constituents to dietary fibre, it has been suggested that they may play key roles in textural attributes such as firmness of canned roots (Baumgardner and Scott, 1963; Sistrunk, 1971) and 'moistness' or 'dryness' of baked roots (Swingle, 1966; Lee, Shin and Ahn, 1985). Hardcore, a disorder induced in raw roots by chilling temperatures and manifested in cooked roots as very hard areas of flesh, has been correlated with content of protopectin and other pectin fractions (Daines et al., 1976; Buescher, Sistrunk and Kasaian, 1976). Permanent hardness after cooking, resulting from submergence of roots in flooded fields, has been ascribed to the reaction of middle lamella pectin with calcium and magnesium ions, released from dead cells, to produce a heat-insoluble network.

The total pectic content of four North American cultivars ranged from 0.73% to 1.3% (fwb) (Reddy and Sistrunk, 1980), contrasting with the 2.6% (fwb) found in a Filipino cultivar (Kawabata et al., 1984) and the 3–5% (fwb) found in eight North American cultivars at harvest (Ahmed and Scott, 1958). The fractionation of pectic substances in one fresh raw sweet potato cultivar produced 1.6% of water-soluble pectin, 10.9% of water-insoluble salts of pectic and pectinic acids and 87.5% of protopectin (Kawabata et al., 1984). The pectic substances in North American samples had degrees of methyl esterification ranging from less than 10% in the acid-soluble fraction to about 30% in the water-soluble fraction (Ahmed and Scott, 1958). A water-soluble fraction in New Zealand sweet potatoes contained 9.7% methoxylated components (Holloway, 1983). The presence of methoxyl groups which render pectin more water-soluble enable it to migrate from the cell wall during heat processing, resulting in decreased firmness of, for example, canned sweet potato roots. On the other hand the proportion of free carboxyl groups present means that calcium salts can be added to canned roots to sustain firmness by cross-linking and formation of relatively insoluble calcium pectate. An observed decrease in pectic content and lowering of the intrinsic viscosity of the pectic fractions of cured and stored roots has also been proposed as a factor in loss of firmness of canned roots, and in the higher degree of 'moistness' in baked roots, compared to those freshly harvested (Ahmed and Scott, 1958). The lower starch, pectin, hemicellulose and cellulose levels of 'Georgia Jet' compared to three other cultivars was thought to explain its softer texture after cooking (Reddy and Sistrunk, 1980). The exact role of pectins in textural properties of sweet potatoes has still not been fully clarified.

Factors influencing fibre content

Significant cultivar differences in pectins, hemicellulose and cellulose were found for four American cultivars grown in one location (Reddy and Sistrunk, 1980). Small (3.8–5.7 cm diameter) roots had significantly higher contents of hemicellulose and cellulose than large (5.7–7.6 cm diameter) roots. Three Korean cultivars described as dry, intermediate and moist types contained similar water-soluble pectin contents, but the dry type contained a higher concentration of acid-soluble pectin (which is partly converted to water-soluble pectin during baking) both before and after baking (Lee et al., 1985). It is possible that the acid-soluble pectin content was in part responsible for the texture of the root type perceived as dry in mouthfeel.

Apart from genetic differences, the factors affecting fibre concentration have been little studied, and then mostly in relation to crude (insoluble) fibre. Applications of nitrogen or phosphorus fertilizer had little effect on crude fibre (Constantin, Hernandez and Jones, 1974); potassium applications up to 140 kg/ha slightly increased it on a dry weight basis, but had no effect on a fresh weight basis (Constantin, Jones and Hernandez, 1977). Supplementary irrigation which increased soil moisture had little or no effect on crude fibre levels (Constantin et al., 1974), nor were they influenced by azide applied at rates designed to control soil rot (*Streptomyces ipomoea* (Person and W.J. Martin) Waks and Henrici)) (Constantin and Hernandes, 1977). Sweet potatoes grown in elevated concentrations of carbon dioxide had decreased levels of insoluble dietary fibre (Lu, Biswas and Pace, 1986).

Flavour components

In many of the major production areas, for example China, sweet potato is used predominantly as a livestock feed rather than as a food for humans. One reason for this may be the dominant flavour of sweet potato which limits its incorporation into locally preferred dishes and its introduction into new geographical areas (Kays and Horvat, 1984). This contrasts with other staples such as rice, cassava, potato etc. which have bland flavours that are easily disguised and adapted to local tastes.

Flavours are perceived as a combination of odour and taste. While taste is basically limited to four sensations of sweet, salty, sour and bitter and is recognized, in an individual, by taste receptors numbering in the thousands, up to 10,000 odours can be distinguished by millions of odour receptors in that same individual (see Figure 2.1). The ability to manipulate the flavour of a food by adding characteristic combinations of flavours can greatly increase the likelihood of a food's acceptance into a particular cuisine (Rozin, 1977). The principal taste sensation of cooked sweet potatoes is sweetness due mainly to the presence of maltose (0.33), glucose (0.74), sucrose (1.0) and fructose (1.73) in increasing order of relative sweetness. The total sugar concentration (see pp.49–50) and the relative amounts of individual sugars vary considerably among cultivars (Kays and Horvat, 1984; and see Table 2.3). Expression of the sugar composition of a cultivar or line as sucrose equivalents can therefore give a meaningful estimate of relative sweetness. A non-sweet or low-sweet staple type often has a sucrose equivalent of 1 to 3 compared to 30 to 45 for normal lines (Kays, 1990).

Foods can be classed into four major groups depending on their characteristic flavour volatiles (Nursten, 1978).

1. Foods with an aroma composed primarily of one character impact compound (CIC).

a

b

Figure 2.1. Tasting different cooked cultivars of sweet potato roots in (a) Rwanda and (b) China. Variations in chemical constituents, e.g. pigments, sugars and flavour compounds, can influence consumer preferences and modes of utilization (M. Iwanaga).

2. Foods whose aroma is due to a mixture of a small number of compounds, one of which may be a CIC.
3. Foods whose aroma is due to a large number of compounds, none of which is a CIC, but which is capable of being reproduced.
4. Foods whose aroma is composed of an unreproducible complex mixture of compounds.

Little research has yet been carried out on the flavour of sweet potatoes. Among the volatile constituents distilled from baked 'Jewel', a North American cultivar, 30 were identified: (the use of (*) after the name indicates that the compound was also found by other researchers as described below) methanol(*), ethanol, acetone(*), diethyl ether, dichloromethane, 2,3-butanedione (diacetyl*), 3-methyl pentane, hexane(*), tetrahydrofuran, methylcyclopentane, 2,3-pentanedione, methylbenzene (toluene*), 2-methyltetrahydrofuran-3-one, furfural(*), dimethyl benzene (xylene*), isobutyronitrile(*), 2-pyrone(*), heptanal, 2-furyl methyl ketone, benzaldehyde(*), 5-methyl-2-furaldehyde, trimethylbenzene (mesitylene), octanal, 2-pentylfuran, phenylacetaldehyde, nonanal, linalool, decanal, beta-ionone(*) and 4-(2,2,3,3-tetramethylbutyl) phenol (Purcell, Later and Lee, 1980).

In addition, the volatile profile chromatograms of three contrasting baked cultivars have been compared and some of the volatiles from two of the three identified (Kays and Horvat, 1984). The three cultivars ('Jewel', 'Morado' and 'Tainung 57') represented North American, Central American and Asian centres of selection respectively. 'Jewel', a moist orange-fleshed type, was described as having a strong, sweet, caramel aroma, whereas 'Morado', a dry, white-fleshed type had very little aroma, described as starchy. 'Tainung 57', a yellow, dry-fleshed cultivar had a slightly stronger aroma than 'Morado' and was described as similar to roasted chestnuts. The volatile profiles of the three cultivars exhibited distinct differences, 'Jewel' having many more compounds than the other two cultivars and also greater concentrations of the constituents common to all three. In addition to the compounds above labelled (*), the following were identified in 'Jewel': pentene-2, pyridine, 2-phenyl-2-methylbutane, 2-acetylfuran, *n*-propylbenzene, limonene, 2,4,6-trimethylpyridine, several monoterpenes and sesquiterpenes. Compounds identified in 'Morado' but not in 'Jewel' included pentane-2-one, 2-pentanol, methylacetate, dimethylbenzene, 2-propenylfuran, 2-methyl-6-ethylpyridine, alpha-terpinol, heptylbenzene, hexadecanol, heptadecanol, 1-isopropyl-4-isopropenylbenzene, tetradecanol, a monoterpene and several sesquiterpenes. A number of the volatiles identified in both cultivars are distinctive aromatic compounds, known to be

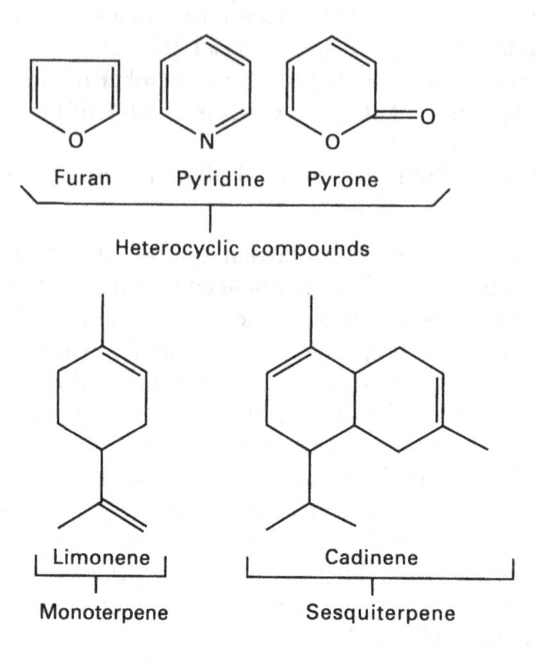

Figure 2.2. Basic structures of some volatile compounds, found in baked sweet potato, which have been associated with the aroma of other foods (from Kays and Horvat, 1984).

involved in the aromas of other foods, for example the furans, pyridines and pyrones, the monoterpene limonene and the sesquiterpene cadinene (see Figure 2.2). However, a CIC was not identified. As the authors found a substantial difference in the volatile compounds and concentrations of sugars of several cultivars, they anticipate an even greater variation among the cultivars in the world gene pool. They suggest that the use of an analytical technique for rapid screening of genetic crosses for flavour would enhance the rate at which flavour is altered, thus increasing the level of acceptance of sweet potato as a human food. Later research (Tiu, Purcell and Collins, 1985) compared the volatiles from a series of baked sweet potato cultivars with olfactory responses elicited by them. Of the 27 volatiles found, five were associated with cultivars having good flavour and eight with cultivars having poor flavour. The study suggested the possibility of specifying baked sweet potato aroma on the basis of a few volatile compounds and thus the selection of cultivars with improved flavours.

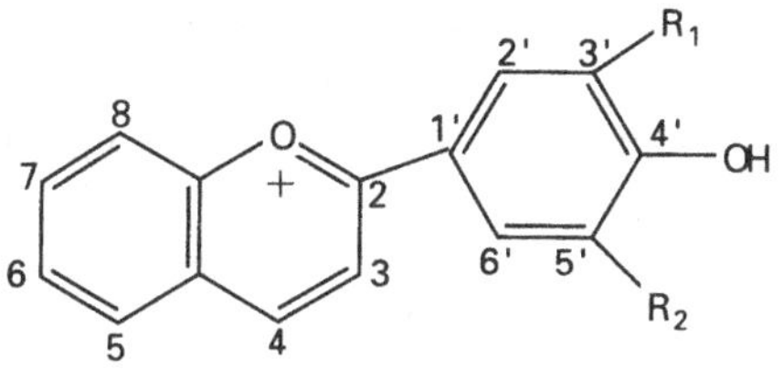

Flavylium cation

Cyanidin $R_1 = OH; R_2 = H$
Peonidin $R_1 = OCH_3; R_2 = H$

Figure 2.3. Basic structure of the anthocyanins found in sweet potato. Acylated glucosides are attached to positions 3′ and 5′.

Pigments

Anthocyanins

The red, purple or blue types of pigmentation found in various parts of the plant are caused by the presence of acylated anthocyanins. In some cultivars, the presence of anthocyanins is so pronounced that the flesh is the colour of beetroot. The major sweet potato anthocyanins are acylated glucosides of cyanidin and peonidin substituted in the 3′- and 5′-carbon positions on the flavylium nucleus (see Figure 2.3); acyl groups of the pigments are caffeic acid, ferulic acid and *p*-hydroxybenzoic acid (Imbert, Seaforth and Williams, 1966; Nozue, Kawai and Yoshitama, 1987). Cyanidin glucoside is present in sweet potato callus in larger amounts than peonidin glucoside (Nozue et al., 1987). Other unidentified anthocyanins are present in minor amounts. The same pigments were found in the stem sap of a West Indian cultivar and in the root periderm or flesh of several other pigmented cultivars (Imbert et al., 1966). The anthocyanins found in different parts of the plant are apparently synthesized *in situ* and are not transferred or moved to other plant organs (Kehr, Ting and Miller, 1955). Sweet potato anthocyanin synthesis has been studied in cell culture with a view to high anthocyanin production for food processing purposes (Nishimaki and Nozue, 1985; Nozue and Yasuda, 1985; Nozue et al., 1987). The concentration of anthocyanins in purple-fleshed cultivars was found to decrease from the periphery to the centre of the root (Cascon et al., 1984) and to depend on both genetic and cultivation factors. Stage of development of the root also affected anthocyanin levels. Large roots of 300–400 g contained about 200 mg/100 g total anthocyanins (dwb), whereas those of the same cultivar weighing only 80–150 g contained about 300 mg/100 g (dwb) anthocyanins. Several cultivars from different growing regions varied in anthocyanin content from 100 mg to 430 mg/100 g (dwb). The possibility therefore exists of selecting cultivars especially for the purpose of extracting their very high levels of anthocyanins. There is an increasing world-wide interest in the use of food colorants from natural sources to replace those made by chemical synthesis. However, many plant-source colorants are less stable than the synthetic colour equivalents. Sweet potato anthocyanins have been found to constitute a source of stable food pigments, suitable for addition to beverages (Bassa and Francis, 1987). The acyl groups were found to lend stability to the anthocyanin pigments. The manufacture of a violet or reddish-violet pigment from sweet potatoes for use in the food industry as a beverage colorant has been described (Japanese Examined Patent, 1981). A method for preparing anthocyanin extract from sweet potatoes is described in Chapter 6.

Carotenoids

Carotenoid pigments are responsible for the cream, yellow, orange or deep orange flesh colour of sweet potato roots. This attribute may (as in the case of red- or purple-fleshed cultivars) be exploited to introduce an attractive colour into foods to which sweet potato is added. The depth of flesh colour is largely a function of the concentration of beta-carotene. The percentage of total carotenoids present as beta-carotene is high in yellow to deep orange-fleshed cultivars (Purcell, 1962; Purcell and Walter, 1968), being 89.9% and 86.4% in 'Goldrush' and 'Centennial', respectively, but decreases as the total carotenoid concentration falls (Ezell and Wilcox, 1946). Some cultivars with white flesh contain no beta-carotene (Wang and Lin, 1969); others contain small quantities (Bradbury and Holloway, 1988). Those with creamy or light yellow flesh, may also contain traces only (Garcia et al., 1970). Carotenoids other than beta-carotene identified in orange-fleshed sweet potatoes (most in amounts contributing only $\leq 1\%$ of the total) include alpha-, gamma- and zeta-carotenes, phytoene, phytofluene, beta-carotene-epoxide, hydroxy-zeta-carotene, and beta-carotene furanoxide (Purcell, 1962; Purcell and Walter, 1968). Either beta-zeacarotene or neurosporene was the principal carotenoid in several white or pale-fleshed lines (Martin, 1983). Phytoene, phytofluene, zeta-carotene and neurosporene are intermediates in the biosynthetic pathway leading to the production of biologically active carotenoids (Bauernfeind, 1972). The chief importance in the diet of beta-carotene and the other nutritionally active carotenoids lies in their provitamin A activity (see Vitamins, below).

The depth of colour in white or pale yellow lines deepens on cooking, and may become unpleasantly green. This greenish discoloration may be due to the epoxides of one or more carotenes (Martin, 1983). Alternatively, it may be due to reactions involving phenolic compounds (see Polyphenolics, below).

Organic acids

Organic acids are important constituents of plant foods, influencing flavour, stability and keeping quality, although with respect to these roles there is at present no information for the sweet potato.

The analysis by HPLC of six different American sweet potato cultivars with a range of dry matter contents, revealed that total concentrations of the major organic acids in the whole raw root varied from 0.46% to 0.59% (fwb) (Picha, 1985a). Individual organic acid concentrations differ widely between cultivars (Picha, 1985a; Holloway et al., 1989). The three most abundant acids found in American and South Pacific cultivars were citric, malic and succinic acids; citric and malic acids were in concentrations higher than that of succinic in the American cultivars (Picha, 1985a), the reverse being found for the South Pacific cultivars (Holloway et al., 1989). No oxaloacetic acid was detected in the American raw root samples and its concentration never exceeded 0.02% in baked roots. There was an inverse correlation between citric and malic acid concentrations among the cultivars tested. The concentrations of all three acids increased during oven baking of the roots, and although this could be partially attributed to water loss, there was some net synthesis (Picha, 1985a).

Free oxalic acid, soluble potassium, sodium and ammonium oxalates and insoluble calcium oxalate are widely distributed in plants. High oxalate levels can cause acute poisoning, resulting in hypocalcaemia, or chronic poisoning in which calcium oxalate is deposited as crystals in the kidneys, causing renal damage. Furthermore, oxalic acid and soluble oxalates can bind calcium, reducing its bioavailability and calcium oxalate itself is poorly utilized by humans. Therefore calcium availability from a plant depends not only on total calcium concentration, but also on the proportion not bound up as oxalate. Less than 2 mg oxalic acid/100 g (fwb) was present in six American cultivars (Picha, 1985a). A range of 32.4 to 144.0 mg oxalic acid/100 g (fwb) was found among 12 lines grown in Hydrabad, India (Geervani et al., unpublished data), compared with much higher concentrations of 317–378 mg/100 g in taro (*Colocasia*) lines. Total oxalates in South Pacific sweet potato roots averaged 94 mg/100 g (fwb) (Solomon Islands) and 59 mg/100 g (fwb) (Solomon Islands and Fiji) (Holloway et al., 1989). These were made up of approximately equal quantities of soluble oxalates and insoluble calcium oxalate. However, two thirds of the total calcium found in the roots was free calcium, not combined as calcium oxalate, and therefore bioavailable. The small quantities of soluble oxalates in the sweet potato samples were similar to those in other root crops examined, but the content of calcium oxalate was ten times less than that of either giant swamp taro, *Cyrtosperma chamissonis* (Schott.) Merr., or elephant foot yam, *Amorphophallus campanulatus* Blume.

The oxalic acid concentration of sweet potato and other roots and tubers was found to increase as a result of infection with various pathogenic microorganisms (Faboya, Ikotun and Fatoki, 1983). The increase in oxalic acid was suggested to aid pathogen penetration by sequestering calcium or magnesium in the middle lamella of cell walls, increasing the susceptibility of pectates to hydrolysis by cell wall-degrading enzymes. Oxalic acid may also lower the pH of root tissue to a level suitable for pathogenic enzymic degradative activity. More information is required about the types and concentrations of organic acids in a wide range of cultivars from different locations and about the role these acids play in quality aspects of sweet potato roots.

Lipids

Although the lipid content of sweet potato is low and nutritionally insignificant, interest in its individual components has been stimulated by their role in the production of off-odours and flavours in dehydrated sweet potato flakes.

Lipid concentration, though low, is variable. It has been reported in various food composition tables to range from 0.1% to 0.8% (fwb) in raw sweet potatoes (Collazos et al., 1974; Leung, Busson and Jardin, 1968; Haytowitz and Matthews, 1984). Highly significant genetic differences were found among nine United States cultivars grown under uniform conditions, with values ranging from 1.21% to 2.55% on a dry weight basis (Boggess, Marion and Dempsey, 1970). The lipid content (dwb) of the inedible portion removed as peel has been found to be higher than that of the flesh (Faboya, 1981; Makki et al., 1986).

The lipids of one American cultivar 'Centennial' have been identified and quantified (Walter, Hansen and Purcell, 1971). The analysis was not carried out immediately after harvest, but after curing for 2 weeks followed by storage at 16°C and 60% relative humidity for 9 months. The lipids found were divided into three classes: neutral lipid (42.1%),

Table 2.5. *Lipid composition of raw 'centennial' sweet potato roots*

Lipid component	% of total lipid
Neutral lipids	42.1
Triglycerides	26.9
Steryl esters	6.1
Diglycerides	3.8
Hydrocarbons	2.8
Sterol (free)	2.5
Glycolipids	30.8
Monogalactosyl diglyceride	13.6
Digalactosyl diglyceride	6.3
Cerebroside	4.7
Esterified steryl glucoside	3.5
Unknown	2.1
Steryl glucoside	0.6
Phospholipids	27.1
Phosphatidyl ethanolamine	7.8
Phosphatidyl choline	7.0
Phosphatidyl inositol	5.1
Unknown	3.0
Cardiolipid	1.6
Phosphatidyl glyceride	1.2
Phosphatidyl serine	1.1
Phosphatidic acid	0.4

Note:
From: Walter et al., 1971. Reprinted from *Journal of Food Science* 1971. **36** (5):796. Copyright © by Institute of Food Technologists. Roots cured, and stored for 9 months.

containing neither sugar nor phosphorus in their molecules; glycolipids (30.8%), containing sugar; and phospholipids (27.1%), which contain phosphorus. The composition of these lipids is shown in detail in Table 2.5. Lipid-soluble carotenoid pigments were isolated mainly in the neutral lipid fraction and amounted to 2.3% of the total lipids. Analysis of the lipid fractions for their fatty acid composition showed that palmitic (16:0) and linoleic (18:2) acids are most abundant in all fractions, comprising 29.3% and 44.7%, respectively, of the total lipids (see Table 2.6). Whilst other studies gave similar results (Boggess, Marion and Woodroof, 1967; Opute and Osagie, 1978), approximately equal

Table 2.6. *Major fatty acids of sweet potato root lipid*

	Fatty acid (%)								
	14:0	15:0	16:0	16:1	17:0	18:0	18:1	18:2	18:3
Centennial[a]	1.7	1.3	29.3	2.0	1.5	6.8	2.0	44.7	8.8
Nigerian Variety[b]	Tr	ND	31.6	ND	ND	6.6	1.8	46.4	12.7

Notes:
ND, not determined; Tr, trace.
[a] From: Walter et al., 1971. Roots cured and stored.
[b] From: Opute and Osagie, 1978. Fresh roots.

amounts of linoleic and oleic (18:1) acids were found in Korean sweet potato lipids (Lee and Lee, 1972), and oleic acid was the most abundant unsaturated acid found in a Nigerian sample (Faboya, 1981). Linolenic acid (18:3) was also scarce in this sample, although it was found in moderate amounts (for example 12.7% and 8.8% of total lipids) by other researchers (Oputie and Osagie, 1978; Walter et al., 1971). Sweet potato lipid is therefore highly unsaturated. The sum of linoleic and linolenic acids accounted for 59.1% of the total fatty acids in a West African sample (Opute and Osagie, 1978) whilst the 'Centennial' study showed that neutral lipids contained 52.4% unsaturated acids and phospho- and glycolipids were more than 60% unsaturated (Walter et al., 1971). Moreover, sweet potatoes are relatively high in lipid-oxidizing activity (Rhee and Watts, 1966). The uptake of oxygen by stored dehydrated sweet potato flakes and consequent autoxidation of the highly unsaturated carotenoid pigments and unsaturated fatty acids in a portion of the total lipid, leading to loss of colour and production of off-odours and off-flavours is referred to further in Chapter 5 (quality aspects of processed products).

Polyphenolics

The sweet potato possesses a number of different compounds known collectively as polyphenolics, the oxidation of which by free oxygen is catalysed by enzymes called polyphenol oxidases. Substrate and enzyme are separated in intact tissues, but if the tissues are disrupted in some way in the presence of oxygen the components mingle and oxidation rapidly occurs. This leads to the production of quinones which either polymerize directly or combine with amino acids and amino groups in proteins to form dark coloured (brown) compounds. The reaction, known as enzymic browning, is part of the plant's defence mechanism against invading parasites such as insects and fungi. It also occurs when sweet potato tissues are bruised, for example during harvest or transport, or wounded, such as during cutting or peeling in the initial stages of processing. This leads to an unpleasant appearance, and consequent loss of quality, of both the fresh root and its processed products. Processing may therefore include a technique for inhibition of browning (for example blanching, lowering of pH, treatment with sulphite etc.).

A green discoloration occurring in sweet potato has also been attributed to a chlorogenic acid–amino group reaction (Uritani, 1953; Uritani, Hoshiya and Takita, 1953) which has been studied in a model system (Matsui, 1981). Green pigments are formed under weakly alkaline conditions when phenolic esters are oxidized and subsequently react with an amino group.

A bitter flavour has been attributed to the phenolic compounds (phenolics) in cereals and vegetables, but little information is available about the effect of these on sweet potato flavour. The presence of both polyphenolics and polyphenol oxidase in sweet potato roots has been studied extensively in order to determine the factors most affecting the darkening of tissues by this phenomenon. Methods for the determination of sweet potato phenolics have been compared (Walter and Purcell, 1979).

The majority of the phenolics are esters formed between quinic acid and caffeic acid. These phenolic esters are chlorogenic acid, isochlorogenic acid and related compounds (see Figure 2.4), with isochlorogenic acid predominating in sweet potato (Kojima, M., personal communication). They accumulate in sweet potato tissue when it is either wounded mechanically (McClure, 1960) or attacked by the black rot fungus *Ceratocystis fimbriata* Ell. and Halst. (Uritani, Uritani and Yamada, 1960). Total phenols and ortho-dihydroxyphenols also greatly increase in concentration in roots infected with *Rhizopus stolonifer* (Ehr. ex Fr.) Lind. (Thompson, 1979). The proposed biosynthetic pathway by which chlorogenic and isochlorogenic acids may form in injured tissues has been outlined (Villegas and Kojima, 1986).

The potential for degree of darkening of root tissue varies between cultivars; phenolic content ranged from 14 to 51 mg/100 g (fwb) in seven cultivars (Walter and McCollum, 1979). Darkening potential also varies year-to-year between cultivars (Walter and Purcell, 1980). Darkening is inhibited by the presence in roots of ascorbic acid, which reduces the quinones, produced as a result of oxidation, back to phenolics before brown pigments can be formed. Once all ascorbic acid is utilized, however, browning proceeds.

The phenolic composition of 14 cultivars has been studied qualitatively directly after harvest and again after curing and 5 months of storage

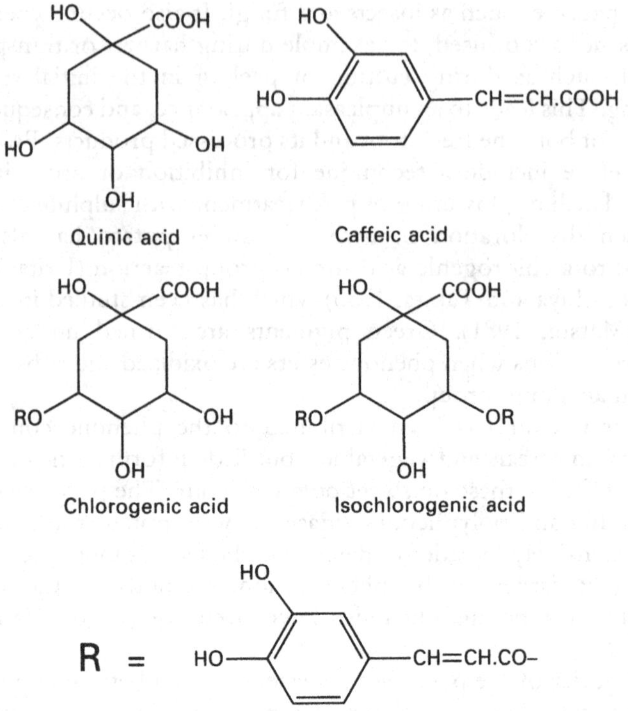

Figure 2.4. Some phenolic compounds found in sweet potato.

at 15°C and 80–85% relative humidity (Thompson, 1981). Chlorogenic acid and isochlorogenic acid were present in all cultivars at harvest and after storage, 4-*o*-caffeoylquinic acid in all but two cultivars at harvest and in all after storage, and neochlorogenic acid in only one (stored) cultivar.

In order to assess rapidly the potential of different cultivars for use in the food processing industry, from the viewpoint of least degree of enzymic browning, a simple method to assess the extent of phenol oxidase action in a disc of sweet potato tissue submerged in a 1% phenol solution was developed (Cereda et al., 1980). However, it has been suggested that genetic selection for minimum browning should be based on a low level of phenolic substrate rather than low phenol oxidase activity, as browning is significantly correlated only with phenolic content (Walter and Purcell, 1980). The phenol oxidase is present in very high concentrations relative to the substrate.

Qualitative histochemical tests showed the phenolics to be located in the periderm, in the approximately 1 mm of tissue directly below the periderm, in the latex of laticifers, in the phloem, in the cambium which separates the secondary phloem from the secondary xylem, in the anomalous secondary cambia of the central core, in the parenchyma cells adjacent to the xylem elements, and in the walls of the xylem elements (Schadel and Walter, 1981). The cross-sectional and end-to-end distribution of the phenolics was further investigated quantitatively, using HPLC (Walter and Schadel, 1981). Cross-sectionally, 78% of the phenolics were found to be concentrated in the outer 5–6 mm of tissue which include the periderm (skin) and the secondary root tissue beneath the periderm and external to the cambium. A similar localization of the phenolics in the outer tissue was found in Korean sweet potatoes (Kim, Hahn and Yoo, 1971). There was little end-to-end variation in phenolics (Walter and Schadel, 1981). The authors concluded that a severe peeling treatment which removed the outer 5 mm of tissue would result in a product less prone to darkening. However, this could be detrimental to the nutritional value (see Factors affecting mineral content, and Distribution of protein within the root, below).

Polyphenols can act as antibiotics, both chlorogenic and isochlorogenic acids being slightly inhibitory to the strain of *C. fimbriata* attacking the sweet potato (Uritani, 1978). Phenolics of unspecified chemical composition were also reported to increase significantly in roots fed upon by adults or larvae of the sweet potato weevil (*Cylas formicarius* Fab.) (Padmaja and Rajamma, 1982). This rise, which took place for up to 14 days of feeding, coincided with the development of an unpleasant aroma and a bitter taste in the roots.

It has been suggested that the marked antioxidative activity displayed by a methanolic extract of sweet potato is due to the synergistic effect of phenolic compounds and free amino acids (Hyase and Kato, 1984). It is further proposed that sweet potato could be a useful source of antioxidant activity, perhaps from the liquid waste produced during the manufacture of sweet potato starch or ethanol.

Phytoalexins

Plants are exposed, under natural conditions, to damage as a result of, for example, high winds and heavy rain and also attacks by fungi, bacteria, viruses and insects. They have therefore acquired defence mechanisms which enable them to resist or minimize the effects of such damage; in response to mechanical injury they form a wound layer and produce abnormal secondary metabolites such as the polyphenolics described in the previous section, or coumarins (Uritani, 1978) which are associated with undesirable greening of sweet potato flesh. When infected by fungi or other microorganisms, plants produce antibiotic substances in the infected regions of tissue or in non-infected tissue close to those regions. Such substances are known as phytoalexins; they are produced not only as a result of mechanical injury or parasitic infection, but also in response to toxic chemical treatments. The first phytoalexin to be isolated was a sesquiterpene (named at the time ipomoeamaron) from *C. fimbriata*-infected sweet potato roots (Hiura, 1943). After several years of investigation, the substance was shown to be a furanoterpene (Kubota and Matsuura, 1953) and renamed ipomeamarone. This compound and other related substances have economic importance as they cause sickness and sometimes death in animals fed on infected sweet potato roots containing them; use of infected high-carbohydrate mash from sweet potatoes for fermentative production of alcohol gives a very bitter and toxic by-product. Although there is no known case of acute human poisoning by ingestion of infected roots, there may be chronic effects, as yet undetected, due to long-term ingestion of diseased roots in times of dire necessity such as a famine. The chemistry of phytoalexins such as ipomeamarone, their occurrence in sweet potatoes and their effects on animal health have been the subject of intensive research, which is reviewed in detail in Chapter 4.

Enzymes

The sweet potato contains many enzyme systems, which catalyse individual synthetic and degradative processes within the tissues. The most important enzymes from the viewpoint of quality in both cooked and processed roots are the amylases. These break down starch to shorter chain molecules. The presence of diastase (the original term for a mixture of alpha- and beta-amylases), with starch-degrading properties was first demonstrated in sweet potatoes in 1920 (Gore, 1920). In 1948, beta-amylase was isolated (Balls, Walden and Thompson, 1948), and is still produced commercially by this method, although a more rapid procedure has recently been suggested (Roy and Hegde, 1985). The entire commercial production of beta-amylase in the United States comes from the sweet potato, and the enzyme finds many applications in the food industry.

In 1966, the importance of alpha-amylase was also demonstrated in sweet potato roots (Ikemiya and Deobald, 1966). Among its unusual characteristics were high temperatures for optimum activity (70–75°C), high thermal stability and low activity at ordinary temperatures. Freshly harvested sweet potatoes contain relatively little alpha-amylase, but the level increases greatly during storage, unlike beta-amylase the initially higher concentration of which changes little and erratically during storage (Walter, Purcell and Nelson, 1975). Activity of both alpha- and

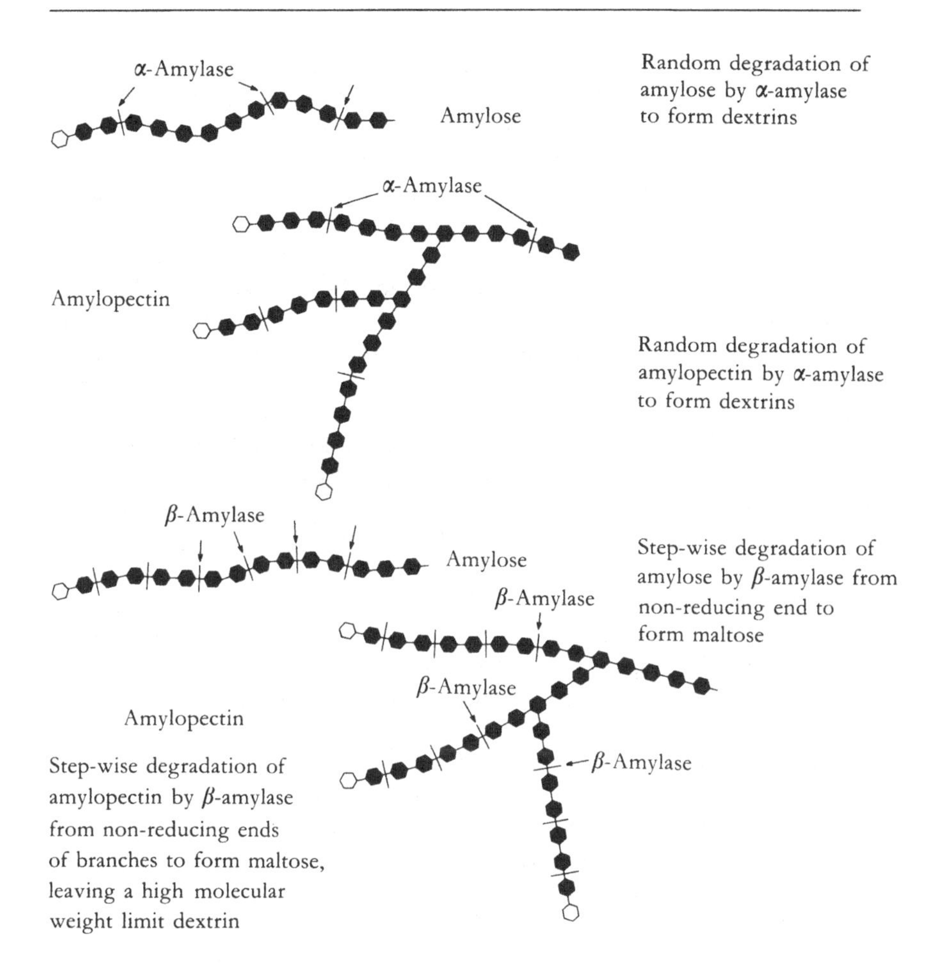

Figure 2.5. Amylose and amylopectin degradation by the amylases.

beta-amylases varies considerably with cultivar (Martin, F.W. and Deshpande, D.S., unpublished data; Walter et al., 1975). Alpha-amylase activity is distributed uniformly throughout the inner tissues of the roots, with an indication of minor concentration in the inner core, whereas beta-amylase is in the highest concentration in the innermost tissues (Ikemiya and Deobald, 1966). The outer cork layer and skin contains low concentrations of both enzymes.

The action of the amylases on starch is illustrated in Figure 2.5. Beta-amylase attacks the alpha-1,4 linkages within the amylose chain in a step-wise fashion, starting at the non-reducing end, to give maltose (and in the case of amylose with an odd number of glucose units, a little maltotriose). Amylopectin is similarily hydrolysed, but as beta-amylase is unable to hydrolyse or by-pass alpha-1,6 links a high molecular weight limit dextrin remains unhydrolysed. Alpha-amylase splits alpha-1,4 links at random to form dextrins, after which these fragments are slowly hydrolysed to maltose. Amylopectin on breakdown gives maltose and polysaccharide fragments called 'limit dextrins' (the enzyme is limited in its ability to hydrolyse starch, being unable to split alpha-1,6 bonds). Both alpha- and beta-amylases appear to contribute to starch breakdown during cooking, and it is probable that by doing so they influence both sweetness and the important quality attribute mouthfeel in the cooked roots. Cultivars with very low or negligible beta-amylase levels immediately after harvest, such as the Japanese cultivar 'Satsumahikari' (Kukimura et al., 1989), do not become sweet on cooking and can be used in the place of potatoes (*Solanum tuberosum*) for many purposes (see Chapter 6). The activity of beta-amylase in the roots, and the maltose content after cooking are respectively 0.11 IU/ml and 0.6%, which may be compared with the corresponding figures of 950 IU/ml and 11.5% for a conventional table cultivar 'Koukei 14'. The mode of inheritance of beta-amylase absence in the storage roots of such cultivars has been studied (Kumagai et al., 1990). However, it has been shown that such cultivars still sweeten during storage (Baba et al., 1987; see Chapter 5).

Mouthfeel is generally interpreted as the degree of 'moistness' or 'dryness' of the cooked root. These terms are used to indicate organoleptic characteristics and are not related to water content. The degradation of starch by the amylases during cooking, particularly alpha-amylase, may influence mouthfeel by lowering the water-binding capacity and viscosity of sweet potatoes this being interpreted by the mouth as increased moistness (S-101 Technical Committee, 1980). It has been suggested (Walter et al., 1975) that mouthfeel in cooked sweet potatoes depends on the amount of starch remaining after hydrolysis, the amounts and sizes of dextrins formed and the amount of sugar present, all of which are influenced by amylolytic activity. The improvement of quality in dehydrated sweet potato flakes when roots have been stored prior to processing has been attributed to the rise in alpha-amylase levels during storage (Ikemiya and Deobald, 1966). The complex changes in carbohydrates taking place during postharvest processes and their relation to quality and consumer acceptance will be discussed further in Chapter 5.

Other enzymes with importance to the quality factors of colour and flavour, namely the polyphenoloxidases and the lipoxidases have been mentioned under Polyphenolics and Lipids, above.

Vitamins

Sweet potatoes are substantial sources of ascorbic acid (vitamin C) and contain moderate amounts of thiamin (B_1), riboflavin (B_2) and niacin as well as pyridoxine and its derivatives (B_6), pantothenic acid (B_5) and folic acid. They have been reported to contain satisfactory quantities of vitamin E. Their chief importance, however, lies in their ability to produce variable and sometimes large quantities of the carotenoids which act as precursors of vitamin A. For this reason the carotenoid content of sweet potatoes will be discussed in some detail. Vitamin contents of roots as determined by various authors are given in Table 2.7.

Carotenoid vitamin A precursors

Importance

Vitamin A deficiency is one of the major public health problems which parts of the developing world are presently facing. It is the main cause of child blindness, and even in its less acute forms, vitamin A deficiency hinders normal growth and development and lowers resistance to infections. The problem is most extensive in some of the Asian countries, but also affects Africa and Latin America. The symptoms of vitamin A deficiency and the extent of its occurrence are discussed in more detail in Chapter 3.

Although the problem of vitamin A deficiency has to be tackled by multiple means, one approach is to encourage the production and consumption of foods rich in provitamin A, such as yellow- or orange-fleshed sweet potatoes. Selection of varieties with a moderately dry mouthfeel, high satiety value, a bland, pleasant taste, and containing provitamin A carotenoids in sufficient quantities to contribute significantly to dietary requirements, is a goal which should be urgently pursued by plant breeders. The AVRDC reported in 1981 that a significant negative correlation between sugar and dry matter contents, coupled with a low positive correlation between sugar and beta-carotene in sweet potato lines with varying sugar concentrations, indicated that it is possible to develop a low-sugar, high dry matter sweet potato which still retains a satisfactory level of beta-carotene.

In addition to their normal provitamin A function, recent research has indicated that some provitamin A compounds, including beta-carotene, have anti-cancer, anti-aging and anti-ulcer properties. Particular interest has been shown in the possible protective effect of beta-carotene against

Table 2.7. *Vitamin content of raw sweet potato (per 100 g (fwb))*

Source	Carotene (μg)	Thiamin (mg)	Riboflavin (mg)	Niacin (mg)	Pyridoxine (mg)	Pantothenic acid (mg)	Total folate (μg)	Ascorbic acid (mg)	Vit. E (mg)
Food composition tables[a]	70[b] (White)	0.1	0.05	0.8	0.26[c]	0.59[c]	14[c]	28	4.56[c]
	90 (Purple)		0.147[c]						
	250[b] (Yellow)								
	4,000[d]								
Authors	400–24,800[e] (White to deep orange)	0.09[f] (0.043–0.123)	0.03[f] (0.019–0.059)	0.6[f] (0.259–0.887)		0.7[g] (0.4–0.9)		23.6[h] (17.3–34.5)	1.75[i] (0.33–4.66)
	30–3,308[j]								
	76[k] (Creamy-yellow)	0.17[l]	Tr[l]	0.5[l]	0.09[l]		17[l]	17[l]	
		0.15[k]	0.035[k]	0.175[k]	0.094[k]		191[m] (dwb)	35[k]	0.75[k]

Notes:

[a] Mean of: Leung and Flores, 1961; Leung et al., 1968, 1972; Pellet and Shadarevian, 1970; Tan, Wenlock and Buss, 1985; Watt and Merrill, 1975.
[b] Collazos et al., 1974.
[c] Haytowitz and Matthews, 1984.
[d] Tan et al., 1985.
[e] Wang and Lin, 1969. Taiwanese samples.
[f] Bradbury and Singh, 1986b. South Pacific samples. Ranges in parentheses.
[g] Junek and Sistrunk, 1978. American samples. Ranges in parentheses.
[h] Bradbury and Singh, 1986a. South Pacific samples. Ranges in parentheses.
[i] Hirahara and Koike, 1989. Japanese samples. Ranges in parentheses.
[j] Bureau and Bushway, 1986. American samples.
[k] Visser and Burrows, 1983. New Zealand samples.
[l] Kwiatkowska et al., 1989. Samples of unspecified origin imported into Britain.
[m] Huq, Abalaka and Stafford, 1983.

cancer. Thirty-three reports were found in the literature of epidemiological studies concerned with the effect of diet, particularly the ingestion of carotenoid-rich vegetables, on the incidence of cancer (Mathews-Roth, 1985). All but four reports suggested that there is an inverse relationship between the ingestion of carotenoid-containing vegetables and the incidence of cancer. Many of these studies have been reviewed (Peto et al., 1981). One possible mechanism for this protective effect is the action of beta-carotene as an antioxidant at low oxygen pressures, such as those found in tissues under physiological conditions, thereby trapping free radicals which would otherwise initiate harmful reactions such as lipid peroxidation which could lead to cancer (Burton and Ingold, 1984). Such research is still being pursued, but if confirmed could arouse renewed interest in yellow- or orange-fleshed sweet potato.

Activity

The carotenoids found in sweet potato roots have already been mentioned under Pigments, above. A review of the chemistry and activity of provitamin A carotenoids, including those found in the sweet potato, has been published (Bauernfeind, 1972). Carotenoids are produced only in the plant world and are converted in the intestinal mucosa of humans to vitamin A, which is exclusively an animal compound. To be converted to vitamin A in mammals, a carotenoid must have at least one unsubstituted beta-ionone ring and a polyene side-chain attached. The other end of the molecule may vary in structure and can be lengthened, but not shortened to less than an 11-carbon polyene fragment. Beta-carotene possesses two beta-ionone rings, one at either end of a long polyene chain, and has the highest provitamin A activity of all the biologically active carotenoids, being assigned a potency of 100%. Alpha- and gamma-carotenes are also biologically active, with approximately half the activity of beta-carotene, and beta-zeacarotene is 20–40% as active as beta-carotene.

The picture is further complicated by the existence of stereoisomers of the various carotenoids, the *cis* forms having a lower potency than the more stable all-*trans* forms. A few *cis* forms can be found in unprocessed foods, but most reported in fresh plant extracts have generally been considered as artifacts produced by extraction and processing (Quackenbush, 1987). Heating of foods can result in the isomerization of some *trans* bonds to *cis*, especially at the 9-, 13- and sometimes 15-carbon positions. The existence of some *cis* forms of beta-carotene was demonstrated, but not quantified, in various vegetables including sweet potatoes, by HPLC (Bushway, 1986). The level of *cis* isomers as a percentage of total carotenoids is very low in fresh sweet potatoes, but

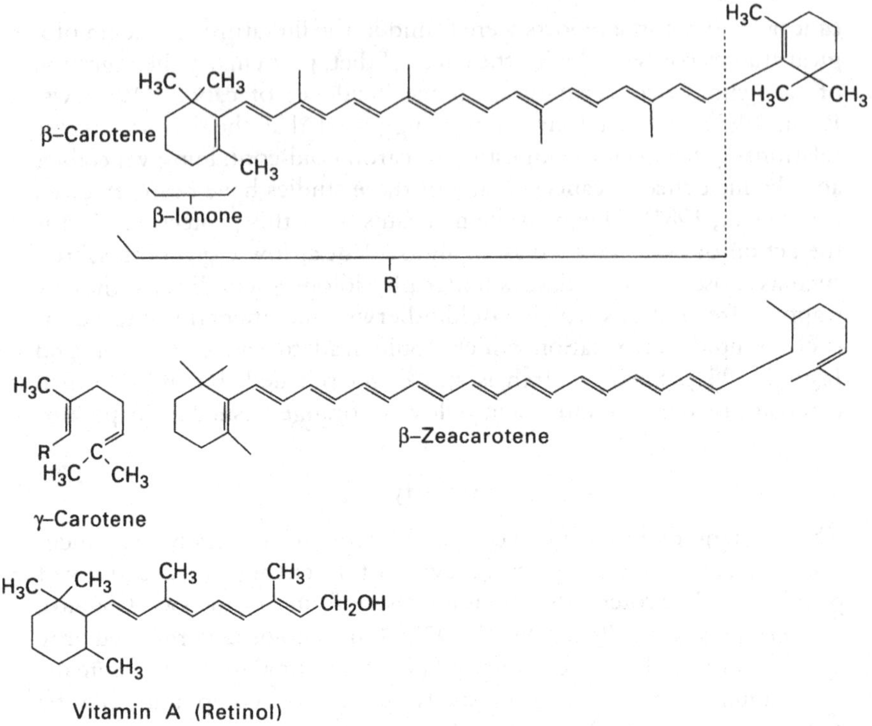

Figure 2.6. Chemical structure of some biologically active carotenoids found in sweet potato and of vitamin A (retinol).

may increase significantly during high temperature processing (Chandler and Schwartz, 1988; Quackenbush, 1987; Sweeney and Marsh, 1971). The isomerization of beta-carotene in sweet potato roots during cooking and processing is discussed in Chapter 6.

The chemical structure of some of the biologically active carotenoids of sweet potato is shown in Figure 2.6. Where levels of total carotenoids are high, as in the orange-fleshed cultivars of sweet potato, beta-carotene is the major carotenoid present. However, the fraction of the total as beta-carotene decreases as the total carotenoid content decreases. At harvest, beta-carotene comprised 89% and 19% of the total in two cultivars with 6.04 and 0.57 mg/100 g (fwb) total carotenoids, respectively (Ezell and Wilcox, 1958). Methods of analysis which measure only total carotenoids, expressed as 'beta-carotene', make no allowance for the difference in activity of the various provitamin carotenoids present which should be determined separately. A total carotenoid determination can overestimate vitamin A potential; on the other hand, pale-fleshed varieties should not automatically be assumed to lack provitamin activity totally. Samples of sweet potato from Papua New Guinea and the Solomon Islands contained respectively an average of 0.084 mg/100 g and 0.048 mg/100 g (fwb) of beta-carotene (Bradbury and Holloway, 1988). Cream to pale yellow sweet potatoes collected in five major American cities during 3 months of a 1 year period contained 0.184 mg/100 g to 0.368 mg/100 g beta-carotene (determined by HPLC) (Bushway, R.J., personal communication). The principal carotenoid in seven lines of cream to yellow-fleshed sweet potatoes was beta-zeacarotene (Martin, 1983). However, the cultivars with deep orange flesh are rich sources of beta-carotene, American samples having been found to range from 3.36 mg/100 g to 19.60 mg/100 g (fwb) of beta-carotene (Bushway, R.J., personal communication).

The biological activity of the carotenoids does not depend solely on their chemical structure or their concentration in the food; digestibility and absorption aspects are important. Factors which affect ultimate utilization include the dietary level of carotenoid administered, the type and sufficiency of dietary fat, the presence of antioxidants such as ascorbic acid or prooxidants such as nitrites, the presence of adequate bile in the dietary tract, the type and adequacy of dietary protein, and the age of, or presence of disease or parasites in, the organism. Thus the efficiency of use of the provitamin A carotenoids in sweet potato and other foods is affected by other factors contributing to malnutrition in the developing world. In addition, various components in a fruit or vegetable food itself can influence bioavailability of provitamin A carotenoids. The presence of chlorophyll and non-provitamin A carotenoids, and various components of dietary fibre or lignin can adversely affect beta-carotene bioavailability (Tsou, 1986, 1987). However, provided other deterrents are not present, beta-carotene has been found to be best utilized at quite low levels in the diet, the efficiency of conversion to vitamin A markedly decreasing as the dietary level increases. Theoretically one molecule of beta-carotene should, in the body, yield two molecules of vitamin A (retinol), but in practice because carotenoid absorption is often poor and conversion to vitamin A is inefficient, ratios of between 4:1 and 8:1 by weight beta-carotene:retinol have been suggested (Bauernfeind, 1972). For public health purposes, a provitamin A unitage three times that of vitamin A is generally adopted. This is a weight ratio of 6 μg beta-carotene to 1 μg retinol, which assumes that the biological activity and digestibility of beta-carotene are about 50% and 33%, respectively, of that of retinol. Thus the bioavailability of pure beta-carotene is 16.7%. Other carotenoids are taken to have a weight ratio of 12 μg to every 1 μg of retinol, on the assumption that their biological activity is 50% that of beta-carotene. Retinol equivalents (vitamin A) are thus calculated as:

$$\frac{\mu\text{g beta-carotene}}{6}+\frac{\mu\text{g other carotenoids}}{12}$$

It is now usual to express the vitamin A value of plant foods in terms of their retinol equivalents. In the past, vitamin A in all foods was often expressed in international units (IU), these being calculated as 1 IU of vitamin A = 0.3 μg retinol and 1 IU of beta-carotene = 0.6 μg beta-carotene or 1.2 μg other carotenoids with vitamin A activity. This has been largely discontinued.

It can be appreciated that the utilization of provitamin A carotenoids from sweet potato is influenced by the nature of the carotenoids present and the type and level of fibrous compounds in the sweet potato itself, and the adequacy of other dietary factors. There has been little research on the bioavailability of carotenoids from sweet potato. When fed to rats at the level of 10 μg beta-carotene equivalents/g feed, two sweet potato cultivars had bioavailabilities of 17.01% and 14.86%, not significantly different from each other, but significantly lower than pure beta-carotene at 21.11% (Tsou, 1986). The presence of fibre in the form of cell wall residues was found to be influential in reducing sweet potato beta-carotene bioavailability. Absorption of beta-carotene has been studied in humans. The average absorbed carotene as the proportion of intake from cooked mashed sweet potato was 28% in six children (Moschette, 1955) and 54% in eight young adults (James and Hollinger, 1954) on intakes of 3.5 mg beta-carotene/day, and otherwise adequate diets. These studies were undertaken with subjects in developed countries. In tropical areas where diets may be deficient in lipids and/or protein and high in fibre, the absorption of carotenoids from sweet potato may be lower than those indicated above. More research is needed to determine the efficiency of absorption and utilization of carotenoids in sweet potato from a variety of diets. This would result in a better understanding of the cultivars required in different localities to provide carotenoid contents sufficient to furnish a high proportion of the daily vitamin A requirement consistent with other consumer preferences.

Determination of carotenoid content

Many of the analyses involving sweet potatoes have determined only total carotenoid levels, but express these as beta-carotene or carotene. This is not very critical from a nutritional standpoint when the cultivars are those whose carotenoids consist chiefly of beta-carotene, such as dark yellow- or orange-fleshed cultivars. In cultivars with lower carotenoid levels, the expression of total carotenoids as beta-carotene alone leads to some overestimation of their provitamin A potential. With the present-day availability of more sophisticated analytical techniques, for example HPLC, which determine individual carotenoids, a more precise knowledge of provitamin A activity and its variation with external factors is possible. This is important if the use of paler-fleshed, bland-tasting cultivars which provide satisfactory provitamin A activity is to be explored. Few studies have so far been reported utilizing HPLC to determine individual carotenoid levels. However, sweet potatoes analysed by this method (Bushway, 1986) were found to contain beta-carotene and small amounts of its *cis* isomers, but not alpha-carotene. The *cis* isomers were not quantified; beta-carotene content of different sweet potato samples both within and between supermarkets was determined. South Pacific samples with white flesh have been analysed for alpha- and beta-carotenes (Singh and Bradbury, 1988). No alpha-carotene was found. Beta-carotene levels were low giving a vitamin A content of only 5.5 to 21.4 μg/100 g (fwb). The presence of other carotenoids was not reported. It is to be hoped that this method will find increasing application for studying the effects of different pre- and postharvest treatments on individual carotenoids in the sweet potato. The section which follows deals mainly with variations in total carotenoid content.

Factors affecting carotenoid content

The mechanism for carotenoid synthesis appears to be a genetic factor, either present or absent in the root. When cultivars high in carotenoids were grafted on to those which contained none, carotene failed to develop in the carotenoid-free cultivars (Miller and Gaafar, 1958). Moreover, synthesis was found to take place in the root itself from precursors formed in the vines (Kehr, Ting and Miller, 1955).

The major factor influencing total carotenoid content (and thereby beta-carotene) is cultivar. The carotenoid content of 17 cultivars grown in Taiwan ranged from 0.400 mg/100 g (fwb) in a local cultivar up to 24.8 mg/100 g (fwb) in a cultivar introduced from the United States (Wang and Lin, 1969). The carotenoids of 26 cultivars grown at Los Baños, Philippines, with flesh colour from white through cream and yellow to carrot-like, varied from traces only to 11.45 mg/100 g (fwb) (Garcia et al., 1970), while the mean carotenoid content of 10 American cultivars with varying flesh colours ranged from 2.55 to 6.73 mg/100 g (fwb) and 5.2 to 26.1 mg/100 g (dwb) (Speirs et al., 1953). The carotenoid level of cultivars or clones can be raised by crosses involving one high carotenoid parent (Wang and Lin, 1969).

However, variations occur in both total carotenoids and beta-carotene not only between but within cultivars. Individual roots of two American cultivars, one with a carotenoid level an order of magnitude greater than the other, varied between 47% and 82% in carotene and 49% and 51% in total carotenoids; means of all roots from single plants varied from those of other plants in the same area by 32% and 85% in carotene and 27% and 40% in total carotenoids (Ezell and Wilcox, 1958). It is interesting to note that the greatest variation took place in beta-carotene in the cultivar with the low carotene content. The stem end of a root has also been found to contain two to three times as much carotene as the root end (Ezell and Wilcox, 1946). All these results indicate that sampling must be carefully carried out if meaningful results are to be obtained.

Variation in carotenoid content is much greater between cultivars than between production sites (Hammett, 1974), but significant differences between locations for a particular cultivar occur (Hammett, 1974; Speirs et al., 1953). Four cultivars all showed much greater carotenoid levels at one location than at another (Speirs et al., 1953). Their mean carotenoid content at harvest was 19 mg/100 g at the first location and 31 mg/100 g (dwb) at the second. The chief difference in the two locations was a mean monthly air temperature 5.6 deg.C higher at the second than at the first. Differences between locations varied from as much as 62% to 93% in three American cultivars (Hammett, 1974).

At a particular location, quite wide seasonal variations in carotenoids occur. Over the course of 6 years, total carotenoids in an American cultivar ranged from 13.5 to 21.3 mg/100 g (dwb) (Speirs et al., 1953) and for two other cultivars the average beta-carotene or total carotenoid levels were twice as high in one season as in a second (Ezell and Wilcox, 1958). However, location effects have been shown to have more influence on carotenoid content than year-to-year effects within one location (Speirs et al., 1953). Overall it can be concluded that cultivars must be tested at the desired production location over a number of years to establish their provitamin A potential.

Variations in planting date within the usual season were not shown consistently to affect carotenoid levels at harvest time (Speirs et al., 1953) in either high or low carotenoid cultivars. Time of harvest of roots from the same planting also showed no consistent effect on carotenoid content. Similarity in the pattern of change in carotenoid content among the cultivars suggested that environmental conditions prior to harvest may have been influencing concentration in the roots. These results are at variance with others (Ezell, Wilcox and Crowder, 1952) which showed increases in both carotene and total carotenoids in four cultivars at one location during the first part of the harvest season, a maximum at about the time of usual commercial harvest and decreases thereafter. A more recent study (Abubakar, 1981) showed that carotenoids were highest at first harvest (125 days after planting) for two out of three cultivars, the third increasing in carotenoid concentration with prolonged harvest date.

One group of workers (Ezell and Wilcox, 1958) hypothesized that final carotenoid content was partially dependent on rate of root growth, the slower growing roots producing the higher carotenoid levels, and cited examples of other research where differences in cultivation practices or climate produced similar trends in both growth and carotenoid levels.

Cultural practices have been little studied for their effects on sweet potato carotenoids. On the whole their influence on total levels appears to be minimal. However, irrigation at high levels was found to decrease total carotenoids (Constantin et al., 1974). Fertilization with up to 140 kg K/ha or 74 kg P/ha did not affect carotenoid levels (Constantin et al., 1977), nor did the carotenoid content of three cultivars fertilized with 33.6, 29.3 and 28.0 kg NPK/ha show any significant difference to unfertilized controls (Constantin et al., 1975). Soil pH values between 4.4 and 7.2 had no effect on carotenoid levels (Constantin et al., 1975). Sweet potatoes grown on soil treated with azide at rates designed to control soil rot contained levels of carotenoids similar to those grown on untreated soil in four out of five seasons. In one season, the sweet potatoes in plots treated with the highest azide level had slightly higher carotenoid content, but were also the lowest yielding.

Diseased sweet potato tissue shows lowered carotenoid levels. Roots infected with the mould *Rhizopus stolonifer* (Ehr. ex. Fr.) Lind contained only 24 mg carotenoids/100 g (fwb) as compared with 50 mg/100 g in uninfected roots (Thompson, 1979).

Vitamin C

Occurrence

Most staples and vegetables contain vitamin C in both the reduced (ascorbic acid) and oxidized (dehydroascorbic acid) forms (see Figure 2.7). Both have full vitamin C activity, but the postharvest oxidation of ascorbic to dehydroascorbic acid, which is reversible, may proceed further irreversibly to produce inactive 2,3-diketogulonic acid with consequent loss of vitamin C activity. Methods for vitamin C determination in sweet potatoes and other crops which measure only ascorbic acid, and not dehydroascorbic acid also, give misleadingly low vitamin C values. Conversely they can imply misleadingly high losses of vitamin C in stored, cooked or processed roots.

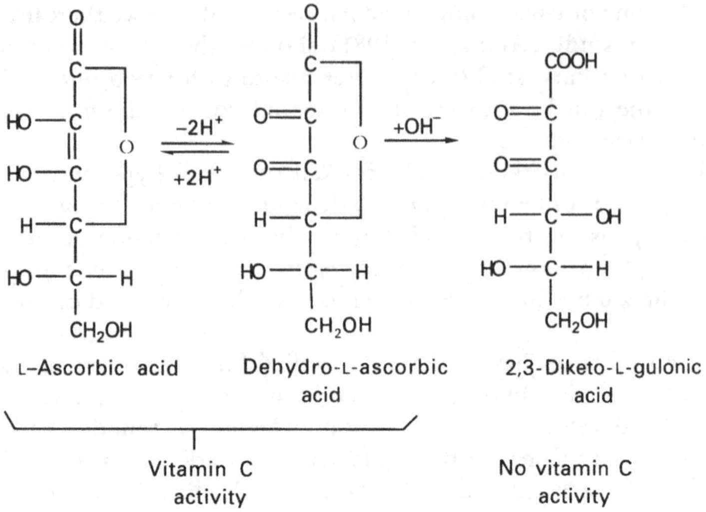

L-Ascorbic acid ⇌ ($-2H^+$ / $+2H^+$) Dehydro-L-ascorbic acid → ($+OH^-$) 2,3-Diketo-L-gulonic acid

Vitamin C activity — No vitamin C activity

Figure 2.7. Structural changes in ascorbic acid.

No dehydroascorbic acid was found in Cuban sweet potatoes (Schmandke and Olivarez Guerra, 1969), but the ascorbic acid content of these roots was low at 9.5 mg/100 g compared with values determined by other workers. A recent study (Bradbury and Singh, 1986a) on a number of South Pacific cultivars freshly harvested, then flown to Australia (1 week of transit time) and subsequently stored for a short period at 15°C, showed that dehydroascorbic acid contributed a high proportion (about 30%) of the total vitamin C content. The mean (with ranges in parentheses) ascorbic, dehydroascorbic and combined acids for 10 cultivars were, in mg/100 g (fwb): 14.3 (9.5–25.0), 9.2 (7.3–13.6), and 23.6 (17.3–34.5), respectively.

Factors affecting vitamin C content

It is common practice to dry root crop samples at 40°C before analysis. However, samples of sweet potato lost about 20% total ascorbic acid during drying at 40°C for 2–3 days (Bradbury and Singh, 1986a); hence preference should be given to determination of ascorbic acid on fresh or freeze-dried samples.

Considerable variability can exist between different sweet potato samples, for example total ascorbic acid in four Solomon Island cultivars varied from 19.8 to 32.9 mg/100 g (fwb) and 57.5 to 94.2 mg/100 g (dwb) (Bradbury and Singh, 1986a). Samples of Papua New Guinea sweet potatoes were reported to contain as much as 64 mg ascorbic acid/100 g edible portion (form of vitamin unspecified) (Farnworth, 1973). This is partially attributable to cultivar differences; the reduced ascorbic acid contents of four cooked cultivars grown in one location in the same year ranged from 10.3 to 30.8 mg/100 g (fwb) (Lanier and Sistrunk, 1979).

Early work (Ezell, Wilcox and Hutchins, 1948) indicated that sweet potato ascorbic acid may vary considerably within as well as between cultivars, there being a range of 160% in 30 individual roots of 'Porto Rico'. For the same cultivar, it was also shown that the centre or stem end portions (which were not significantly different) contain more ascorbic acid than does the root end (Spiers et al., 1945).

For a single cultivar, 'Unit 1 Porto Rico', significant differences in ascorbic acid between locations and between different years at the same location were demonstrated (Spiers et al., 1953), and five cultivars showed consistently higher ascorbic acid at a Virginia location than at one in Louisiana. Even at a single location higher levels of ascorbic acid were found in relatively young roots from late plantings than in older roots planted earlier. Roots of the former type were taken from Virginia, and of the latter from Louisiana. The possible cause of much of the location effect, therefore, was attributed to differing dates of planting and harvesting. However, since the magnitude of planting or harvesting effects varied, there must also have been environmental influences contributing to the differences in ascorbic acid between locations. Another study found an insignificant increase in ascorbic acid with later harvest date in three cultivars (Abubakar, 1981). Raising fertilizer levels from 62 to 186.7 kg K/ha significantly lowered ascorbic acid content (Sharfuddin and Voican, 1984).

The ascorbic acid level in sweet potato roots infected with R. *stolonifer* was only 20 mg/100 g compared to 32 mg/100 g (fwb) in those with healthy tissue (Thompson, 1979), a finding attributed to host–parasite interaction.

Other vitamins

The ranges of thiamin, riboflavin and niacin contents in some South Pacific islands cultivars were 0.04–0.12, 0.02–0.06 and 0.26–0.89 mg/100 g (fwb) respectively (Bradbury and Singh, 1986b; Table 2.7). There does not appear to be any published information on the factors contributing to such variations in these or other B group vitamins in sweet potatoes. The total vitamin B_6 content of raw sweet potatoes was determined in Britain as 0.09 mg/100 g (fwb) (Kwiatkowska, Finglas and Faulks, 1989), a figure similar to that found for New Zealand samples, but much lower than that for American samples (see Table 2.7). Total B_6 was made up of 33% pyridoxine, 44% pyridoxal and 23% pyridoxamine (Kwiatkowska et al., 1989).

Tocopherol (vitamin E) has been recorded as ranging from 0 to 10 mg/100 g (fwb) from the work of different researchers (Hirahara and Koike, 1989). In all but one of these analyses the nutritionally most potent form, alpha-tocopherol, was the predominant component, with other tocopherols undetected or present only in traces. However, a North American sweet potato baby food was found to contain alpha-, beta- and delta-tocopherols (Davis, 1973). Raw New Zealand sweet potato roots were reported to contain 0.75 mg tocopherol/100 g (fwb) (Visser and Burrows, 1983). A recent analysis by HPLC of nine cultivars grown in different locations and at different times in Japan showed a range in alpha-tocopherol of 0.333–4.660 mg/100 g (fwb) (Hirahara and Koike, 1989). These differences were more attributable to the effect of growth location than genetic variation. Alpha-tocopherol distribution within the root, investigated in one cultivar, was in the order centre > upper part > lower part, but this difference was not significant.

Minerals

Occurrence

The ash (non-volatile inorganic residue) content of sweet potatoes averages approximately 1% of the fresh root weight (about 3–4% of the dry weight). Variation in the ash content of South Pacific samples is shown in Table 2.2. Ash contains a variety of minerals and trace elements, some of which have a function in the life of the plant. Others are absorbed in varying quantities depending on their concentrations in the soil in which the roots are grown, or are derived from fertilizers or sprays employed during cultivation. The ash content of sweet potato peel is much higher than that of the flesh. The average ash concentrations in the peel and flesh of two Egyptian cultivars were respectively 14.1% and 4.6% (dwb) (Makki et al., 1986).

In general K is the element present in the greatest concentration, followed by P and Ca or Mg (Lopez, Williams and Cooler, 1980; Monro, Holloway and Lee, 1986; Ohtsuka et al., 1984; Picha 1985b; see also Table 2.2). Na may be in a higher (Lopez et al., 1980) or lower (Monro et al., 1986) concentration than Ca. The ratio of K to Na is very high. Other elements always present as shown by various analyses are Fe, Cu, Mn, Zn, S and Cl (Lopez et al., 1980; Monro et al., 1986; Ohtsuka et al., 1984). In addition, B (Monro et al., 1986), Cd, Ni and Pb (Furr et al., 1981; Kim et al., 1981), Hg (Kim et al., 1981), Se (Wolinsky et al., 1988) and Si (Lopez et al., 1980) may be present. Root mineral and trace element concentrations are shown in Tables 2.2, 2.8 and 2.9.

Table 2.8. *Mineral content of raw sweet potato roots (mg/100 g (fwb))*

Source	Ca	K	P	Mg	Fe	Na
Various authors[a]						
Mean	24	396	41	20	0.69	21
(Ranges)	(17–34)	(342–488)	(28–54)	(14–23)	(0.59–0.86)	(13–30)
Food composition tables						
Mean	30[b]	269[c]	41[d]	—	1.2[b]	24[c]
(Ranges)	(22–41)	(210–320)	(31–56)	—	(0.7–2.5)	(10–36)

Notes:

[a] Mean of values from: Lopez et al., 1980; Monro et al., 1986; Ohtsuka et al., 1984; Picha, 1985b.

[b] Mean of values from: Collazos et al., 1974; Leung and Flores, 1961; Leung et al., 1968, 1972; Tan et al., 1985; Pellet and Shadarevian, 1970; Platt, 1962; Watt and Merrill, 1975.

[c] Mean of values from: Leung et al., 1972; Tan et al., 1985; Watt and Merrill, 1975.

[d] Mean of values from: Collazos et al., 1974; Leung and Flores, 1961; Leung et al., 1968, 1972; Pellet and Shadarevian, 1970; Watt and Merrill, 1975.

Factors affecting mineral content

There is little information on this topic. Within cultivars, minerals are not evenly distributed throughout the root. Contents of some minerals in undefined peel fractions from two Egyptian sweet potato cultivars were higher than in the flesh averaging (mg/100 g (dwb)) 58.7 and 6.84, 4.03 and 0.41, 5.58 and 3.53, 0.60 and 0.30, 0.30 and 0.13 for Ca, Fe, Mg, Zn and Mn, respectively (Makki et al., 1986). In one variety Cu content of the flesh was higher than the peel, but the contrary was true of the other variety. Se was shown to be present entirely in the skin section of North American sweet potatoes (Wolinsky et al., 1988). The results indicate that peeling of sweet potatoes may significantly reduce levels of nutritionally important minerals such as Ca and Fe. The extent of this reduction may depend on whether the root is peeled before or after cooking, and the percentage of peel which is removed.

Significant differences in K, P, Ca and Mg were shown in six American cultivars (grown in one location under identical conditions of cultivation), analysed immediately after harvest (Picha 1985b). Variations, due no doubt to both cultivar and location effects, were found in P, Ca, Mg,

Table 2.9. *Trace and heavy-metal content of raw sweet potato roots (mg/100 g (fwb))*

Element	Content	
Boron[a]	0.64	
Cadmium[b (c)]	0.0045	(0.006)
Chromium[d]	<0.014	
Cobalt[d (e)]	<0.041	(0.022)
Copper[f (e)]	0.16	(0.26)
Fluorine[e]	0.86	
Iodine[e]	0.0045	
Lead[b (c)]	0.089	(0.04)
Molybdenum[d (e)]	<0.24	(0.057)
Manganese[f]	0.24	
Mercury[b]	0.0015	
Nickel[b (c)]	Tr	(0.02)
Selenium[g]	0.002	
Silicon[d]	0.99	
Tin[d]	<0.61	
Zinc[f (e)]	0.24	(2.0)

Notes:
Tr, trace.
[a] Monro et al., 1986.
[b] Kim et al., 1981.
[c] Furr et al., 1981 (dwb).
[d] Lopez et al., 1980.
[e] Rao and Polacchi, 1972.
[f] Mean of values from: Monro et al., 1986; Lopez et al., 1980; Ohtsuka et al., 1984; Visser and Burrows, 1983.
[g] Wolinsky et al., 1988 (baked).

K, Na and Fe among a number of cultivars grown in three areas of Papua New Guinea (Farnworth, 1973).

The concentrations of P and K in roots analysed at intervals, during a 14-week period beginning 2 weeks after planting, decreased slightly with increasing growth (Scott and Bouwkamp, 1974). There were no definite seasonal trends in Ca, Mg, Fe, Mn or B.

The application of K fertilizer at levels of up to 150 kg/ha did not significantly affect the content of ash and its components Fe, Mn, Mg, Ca, K and P in sweet potato roots grown on a red oxisol soil in Tanzania (Uriyo, 1974), although Na content was significantly depressed by 50 kg K/ha. At a level of 240 kg/ha, applied K lowered the ash content of

'Haiti' cultivar, planted in a red soil with an elevated Mg content, from 1.03% to 0.84% (Caraballo Llosas, 1974). The levels of K increased, but those of Ca and Mg decreased, P being only slightly affected. Na concentration of roots was greater when $NaNO_3$ was applied as fertilizer than with the use of NH_4NO_3 or $Ca(NO_3)_2$ fertilizers. The concentration of Na in roots fertilized with $NaNO_3$ rose from 86 mg/100 g to 134 mg/100 g (fwb) when the nitrogen application was increased from 101 kg/ha to 202 kg/ha (Hammett et al., 1984). This compared to a concentration of 15 mg/100 g reported in tables (Marsh, Klippstein and Kaplan, 1980). However, as 1100 to 3300 mg Na/day is considered safe and adequate for an adult, even at the higher rate of nitrogen application, sweet potato roots should not be considered a high Na food (Hammett et al., 1984).

There is increasing interest in the use of sewage sludges as soil conditioners or fertilizers in agriculture, but at the same time concern that toxic, heavy metals, known to be present especially in municipal sludges, may be taken up by crops grown on sludge-amended soil. A general increase in Cd, Cu, Ni, Pb and Zn contents of sweet potato roots took place when they were cultivated on either acid or neutral sewage sludge-amended soils (Furr et al., 1981), but the quantities present were still low compared to those in leafy vegetables such as lettuce, spinach etc. This indicates that analysis of such elements should also be carried out on sweet potato leaf tips and shoots grown for vegetable use, if sewage sludge is to be used as a fertilizer. The heavy metals Cu, Pb, Zn, Hg, and Cd were also present in sweet potatoes collected from the Jinju district of South Korea (Kim et al., 1981). Ni was present in trace amounts only.

Nitrogenous constituents

Determination

Although the nitrogenous compounds, collectively referred to normally as crude protein and here-in called total protein, comprise a quantitatively minor fraction of sweet potato dry matter compared with the carbohydrates, there has been a considerable amount of recent literature about them. This undoubtedly reflects the concern of researchers to identify ways of increasing the rather low protein content of sweet potato roots, given their importance in the diets of the poorest sections of some developing country populations, coupled with the desire to promote general dietary utilization of sweet potato as a local alternative to cereal importations. Moreover the overall picture of the sweet potato crop as an important world protein source is impressive. It has been estimated (Walter et al., 1984) that sweet potato yields on average 184 kg protein/ha, comparing favourably with the estimated average yields for wheat (200 kg/ha) and rice (168 kg/ha), and that as one of the major global crops it has the potential to provide about 2 million tonnes of protein world-wide.

However, to consumers of cooked or processed sweet potatoes, the concentration of protein in the food as eaten is of first concern. The composition of most roots and tubers changes little on boiling, whereas that of cereals in preparations which absorb large quantities of water (e.g. boiled rice, maize porridge) changes significantly (Woolfe, 1987) as nutrients are diluted. Comparison of the protein content of raw sweet potato and foods such as cereals is therefore less misleading if carried out on a dry weight basis. Much of the information pertaining to sweet potato protein is therefore given as dwb or both fwb and dwb.

Total protein in sweet potato is usually determined as Kjeldahl N × 6.25. This and other methods of protein determination such as auto-analysis, dye-binding or spectroscopy are unsuitable for mass field screening which could be used by plant breeders immediately after harvest. The AVRDC (Anon., 1985a) have reported on the use of a tetrabromophenol blue indicator paper which can be used to indicate the intensity of water-soluble protein in a large number of samples. This leads to some error in estimating total protein, as the percentage of water-soluble protein in total protein varies from one sweet potato line to another. However, it appears to be most accurate when high dry matter lines are tested. Since high dry matter may also be an important consideration in nutritional value assessment of sweet potatoes, lines could first be screened for dry matter and then for protein content.

Nature of the nitrogenous constituents

The total protein content of sweet potato (calculated as above) is on average about 5% (dwb) or 1.5% (fwb). This includes all nitrogenous compounds present in the analysate. At harvest these are made up of approximately 75% true (coagulable) protein and 25% non-protein nitrogen (NPN). Wide variations occur in total protein content and in the ratios of its constituent parts (see Table 2.2 and next section).

A large percentage of the true soluble protein is thought to function, in the plant, in a storage capacity as it has been found in a high concentration in fresh roots, but only at low levels in the stem, petioles and leaves or sprouted roots (Maeshima, Sasaki and Asahi, 1985) and was almost absent in roots stored for 1 year (Li and Oba, 1985). This major protein, sporamin, accounting for more than 80% of the total soluble protein in roots, can be separated into two closely related proteins: sporamins A and B. These have similar molecular weights (~25,000) and similar, but not identical, amino acid profiles according to one group of researchers (Maeshima et al., 1985). The chemical score of sporamin A at 77%, with lysine as the limiting amino acid, was higher than that of sporamin B at 55%, with methionine limiting. Other workers (Li and Oba, 1985) found one major protein with a molecular weight of 25,000 apparently existing in two isomeric forms, which accounted for 60–70% of the total soluble protein. This major protein was suggested as a useful marker for breeding sweet potatoes with higher protein contents. It seems likely that their major protein is in fact the same as sporamin. The genes coding for sporamins A and B have recently been isolated and identified (Hattori and Nakamura, 1988). Experimental evidence suggests that accumulation of sucrose transported from the leaves may play an important part in the expression of these genes during root development in the field (Nakamura, 1989, and personal communication).

Part of the protein present acts as an inhibitor of the proteolytic digestive enzyme trypsin (Sohonie and Bhandarkar, 1954). The relationship between the concentrations of trypsin inhibitor and total protein and the significance of this inhibitor in sweet potato roots will be discussed in Chapter 4.

The NPN of sweet potato includes peptides too small to be precipitated by reagents which coagulate true protein, free amino acids, amides and other non-polymeric nitrogen compounds. The composition of NPN immediately after harvest has not been published. Its main components after 107 days of storage were asparagine (61%), aspartic acid (11%), glutamic acid (4%), serine (4%) and threonine (3%)(Purcell and Walter, 1980). The remaining 5.5% of recovered nitrogen consisted of small amounts of other amino acids and ammonia. A further 11.5% N remained unrecovered and unidentified. NPN is undoubtedly part of a metabolically active pool containing large amounts of asparagine which are readily available for synthesis of amino acids as demanded by the root (Purcell and Walter, 1980). Aspartic and glutamic acids would be available for transamination reactions. From a nutritional viewpoint NPN would be able to supply total utilizable nitrogen to the body, but provides little in the way of essential amino acids. It therefore acts as a diluent of the amino acids in the true protein and hence alters the nutritional value of sweet potato total protein according to its relative concentration. The determination of total protein alone may therefore overestimate protein nutritional value. The amino acid content of protein and NPN determined in American sweet potatoes is shown in Table 2.10.

Distribution of protein within the root

There is evidence from work with cultivars grown in the United States and Taiwan (Purcell, Walter and Giesbrecht, 1976b; Li, 1982) that total

Table 2.10. *Amino acids in protein and non-protein nitrogen of raw sweet potato roots*

	g amino acid/16 g N	
	NPN[a]	True protein[b]
Asp	10.79	16.7[c]
Thr	2.97	5.9
Ser	4.36	5.8[c]
Glu	3.93	9.7[c]
Pro	0.18	
Gly	0.68	5.1[c]
Ala	2.14	5.6[c]
Val	0.78	7.6
Met	0.15	2.7
½-cystine	0.09	0.6
Ile	0.17	6.1
Leu	0.12	8.4
Tyr	0.46	5.6
Phe	0.97	7.1
Lys	0.08	5.8
His	1.08	1.7[c]
Try	ND[d]	1.0
Arg	0.68	5.5[c]
As	61.48	ND
NH_3	0.21	

Notes:
ND, not determined.
[a] Purcell and Walter, 1980. NPN, non-protein nitrogen fraction from cured and stored roots.
[b] Purcell et al., 1972. Fresh roots. Protein coagulated with trichloroacetic acid.
[c] Calculated from results of Purcell et al., (1972).

protein is not evenly distributed throughout the root. End-to-end gradients of protein are small but significant with higher concentrations at the proximal (stem) end than at the distal (root) end. A higher concentration of total protein has been reported in the outer layer of flesh, close to the skin, which is removed by peeling (Purcell et al., 1976b; Makki et al., 1986; Kimber, 1976) than in the rest of the flesh. One group of workers found that peel removed by scraping (2.5% of the total weight) or deep peeling (8.9% of the total weight) contained respectively 87% and 47% more protein per unit weight than the peeled material

(Bradbury et al., 1984). Deep peeling would therefore reduce the protein content of the remaining material by about 12% and scraping by about 4%. Other work (Kimber, 1976), however, indicated that when roots were peeled by cutting off the skin and about 1–2 mm of adhering flesh, the material remaining after peeling contained only about 0.1–0.2% less protein than the whole root. Although the protein-rich tissue has a relatively high protein concentration it is present in very small quantities. Radial differences in total protein have been found to be insignificant (Purcell et al., 1976b). Results suggest that proper sampling for total protein analysis should entail use of longitudinal sections of tissue.

Factors influencing protein concentration

Considerable genetic variability seems to exist for both total and true protein, being a major factor in observed differences in sweet potato protein content. However, variations due to environmental effects and differences in cultural management practices have also been noted.

A range of total protein from 1.27% to 10.07% (dwb) was found among 300 lines grown in Taiwan under similar cultural management in a single season, with the majority containing 4–5% protein (Li, 1974). Total protein of 100 seedlings from seven parental clones grown in a single American location in one season ranged from 4.38% to 8.98% (dwb) with a mean of 6.29% (Dickey et al., 1984). There is thus a strong possibility of enhancing protein content by breeding or selection of high protein cultivars. A number of new, improved cultivars from Peru had high protein contents of 8.9% to 14.9% (dwb) (Carpio Burga, 1985). On a fresh weight basis, the total protein of six American cultivars grown under identical conditions and analysed immediately after harvest ranged from 1.36% to 2.13% (Picha, 1985b). Ten cultivars from Papua New Guinea ranged in mean (for a number of different locations and seasons) fresh weight total protein from 1.29% to 1.81% (Bradbury et al., 1985b).

Variations in total protein content within cultivars have been studied by several groups. One group (Purcell et al., 1978) found that variation was slightly less, and another (Bradbury et al., 1985b) that it was as great among roots from the same plant as between plants. High field-to-field and location (Purcell et al., 1978) variability has also been noted within cultivars.

Environment (which may vary according to location, season or year of growth and entails changes due to climate, soil, incidence of pests and diseases etc.) may exert a significant influence on total protein, affecting some cultivars more than others. Analysis of variance of the data gathered for six genotypes grown for 3 years at six locations (18 environments) showed that total protein varied by genotype, environment and interaction of genotype × environment ($p < 0.01$) (Collins and Walter, 1982).

Differences in total protein content of roots of a number of Taiwanese cultivars were significant ($p < 0.01$) among cultivars, locations, years and seasons (Li, 1975); ranges were 3.2–8.3%, 2.8–6.5%, 4.3–5.8% and 3.4–6.1% (dwb), respectively. Cultivar × location interaction was also highly significant, indicating the importance of testing in different locations, but cultivar × year interaction was small and insignificant, indicating that cultivars ranked similarly in each year.

Genotype × environment interactions can be used to produce a graph which pinpoints cultivars combining relatively high protein content with low environmental protein variability, as illustrated with Papua New Guinea sweet potatoes (Bradbury et al., 1985b).

Cultural management techniques, which include irrigation, fertilization, plant spacing and variations in planting and harvest time, may also influence total protein content. The increase in total protein with increasing application of nitrogen in fertilizer has been well documented (Constantin et al., 1974; Gonzales, Cadiz and Bugawan, 1977; Kimber, 1976; Li, 1975; Purcell et al., 1982; Yeh, Chen and Sun, 1981). Increasing nitrogen applications from 0 to 136 kg/ha increased the total protein (dwb) of 'Serenta', a cultivar widely grown in the Papua New Guinea highlands, from 2.72% to 4.35% (Kimber, 1976). Total root nitrogen (dwb) of 'Jewel' increased from 1.12% to 1.46%, without any alteration of NPN:total nitrogen ratio by increased nitrogen application in a sandy loam soil of 0 to 112 kg/ha (Purcell et al., 1982). Root protein yields per plot (mean of three cultivars) more than doubled on a soil of poor fertility in the Philippines when 150 kg N/ha was applied (Gonzales et al., 1977). There are indications that there may be an optimum rate of nitrogen application for maximum production of root nitrogen, as high levels of applied nitrogen may be less efficiently utilized for root nitrogen production (Kimber, 1976) and may bring about decreased root (Yeh et al., 1981) or dry matter (Li, 1975) yields.

Potassium fertilization has been reported to have little or no effect on total protein content of roots (Li, 1975; Yeh et al., 1981; Purcell et al., 1982), or to decrease it (Constantin et al., 1977; Caraballo Llosas, 1974). Potassium applications of 140 kg/ha (Constantin et al., 1977) or 240 kg/ha (Caraballo Llosas, 1974) caused total protein to decrease significantly. Total protein yield of the crop may increase as a result of increased yield of roots with potassium applications (Purcell et al., 1982; Yeh et al., 1981).

Protein content (% dwb) was decreased by increased rate of phosphorus application (Constantin et al., 1977; Hammett et al., 1982), and fertilization with sulphur had no effect (Purcell et al., 1982; Bradbury et al., 1985b). The use of shredded waterhyacinth (25 tonnes/ha) in combination with bonemeal was reported to increase root total protein in Indian sweet potatoes to 5.45% (dwb) from 3.8% in untreated controls or 4.39% with bonemeal alone, and to produce the highest yields (Maurya and Dhar, 1976).

Supplemental irrigation has been reported to bring about a change in total protein content (Constantin et al., 1974). Percentage protein decreased as the moisture content of the soil increased. Alteration of plant population from 35,000 to 50,000 plants/ha did not significantly affect root protein levels (Li, 1975).

An investigation of the effect of length of growing season on protein content of 16 American cultivars showed that percentage protein (dwb) decreased significantly when harvest date was prolonged from 102 to 165 days after planting (Purcell, Pope and Walter, 1976a), but the authors concluded that selection of high protein cultivars offers a greater chance of obtaining high protein sweet potatoes than does utilizing early harvest, which in terms of yield reduction may in any case not be economically feasible. Moreover these results were in contrast to those found for six Brazilian cultivars which had the highest protein contents when harvested 8 months after planting than after 4 or 6 months (Menezes et al., 1976).

Inoculation of the root areas of transplants of 'Centennial' with 5 ml suspensions (10^9 cells/ml) of selected strains of the bacterium *Azospirillum* has been found significantly to increase root nitrogen concentration over that of untreated controls (Crossman and Hill, 1984). Growing 'Georgia Jet' sweet potatoes in elevated concentrations of carbon dioxide decreased their total protein content (Lu et al., 1986).

The positive correlation between high protein content and orange flesh colour found among seedlings in Taiwan (Li, 1982) was not confirmed in Papua New Guinea cultivars (Bradbury et al., 1985b), in which neither skin nor flesh colour was related to total protein content.

There are indications in the literature that protein content is not strongly negatively correlated with either yield (Li, 1982; Kimber, 1976) or dry matter content (Purcell et al., 1976a). Therefore the chances of obtaining combinations of high yield, high dry matter and high protein appear likely.

The fact that a percentage of sweet potato total nitrogen is present as NPN (which has very low concentrations of essential amino acids) means that investigations into the sources of variation in true protein and NPN levels provide a better understanding of the protein nutritive value than total nitrogen determinations alone. Coagulable true protein varied from 1.73% to 9.14% (dwb) and 0.49% to 2.24% (fwb) among

100 entries in a root collection in one location under identical conditions of cultivation (Purcell, Swaisgood and Pope, 1972). Most cultivars contained between 4.5% and 7% protein (1.5–2% fwb). Therefore true protein appears to vary as much among cultivars as does total protein.

In a selection of 10 high protein seedlings out of 100 seedlings grown from seven parents (Dickey et al., 1984), the percentage of NPN in total nitrogen ranged from 22.1% to 37.7% (mean 29.6%). The parents ranged from 16.6% to 33.7% (mean 25.8%). There was no significant difference between the means of parents and selections. Total protein and NPN levels were not significantly correlated ($r = 0.3$) and hence selection for high total protein need not result in appreciable increases in NPN levels. Furthermore, when increased nitrogen fertilizer application resulted in an increase in root total nitrogen, there was no change in NPN:total protein ratio (Purcell et al., 1982). In other words, as total nitrogen increased so did protein nitrogen. Thus nitrogen fertilization may increase protein nutritive value of sweet potato. Potassium fertilizer application did not alter the NPN:total nitrogen ratio. Undefined conditions relating to location, year and environment which affected NPN:total nitrogen ratio have still to be sought.

Variations in individual amino acids have been noted not only between, but also within cultivars (Bradbury et al., 1984, 1985b; Purcell and Walter, 1982). Amino acid variations occur even between roots from the same plant and are high for a cultivar grown in several different environments (Bradbury et al., 1985b). However, because there is much greater variability in protein content than in protein quality, improvements may be made in the former rather than the latter by selection/breeding. Amino acid content of sweet potato relating to nutritional value is discussed in greater detail in Chapter 3. Recent developments in the improvement of sweet potato protein quality by genetic engineering are also described.

Tops

In those areas of the world where sweet potato tops are consumed, cultivars may be grown which are suitable for harvesting of both roots and leaves, or for harvesting of leaves alone. An alternative title for this section could have been Greens as in fact different parts of the sweet potato top or vine are eaten in different countries. The most common form in which they are found is the 10–15 cm apical tip of the vine, including both the stem, petioles and tender leaves. Tips may be sold in the market either by number (in bundles) or by weight. Cultivars have been shown to vary in yield as judged by weight or number of tips produced per unit area, and could therefore be selected by farmers depending on how they are to be sold (Villareal et al., 1979b). In some places, for example Korea and Japan, the petioles alone are eaten instead of the tips, whereas the Haka people of Taiwan eat the mature leaves and petioles (Villareal et al., 1985).

The cutting off of tips, or leaves and stems at regular intervals is known as topping. Where sweet potatoes are to be grown for both roots and tops, topping has to be carried out in such a way as to optimize the yield of both products. Topping carried out too frequently or at the time of rapid root formation and development results in low root yield. However, cultivars exist which are grown only for their tops (for example in the Philippines) and they produce no enlarged roots worth harvesting (Villareal et al., 1979b).

There appear to be differences in the percentage distribution of leaves, stems and petioles between cultivars (Anon. 1985b). The tops of line CN 1508-93 grown in Taiwan had long petioles constituting 38% of the edible portion, a much higher percentage than that in three other lines (Table 2.11). Such a line might be particularly suitable for areas where only petioles are consumed.

In contrast to sweet potato roots, there is little information about the chemical composition of sweet potato greens, most of the work to date having been carried out on the tender tips, and some components having been little explored. There is necessarily also a scarcity of information about the factors causing variations in chemical and nutritional composition and quality. This reflects the neglect, until recently, of research into the usefulness in the human diet of sweet potato tips and other such underexploited green vegetables. Most of the information available is concerned with the use of sweet potato vines as animal feed. This is discussed in Chapter 7.

Dry matter

As might be expected, the level of dry matter in sweet potato greens is lower than that in the roots, being on average about 12–14%. However, a pooled sample of Singapore market sweet potato leaves purchased over 2–3 months contained only 5.9% dry matter (Candlish, Gourley and Lee, 1987). Table 2.11 shows the considerable genetic differences which exist in dry matter. Young leaves of Nigerian samples were reported to contain 12.5% dry matter (Oyenuga, 1968). The dry matter in 10 cultivars grown in Taiwan ranged from 12.6% to 16.2% (mean 14.5%) for 15 cm tips (Villareal et al., 1979b). Young leaf tips, taken from the apical 10 cm of American-grown vines, with 17.3% dry matter were not significantly different from older leaves, taken from the next 10 cm down and containing 18.8% dry matter (Pace et al., 1985b). However, the leaf

Table 2.11. *Distribution and proximate composition of dry matter in sweet potato tops*

Portion	Weight distribution (%)	DM (%FW)	Protein (%DM)	Crude fibre (%DM)	Starch (%DM)	Sugar (%DM)	Ash (%DM)
Total[a]		12.52	20.90	14.92	3.71	8.82	13.52
Stem[a]	26.0	9.63	13.64	20.69	3.91	12.64	13.79
Petiole[a]	23.9	7.91	12.74	16.64	4.47	11.03	18.35
Leaf[a]	50.1	16.26	28.56	11.10	3.25	5.98	10.94
Total[b]		6.62	20.67	15.00	3.63	12.99	16.67
Stem[b]	24.4	7.92	14.06	20.79	3.37	16.84	14.92
Petiole[b]	37.9	5.92	13.25	17.06	4.00	15.68	22.27
Leaf[b]	37.7	6.47	32.42	9.19	3.43	7.79	12.18

Notes:
DM, dry matter; FW, fresh weight.
From: Anon, 1985b.
[a] Average of three lines with very similar DM contents.
[b] Line CN1508–93 with low DM content.

samples contained significantly higher levels of dry matter than did the corresponding young or old stems/petioles which had 12.1% and 12.8% dry matter, respectively. This differential in dry matter between the leaves and stems/petioles was also found in three lines containing approximately 12% dry matter in Taiwan, but not in a low dry matter line (Table 2.11).

The proximate composition of dry matter in sweet potato tops and their constituent parts, as determined for several lines grown under similar conditions in Taiwan (Anon. 1985b), is shown in Table 2.11. The major component was found to be non-starch polysaccharides which accounted for about 60% of sweet potato tip dry matter (Anon. 1985b). This appears to be the only such detailed analysis carried out so far. More up-to-date analyses are clearly needed if this neglected vegetable is to receive the attention it deserves.

Pigments

Sweet potato leaves and stems or petioles may be entirely green, green with varying degrees of purple pigmentation, predominantly purple, or less commonly, yellow. Varieties with some purple pigment may be preferred over wholely green types as they add colour and attractiveness to the table (Villareal et al., 1979a). The green colour is of course due to chlorophyll, this being partly or wholely masked by anthocyanins where purple pigments are present. Yellow leaves no doubt have lower levels of chlorophyll, allowing carotenoids to predominate. There appears to be little information about individual carotenoids in sweet potato leaves. It seems likely, however, that beta-carotene is the major provitamin A carotenoid present, but other pigments including xanthophylls can also be important (Tsou, 1986). Thus dried sweet potato vines analysed at varying stages of field maturity were shown to contain beta-carotene at approximately 30–50% by weight of xanthophylls (Garlich et al., 1974). A Japanese study showed that fresh vines of varying ages contained 14–37 μg carotene/g, but only 0.3–0.6 μg cryptoxanthine/g (Sutoh, Uchida and Kaneda, 1973). The same study revealed a progressive decrease in carotenoid levels in fresh vines, with an increase in their age at harvest. This was confirmed by others who showed that beta-carotene and xanthophylls respectively declined from 21.8 mg/100 g and 31.7 mg/100 g (90% dry matter basis) in 'Centennial' vines harvested at an early stage in July to 7.6 mg/100 g and 16.1 mg/100 g when harvested late, in October (Garlich et al., 1974). Sweet potato foliage xanthophylls have been investigated as a potentially useful source of broiler skin and egg yolk pigmentation when included in chicken diets (see Chapter 7).

Oxalic acid

The only organic acid to be investigated in sweet potato tops, because of its implications for their nutritional value, is oxalic acid. The oxalate content of tips of 10 cultivars grown in the same location in Taiwan differed one from another, ranging from (fwb) 280 mg/100 g to 450 mg/100 g (mean 370 mg/100 g) (Villareal et al., 1979b). Three cultivars, representing broad, medium and fine leaf types contained 490, 480 and 340 mg oxalate/100 g, respectively. Frequency of harvesting did not affect oxalate content, but it decreased from 460 mg/100 g to 420 mg/100 g (means of the three cultivars) when the level of nitrogen fertilization increased from 0 to 120 kg/ha. This investigation revealed that it is possible to combine lower oxalate content with other desirable characteristics in the leaves. The fine leafed type 'Kinangkong' from the Philippines, for example, contained the lowest oxalate concentration, combined with relatively high protein and dry matter contents.

The distribution of oxalate in leaves at different positions along the stem has been investigated in two cultivars (AVRDC, 1976). The tips consisting of the vine and leaves in the first 15 cm from the apex, and then the remaining leaves separated into groups of 15 cm intervals along the stem, were analysed. Oxalate concentration was found, in both cultivars, to increase significantly from the tip to the base. In one cultivar the oxalate content was three times greater in the oldest leaves than in the tips. This finding plus the fact that protein concentration was also highest in the tips (see below) was used to support a recommendation that only sweet potato leaf tips should be eaten.

On a fresh-cooked basis, sweet potato leaves and other tropical leafy vegetables have been shown (Evenson and Standal, 1984) to contain much higher levels of oxalic acid than do common temperate climate vegetables, with the exception of spinach. The fact that some developing-country diets are deficient in calcium and other minerals, bound and rendered unavailable to the body by oxalic acid, makes selection of low oxalate types an important goal for researchers seeking to popularize sweet potato leaves.

Fibre

The determination of total fibre and its individual constituents in sweet potato tops is important not only from a nutritional, but also from a quality viewpoint. As providers of dietary fibre, sweet potato tops make a positive contribution to certain physiological functions. Fibre levels in tops may also, however, adversely influence bioavailability of provitamin A carotenoids (see below) and absorption of minerals. Sweet potato tops are often rejected by consumers as being fibrous, tough and less palatable than other green leafy vegetables. One reason for the consumption of tips rather than older leaves is the tender nature of the young leaves which occur in the apices of the vines. The fibre content of tips themselves has also been shown to increase with increased age of tips at time of harvest (AVRDC, 1976). The mean fibre contents (% fwb) of tips from 10 cultivars harvested 40, 79 and 118 days after planting were 1.96, 1.98 and 2.07, respectively.

Total dietary fibre was found to average 2.4 g/100 g fresh weight (40.6% of dry matter) in a number of samples of sweet potato tops purchased in Singapore (Candlish et al., 1987). This quantity consisted of 1.6 g (66%) non-cellulosic polysaccharides (uronic acids, pentoses and hexoses) and 0.8 g (34%) cellulose, a distribution similar to that in four Taiwanese lines (Anon., 1985b). In addition lignin, which was simply the residue left at the end of a multiple-stage procedure for determining other constituents, averaged 0.36 g/100 g fresh tops, but had a very high coefficient of variation of ± 0.4.

Non-starch polysaccharides, which occurred in the forms water-soluble to water-insoluble 1:5, made up the major component (60%) of sweet potato tip dry matter in four Taiwanese lines (Anon., 1985b). About 80% of the total edible fibre consisted of pectic substances, cellulose and hemicellulose with the remainder as gums and lignin. The percentages of individual components (excluding gums and lignin) were determined in soluble and insoluble non-starch polysaccharide fractions for complete tips, as well as leaves, stems and petioles. Hemicellulose was the major component in both soluble and insoluble non-starch polysaccharide and in all plant parts. Since the soluble form is extracted during cooking, the author concluded that the insoluble form is probably the main constituent affecting the texture of sweet potato greens. It was thought that a relatively high percentage of a pectic substance in the insoluble non-starch polysaccharide of one of the four lines analysed might be a factor contributing to this line's greater tenderness. In general there appeared to be few differences in content of total fibre or its individual components between the tops of sweet potato and water convolvulus (*Ipomoea aquatica*), which is often preferred as a green vegetable (Candlish et al., 1987; Anon., 1985b).

Since lignin content is also likely to influence toughness in sweet potato tops, it should be determined more accurately in investigations of the level of fibre contributing to toughness, the site of greatest concentration of fibre (in stems, petioles or leaves), and the factors which cause variations in fibre content.

Nitrogenous constituents

Content

Leaf nitrogenous constituents have been determined as total protein (N × 6.25) by most authors to date. As in the roots, part of the total nitrogen in leaves exists as NPN. Only one source was found which reported concentrations of true as well as total protein, and the ratio of true to total protein, in the leaves of two cultivars (Ruinard, 1969). These can be seen in Table 2.12, and indicate that true protein represents a high percentage of the total in leaves. There was an indication in cultivar 'Okinawa 2' that the ratio decreased when leaves were harvested at 4 or 5 months rather than at 3 months, but it remained consistently higher than the corresponding ratio in the roots.

The tops of sweet potatoes are superior to the roots in total protein content, averaging about 3% (fwb) and 20% (dwb). Considerable variations do exist: total protein (dwb) of Nigerian sweet potato leaves were reported as 24.6% (Oyenuga, 1968), but that of two lines grown in Taiwan was only 10.8% and 13.2% (Cheng, 1978), whereas two cultivars from the United States averaged 29.9% over three harvest times and two seasons (Pace, Dull and Phills, 1985a). The reasons for such variations have been studied in more detail than for other constituents of the dry matter and are due to cultivar and other effects.

Table 2.12. *Comparative composition of leaves and roots in two cultivars (% dry matter)*

	'Genjem 2'		'Okinawa 2'	
	Leaves[a]	Roots	Leaves[a]	Roots
Starch	4.2	ND	10.1	ND
Sugar	1.1	ND	0.2	ND
Total protein	30.8	5.1	31.7	6.0
True protein	25.8	4.7	30.6	4.8
Ash	?	3.5	?	3.8
K	3.37	1.29	3.03	1.6
P	0.62	0.15	0.46	0.11
Mg	0.30	0.09	0.23	0.07
Ca	1.0	0.34	0.75	0.30
Na	0.06	0.04	0.06	0.05
True: total protein	0.84	0.92	0.97	0.80
Carotene (mg/100 g DM)	72.1	1.5	62.0	0.70
% DM	14.1	30.4	15.5	26 ?
Flesh colour	Yellow		White	

Notes:
ND, apparently not determined; DM, dry matter.
From: Ruinard, 1969.
[a] Top shoots (30 cm) harvested after 3 months growth.

Factors affecting protein content

The major source of variation in leaf protein content is cultivar. The leaves (40 cm tips) of 300 lines grown in one location in Taiwan under similar cultivation techniques and in a single season ranged from 12.1% to 25.7% total protein (dwb) (Li, 1974). The most frequently observed contents were 19.5% to 20.5%.

Variations in protein content do not exist just between cultivars, but are significant between plant parts within a cultivar. The protein content of leaves was significantly higher than that of the stems or petioles in four lines grown in Taiwan (Anon., 1985b; and see Table 2.11). Furthermore, when the variation in protein content of leaves according to their position on the stem was investigated (AVRDC, 1976), the tip, representing the first 15 cm from the apex, was found to have the highest protein content and the oldest leaves furthest from the tip the lowest. For two cultivars, protein contents fell from about 2.4% and 2.8% in tips to 1.3% and 1.5% in leaves 75–90 cm from the apex. For these reasons the consumption of whole tips ensures a higher protein intake than if older leaves or only the petioles are eaten.

Cultivation management and harvesting techniques also affect protein content of tops. Highly significant differences in leaf protein between plants grown in spring, winter and autumn seasons was found for two cultivars in Taiwan (Li, 1975).

The significant increase in total protein content of leaves or tips which takes place with increasing nitrogen fertilization has been confirmed by several researchers (Gonzales et al., 1977; Li, 1975; Villareal et al., 1979b; Yeh et al., 1981). Protein contents increased with nitrogen applications of 60 kg/ha or more. Furthermore, because tip yield is also increased by nitrogen application, the total protein harvested also increases. An increased percentage of the amino acid lysine has also been reported in sweet potato leaves harvested from 100 kg N/ha fertilized plants (Yeh et al., 1981). According to one author, potassium fertilization has little effect on the protein content of vines (Li, 1975), but other authors produced increases in both protein and lysine contents of a cultivar used especially for its leaves and stems (Yeh et al., 1981) by application of potassium fertilizer at 120 kg/ha.

Protein content of tips varies with harvesting date, and its interaction with cultivar. Three types of response have been noted amongst cultivars when 15 cm tips were harvested 40, 80 and 120 days after planting (Villareal et al., 1979b). Some cultivars showed a steady reduction in protein content from the first to the third harvesting; others increased in protein between 40 and 80 days after planting and thereafter decreased but still retained a good protein production; yet another sustained its productivity over the whole period with only a small reduction between 80 and 120 days after planting. The first response was expected, since the nitrogen available for protein synthesis could be drawn on by both roots and tips. Thus 40 day samples had the highest protein and the 120 day ones the least; 40 day harvesting coincided with root initiation, whereas 80 day harvesting coincided with root enlargement, and it is possible that the second and third responses were due to competition between roots and tips for available nitrogen. Presumably the tips which increased in protein content were able to take more of the nitrogen for protein synthesis. Others have shown similar results for the production of total protein per plot from sweet potato leaves (Gonzales et al., 1977). Recent results for two American cultivars, harvested in what was described as the vegetative, mid-vegetative and root-forming stages of growth, showed significant decreases of protein content in 10 cm tips between the first and second harvests with a stabilization or very slight increase thereafter (Pace et al., 1985a). Frequency of harvesting (either weekly or bimonthly) has been reported not to affect protein content (Villareal et al., 1979b). If tips are harvested frequently a greater total yield of tips and tip nutrients, including protein, will result. However, if roots are to be harvested also, a greater frequency of topping will result in decreased root yields (Bartolini, 1982; Gonzales et al., 1977). Therefore when tops are harvested it is important to time harvesting so that the production of roots is affected as little as possible.

Vitamins

Sweet potato leaves are very good sources of several vitamins: beta-carotene (provitamin A), thiamin (B_1), riboflavin (B_2), folic acid and ascorbic acid (C). They also contain niacin and pyridoxine (B_6). However, few up-to-date analyses are available for sweet potato leaves, and there is a lack of information about variations in vitamin content and the factors which affect concentrations, both pre- and postharvest.

Little attempt appears to have been made so far to analyse tops for their individual carotenoid content and some authors have expressed the vitamin A equivalent in IU (see p. 76). These are probably based on analyses for total carotenoids and may therefore be inaccurate. The carotenoid content of sweet potato leaves has been reported to be on average 3610 μg/100 g (fwb) (Kay, 1987). Fresh vines (leaves and stems) ranging in age at harvest from 89 to 147 days were reported to contain from 3777 μg/100 g to 1597 μg/100 g provitamin A carotenoids (Sutoh et al., 1973), showing that carotenoid content declines as the age of vines increases. A range of vitamin A activity of 3520 to 8320 IU/100 g has been reported for the fresh 15 cm tips of 10 cultivars grown in the same location in Taiwan (Villareal et al., 1979b), but the difference between cultivars was not significant. The average vitamin A activity was 5580 IU/100 g. This is similar to the figure of 5256 IU/100 g found for freshly harvested young sweet potato shoots grown in the Philippines (Oñate et al., 1970). Sweet potato leaves and tender tips are reported to contain 2700 μg/100 g (fwb) of beta-carotene equivalents in food composition tables (Leung, Butrum and Chang, 1972).

The beta-carotene content of tips, leaves, stems and petioles has been determined in four lines grown under similar conditions in Taiwan (Anon., 1985b). Tips varied from 0.61 to 2.66 mg beta-carotene/100 g fresh weight among the four lines. The highest concentration of beta-carotene in all four lines was in the leaves (range 1.01–5.53 mg/100 g) with petioles containing only between 0.14 and 0.47 mg/100 g.

The bioavailability of provitamin A carotenoids in sweet potato tops has not been determined directly. However, it seems likely to be similar to that of other green leafy vegetables. In a rat-feeding experiment, the provitamin A carotenoid bioavailabilities of spinach, water convolvulus and field mustard were not significantly different and averaged 12.2%. This figure was significantly lower than that for sweet potato roots (Tsou, 1986) determined in the same experiment. Chlorophyll, non-provitamin A carotenoids and fibre were all found to lower the bioavailability of pure beta-carotene in rats and were postulated to be the major factors responsible for the low bioavailability of provitamin A in green leafy vegetables (Tsou, 1986; 1987). If the bioavailability of beta-carotene in sweet potato tops proves to be about 12% then it would be more accurate to calculate retinol equivalents by a conversion factor of 8 μg beta-carotene (or 16 μg other provitamin A carotenoids) to 1 μg retinol.

Generally speaking, if the leaves or whole tops rather than just the petioles are eaten, sweet potato tops appear to be fairly good sources of provitamin A, unlike the roots of some cultivars which contain little or no provitamin A activity; but a great deal more and detailed information is required on this topic.

The existence of widespread riboflavin deficiency in Southeast Asia has prompted some interest in the riboflavin content of sweet potato leaves. A range of 0.29 to 0.41 mg riboflavin/100 g (fwb) occurred among 10 cultivars grown under the same conditions in Taiwan (Villareal et al., 1979b). The riboflavin content of samples taken directly from the field in Malaysia ranged from 0.26 to 0.35 mg/100 g (mean 0.31 mg/100 g (fwb)), whereas other samples bought in the market contained less – 0.16 to 0.24 mg/100 g (mean 0.19 mg/100 g) (Caldwell and Enoch, 1972). Riboflavin content was not found to be affected by either leaf type (fine, medium or broad) or by increasing levels of nitrogen fertilization up to 120 kg/ha (Villareal et al., 1979b). More information on the effects of postharvest treatment of leaves such as the effects of bruising, drying, wilting, exposure to direct sunlight etc. on the content of riboflavin and other vitamins should be urgently sought if leaves are to be useful in alleviating existing deficiency states.

Considerable differences in ascorbic acid content of different leaf samples have been reported in food composition tables and in the literature. However, there are indications that freshly harvested leaves are generally a good source of ascorbic acid and in some cases extremely good. There is no direct information as to the forms of ascorbic acid which occur in fresh raw leaves, nor how these change with the time which may elapse between harvest and cooking. However, ascorbic, dehydroascorbic and diketogulonic acids were found to account for approximately 83%, 13% and 4%, respectively, of six other types of fresh, raw, green leafy vegetable (Pasricha, 1967). The ascorbic acid content of 10 cultivars grown in Taiwan was reported to range from 32 to 73 mg/100 g (fwb) (Villareal et al., 1979b), that of market samples

Table 2.13. *Vitamins in raw sweet potato leaves (per 100 g (fwb))*

Vitamin	Content
β-Carotene	
leaves (mg)	3.44[a] (1.01–5.53)
petioles (mg)	0.24[a] (0.14–0.47)
stems (mg)	0.65[a] (0.27–1.63)
Thiamin (mg)	0.1[b] (0.03–0.12)
Riboflavin (mg)	0.35[c] (0.29–0.41)
Niacin (mg)	0.9[b] (0.6–1.0)
Pyridoxine (mg)	0.21[d]
Folic acid (μg)	88.4[d]
Ascorbic acid (mg)	55[e] (11–136)

Notes:
Ranges in parentheses.
[a] Anon., 1985b; data for four cultivars grown at AVRDC.
[b] Leung et al., 1968.
[c] Villareal et al., 1979b.
[d] Rao and Polacchi, 1972.
[e] Mean of data from: Leung et al., 1972; Caldwell, 1972; Watson, 1976; Villareal et al., 1979b; Haytowitz and Matthews, 1984.

purchased in Malaysia from 71 to 136 mg/100 g (fwb) (Caldwell, 1972) and that of Ghanaian sweet potato leaves from 83 to 121 mg/100 g (Watson, 1976). Others have reported leaves to contain about the same quantity of ascorbic acid as the roots, i.e. 20–25 mg/100 g (fwb) (Kay, 1987; Evensen and Standal, 1984). North American leaves have been reported to contain only 11 mg/100 g (Haytowitz and Matthews, 1984).

Contents of niacin, pyridoxine and folic acid have been reported in food composition tables, but more up-to-date and detailed information is required about them. Vitamins in sweet potato leaves as determined by various authors to date are shown in Table 2.13.

Table 2.14. *Minerals and trace elements in raw sweet potato leaves (dwb)*

mg/100 g	
Ca	1351
Fe	20
K	3018
Mg	432
Na	38[a]
P	264
p.p.m.	
Al	328
B	55
Ba	160
Cu	12
Mn	210
Mo	2
Sr	111

Notes:
'Centennial' cultivar only. Calculated from Paterson & Speights (1971).
[a] Leung et al., 1968.

Minerals

Content

The ash content of sweet potato leaves is about 1.6% (fwb) or 12% (dwb). It contains a wide variety of minerals and trace elements, including relatively substantial quantities of Fe and Ca as well as P, Mg, Zn and Cu. The K:Na ratio is very high. The contents of minerals and trace elements determined in 'Centennial' cultivar sweet potato leaves is shown in Table 2.14. On a dry weight basis, the concentrations of minerals are much greater in foliar than in root tissue (Hammett et al., 1984; Scott and Bouwkamp, 1974 and Table 2.12).

Factors affecting mineral contents

Cultivar may play a minor part in affecting concentrations of some minerals compared with external factors. No significant differences in Ca or Fe contents were noted for 10 cultivars grown under the same conditions in one Taiwanese location (Villareal et al., 1979b). However,

Table 2.15. *Variations in Ca, Fe and Zn between leaves and stems in 'Jewel' sweet potato (mg/100 g)*

	Ca		Fe		Zn	
	dwb	fwb	dwb	fwb	dwb	fwb
Leaf tips	836^{A}[a]	150^{D}	10.9^{A}	1.8^{D}	2.9^{A}	0.5^{E}
Older leaves	1144^{B}	217^{F}	13.4^{B}	2.4^{E}	2.5^{B}	0.5^{E}
Stem – petioles from tips	931^{A}	118^{E}	4.7^{C}	0.5^{F}	1.6^{C}	0.2^{F}
Older stem – petioles	852^{A}	109^{E}	4.6^{C}	0.6^{F}	1.2^{D}	0.2^{F}

Notes:
Adapted from Pace et al., 1985b.
[a] Means in same column but with different superscripts differ significantly ($p<0.05$).

recent studies showed that contents of Ca, Fe and Zn can vary in different parts of the green tops (Pace et al., 1985b; Anon., 1985b). Some results are shown in Table 2.15. Older leaves of 'Jewel' contained significantly more Ca and Fe than did young leaf tips. Both young and older stem-petioles contained significantly less of all three minerals than did the young or older leaves. The Fe concentration in leaf tips of four Taiwanese lines varied from 1.53 to 3.35 mg/100 g (fwb) (Anon., 1985b). Individual analyses of leaves, stems and petioles showed the highest Fe concentration to be in the leaves.

Applications of foliar sprays containing minerals were found to enhance the concentrations of some minerals and trace elements in 'Centennial' sweet potato leaves grown in the same field (Paterson and Speights, 1971). Use of a spray containing Zn significantly increased the leaf Zn content, whereas applied Fe significantly reduced leaf Ca and B content. Application of B significantly increased B and decreased Zn in the leaves. Use of sprays containing Zn, Fe, B or Mn or combinations of these had no effect on the P, K, Mo, Mn, Cu, Sr or Ba contents of leaves.

It seems likely that the mineral content of the soil in which sweet potato plants are grown will influence leaf mineral contents, but so far little information appears to be available on this topic.

Summary

Some of the compositional advantages of sweet potato tops when compared to roots can be seen in Table 2.12. The dry matter of tops contains greater quantities of protein and minerals than that of the roots. In addition, the tops of many cultivars have more favourable ascorbic acid and riboflavin concentrations than do the roots, and frequently maintain a fairly good supply of provitamin A carotenoids, which in the roots may be absent. Several investigations have shown that in general the concentrations of major nutrients in tops are higher in the leaves than in the stems or petioles, and for this reason it is preferable for people of low nutritional status to be encouraged to eat the leaves or whole tips rather than the petioles alone.

Research into the chemical composition of little known and under-utilized green vegetables such as sweet potato leaves is only in its initial stages and information is patchy and inadequate. It is to be hoped that complete analyses of a wide variety of leaf types will be published as interest develops in promoting their use in human diets, by selection of cultivars to suit local requirements, improvement of quality characteristics and optimization of desirable nutritional components.

References

Abubakar, M. 1981. Effects of harvest dates and intra-row spacing on yield and quality of three sweet potato (*Ipomoea batatas* (L.) Lam) cultivars. *Diss. Abstr. Int., B*, **41** (12): 4335.

Ahmed, E.M. and Scott, L.E. 1958. Pectic constituents of the fresh roots of the sweet potato. *Proc. Am. Soc. Hort. Sci.* **71**: 376–87.

Anon. 1981. *AVRDC Progress Report for 1980*. AVRDC, Shanhua, T'ainan, p. 71.

1985a. *Mass screening techniques for sweet potato protein determination*, AVRDC Progress Report for 1983. AVRDC, Shanhua, T'ainan, pp. 301–4.

1985b. *Composition of edible fiber in sweet potato tips*, AVRDC Progress Report 1985. AVRDC, Shanhua, T'ainan, pp. 310–13.

AVRDC. 1976. *Nutritional progress and potential for sweet potatoes*, Sweet Potato Report. AVRDC, Shanhua, T'ainan, pp. 37–40.

Azhar, A. and Hamdy, M.K. 1981a. Alcohol fermentation of sweet potato. I. Acid hydrolysis and factors involved. *Biotechnol. Bioeng.* **23**: 879–86.

1981b. Alcohol fermentation of sweet potato. Membrane reactor in enzymatic hydrolysis. *Biotechnol. Bioeng.* **23**: 1297–307.

Baba, T., Nakama, H., Tamaru, Y. and Kono, T. 1987. [Development of snack foods produced from sweet potatoes. V. Changes in sugar and starch contents during storage of new type sweet potato (low beta-amylase activity in roots)] Japanese. *J. Jap. Soc. Food Sci. Technol.* **34** (4): 249–53.

Balls, A.K., Walden, M.K. and Thompson, R.R. 1948. Crystalline beta-amylase from sweet potatoes. *Biol. Chem.* **173** (9): 9–19.

Bartolini, P.U. 1982. Timing and frequency of topping sweet potato at varying levels of nitrogen. In: Villareal, R.L. and Griggs, T.D. (eds.), *Sweet potato*, Proceedings of the First International Symposium. AVRDC, Shanhua, T'ainan, pp. 209–14.

Bassa, I.A. and Francis, F.J. 1987. Stability of anthocyanins from sweet potatoes

in a model beverage. *J. Food Sci.* **52** (6): 1753–4.
Bauernfeind, J.C. 1972. Carotenoid vitamin A precursors and analogs in food and feeds. *J. Agric. Food Chem.* **20** (3): 456–73.
Baumgardner, R.A. and Scott, L.E. 1963. The relation of pectic substances to firmness of processed, sweet potatoes. *Proc. Am. Soc. Hort. Sci.* **83**: 629–40.
Beiqing, Z. 1983. [Alcoholic fermentation of uncooked sweet potato] Chinese. *Food Ferment. Ind.* **6**: 59–63.
Bertoniere, N.R., McLemore, T.A. and Hasling, V.C. 1966. The effect of environmental variables on the processing of sweet potatoes into flakes and on some properties of their isolated starches. *J. Food Sci.* **31**: 574–82.
Boggess, T.S., Marion, J.E. and Dempsey, A.H. 1970. Lipid and other compositional changes in 9 varieties of sweet potatoes during storage. *J. Food Sci.* **35** (3): 306–9.
Boggess, T.S., Marion, J.E. and Woodroof, J.G. 1967. Changes in lipid composition of sweet potatoes as affected by controlled storage. *J. Food Sci.* **32** (5): 554–8.
Bradbury, J.H., Baines, J., Hammer, B., Anders, M. and Millar, J.S. 1984. Analysis of sweet potato (*Ipomoea batatas*) from the highlands of Papua New Guinea: relevance to the incidence of *Enteritis necroticans*. *J. Agric. Food Chem.* **32** (39): 469–73.
Bradbury, J.H., Beatty, R.E., Bradshaw, K., Hammer, B., Holloway, W.D., Jealous, W., Lau, J., Nguyen, T. and Singh, U. 1985a. Chemistry and nutritive value of tropical root crops in the South Pacific. *Proc. Nutr. Soc. Austr.* **10**: 185–8.
Bradbury, J.H., Bradshaw, K., Jealous, W., Holloway, W.D. and Phimpisane, T. 1988. Effect of cooking on nutrient content of tropical root crops from the South Pacific. *J. Sci. Food Agric.* **43** (4): 333–42.
Bradbury, J.H., Hammer, B., Nguyen, T., Anders, M. and Millar, J.S. 1985b. Protein quantity and quality and trypsin inhibitor content of sweet potato cultivars from the highlands of Papua New Guinea. *J. Agric. Food Chem.* **33** (2): 281–5.
Bradbury, J.H. and Holloway, W.D. 1988. *Chemistry of tropical root crops: significance for nutrition and agriculture in the Pacific*, ACIAR Monograph Ser. No. 6, Canberra.
Bradbury, J.H. and Singh, U. 1986a. Ascorbic and dehydroascorbic acid content of tropical root crops from the South Pacific. *J. Food Sci.* **51** (4): 975–8.
1986b. Thiamin, riboflavin and nicotinic acid contents of tropical root crops from the South Pacific. *J. Food Sci.* **51** (6): 1563–4.
Buescher, R.W., Sistrunk, W.A. and Kasaian, A.E. 1976. Induction of textural changes in sweet potato roots by chilling. *J. Am. Soc. Hort. Sci.* **101** (5): 516–19.
Bureau, J.L. and Bushway, R.J. 1986. HPLC determination of carotenoids in fruit and vegetables in the United States. *J. Food Sci.* **51** (1): 128–30.
Burton, G.W. and Ingold, K.V. 1984. Beta-carotene: an unusual type of lipid antioxidant. *Science* **224**: 569–73.
Bushway, R.J. 1986. Determination of alpha- and beta-carotene in some raw

fruits and vegetables by high-performance liquid chromatography. *J. Agric. Food Chem.* **34** (3): 409–12.

Caldwell, M.J. 1972. Ascorbic acid content of Malaysian leaf vegetables. *Ecol. Food Nutr.* **1** (4): 313–17.

Caldwell, M.J. and Enoch, I.C. 1972. Riboflavin content of Malaysian leaf vegetables. *Ecol. Food Nutr.* **1** (4): 309–12.

Candlish, J.K., Gourley, L. and Lee, H.P. 1987. Dietary fiber and starch contents of some Southeast Asian vegetables. *J. Agric. Food Chem.* **35**: 319–21.

Caraballo Llosas, N. 1974. The effect of potassium application on the magnesium content of sweet potato (*Ipomoea batatas*). *Centro Agricola* **1** (2): 73–83.

Carpio Burga, R. del. 1985. [Sweet potatoes of high protein value] Spanish. *Agricultura de las Americas*, January: 24–25.

Cascon, S.C., Carvalho, M.P.M., Moura, L.L., Guimaraes, I.S.S. and Philip, T. 1984. [Pigments of purple sweet potatoes for use in foods] Portuguese. *EMBRAPA Bull. Pesq.* No. 9, EMBRAPA Centro Tecnologia Agricola e Alimentar, Rio de Janeiro.

Cereda, M.P., Cagliari, A.M., Heezen, A.M. and Fioretto, R.B. 1980. [Evaluation of the enzymatic action of the phenol oxidase in sweet potato pulp (*Ipomoea batatas*)]. Portuguese. *Turrialba* **30** (2): 147–51.

Cereda, M.P., Conceição, F.A.D., Cagliari, A.M., Heezen, A.M. and Fioretto, R.B. 1982. [Comparative study of sweet potato (*Ipomoea batatas*) varieties to estimate their utilization in the food industry] Portuguese. *Turrialba* **32** (4): 365–70.

Cerning-Beroard, J. and Le Dividich, J. 1976. [Feeding value of some starchy tropical products: *in vitro* and *in vivo* study of sweet potato, yam, malanga, breadfruit and banana] French. *Ann. Zootech.* **25** (2): 155–68.

Chandler, L.A. and Schwartz, S.J. 1988. Isomerization and losses of *trans-β*-carotene in sweet potatoes as affected by processing treatments. *J. Agric. Food Chem.* **36**(1): 129–33.

Chang, S.M. 1983. The fine structure of the amyloses from some tuber starches and their noodle quality. *Proc. 6th Int. Congr. Food Sci. Technol.* **1**: 111–12.

Cheng, H-H. 1978. [Protein content and amino acid composition in tubers and stems and leaves of sweet potato cultivars in Taiwan] Chinese. *J. Agric. Res. China* **27** (4): 291–5.

Collazos, C. and 15 others. 1974. [*The composition of Peruvian foods*] Spanish. 4th edn. Ministry of Health, Lima.

Collins, W.W. and Walter, W.M. 1982. Potential for increasing nutritional value of sweet potatoes. In: Villareal, R.L. and Griggs, T.D. (eds.), *Sweet potato*, Proceedings of the First International Symposium, AVRDC, Shanhua, T'ainan, pp. 355–63.

Constantin, R.J. and Hernandez, T.P. 1977. Effects of azide soil treatments on quality and yield of sweet potatoes. *HortScience* **12** (5): 457–8.

Constantin, R.J., Hernandez, T.P. and Jones, L.G. 1974. Effects of irrigation and nitrogen fertilization on quality of sweet potatoes. *J. Am. Soc. Hort. Sci.* **99** (4): 308–10.

Constantin, R.J., Jones, L.G. and Hernandez, T.P. 1975. Sweet potato quality as affected by soil reaction (pH) and fertilizer. *J. Am. Soc. Hort. Sci.* **100** (6): 604–7.

1977. Effects of potassium and phosphorus fertilization on quality of sweet potatoes. *J. Am. Soc. Hort. Sci.* **102** (6): 779–81.

Crossman, S.M. and Hill, W.A. 1984. Sweet potato yield and nitrogen content in response to inoculation with several *Azospirillium* isolates. *HortScience* **19** (2): 206.

Daines, R.H., Hammond, D.F., Haard, N.F. and Ceponis, M.J. 1976. Hardcore development in sweet potatoes: a response to chilling and its remission as influenced by cultivar, curing temperature and time and duration of chilling. *Phytopathology* **66**: 582–7.

Davis, K.C. 1973. Vitamin E content of selected foods. *J. Food Sci.* **38** (3): 442–6.

Dickey, L.F., Collins, W.W., Young, C.T. and Walter, W.M. 1984. Root protein quantity and quality in a seedling population of sweet potatoes. *HortScience* **19** (5): 689–92.

Doremus, G.L., Crenshaw, F.A. and Thurber, F.H. 1951. Amylose content of sweet potato starch. *Cer. Chem.* **28**: 308–17.

Englyst, H.N., Bingham, S.A., Runswick, S.A., Collinson, E. and Cummings, J.H. 1988. Dietary fibre (non-starch polysaccharides) in fruit, vegetables and nuts. *J. Hum. Nutr. Dietet.* **1** (4): 247–86.

1989. Dietary fibre (non-starch polysaccharides) in cereal products. *J. Hum. Nutr. Dietet.* **2** (4): 253–71.

Evensen, S.K. and Standal, B.R. 1984. *Use of tropical vegetables to improve diets in the Pacific region.* HITAHR Res. Ser. 028, HITAHR, University of Hawaii.

Ezell, B.D. and Wilcox, M.S. 1946. The ratio of carotene to carotenoid pigments in sweet potato varieties. *Science* **103**: 193–4.

1958. Variation in carotene content of sweet potatoes. *J. Agric. Food Chem.* **6** (1): 61–5.

Ezell, B.D., Wilcox, M.S. and Crowder, J.N. 1952. Pre- and post-harvest changes in carotene, total carotenoids and ascorbic acid content of sweet potatoes. *Plant Physiol.* **27**: 355–69.

Ezell, B.D., Wilcox, M.S. and Hutchins, M.C. 1948. Effect of variety and storage on ascorbic acid content of sweet potatoes. *Food Res.* **13**: 116–22.

Faboya, O.O.P. 1981. The fatty acid composition of some tubers grown in Nigeria. *Food Chem.* **7** (2): 151–4.

Faboya, O.O.P., Ikotun, T. and Fatoki, O.S. 1983. Production of oxalic acid by some fungi infected tubers. *Z. Allg. Mikrobiol.* **23** (10): 621–4.

Farnworth, E.R. 1973. The composition of Papua New Guinea foods. *Sci. New Guinea* **1** (3–4): 21–41.

Furr, A.K., Parkinson, T.F., Elfving, D.C., Bache, C.A., Gutenmann, W.H., Doss, G.J. and Lisk, D.J. 1981. Elemental content of vegetables and apple trees grown on Syracuse sludge-amended soils. *J. Agric. Food Chem.* **29** (1): 156–60.

Garcia, E.H., Querido, I.B. and Cahanap, A.C. 1970. The relation of carotene and starch content of some twenty-six varieties of sweet potatoes (*Ipomoea batatas*). *Phil. J. Plant Ind.* **35** (3–4): 203–13.

Garlich, J.D., Bryant, D.N., Covington, H.M., Chamble, D.S. and Purcell,

A.E. 1974. Egg yolk and broiler skin pigmentation with sweet potato vine meal. *Poult. Sci.* **53** (2): 692–9.

Gonzales, F.R., Cadiz, T.G. and Bugawan, M.S. 1977. Effect of topping and fertilization on the yield and protein content of three varieties of sweet potato. *Phil. J. Crop Sci.* **2**: 97–102.

Gore, H.C. 1920. Occurrence of diastase in sweet potato in relation to the preparation of sweet potato syrup. *J. Biol. Chem.* **44** (1): 19–20.

Goshima, G., Kubo, K., Ohashi, K. and Tsuge, H. 1984. Classification of various starch granules by pasting characteristics before and after defatting. *J. Jap. Soc. Food Sci. Technol.* **31** (7): 429–43.

Hall, M.R. and Smittle, D.A. 1983. *Industrial-type sweet potatoes: a renewable energy source for Georgia.* Univ. Georgia, Coll. Agric. Exp. Stations Res. Rep. No. 429, Tifton, GA.

Hammett, H.L. 1974. Total carbohydrate and carotenoid content of sweet potatoes as affected by cultivar and area of production. *HortScience* **9** (5): 467–8.

Hammett, H.L. and Barrantine, B.F. 1961. Some effects of variety, curing and baking upon the carbohydrate content of sweet potatoes. *Proc. Am. Soc. Hort. Sci.* **78**: 421–6.

Hammett, H.L., Constantin, R.J., Jones, L.G. and Hernandez, T.P. 1982. The effect of phosphorus and soil moisture levels on yield and processing quality of 'Centennial' sweet potatoes. *J. Am. Soc. Hort. Sci.* **107** (1): 119–22.

Hammett, L.K., Miller, C.H., Swallow, W.H. and Harden, C. 1984. Influence of N source, N rate, and K rate on the yield and mineral concentration of sweet potato. *J. Am. Soc. Hort. Sci.* **109** (3): 294–8.

Hattori, T. and Nakamura, K. 1988. Genes encoding for the major tuberous root protein of sweet potato: identification of putative regulatory sequence in the 5′ upstream region. *Plant Mol. Biol.* **11**: 417–26.

Haytowitz, D.B. and Matthews, R.H. 1984. *Composition of foods: vegetables and vegetable products*, Human Nutrition Information Series, USDA Agric. Handbook No. 8–11, Washington, DC.

Hirahara, F. and Koike, Y. 1989. [Tocopherol content in sweet potato tubers of different cultivars, places harvested and cooking methods] Japanese. *Jap. J. Nutr.* **47** (2): 85–91.

Hiura, M. 1943. [Studies on storage and rot of sweet potato] Japanese. *Rep. Gifu Agric. Coll.* **50**: 1–5.

Holloway, W.D. 1983. Composition of fruit, vegetable and cereal dietary fibre. *J. Sci. Food Agric.* **34** (11): 1236–40.

Holloway, W.D., Argall, M.E., Jealous, W.T., Lee, J.A. and Bradbury, J.H. 1989. Organic acids and calcium oxalate in tropical root crops. *J. Agric. Food Chem.* **37** (2): 337–41.

Holloway, W.D., Monro, J.A., Gurnsey, J.C., Pomare, E.W. and Stace, N.H. 1985. Dietary fiber and other constituents of some Tongan foods. *J. Food Sci.* **50** (6): 1756–7.

Huq, R.S., Abalaka, J.A. and Stafford, W.L. 1983. Folate content of various Nigerian foods. *J. Sci. Food Agric.* **34** (4): 404–6.

Hyase, F. and Kato, H. 1984. Antioxidative components of sweet potatoes. *J.*

Nutr. Sci. Vitaminol. **30** (1): 37–46.

Ikemiya, M. and Deobald, H.J. 1966. New characteristic alpha-amylase in sweet potatoes. *J. Agric. Food Chem.* **14**: 237–41.

Imbert, M.P., Seaforth, C.E. and Williams, D.B. 1966. The anthocyanin pigments of the sweet potato. *Proc. Am. Soc. Hort. Sci.* **88**: 481–5.

James, W.H. and Hollinger, M.E. 1954. The utilization of carotene. II. From sweet potatoes by young human adults. *J. Nutr.* **54** (1): 65–74.

Japanese Examined Patent. 1981. 5 617 061. [Violet pigment] Japanese. San Ei Chemical Industry KK.

Junek, J. and Sistrunk, W.A. 1978. Sweet potatoes high in vitamin content but content is affected by variety and cooking method. *Ark. Farm Res.* **27** (5): 7–8.

Kainuma, K. 1984. Uses of sweet potato starch. *Farming Japan* **18** (5): 36–40.

Kawabata, A., Sawayama, S., del Rosario, R.R. and Noel, M.G. 1984. Effect of storage and heat treatment on the sugar constituents of tropical root crops. In: Uritani, I. and Reyes, E.D. (eds.), *Tropical root crops: postharvest physiology and processing.* Japan Scientific, Tokyo, pp. 243–58.

Kay, D.E. (Revised by Gooding, E.G.B.). 1987. *Root crops*, Crop and Product Digest No.2, 2nd edn. Tropical Development and Research Institute, London (now Natural Resources Institute, Chatham Maritime).

Kays, S.J. 1990. Strategies for selecting conventional and new flavor types of tropical root and tuber crops to increase consumer acceptance and use. In: *Tropical root and tuber crops changing role in a modern world.* Proceedings of the Eighth Symposium of the International Society for Tropical Root Crops (in press).

Kays, S.J. and Horvat, R.J. 1984. A comparison of the volatile constituents and sugars of representative Asian, Central American and North American sweet potatoes. *Proc. 6th Symp. Int. Soc. Trop. Root Crops*, pp. 577–86.

Kehr, A.E., Ting, Y.C. and Miller, J.C. 1955. The site of carotenoid and anthocyanin synthesis in sweet potatoes. *Proc. Am. Soc. Hort. Sci.* **65**: 396–8.

Kim, K.J., Hahn, Y.S. and Yoo Y.J. 1971. [Studies on utilization of the sweet potato (Part 2)] Korean. *Kungnip Kongop Yonguso Pogo* **21**: 169–74.

Kim, K. and Hamdy, M.K. 1987. Depolymerization of starch by high pressure extrusion. *J. Food Sci.* **52** (5): 1387–90.

Kim, M.C., Sung, N.K., Shim, K.H., Lee, M.H. and Lee, I. 1981. [The contents of heavy metal in fruits and vegetables collected from Jinju district] Korean. *Korean J. Food Sci. Technol.* **13** (4): 299–306.

Kimber, A.J. 1976. Some factors influencing the protein content of sweet potato. In: Wilson, K. and Bourke, R.M. (eds.), *1975 Papua New Guinea Food Crops Conference Proceedings*, Department of Primary Industry, Port Moresby, pp. 63–74.

Kubota, T. and Matsuura, T. 1953. [Chemical studies on the black rot disease of sweet potato. VI. Chemical constitution of ipomeamarone] Japanese. *J. Chem. Soc. Japan* **74**: 248–51.

Kukimura, H., Yoshida, T., Komaki, K., Sakamoto, S., Tabuchi, S., Ide, Y. and Yamakawa, O. 1989. 'Satsumahikari': a new sweet potato cultivar. *Bull. Kyushu Nat. Agric. Exp. Stn* **25** (3): 250.

Kumagai, T., Umemura, Y., Baba, T. and Iwanaga, M. 1990. The inheritance of β-amylase null in storage roots of sweet potato, *Ipomoea batatas* (L.) Lam. *Theor. Appl. Genet.* **79** (3): 369–76.

Kwiatkowska, C.A., Finglas, P.M. and Faulks, R.M. 1989. The vitamin content of retail vegetables in the UK. *J. Hum. Nutr. Dietet.* **2** (3): 159–72.

Lanier, J.J. and Sistrunk, W.A. 1979. Influence of cooking method on quality attributes and vitamin content of sweet potato. *J. Food Sci.* **44** (2): 374–80.

Lee, K.Y. and Lee, S.R. 1972. [A study on the systematic analysis of lipids from sweet potatoes] Korean. *Korean J. Food Sci. Technol.* **4** (4): 309–16.

Lee, K.A., Shin, M.S. and Ahn, S.Y. 1985. The changes of pectic substances in sweet potato cultivars during baking. *Korean J. Food Sci. Technol.* **17** (6): 421–5.

Leung, W-T.W., Busson, F. and Jardin, C. 1968. *Food composition table for use in Africa.* US Department of Health, Education and Welfare Public Health Service, Bethesda, MD; FAO, Rome.

Leung, W-T.W., Butrum, R.R. and Chang, F.H. 1972. *Food composition table for use in East Asia.* Part I. *Proximate composition, mineral and vitamin contents of East Asian foods.* US Department of Health, Education and Welfare, Bethesda, MD; FAO, Rome.

Leung, W-T.W. and Flores, M. 1961. *Food composition table for use in Latin America.* INCAP, Guatemala; Interdepartmental Committee on Nutrition for National Defense, Bethesda, MD.

Li, L. 1974. [Variation in protein content and its relation to other characters in sweet potatoes (*Ipomoea batatas* L.)] Chinese. *J. Agric. Assoc. China* **88**: 17–22.

1975. [Studies on the influence of environmental factors on the protein content of sweet potatoes] Chinese. *J. Agric. Assoc. China* **92**: 64–72.

1982. Breeding for increased protein content in sweet potatoes. In: Villareal R.L. and Griggs, T. (eds.) *Sweet potato*, Proceedings of the First International Symposium. AVRDC, Shanhua, T'ainan, pp. 345–54.

Li, L. and Liao, C.H. 1983. [Variation in crude starch percentage of sweet potato (*Ipomoea batatas* (L.) Lam.) (Varietal differences)] Chinese. *J. Agric. Res. China* **32** (4): 325–35.

Li, H.S. and Oba, K. 1985. Major soluble proteins of sweet potato roots and changes in proteins after cutting, infection or storage. *Agric. Biol. Chem.* **49** (3): 737–44.

Lii, C-Y. and Chang, S-M. 1978. Studies on the starches in Taiwan. 1. Sweet potato, cassava, yam and arrowroot starches. *Proc. Natl. Sci. Counc. Republic of China* **2** (4): 416–23.

Lopez, A., Williams, H.L. and Cooler, F.W. 1980. Essential elements in fresh and canned sweet potatoes. *J. Food Sci.* **45** (3): 675–8.

Losh, J.M., Phillips, J.A., Axelson, J.M. and Schulman, R.S. 1981. Sweet potato quality after baking. *J. Food Sci.* **46** (1): 283–6.

Lu, J.Y., Biswas, P.K. and Pace, R.D. 1986. Effect of elevated CO_2 growth conditions on the nutritive composition and acceptibility of baked sweet potatoes. *J. Food Sci.* **51** (2): 358–9.

Lund, E.D. and Smoot, J.M. 1982. Dietary fiber content of some tropical fruits and vegetables. *J. Agric. Food Chem.* **30** (6): 1123–7.

Madamba, L.S.P., Bustrillos, A.R. and San Pedro, E.L. 1975. Sweet potato starch: physicochemical properties of the whole starch. *Philipp. Agric.* **58**: 338–50.

Maeshima, M., Sasaki, T. and Asahi, T. 1985. Characterization of major proteins in sweet potato tuberous roots. *Phytochemistry* **24** (9): 1899–1902.

Makki, H.M., Abdel-Rahmann, A.Y., Khalil, M.K.M. and Mohamed, M.S. 1986. Chemical composition of Egyptian sweet potatoes. *Food Chem.* **20** (1): 39–44.

Marsh, A.C., Klippstein, R.N. and Kaplan, S.D. 1980. The sodium content of your food. *Home and Garden Bull.* No. 233, USDA, Washington, DC.

Martin, F.W. 1983. The carotenoid pigments of white-fleshed sweet potatoes – reference to their potential value as sources of vitamin A activity. *J. Agric. Univ. Puerto Rico* **67** (4): 494–500.

Martin, F.W. and Deshpande, S.N. 1985. Sugars and starches in a non-sweet sweet potato compared to those of conventional cultivars. *J. Agric. Univ. Puerto Rico* **69** (3): 401–6.

Mathews-Roth, M.M. 1985. Carotenoids and cancer prevention – experimental and epidemiological studies. *Pure Appl. Chem.* **57** (5): 717–22.

Matsui, T. 1981. Greening pigments produced by the reaction of ethyl caffeate with methylamine. *J. Nutr. Sci. Vitaminol.* **27** (6): 573–82.

Maurya, K.R. and Dhar, N.R. 1976. Influence of phosphatic fertilizers alone and in combination with waterhyacinth on the yield and composition of sweet potato (*Ipomoea batatas* Poir). *Mysore J. Agric. Sci.* **10** (3): 387–96.

McClure, T.T. 1960. Chlorogenic acid accumulation and wound healing in sweet potato roots. *Am. J. Bot.* **47** (4): 277–80.

Menezes, D.M. de, Rego, M.M. do, Nobre, A. and Meneguelli, C.A. 1976. [Effect of harvest time on the sugar and protein content of sweet potatoes] Portuguese. *Pes. Agropec. Bras., Agron.* **11** (12): 49–52.

Miller, J.C. and Gaafar, A.K. 1958. A study of the synthesis of carotene in the sweet potato plant and root. *Proc. Am. Soc. Hort. Sci.* **71**: 388–90.

Monro, J.A., Holloway, W.D. and Lee, J. 1986. Elemental analysis of fruit and vegetables from Tonga. *J. Food Sci.* **51** (2): 522–3.

Moschette, D.S. 1955. Metabolic studies with pre-adolescent girls. 1. Utilization of carotene. *J. Am. Dietet. Assoc.* **31** (1): 37–44.

Nakamura, K. 1989. Regulation of expression of underground tuberous organ storage protein genes. *First United States-Japanese Symposium on Biotechnology*, FL, p. 9.

Nishimaki, T. and Nozue, M. 1985. Isolation and culture of protoplasts from high anthocyanin-producing callus of sweet potato. *Plant Cell Rep.* **4**: 248–51.

Nozue, M., Kawai, J. and Yoshitama, K. 1987. Selection of a high anthocyanin-producing cell line of sweet potato cell cultures and identification of pigments. *J. Plant Physiol.* **129**: 81–8.

Nozue, M. and Yasuda, H. 1985. Occurrence of anthocyanoplasts in cell suspension cultures of sweet potato. *Plant Cell Rep.* **4**: 252–5.

Nursten, H.E. 1978. Why flavour research? How far have we come since 1975 and where now? In: Land, D.G & Nursten, H.E. (eds.), *Progress in flavour research*, Proceedings of the 2nd Weurman Flavour Research Symposium. Applied Science Publishers, London, pp. 337–55.

Ohtsuka, R., Kawabe, T., Inaoka, T., Suzuki, T., Hongo, T., Akimichi, T. and Sugahara, T. 1984. Composition of local and purchased foods consumed by the Gidra in lowland Papua. *Ecol. Food. Nutr.* **15**: 159–69.

Oñate, L.U., Arago, L.L., Garcia, P.C. and Abdon, I.C. 1970. Nutrient composition of some raw and cooked Philippine vegetables. *Phil. J. Nutr.* **23** (3): 33–44.

Opute, F.I. and Osagie, A.U. 1978. Fatty acid composition of total lipids from some tropical storage organs. *J. Sci. Food Agric.* **29** (11): 959–62.

Oyenuga, V.A. 1968. *Nigerian foods and feeding stuffs.* Ibadan University Press, Ibadan.

Pace, R.D., Dull, G.G. and Phills, B.R. 1985a. Proximate composition of sweet potato greens in relation to cultivar, harvest date, crop year and processing. *J. Food Sci.* **50** (2): 537–8.

Pace, R.D., Sibiya, T.E., Phills, B.R. and Dull, G.G. 1985b. Ca, Fe and Zn content of 'Jewel' sweet potato greens as affected by harvesting practices. *J. Food Sci.* **50** (4): 940–1.

Padmaja, G. and Rajamma, P. 1982. Biochemical changes due to weevil (*Cylas formicaricus* Fab.) feeding on sweet potato. *J. Food Sci. Technol., India* **19**: 162–3.

Pasricha, S. 1967. Effect of different methods of cooking and storage on the ascorbic acid content of vegetable. *Ind. J. Med. Res.* **55** (7): 779–84.

Paterson, D.R. and Speights, D.E. 1971. Effects of foliar applications of iron, manganese, zinc and boron on crop yield and mineral composition of sweet potato leaf tissue. *J. Rio Grande Valley Hort. Soc.* **28**: 86–90.

Pellet, P.L. and Shadarevian, S. 1970. *Food composition tables for use in the Middle East*, 2nd edn. American University of Beirut, Beirut.

Peto, R., Doll, R., Buckley, J.D. and Sporn, M.B. 1981. Can dietary beta-carotene materially reduce human cancer rates? *Nature (London)* **290**: 201–8.

Picha, D.H. 1985a. Organic acid determination in sweet potatoes by HPLC. *J. Agric. Food Chem.* **33** (4): 743–5.

1985b. Crude protein, minerals and total carotenoids in sweet potatoes. *J. Food Sci.* **50** (6): 1768–9.

1985c. HPLC determination of sugars in raw and baked sweet potatoes. *J. Food Sci.* **50** (4): 1189–90.

Platt, B.S. 1962. *Tables of representative values of foods commonly used in tropical countries.* Medical Research Council Special Report Ser. No. 302. HMSO, London.

Prabhuddham, S., Tantidham, K., Poonperm, N., Lertbawornwongsa, C. and Tongglad, C. 1987. A study of sweet potato quality and processing methods. Paper presented during Training Course on Technology of Sweet Potato Production, 14 July 1987, Pichit Horticultural Research Center, Thailand [Mimeo].

Purcell, A.E. 1962. Carotenoids of Goldrush sweet potato flakes. *Food Technol.* **16** (1): 99–102.

Purcell, A.E., Later, D.W. and Lee, M.L. 1980. Analysis of the volatile constituents of baked 'Jewel' sweet potatoes. *J. Agric. Food Chem.* **28** (5): 939–41.

Purcell, A.E., Pope, D.T. and Walter, W.M. 1976a. Effect of length of growing season on protein content of sweet potato cultivars. *HortScience* **11** (1): 31.

Purcell, A.E., Swaisgood, H.E. and Pope, D.T. 1972. Protein and amino acid content of sweet potato cultivars. *J. Am. Soc. Hort. Sci.* **97** (1): 30–3.

Purcell, A.E. and Walter, W.M. 1968. Carotenoids of Centennial variety sweet potato, *Ipomoea batatas*, L. *J. Agric. Food Chem.* **16** (5): 769–70.

1980. Changes in composition of the non-protein-nitrogen fraction of 'Jewel' sweet potatoes (*Ipomoea batatas* (Lam.)) during storage. *J. Agric. Food Chem.* **28** (4): 842–4.

1982. Stability of amino acids during cooking and processing of sweet potatoes. *J. Agric. Food Chem.* **30** (3): 443–4.

Purcell, A.E., Walter, W.M. and Giesbrecht, F.G. 1976b. Distribution of protein within sweet potato roots (*Ipomoea batatas* L.). *J. Agric. Food Chem.* **24** (1): 64–6.

1978. Root, hill and field variance in protein content of North Carolina sweet potatoes. *J. Agric. Food Chem.* **26** (2): 362–4.

Purcell, A.E., Walter, W.M., Nicholaides, W.W., Collins, W.W. and Chancey, H. 1982. Nitrogen, potassium, sulfur fertilization and protein content of sweet potato roots. *J. Am. Soc. Hort. Sci.* **107** (3): 425–7.

Quackenbush, F.W. 1987. Reverse phase HPLC separation of *cis*- and *trans*-carotenoids and its application to β-carotenes in food materials. *J. Liquid Chromatogr.* **10** (4): 643–53.

Rao, M.N. and Polacchi, W. 1972. *Food composition table for use in East Asia.* Part II. *Amino acid, fatty acid, certain B-vitamin and trace mineral content of some Asian foods.* US Department of Health, Education and Welfare, Bethesda, MD; FAO, Rome.

Rasper, V. 1969a. Investigations on starches from major starch crops grown in Ghana. I. Hot paste viscosity and gel-forming power. *J. Sci. Food Agric.* **20** (11): 165–71.

1969b. Investigations on starches from major starch crops grown in Ghana. II. Swelling and solubility patterns: amyloclastic susceptibility. *J. Sci. Food Agric.* **20** (11): 642–6.

Reddy, N.N. and Sistrunk, W.A. 1980. Effect of cultivar, size, storage and cooking method on carbohydrates and some nutrients of sweet potatoes. *J. Food Sci.* **45** (3): 682–4.

Rhee, K.S. and Watts, B.M. 1966. Evaluation of lipid oxidation in plant tissues. *J. Food Sci.* **31**: 664–8.

Rosario, R.R. del and Pontiveros, C.R. 1983. Retrogradation of some starch mixtures. *Starch/Staerke* **35** (3): 86–92.

Roy, F. and Hegde, M.V. 1985. Rapid method for purification of beta-amylase from *Ipomoea batatas. J. Chromatogr.* **324** (2): 489–94.

Rozin, P. 1977. The use of characteristic flavorings in human culinary practice. In: Apt, C.M. (ed.), *Flavor: its chemical, behavioral and commercial aspects*, Proceedings of the A.D. Little Flavor Symposium. Westview Press, Boulder, CO, pp. 101–27.

Ruinard, J. 1969. Notes on sweet potato research in West New Guinea (West Irian). In: Tai, E.A., Charles, W.B., Haynes, P.H., Iton, E.F. and Leslie, K.A. (eds.), *Proceedings First International Symposium on Tropical Root Crops*, University of the West Indies, St Augustine, Trinidad, Vol.1, Section III, pp. 88–108.

S-101 Technical Committee. 1980. *Sweet potato quality*. S. Coop. Ser. Bull. 249, Horticultural Crops Laboratory, Athens, GA.

Sakamoto, S. and Bouwkamp, J.C. 1985. Industrial products from sweet potatoes. In: Bouwkamp, J.C. (ed.), *Sweet potato products: a natural resource for the tropics*. CRC Press, Inc., Boca Raton, FL, pp. 219–33.

Schadel, W.E. and Walter, W.M. 1981. Localization of phenols and polyphenol oxidase in 'Jewel' sweet potatoes (*Ipomoea batatas* 'Jewel'). *Can. J. Bot.* **59** (10): 1961–7.

Schmandke, H. and Olivarez Guerra, O. 1969. [The carotene, L-ascorbic acid, dehydroascorbic acid and tocopherol content of Cuban vegetable products] German. *Nahrung* **13** (6): 523–30.

Scott, L.E. and Bouwkamp, J.C. 1974. Seasonal mineral accumulation by the sweet potato. *HortScience* **9** (3): 233–5.

Shanmugan, A. and Venugopal, K. 1975. Starch content of sweet potato (*Ipomoea batatas* Lamb.) varieties. *Sci. Cult.* **41** (10): 504–5.

Sharfuddin, A.F.M. and Voican, V. 1984. Effect of plant density and NPK dose on the chemical composition of fresh and stored tubers of sweet-potato. *Indian J. Agric. Sci.* **54** (12): 1094–6.

Shin, M.S. and Ahn, S.Y. 1983. [Physicochemical properties of sweet potato starches] Korean. *J. Korean Agric. Chem. Soc.* **26** (2): 137–42.

Singh, U. and Bradbury, J.H. 1988. HPLC determination of vitamin A and vitamin D_2 in South Pacific root crops. *J. Sci. Food Agric.* **45** (1): 87–94.

Sistrunk, W.A. 1971. Carbohydrate transformation, color and firmness of canned sweet potatoes as influenced by variety, storage, pH and treatment. *J. Food Sci.* **36**: 39–42.

Sohonie, K. and Bhandarkar, A.P. 1954. Trypsin inhibitors in Indian foodstuffs: Part 1 – Inhibitors in vegetables. *J. Sci. Ind. Res.* **13B**: 500–3.

Speirs, M., Cochran, H.L., Peterson, W.J., Sherwood, F.W. and Weaver, J.G. 1945. *The effects of fertilizer treatments, curing, storage and cooking on the carotene and ascorbic acid content of sweetpotatoes*. S. Coop. Ser. Bull. 3.

Speirs, M. and 18 others. 1953. *The effect of variety, curing, storage and time of planting and harvesting on the carotene, ascorbic acid, and moisture content of sweetpotatoes grown in six southern states*. S. Coop. Ser. Bull. 30.

Sutoh, H., Uchida, S. and Kaneda, K. 1973. [Studies on silage making (XXII) The nutrient content of sweet potato (*Ipomoea batatas* L. var. edulis) at the different stages and the quality of sweet potato vine silage] Japanese. *Sci. Rep. Fac. Agric. Okayama Univ.* **41**: 61–8.

Sweeney, J.P. and Marsh, A.C. 1971. Effect of processing on provitamin A in vegetables. *J. Am. Dietet. Assoc.* **59**: 238–45.

Swingle, H.D. 1966. The relation of pectic substances and starch to consistency and moistness of sweet potatoes. Ph.D. thesis, Louisiana State University.

Szylit, O., Durand, M., Borgida, L.P., Atinkpahoun, H., Prieto, F. and Delort-Laval, J. 1978. Raw and steam-pelleted cassava, sweet potato and yam *cayenensis* as starch sources for ruminant and chicken diets. *Anim. Feed Sci. Technol.* **3** (1): 73–87.

Tamate, J. and Bradbury, J.H. 1985. Determination of sugars in tropical root crops using ^{13}C n.m.r. spectroscopy: comparison with the h.p.l.c. method. *J. Sci. Food Agric.* **36** (12): 1291–302.

Tan, S.P., Wenlock, R.W. and Buss, D.H. 1985. *Immigrant foods.* Second supplement to *McCance and Widdowson's The composition of foods.* HMSO, London.

Thompson, D.P. 1979. Phenols, carotene and ascorbic acid in sweet potato roots infected with *Rhizopus stolonifer. Can. J. Plant Sci.* **59** (4): 1177–9.

1981. Chlorogenic acid and other phenolic compounds in fourteen sweet potato cultivars. *J. Food Sci.* **46** (3): 738–40.

Tiu, C.S., Purcell, A.E. and Collins, W.W. 1985. Contribution of some volatile compounds to sweet potato aroma. *J. Agric. Food Chem.* **33** (2): 223–6.

Truong, V.D., Bierman, C.J. and Marlett, J.A. 1986. Simple sugars, oligosaccharides, and starch determination in raw and cooked sweet potato. *J. Agric. Food Chem.* **34** (3): 421–5.

Tsou, S. 1986. Bioavailability of provitamin A in vegetables and fruits, *AVRDC Annual Report 1986.* AVRDC, Shanhua, T'ainan.

1987. Effect of dietary compositions on bioavailability of provitamin A, *AVRDC Annual Report 1987.* AVRDC, Shanhua, T'ainan.

Uritani, I. 1953. [Phytopathological chemistry of black-rotted sweet potato. Part 10. The mechanism for greening occurring in the sound part next to the injured tissue, when dipped in sodium bicarbonate solution (I)] Japanese. *J. Agric. Chem. Soc. Jap.* **27**: 781–5.

1978. Biochemistry of host response to infection. In: Reinhold, L., Harborne, J.B. and Swain, T. (eds.), *Progress in phytochemistry*, Vol.5. Pergamon Press, Oxford, pp. 29–63.

Uritani, I., Hoshiya, I. and Takita, S. 1953. [Phytopathological chemistry of black-rotted sweet potato. Part 11. The mechanism for greening occurring in the sound part next to the injured tissue (II)] Japanese. *J. Agric. Chem. Soc. Jap.* **27**: 785–9.

Uritani, I., Uritani, M. and Yamada, R. 1960. Similar metabolic alterations induced in sweet potato by poisonous chemicals and by *Ceratocystis fimbriata. Phytopathology* **50**: 30–4.

Uriyo, A.P. 1974. Response of sweet potatoes (*Ipomoea batatas*) to potassium fertilization on a red oxisol soil in Tanzania. Lecture, Department of Soil Science and Agricultural Chemistry, University of Dar es Salaam, Tanzania.

Villareal, R.L., Lin, S.K., Chang, L.S. and Lai, S.H. 1979a. Use of sweet potato (*Ipomoea batatas*) leaf tips as vegetables. I. Evaluation of morphological traits. *Exptl. Agric.* **15** (2): 113–16.

Villareal, R.L., Tsou, S.C.S., Lin, S.K. and Chiu, S.C. 1979b. Use of sweet potato (*Ipomoea batatas*) leaf tips as vegetables. II. Evaluation of yield and nutritive quality. *Exptl. Agric.* **15** (2): 117–22.

Villareal, R.L., Tsou, S.C., Lo, H.F. and Chiu, S.C. 1985. Sweet potato vine tips as vegetables. In: Bouwkamp, J.C. (ed.), *Sweet potato products: a natural resource for the tropics.* CRC Press, Inc., Boca Raton, FL, pp. 175–83.

Villegas, R.J.A. and Kojima, M. 1986. Purification and characterization of hydroxycinnamoyl D-glucose: quinate hydroxycinnamoyl transferase in the root of sweet potato, *Ipomoea batatas* Lam. *J. Biol. Chem.* **261** (19): 8729–33.

Visser, F.R. and Burrows, J.K. 1983. *Composition of New Zealand foods.* 1. *Characteristic fruits and vegetables.* Science Information Publishing Centre, Wellington, NZ, pp. 31–2.

Walter, W.M., Collins, W.W. and Purcell, A.E. 1984. Sweet potato protein: a review. *J. Agric. Food Chem.* **32** (4): 695–9.

Walter, W.M., Hansen, A.P. and Purcell, A.E. 1971. Lipids of cured Centennial sweet potatoes. *J. Food Sci.* **36** (5): 795–7.

Walter, W.M. and McCollum, G.K. 1979. Use of high-pressure liquid chromatography for analysis of sweet potato phenolics. *J. Agric. Food Chem.* **27** (5): 938–41.

Walter, W.M. and Purcell, A.E. 1979. Evaluation of several methods for analysis of sweet potato cultivars phenolics. *J. Agric. Food Chem.* **27** (5): 942–6.

1980. Effect of substrate levels and polyphenol oxidase activity on darkening in sweet potato cultivars. *J. Agric. Food Chem.* **28** (5): 941–4.

Walter, W.M., Purcell, A.E. and Nelson, A.M. 1975. Effects of amylolytic enzymes on "moistness" and carbohydrate changes of baked sweet potato cultivars. *J. Food Sci.* **40** (4): 793–6.

Walter, W.M. and Schadel, W.E. 1981. Distribution of phenols in "Jewel" sweet potato (*Ipomoea batatas* (L.) Lam.) roots. *J. Agric. Food Chem.* **29** (5): 904–6.

Wang, H. and Lin, C.T. 1969. [The determination of the carotene content of sweet potato parental varieties and their offspring] Chinese. *J. Agric. Assoc. China* **65**: 1–5.

Watson, J.D. 1976. Ascorbic acid content of plant foods in Ghana and the effects of cooking and storage on vitamin content. *Ecol. Food Nutr.* **4** (4): 207–13.

Watt, B.K. and Merrill, A.L. 1975. *Handbook of the nutritional contents of foods.* Dover Publications, Inc., NY.

Winaro, F.G. 1982. Sweet potato processing and by-product utilization in the tropics. In: Villareal, R.L. and Griggs, T.D. (eds.), *Sweet potato*, Proceedings of the First International Symposium. AVRDC, Shanhua, T'ainan, pp. 373–84.

Wolinsky, I., Lane, H.W., Warren, D.C. and Whaley, B.S. 1988. Zinc and selenium levels in selected and ethnic/regional foods. *J. Agric. Food Chem.* **36** (4): 749–52.

Woolfe, J.A. 1987. *The potato in the human diet.* Cambridge University Press, Cambridge, England.

Yeh, T.P., Chen, Y.T. and Sun, C.C. 1981. [The effects of fertilizer application on the nutrient composition of high protein cultivars of sweet potatoes on the protein and lysine production] Chinese. *J. Agric. Assoc. China* **113**: 33–40.

CHAPTER 3

The nutritional value of sweet potato roots and leaves

The status which sweet potato enjoys in different parts of the world varies widely. In some areas of one of the richest and most powerful nations on earth, the United States, sweet potato roots are a festive food without which no table for the annual Thanksgiving meal would be complete. Furthermore United States' investigations into the use of sweet potato as a food to be grown in orbiting space stations have been cited (Hill, 1990) as an example of increasing its prestige with the younger age groups! Its high nutritional value has been recognized by its inclusion in school lunch programmes (Harris, 1963; University of New Hampshire, 1979) and in menus for the elderly (Unklesbay, 1978). In other lands, sweet potatoes are sometimes scorned as a 'poor man's food' fit only for those who can afford nothing better. The sweet potato is not, however, invariably assigned a lowly status in tropical countries. In Tonga, it is an important ingredient of traditional feasts, being acceptable for presentation to nobility and persons of high rank, whereas yautia (*Xanthosoma* spp.), cassava and bananas are not acceptable in this role (Taufatofua and Pole, 1987). Moreover, it is used in India for religious ceremonies such as 'Shradth' in which only indigenous vegetables are permissible, the sweet potato being of such ancient cultivation that it is now regarded by the people as indigenous, even though this is not the case (Nair et al., 1987).

Some nations have cause to be thankful to the sweet potato for saving many lives during famine, for example at various times in Japanese history when damage from climatic changes, diseases or insects produced poor harvests of other major crops (Sakamoto, 1984), or in the Philippines, where its use as a backyard crop and a food for prisoners-of-war helped many to survive malnutrition during the Second World War

(Villareal et al., 1985). Past dependency on the sweet potato as a famine food has not improved the esteem in which it is held by consumers in some areas of the far East, namely most parts of the Philippines and in Taiwan (Tsou and Villareal, 1982). However, in Japan, the sweet potato, previously stigmatized as a survival food, is enjoying a comeback, especially among young people, during the present period of prosperity and boom in gourmet food consumption (Duell, 1990). Sweet potato is even being introduced to consumers in temperate areas where it was previously unknown (see Figure 8.1, p. 478). This occurred initially through importation for immigrants from the tropics.

In contrast to the limited use of sweet potato as an exotic or festive food, as one to fall back on when other crops have failed, or to its moderate use as a co-staple or vegetable, the sweet potato is the major staple food of a very few locations. These include Orchid Island off the coast of Taiwan, some parts of the Philippines (Ifugao, Nueva Viscaya, Batanes and Mindoro), some Pacific islands such as Tonga and the Solomons, Irian Jaya in Indonesia, and most notably Papua New Guinea, where it has been described as the 'staff of life of the highland Papuan' (Oomen et al., 1961). In many parts of highland Papua New Guinea sweet potato roots still account for a large percentage of the daily diet, and are now also replacing more traditional staples such as taro (*Colocasia* spp.) in the lowlands (Bourke, 1985).

In being held in low esteem in some countries sweet potatoes are not alone, for legumes too throughout history, at least in Europe and the near East, have been looked on as 'poor man's meat', a socially inferior food (Aykroyd and Doughty, 1964, p. 9). With the popularization of so-called healthy eating and increased interest in vegetarianism the status of legumes, and consequently their consumption, has risen in wealthy countries which previously almost ignored their potential dietary role. The status of sweet potato too will improve, if strenuous efforts are made to demonstrate the valuable part it can play in the diet, when it is readily available all year round, reasonably priced, of good quality, of a type adapted to incorporation in local dishes and sold in some tasty, interesting (and more prestigious) forms as a nutritious snack or convenience food. Every effort should be made to change the present unfortunate name of 'poor man's food' to a badge of honour, recognizing the considerable contribution which sweet potato can make to enhancing the nutritional status of many of the world's great populations. At the same time, research must be directed towards the goal of ensuring that the sweet potato remains a food available to the poor.

Chapter 2 emphasized the wide ranges in nutrient content brought about by genetic, environmental and cultural factors. While keeping in mind these ranges (which similarly exist in other food crops), the present chapter endeavours to highlight the nutritional advantages of sweet potato roots and tops by comparing and contrasting their average nutrient composition with that of other staples and vegetables. The use of yellow-orange sweet potato roots as table desserts or as processed forms similar to dried fruits such as apricots, peaches etc. (see Chapter 6) prompted the further comparison of their nutritional value with a variety of fruits also. A table showing the composition of different forms of cooked and processed sweet potato is included in this chapter for the convenience of readers who wish to make comparisons between these forms and other plant foods; postharvest nutritional changes are discussed in more detail in Chapter 6.

Tables of recommended, or safe levels of, daily nutrient intakes for different age groups have been used to calculate the average percentage contributions to nutrient needs which can be made by 100 g of cooked (boiled) roots or 85 g of cooked green tops; these are intended only as guides to actual or potential contributions. A 100 g portion of the roots was chosen as a modest quantity of sweet potato which even a small child could consume without difficulty at a single main meal or as a snack. In some societies, many times this amount may be eaten at a single meal; in others, where sweet potato is a minor vegetable accompaniment to a meal, less than 100 g may be used. There has been little study of the quantities of leaves, petioles or tips which are consumed. The quantity of 85 g was suggested by one reseacher to be an average serving of leaves. Moreover, the contributions which tops make to the daily requirements for some vitamins await reappraisal in the light of more complete, and up-to-date, analyses.

Techniques which are already in use or are under consideration for improvement of nutritional value in the sweet potato are also discussed. Increasing the quantity and quality of protein is a major field of interest.

For purposes of simplicity the nutritional contributions of roots and tops are mainly treated separately. However, it should be remembered that some societies, particularly those based on subsistence farming, may consume both the roots and tops at a single meal. Their value will therefore be reinforced in terms of nutrients contained in common; alternatively one form of sweet potato can supply a nutrient lacking in the other.

Some societies have traditionally used sweet potato for medicinal as well as dietary purposes (see Chapter 8, China, p. 487) and potential pharmaceutical applications are still being explored. There is an exciting new development in the field of oral rehydration therapy that is included here because of its links with the nutritional status of children.

Roots

Energy

The major role most consumers would undoubtedly expect sweet potato roots to play in their diet is that of energy provider. While this is one important aspect of the sweet potato's nutritional contribution, it is certainly not the only one. Both roots and leaves supply valuable quantities of many other nutrients as we shall see in the following sections.

Sweet potato roots are not alone, of course, among the staple foods in supplying energy. However, they are somewhat better sources of energy than some of the other root and tuber staples, as can be seen in Table 3.1. The roots and tubers shown are representative of those grown in different parts of the world. Their proximate composition is given on a raw basis because when they are boiled, as is frequently the case, their composition remains virtually unchanged. All roots and tubers have a negligible lipid content (for nutritional purposes); therefore they are not outstandingly rich sources of energy. The energy content of sweet potato (and other crops) depends on the dry matter content which can vary widely (see Chapter 2). Although it supplies less energy than cassava, sweet potato (with about 465 kJ (111 kcal) per 100 g (fwb)) has an energy value similar to those of yam, taro and plantain, and nearly one and a half times that of potatoes, some types of yam, giant swamp taro, oca and ullucu.

Although sweet potato and other roots and tubers are often considered to be poor energy sources compared to cereal staples or legumes, Table 3.2 shows that this is not invariably the case. If sweet potatoes are compared on a cooked basis with beans and cereals (the energy density of which must change on addition of water), it can be seen that boiled or baked sweet potato roots compare favourably with boiled beans, boiled rice, and cereals cooked as porridges or made into noodles. They compare less favourably on a weight to weight basis with drier products such as bread, tortillas or chapathis. Sweet potato flour, however, has an energy content similar to that of cassava flour or maize meal, both of which are in common use in some countries.

The sweet potato has frequently been mentioned as a source of energy in the diet in cases where it invariably forms the major staple or, more often, when it replaces rice or wheat in times of their scarcity or high price. One hundred grams of sweet potato would supply 10%, 8%, and 7% of the daily energy requirements (FAO/WHO/UNU, 1985) of a 1–2, 2–3 and 3–5 year-old child, respectively, and 3–8% of that of an adult, depending on age, sex and activity. However, in parts of Irian Jaya,

Table 3.1. *Proximate composition of sweet potato roots and other root and tuber staples (per 100 g edible portion)*

Staple	Moisture (%)	Energy (kcal)	Energy (kJ)	Protein (g)	Lipid (g)	Total carbohydrate (g)	Dietary[a] fibre (g)	Ca (mg)	P[b] (mg)	Fe (mg)
Sweet potato[c] (*Ipomoea batatas*)	70	111	465	1.5	0.3	26.1	3.9 (0.4)	32	39	0.7
Arrowroot[d] (*Maranta arundinacea*)	61	136	569	2.0	0.1	33.2		17	21	2.5
Cassava[e] (*Manihot esculenta*)	63	141	590	1.0	0.3	32.4	4.4 (0.4)	39	41	1.1
Elephant foot yam[f] (*Amorphophallus* spp.)	77	89	372	1.5	0.1	20.5	1.5[g]	13	60	1.2
Giant taro[g] (*Alocasia macrorrhiza*)	70	111[h]	464[h]	2.2	0.1	27.0[h]	1.9	38	44	0.8
Giant swamp taro[g] (*Cyrtosperma chamissonis*)	75	83	348	0.5	0.2	20.6	2.8	182	16	0.6
Oca[i] (*Oxalis tuberosa*)	84	61	255	1.0	0.6	14.3		22	36	1.6
Ullucu[i] (*Ullucus tuberosus*)	84	62	259	1.1	0.1	15.1		3	28	1.1
Plantain, green[j] (*Musa paradisiaca*)	63	91	381	0.8	0.1	24.3	5.4 (0.7)	6		0.6
Potato[e] (*Solanum tuberosum*)	78	80	335	2.1	0.1	18.5	2.1[k]	9	50	0.8
Taro[l] (*Colocasia esculenta*)	72	103	432	1.7	0.2	23.1	4.0 (0.3)	35	65	1.2
Yam, Chinese[m] (*Dioscorea esculenta*)	73	104	434	2.4	0.1	23.4	1.2[g]	8[g]	42	0.8[g]
Yam, winged[n] (*D. alata*)	76	88	368	1.8	0.1	21.0[o]	2.3 (0.3)	11[p]	41	0.5[p]
Yautia[q] (*Xanthosoma* spp.)	66	125	523	1.7	0.2	29.2	3.4 (0.2)	14	52	0.7

Notes:

References: (1) Leung and Flores, 1961; (2) Leung, Busson and Jardin, 1968; (3) Leung, Butrum and Chang, 1972; (4) Carribean Food and Nutrition Institute, 1974; (5) Bradbury and Holloway, 1988; (6) Haytowitz and Matthews, 1984; (7) Ohtsuka et al., 1984; (8) Platt, 1962; (9) Collazos et al., 1974; (10) Paul and Southgate 1978; (11) Holloway et al., 1985.

[a] Calculated from (11). Excludes lignin, which is given in parentheses. See also notes *g* and *k*.

[b] Excludes (4) and (5). See also notes *c* to *g*, *i*, *j*, *l*, *m*, *n* and *q*.

[c] Average figures from references (1) to (6).

[d] (1), (3) and (4).

[e] Except where noted, the data are average figures from references (1) to (4).

[f] (7).

[g] (5), dietary fibre figures include lignin.

[h] (8).

[i] (9).

[j] (4).

[k] (10).

[l] (1) to (5) and (7), includes lignin.

[m] (2), (3) and (5).

[n] (3), (5) and (7).

[o] (3) and (7).

[p] (5) and (7).

[q] (1), (2), (4) and (5).

Table 3.2. *Proximate composition of sweet potato roots and other plant foods (per 100 g)*

Food	Moisture (%)	Energy (kcal)	Energy (kJ)	Protein (g)	Lipid (g)	Total carbohydrate (g)	Dietary fibre (g)	Ca (mg)	P (mg)	Fe (mg)
Sweet potato										
Boiled[a]	71	114	477	1.7	0.4	26.3	2.4[b]	32	47	0.7
Baked[a]	64	141	590	2.1	0.5	32.5		40	58	0.9
Flour[c]	12	336	1406	2.4	0.7	79.2		70	98	3.2
Cassava										
Flour[d]	13	341	1427	1.5	0.5	83.4		99	92	3.3
Rice										
Boiled, white[e]	68	135	565	2.3	0.3	28.0	0.8[f]	8	36	0.3
Flour[g]	12	365	1527	6.8	0.7	80.0		17	135	1.6
Noodles[h] cooked	79	88	368	1.0		20.3		7	7	0.6
Maize										
Porridge[e]	81	76	318	1.8	0.8	15.6		4		0.6
Tortilla[e]	48	210	879	4.6	1.8	45.3		196	138	2.6
Meal[i]	12	354	1481	9.3	3.9	73.6		19	237	3.3
Wheat										
Chaphati[j]	46	202	860	7.3	1.0	43.7	3.4	60		2.1
Bread[e]	33	278	1163	8.7	1.6	55.7	2.7[f]	24	98	1.3
Noodles[h], cooked	75	108	452	2.7	2.1	19.4		21	25	0.8
Pasta[e], cooked	66	132	552	4.1	0.7	26.7		8	59	0.5
Sorghum										
Porridge[k]	80	85	356	2.7	0.5	17.0		4	31	1.7
Meal[k]	11	343	1435	9.5	2.8	75.5		28	238	10.0

Table 3.2. *(cont.)*

Beans (*Phaseolus vulgaris*)										
Boiled[e]	69	118	494	7.8	0.5	21.4	7.4[f]	38	140	2.4

Notes:

References: (1) Leung and Flores, 1961; (2) Leung, Busson and Jardin, 1968; (3) Leung, Butrum and Chang, 1972; (4) Carribean Food and Nutrition Institute, 1974; (5) Watt and Merrill, 1975; (6) Ashida, 1982; (7) Woolfe, 1987; (8) Tan et al., 1985; (9) Englyst et al., 1988; (10) Paul and Southgate, 1978.

[a] (5).
[b] (9), Peeled, boiled.
[c] (1), (2), (3) and (6).
[d] (1), (2), (3) and (4).
[e] Averages calculated in (7).
[f] (10), includes lignin.
[g] (3) and (4).
[h] (3).
[i] (2) and (3).
[j] (8).
[k] (2).

Figure 3.1. Selling sweet potato in the Papua New Guinea Highlands where most families still obtain a high proportion of their energy intake from sweet potato (courtesy of J. Earland).

Indonesia, it has been reported (Karafir, 1987) that a researcher has estimated adult consumption at about 3 kg of sweet potato per day, a quantity which would supply about 14.3 MJ (3400 kcal) of energy, or more than 100% of the daily requirement of most adults. In another area of Irian Jaya, where a varied rather than a monotonous menu pertains, a report noted (Karafir, 1987) that adults and children had been found to consume an average of 0.4 kg and 0.27 kg, respectively of sweet potato, which would represent 11–30% and 15–20% of their respective energy requirements. Consumption in the Papua New Guinea Highlands (see Figure 3.1) of up to 2 kg/day is not unusual (Oomen, 1972; Luyken, Luyken-Koning and Pikaar, 1964), supplying 80–90% of adult daily energy requirements. The disadvantage of such heavy reliance on sweet potato (or other roots and tubers) with their high moisture content and consequently rather low energy density is more apparent in the feeding of small children than that of adults. In one day, a 1–2 year-old child would have to eat 1 kg of boiled sweet potato (which has an energy density of approximately 4.2 kJ (1 kcal)/g, only half that of the diet of a Western child) to supply its energy needs (Church, 1979). This is a quantity too bulky for most to ingest and they would therefore require energy-rich supplements. In this respect baked roots have an advantage over boiled roots in that baking results in a moisture loss sufficient to

Table 3.3. *Proximate composition of sweet potato roots cooked or processed (per 100 g)*

	Energy		Moist.	Protein	Fat	Total carb.	Dietary fibre	Ash
	(kJ)	(kcal)	(%)	(g)	(g)	(g)	(g)	(g)
Boiled (in skin)	477	114	70.6	1.7	0.4	26.3	2.4[a]	1.0
Baked (in skin)	590	141	63.7	2.1	0.5	32.5		1.2
Flour[b]	1410	337	13.2	3.3	0.6	78.3		2.7
Candied	703	168	60.0	1.3	3.3	34.2		1.2
Canned								
(in syrup)	477	114	70.7	1.0	0.2	27.5		0.6
(vacuum pack)	452	108	71.9	2.0	0.2	24.9		1.0
Flakes (dry)	1586	379	2.8	4.2	0.6	90.0		2.4
(reconstd)	398	95	75.7	1.0	0.1	22.6		0.6
Frozen[c]	399	96	74.9	1.7		22.2		1.0

Notes:
Moist., moisture; carb., carbohydrate; reconstd, reconstituted.
Unless otherwise indicated, data from: Watt and Merrill, 1975, p. 61.
[a] Englyst et al., 1988; boiled 5 min, flesh only.
[b] Ashida, 1982, p. 48.
[c] Haytowitz and Matthews, 1984.

raise the energy density significantly (Bradbury et al., 1988; and see Table 3.3).

The energy value of a particular sample of sweet potato can be rapidly and simply determined in developing countries by laboratories which are unequipped to carry out standard analyses, for example bomb calorimetry, by a method devised in Australia (Bradbury, 1986). A study showed that for foods, such as root crops, containing small amounts of fat or dietary fibre ($\leq 3\%$) the available energy E (kJ/100 g) is negatively correlated with the percentage moisture M by the equation $E = -17.38\,M + 1699$; $r = -0.998$ over the range $M = 0$–96%. The difference between energy values found by standard methods (E_s) and those determined from the above empirical equation (E) for two sets of sweet potato samples were calculated. The mean difference, $E_s - E$ (kJ/100 g), (and its percentage of the total mean energy E_s) for 38 sweet potato samples with a range of moisture content 58–78% and for an additional 84 samples with moistures 63–89% was -6.3 (-1.2%) and -9.3 (-2.1%), respectively. On average, therefore, the use of the equation overestimates the energy somewhat, though the standard deviations of 12 and 26 obtained for the two sets of samples indicated that there is a variability in $E_s - E$ from positive to negative values. It was concluded that careful moisture determination, made by drying samples to constant weight in an oven at 100°C should allow the calculation of the energy content with an accuracy of 5–10%. This could prove valuable in areas where it is desired to promote sweet potato as a weaning food.

The starch, which makes up a high proportion of sweet potato dry matter and therefore furnishes much of the energy, is easily digestible. The digestibility of the isolated starch from 13 Taiwanese sweet potato lines, determined using rats, varied from 90% to 99% and was not significantly different from that of corn starch (Anon., 1985).

The process of preparing a drum-dried weaning food based on a blend of sweet potato, full-fat soy flour, minerals, vitamins and methionine has been described (Lee, 1970). The overall nutritive value of this food (in terms of feed efficiency, body composition and protein retention for rats) was found to be comparable to milk or to another food based on corn starch in place of sweet potato.

Protein

The protein content of diets among the low income groups in developing countries is derived mostly from foods of vegetable origin. This situation is unlikely to change in the foreseeable future. It is important, therefore, to examine the content and quality of the protein in vegetable foods such as the sweet potato to determine their actual and potential role in dietary protein adequacy. It must be borne in mind, however, that total dietary energy must be adequate in order for protein to be used in growth and tissue maintenance and repair rather than as an energy source.

Quantity

The average total protein (N × 6.25) content of sweet potato is low at 1.5% (fwb) and 5% (dwb). Even so, it is still superior to some other roots and tubers (see Table 3.1) such as cassava, plantain, the giant swamp taro of the South Pacific or the oca of South America. It is somewhat inferior to potato and yams and to cereals, even those cooked as porridges (see Table 3.2). However, a wide range of protein contents exists among clones (see Table 3.5) and those with higher than average, environmentally stable levels can be identified. The protein content of sweet potato is a little higher (2–3%) if it is used in a form where some of the moisture has been removed during cooking, for example roasted or as a dry flour (Table 3.3). However, some 25% of the total nitrogen of fresh sweet potato roots occurs as non-protein nitrogen (NPN; see Chapter 2), so that total nitrogen determinations overestimate protein content.

Table 3.4. *NDpE% of sweet potatoes and other staple foods compared to that of breast milk*

Food	NDpE%
Breast milk	8.0
Oats	7.0
Potatoes	6.0
Wheat	6.0
Sorghum	4.9
Rice	4.9
Yam	4.8
Maize	4.5
Sweet potato	3.4
Plantain	1.5
Cassava	<1

Notes:
Data from: Cameron and Hofvander, 1976.
NDpE% = Net dietary protein energy percentage, or the percentage of total energy provided by protein. (Note: requirement for adult = 4.0, for 1 year old = 6.0.)

In the boiled form in which it is most commonly eaten, however, sweet potato is not a well-balanced food in terms of energy and protein. The net dietary protein energy percentage (NDpE%), which is calculated as:

$$\frac{(\text{g protein}/100\ \text{g food}) \times \text{chemical score} \times 16.8 \times 100}{\text{total kJ}/100\ \text{g food}}$$

provides a useful estimate of the quality of a food or diet by showing the percentage of the total energy supplied by protein. The NDpE% of breast milk is approximately 8, which meets the known requirements of infants. As children grow, energy requirements increase and the percentage drawn from protein decreases until, at 1 year old, the child needs an NDpE% of about 6. This decreases to 4 in an adult. Table 3.4 shows the NDpE% of some staple foods; that of sweet potato is below even the adult requirement, signifying that even if enough is eaten to supply energy needs, the protein requirement will not be fulfilled.

NDpE% is calculated using both the quantity and quality of a protein source. The range of protein contents reported by different authors (Table 3.5) shows that a wide variation exists and it may be possible to select, or breed for, higher protein lines. To determine the value of this procedure, it is worth examining the quality of sweet potato protein as

Table 3.5. *Reported protein contents of sweet potato roots from different sources*

		% (dwb)		% (fwb)	
Ref.[a]	Source	Range	Mean	Range	Mean
1	PNG highlands	2.2–3.8			
1	PNG lowlands	4.8–9.2			
2	PNG	0.6–2.6			
3	Taiwan	2.51–7.46			
4	Pacific			0.63–0.81	
5	USA	1.73–9.14	5.76	0.49–2.56	
6	PNG			0.50–2.06	1.34
7	PNG	1.6–5.3	3.14	0.50–1.82	1.00
8	USA	4.38–8.98	6.29		

Notes:
PNG, Papua New Guinea.
[a] References: (1) Oomen et al., 1961; (2) Norgan, Durnin and Ferro-Luzzi, 1979; (3) Yang et al., 1967; (4) Peters, 1958; (5) Purcell et al., 1972; (6) Bradbury et al., 1984; (7) Goodbody, 1984; (8) Dickey et al., 1984.

determined by chemical procedures, animal feeding experiments and human feeding trials.

Amino acid analyses and scores

Complete amino acid analyses have been carried out by several researchers. In addition to those shown in Table 3.6, others have also been reviewed (Walter, Collins and Purcell, 1984). If, from these determinations, an average essential amino acid profile is calculated, it can be seen (Table 3.7) that in general sweet potato has more lysine, but a little less of the sulphur-containing (S-containing) amino acids than does maize or rice protein but less lysine and more S-containing amino acids than does legume (*P. vulgaris*) protein. Although sweet potato protein is shown as similar in amino acid pattern to other roots and tubers in Table 3.7, other authors (Martin and Splittstoesser, 1975) found cassava to be much lower in the essential amino acids than sweet potato.

Wide ranges in all sweet potato essential amino acids have been found. These are due to varietal, environmental and cultural effects and postharvest treatment of roots. They may also reflect differences in analytical techniques which are notoriously difficult to reproduce between laboratories even for a single sample. However, ranges found

Table 3.6. *Amino acids in sweet potato roots as determined by various authors (g/16 g N)*

	1	2	3	4	5	6
Essential						
Lys	3.8	3.42	4.6	4.42	3.98 (2.52–5.22)	3.72
Ile	3.7	3.68	4.8	7.31	3.72 (3.28–4.50)	3.23
Leu	5.9	5.44	7.7	5.57	5.94 (4.98–6.86)	5.14
Met	0.7	1.70	1.5	3.05	2.03 (1.41–2.83)	1.55
Cys	0.8	1.10	0.0	0.92	0.04 (ND–0.42)	1.85
Phe	5.0	3.86	4.5	5.74	4.88 (4.13–5.69)	4.22
Tyr	2.5	2.34	1.2	1.96	3.04 (2.55–3.72)	2.42
Trp	1.4	ND	ND	ND	1.66 (1.21–2.19)	1.09[a]
Thr	5.0	3.78	4.8	3.90	4.26 (3.67–5.22)	4.37
Val	5.5	4.53	6.9	8.17	5.22 (4.36–6.31)	4.78
His	1.8	1.34	ND	2.15	1.66 (1.42–1.91)	1.76
Non-essential						
Asp	14.8	13.20	ND	18.17	14.26 (11.36–19.09)	13.68
Glu	12.4	8.66	ND	12.65	10.30 (8.08–12.82)	7.65
Pro	6.1	3.50	ND	3.73	3.92 (3.15–4.76)	3.37
Gly	3.6	3.74	ND	4.86	4.08 (3.17–4.79)	3.79
Ala	6.9	4.77	ND	5.08	5.30 (4.20–6.49)	4.49
Arg	2.8	4.91	ND	3.97	3.73 (2.59–5.90)	3.68
Ser	5.9	4.08	ND	5.79	4.33 (3.31–5.83)	4.35

Notes:
ND, not determined.
1 Taira and Taira, 1963; average of two cultivars, Japan.
2 FAO, 1970.
3 Martin and Splittstoesser, 1975; average of three cultivars, Puerto Rico.
4 Cheng, 1978; average of seven cultivars, Taiwan.
5 Goodbody, 1984; average (ranges in parentheses) of 15 cultivars, Papua New Guinea.
6 Bradbury and Holloway, 1988; average of eight cultivars from the Solomon Islands and Tonga.
[a] Average of four cultivars from the Solomon Islands only.

by a single author among some Papua New Guinea cultivars are shown in Table 3.6. Lysine contents have been found to range from 2.2 g/16 g N (Yang et al., 1975) to as much as 7.2 g/16 g N (Purcell, Swaisgood and Pope, 1972). The S-containing amino acid cystine, which spares methionine, has been found to be present in some cultivars and completely absent in others. Hence total S-containing amino acids are also very variable, being as high as 3.9 g/16 g N in a cultivar grown in Taiwan (Li,

Table 3.7. *The essential amino acid composition of sweet potato roots and leaves and other plant foods (g/16 g N)*

Food	His	Ile	Leu	Lys	Met + Cys	Phe + Tyr	Thr	Trp	Val
Sweet potato roots[a]	1.7	4.4	6.0	4.0	2.5	6.9	4.4	1.4	5.9
Potato tubers[b]	2.0	3.9	5.9	6.0	3.0	7.8	3.9	1.4	5.1
Cassava roots	2.1	2.8	4.0	4.1	2.7	4.1	2.6	1.2	3.3
Yam tubers	1.9	3.7	6.5	4.1	2.8	8.0	3.6	1.3	4.7
Maize meal	2.7	3.7	12.5	2.7	3.5	8.7	3.6	0.7	4.9
Rice, white	2.3	4.2	8.2	3.6	3.7	8.1	3.3	1.3[b]	5.8
Beans, dry (*Phaseolus vulgaris*)	2.8	4.2	7.6	7.2	1.9	7.7	4.0	1.0[c]	4.6
Leaves									
Sweet potato[d]	3.4	4.8	8.1	3.8	2.9	7.8	4.5	0.88[e]	9.1
Cassava	2.2	4.9	8.6	6.2	2.8	9.4	4.7	1.5	5.7
Horseradish tree (*Moringa oleifera*)	2.2	4.7	8.4	5.8	3.8	5.9[f]	4.5		6.0
Tamarind (*Tamarindus* spp.)	2.3	5.3	9.3	5.9	1.6	9.7	4.6		5.8
Spinach (*Spinacia oleracea*)	2.5	4.8	9.5	7.3	3.7	11.2	5.3		6.1
Water convolvulus (*I. aquatica*)	1.9	2.9	5.2	3.6	1.5[g]	7.6	3.3		4.0

Notes:
Unless otherwise noted, figures calculated from: FAO, 1970.
[a] Average of values given in Table 3.6.
[b] Average values calculated by Woolfe (1987).
[c] Calculated from Paul and Southgate (1978).
[d] Cheng, 1978; average of two cultivars.
[e] Haytowitz and Matthews, 1984.
[f] Phe only.
[g] Met only.

1982). Tryptophan has been found at a very low level by some authors, notably 0.1 g/100 g protein (Splittstoesser and Martin, 1975) and 0.28 g/100 g protein (Dickey et al., 1984). Another author (Goodbody, 1984) has found adequate tryptophan levels . In one study, tyrosine was found to be low in tropical roots and tubers including sweet potato (Martin and Splittstoesser, 1975), but other workers have generally noted adequate amounts (Purcell et al., 1972; Dickey et al., 1984). Some of the differences in amino acids among samples of total sweet potato protein

Table 3.8. *Amino acid scores[a] of sweet potato roots and leaves for various human age groups*

	Infant		Pre-school child		School child		Adult	
	Score	LAA[b]	Score	LAA	Score	LAA	Score	LAA
Roots	60	S-amino acids/Lys	70	Lys	92	Lys	>100	—
Leaves[c]	52?	Trp?	66	Lys	86	Lys	>100	—

Notes:

[a] Calculated by:

$$\text{Score} = \frac{\text{mg amino acid in sweet potato protein} \times 100}{\text{mg amino acid in requirement pattern}}$$

using amino acid contents of sweet potato given in Table 3.7 and human amino acid requirements according to FAO/WHO/UNU (1985).

[b] LAA, limiting amino acid.

[c] Calculated using single figure for tryptophan in sweet potato leaves given by Haytowitz and Matthews (1984); Trp and Lys almost equally limiting.

may be due to variations in the true protein to NPN ratio. Table 2.10 (p. 88) shows the much higher levels of essential amino acids which are present in true protein as compared to NPN. The higher the percentage of NPN in total protein, therefore, the greater the 'dilution' effect on the amino acids of the true protein. The quantities of all essential amino acids (g/16 g N) were higher in 'Centennial' protein with 24% NPN than in 'Jewel' protein with 34% NPN (Walter et al., 1983) and the quality of the protein remaining rose, after extraction of the NPN, in terms of all essential amino acids in both cultivars. Approximately 90% of the NPN is amino or amide nitrogen and as such is able to satisfy some of the requirements for amino acid synthesis *in vivo* (Walter et al.,1984).

Many authors have previously assessed the quality of sweet potato protein by comparing it with a hypothetical reference protein (FAO/ WHO 1973). In this way, the S-containing amino acids have generally been found to be limiting in sweet potato protein (Bradbury et al., 1984), the mean chemical score being calculated as 65% (range 34–116%) from 33 analyses. Within single cultivars from Papua New Guinea, the first limiting amino acid varied between lysine (34% of samples), leucine (32%), S-containing amino acids (21%) and isoleucine (11%) for different locations and seasons (Bradbury et al., 1985b). The mean chemical score of the samples was 73%. Results also showed that the first limiting amino acid varies even between roots from the same plant, so that amino acids within different roots eaten at the same meal would complement one another and therefore raise the chemical score above the value for a single root.

In 1985 it was indicated (FAO/WHO/UNU, 1985) that the amino acid scoring procedure, by which the capacity of a protein or mixture of proteins to meet the essential amino acid and nitrogen requirements of a human being is evaluated, should be based on a knowledge of those requirements. Table 3.8 shows the amino acid scores and first limiting amino acids of sweet potato protein for different age groups, calculated from the latest estimates of human requirements and the average amino acid content of sweet potato. It can be seen that the quality of sweet potato protein is lowest for the infant, but moderate to high for all other age groups. The apparently high quality of this protein for the upper age groups should, however, be treated with caution in view of recent criticisms of the low FAO/WHO/UNU (1985) estimates of essential amino acid needs of adults (Pellett and Young, 1988). A suggested alternative amino acid pattern for all age groups above that of the infant (Pellett and Young, 1988) would give sweet potato protein an amino acid score of 80 for these age groups. For the infant the score is higher at 60 (methionine + cystine and lysine equally limiting) than that of maize at 41 (lysine and tryptophan equally limiting) and beans at 45 (methionine and cystine limiting). The moderate values of lysine and methionine which sweet potato protein possesses in comparison with maize (lower lysine) and beans (lower S-containing amino acids) suggest that it could act as a complement to the other two types of food, especially if its total protein content could be raised. At the same time the deficiencies of S-containing amino acids and lysine for the most nutritionally vulnerable age groups suggest that an improvement in both the quality and quantity of protein would be beneficial in those societies relying partly or heavily on sweet potato as a source of protein.

Animal experiments

Sweet potato has long been used to feed animals which are farmed for meat, such as pigs, cattle and poultry. Its present and potential use in animal feedstuffs are discussed in Chapter 7. There are few published examples of experiments using laboratory animals to investigate the nutritional value of sweet potato protein. Moreover, these have not given consistent results. This probably reflects differences in methodology as much as variations in the actual quality and quantity of protein in samples used by researchers in different parts of the world. On the whole experiments have indicated that the quality of sweet potato protein is moderately good and have demonstrated its capacity for complementing the proteins in other foods.

A group of workers in Taiwan (Yang et al., 1975; Yang, 1982), describing a series of experiments carried out with rats between 1965 and 1975, observed that partial substitution of rice or wheat flour (equivalent to 25% of the cereal protein) with sweet potato flour increased the biological value of the dietary protein from 72 to 80 and 65 to 71 in male and female rats, respectively, and prolonged their lifespan. This result was attributed to the lysine values found in sweet potato protein being higher than in rice or wheat proteins. Replacement of 5%, 10% or 20% of pre-cooked dried rice with pre-cooked sweet potato flour in a diet originally consisting of 94% rice, enhanced the performance of rats in terms of growth rate, feed efficiency, protein efficiency ratio (PER*) and longevity (Yang and Blackwell, 1966). Indian researchers reported in 1954 (Subrahmanyan, Murthy and Swaminathan, 1954) that partial replacement of rice or wheat, which is eaten by poor South Indian vegetarians, with sweet potato flour, led to an improvement in growth in rats compared to those maintained on an exclusively rice diet. Further Indian experiments (Majumdar, Sharma and Kehar, 1960) in which rats were fed for 8 weeks on diets composed of 78.5% sweet potato, tapioca, rice or wheat flour showed that sweet potato or tapioca alone were unable to sustain life. Substitution of 25% of the rice by sweet potato caused no change in rat growth compared to rice alone, but sweet potato substituting for 25% of the wheat caused a reduction in rat growth compared to the group on wheat alone. In contrast the inclusion of tapioca at the same level of substitution improved rat growth. Addition of groundnut cake to the rice/sweet potato or wheat/sweet potato diets increased rat weight gains above those of rats on wheat or rice alone, which suggests that the mixed diets may have been deficient in protein.

A much more recent experiment (Bressani, Navarrete and Elias, 1984) also apparently demonstrated the inferiority of sweet potato protein compared to other roots and tubers in feeding trials carried out to determine the minimum quantity of *P. vulgaris* beans with and without methionine supplementation needed to obtain positive weight gains in rats fed cassava, plantain, potato or sweet potato flours and the PER values of mixed staple/bean diets with 40% bean substitution. Sweet potato performed the most poorly of the roots and tubers tested, needing the greatest quantity of beans to sustain rat body weight and having the

* Protein efficiency ratio (PER): weight gain per weight of protein eaten (usually measured in rats with the test food supplying all the protein at a level of 10% in the diet).

lowest PER even when supplemented with methionine. However, the cassava roots, potatoes and beans were all cooked before being dried and ground to a flour. The sweet potato was described as simply peeled, washed, dried and ground. It is therefore quite possible that the trypsin inhibitor present in raw sweet potato roots (see Chapter 4) had not been destroyed and was interfering with protein digestibility. A more useful experiment would have employed properly cooked sweet potato in which all trypsin inhibitor was known to have been inactivated.

The PER values of two American cultivars, 'Centennial' and 'Jewel' (previously washed, peeled, cut into strips, cooked in steam, dried at 60°C in a forced-draught oven and ground to a flour), were determined in rats and compared with that of casein (Walter et al., 1983). All diets contained 1.6% N. After harvest the roots had first to be stored until sufficient starch had been metabolized to increase nitrogen levels of the prepared flours to more than 1.8%. This resulted in considerably increased NPN contents compared to those of freshly harvested roots, especially in 'Jewel' (see p. 133). The PER of 'Centennial', 2.22, was significantly ($p<0.05$) higher than that of 'Jewel', 2.00 (corrected to casein = 2.50). This result was reflected by the lower percentage of NPN and generally higher levels of all essential amino acids (g/16 g N) in 'Centennial' protein. Sweet potato total protein appeared to be of fairly high quality from this experiment.

True sweet potato protein, separated as either a concentrate or an isolate by precipitation from heated calcium chloride solution, was shown to have high nutritional value; when fed to rats the fractions had PERs and NPRs* numerically superior (though not significantly different) to that of casein (Walter and Catignani, 1981). It has been suggested that such protein fractions, resulting from by-products of industries producing starch (Purcell, Walter and Giesbrecht, 1978) or alcohol (Walter and Catignani, 1981) could be used to feed humans. However, such a use is likely to benefit developed countries such as the United States and Japan rather than poorer nations.

The utilization of a protein in the body does not depend only on its amino acid composition, but also the degree to which it is digested. The digestibility of sweet potato protein is not very high, as judged from human feeding trials (see p. 138) and from studies with pigs (Tsou, Kan and Wang, 1987). It has been suggested that this may be due to the low digestibility of sweet potato carbohydrate, as root and tuber starches are

* Net protein ratio (NPR): weight gain of a test animal plus weight loss of a control animal per gram of protein consumed. Thus:

$$\text{NPR} = \frac{\text{Wt gain of test animal} + \text{Average wt loss of animal fed basal (non-protein) diet}}{\text{Protein (N} \times 6.25\text{) consumed by animal}}$$

known to be less digestible than, and inferior in their effect on protein utilization to, cereal starches. The effect of sweet potato carbohydrate, both in an isolated form as starch and in a complete form as a dried flour, on protein digestibility has been investigated in rats using soy protein or casein. The isolated starch from 13 sweet potato lines grown in Taiwan was highly (90–99%) digestible. The true digestibility of soy protein (75.5–80% for the 13 lines) was numerically below, but not significantly different from, that (84%) fed together with corn starch (Anon., 1985). However, the levels of faecal carbohydrate produced did vary from 4.33 to 49.91 mg/hour. The apparent digestibilities of both casein and soy protein when fed along with dried whole sweet potato were significantly lower than those when fed along with dried cassava, though they were higher than those fed with dried taro (Suzuki et al., 1983). It would appear therefore that some constituent of whole sweet potato is capable of affecting protein digestibility. The nature of this component is not known; it would seem, however, that unavailable carbohydrate in the form of dietary fibre or resistant starch might be involved. When human beings eat large quantities of sweet potato, it has been noted that they produce extremely voluminous faeces (Oomen, 1970; Huang, Lee and Chen, 1979; and see p. 143).

Human feeding

A limited number of experiments have been carried out to determine the capacity of sweet potato root protein to maintain nitrogen balance in humans. In addition such studies have been facilitated by the almost exclusive dependence on sweet potato which exists in societies such as those of the Papua New Guinea Highlands, where even superficial observations would lead to the conclusion that sweet potato has supported life for generations.

Results of feeding experiments or observations of dietary studies seem to indicate that sweet potato protein alone is not adequate to support normal growth in children, but maintains adults at a marginal level of protein adequacy. When sweet potato provides a high proportion of the dietary protein, a subject can be maintained in positive nitrogen balance as long as the level of nitrogen fed is above a certain limit. As has been pointed out (Oomen, 1971) it is not difficult for a person subsisting on a diet consisting almost entirely of sweet potato to achieve sufficient energy per day. Whether they achieve an intake of 10, 20 or 30 g total protein/day then depends on the varieties of sweet potato consumed.

An early paper (Adolph and Liu, 1939) reported that sweet potato could maintain nitrogen balance in human subjects. Much more recently, seven teenage and two adult males from Orchid Island

(Taiwan) were fed diets in which steamed sweet potato supplied 80–90% of their total energy and 75–91% of their total protein (Huang et al., 1979), for either 32 (trial 1) or 53 (trial 2) days, and their nitrogen balance determined. Teenage protein intakes were 0.67 and 0.71 g/kg body weight in trials 1 and 2 respectively and 0.63 g/kg body weight for the adults. At the end of the trial periods, the teenagers were in negative nitrogen balance; the mean result for trial 1 was −0.5 mg N/kg per day and for trial 2 −3.2 mg N/kg per day. Apparent and true protein digestibilities were calculated as 63% and 76%, respectively, from data for all subjects. Plasma urea nitrogen of both teenagers and adults decreased from between 8 and 11 mg/100 ml at the start of trial 2 to between 2 and 3 mg/100 ml after 53 days, and the teenage plasma free amino acid pattern showed some abnormality in that valine, isoleucine and leucine decreased significantly. Their essential/non-essential amino acid ratio decreased from 0.71 to 0.58. These results indicated a degree of protein depletion in the teenagers. The two adults averaged +6.0 mg N/kg per day, although one of them had a slightly negative cumulative nitrogen balance. Adults may thus be able to fulfil their minimum protein requirement when fed about 2.5 kg sweet potato/day, supplemented by a small amount of protein-rich food such as fish and some vegetables. There was no apparent sign of protein deficiency even after 2 months, but subjects showed a tendency to tire more easily during physical exercise.

The supplementary value of sweet potato protein for cereals has also been shown for humans. Adult Taiwanese male students when fed on a high rice diet showed more tendency to be in negative nitrogen balance than when 11.5% of the calories provided by the rice were replaced by cooked sweet potato (Yang et al., 1967).

The Papua New Guinea Highlands have been called a 'man–pig–*Ipomoea*' ecosystem, the sweet potato diet existing there being very little supplemented with other foods (Oomen, 1971). Sweet potato is the first solid food given to babies and from childhood onward often provides more than 90% and 75% of dietary energy and protein, respectively (Oomen, 1971; Oomen et al., 1961). In spite of evidence suggesting the occurrence of protein deficiency among Papua New Guinea Highland populations (Sinnett, 1975), various authors have also noted that the Highlands are densely populated and that the adults are healthy and capable of heavy physical work (Luyken et al., 1964; Hipsley and Kirk, 1965; Oomen, 1970). Negative nitrogen balances were observed in adult and 10–14 year-old Highland Papuans for whom 70% to 100% of the diet was always sweet potato roots (Oomen, 1970). Nitrogen intake for the adults (average weight 51 kg) and children (average weight 25 kg), respectively, was 2.5–3.8 g/day and 1.2–2.0 g/day. In another area of the

Papua New Guinea Highlands, where both sweet potato roots and leaves are commonly consumed at a single meal, nitrogen balances were positive in boys (mean weight 30 kg) where sweet potatoes (2000 g of roots and 200 g of leaves) provided 5 g N or more per day, and negative at lower nitrogen intakes (Luyken et al., 1964). The higher nitrogen intakes were achieved using roots from a cultivar both relatively high in total protein and with a high percentage of the total protein occurring as true protein, and the use of leaves with a high mean protein content of 4.8% (fwb). Supplementing the lower nitrogen intake diets with methionine or lysine did not improve nitrogen retention. Adding corn or corn and methionine did not improve retention above that on the higher nitrogen intake sweet potato diets. Supplementing sweet potatoes with 50 g of peanuts or pigeon peas significantly raised nitrogen intakes and retentions. The authors concluded that sufficient nitrogen for growth is retained by school children 7–15 years old on sweet potato diets providing 5 g N or more per day. However, the experiments did not give information on the level of nitrogen retention in the case of infections or heavy work. Moreover, it has been found that leaves are infrequently consumed in the Highlands of Papua New Guinea and that the protein content of Highland cultivars is often low (Oomen et al., 1961; Oomen, 1971). A summary of published data of Highland diets in which sweet potato is the major protein source shows their $ND_pE\%$ to range from 3.11 to 4.98 (Heywood and Nakikus, 1982). Most of these diets therefore even if eaten in sufficient quantity to supply energy needs would not fulfil protein requirements. The growth of children living on diets composed mainly of sweet potato or of a mixture of sweet potato and taro has been described as among the slowest of any reported population in the world (Malcolm, 1970).

However, the percentage of the world population who live on a mainly sweet potato diet is very low indeed. Most societies, even those living on a wholely vegetarian diet, have a mixture of proteins which in practice complement one another. The experiments and studies described above serve to illustrate that, in contrast to the assumption that sweet potato is solely a source of starch, it is also a source of fairly good quality protein for human feeding. Those societies obtaining part or a high percentage of their protein intake from sweet potato would benefit from an increase in quality and especially quantity of the protein.

Strategies for improvement of sweet potato protein

The wide range of total protein content present among sweet potato clones (see Chapter 2 and Table 3.5) points to the possibility of producing sweet potato lines with higher than average protein contents by either selection or breeding. In addition, genotype × environment interactions affecting protein content could be exploited by selecting genotypes which consistently produce high levels of protein in any particular environment. Cultural manipulations such as fertilization with nitrogen or variation of time between planting and harvest could be applied (see pp. 90–1).

A further possibility to improve nutritional utilization of sweet potato protein is to improve its quality. The deficiencies of sweet potato protein in some of the essential amino acids has already been described. Several potential methods exist to achieve quality improvement. These include conventional plant breeding techniques, or manipulation of existing storage protein genes to increase levels of essential amino acids. A recent and more novel approach utilizes genetic engineering or recombinant DNA technology to construct synthetic genes which encode proteins rich in essential amino acids, and introduce them into the plant. At present, insertion of single genes is all that has been accomplished in plants, but sweet potato is a prime candidate for this type of approach due to its genetic complexity (hexaploidy). Plant breeding may perhaps be rejected due to the large amount of screening and effort involved for rather limited improvements. Difficulties are compounded by the fact that sweet potato is deficient in more than one essential amino acid. Genetic engineering using synthetic genes is flexible enough to produce proteins tailored to supplement optimally a particular plant protein such as that of the sweet potato, and also offers the possibility of introducing the single trait of high quality protein without altering other agronomically important characters of the plant.

The production of a synthetic gene encoding a protein composed of about 80% essential amino acids (normal sweet potato protein is only about 45% essential amino acids), the development of a technique utilizing the bacterium *Agrobacterium rhizogenes* or *Agrobacterium tumifaciens* as a vector for introducing the synthetic gene into plants and the specific transformation in this way of the sweet potato has been amply described by the researchers responsible (Jaynes et al., 1986; Dodds, J.H. and Jaynes, J.M., personal communication; Espinoza et al., 1987). The following is an extremely brief summary of what has been achieved so far.

A. rhizogenes and *A. tumifaciens* possess plasmids (small, circular, self-replicating DNAs) which, when the bacteria infect an area of wounded plant tissue, become integrated into the genome (chromosomes) of the plant cells. This initiates activity which leads to tumour production (*A. tumifaciens*) or 'hairy root' syndrome (*A. rhizogenes*). This is an example in nature of the insertion of new information into a plant chromosome. For this reason *Agrobacterium* plasmids have been called 'nature's genetic

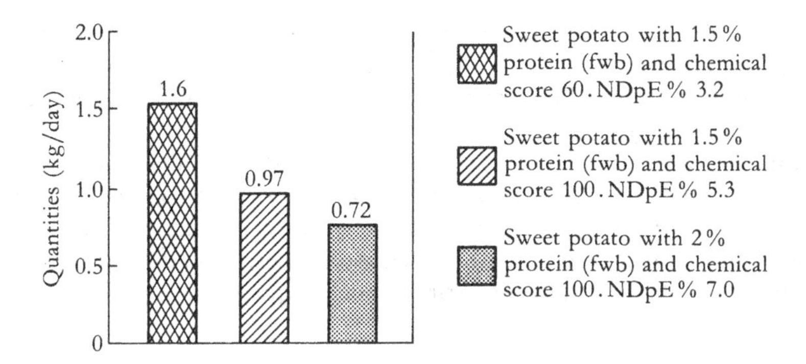

Figure 3.2. Theoretical improvements in the nutritive value of sweet potato protein attainable by genetic engineering. The histogram shows the quantities of 'normal' and 'improved' sweet potato which would have to be eaten by a 2-year-old child to satisfy 100% of its essential amino acid requirements. Calculations assume 100% availability and so represent lower limits.

engineers'. Gene transfer by the plasmids offers great potential for altering the genetic composition of plants in a predictable and desirable way. Any desired gene can be inserted in to the *Agrobacterium* plasmid which acts as a vector for gene transfer.

To date, reproducible methods for isolating and purifying desirable plant genes are still not available. An alternative approach to the use of pure plant genes could be to synthesize genes with required characteristics, for example with high levels of those essential amino acids normally deficient in plant proteins. This can be done mechanically, and in such a way, that the gene is symmetrical enough to produce a high quality protein no matter which way it is read by the cells' protein synthetic machinery. Lysine can also be inserted at frequent intervals to provide numerous sites for proteolytic attack by trypsin, thus promoting bioavailability. To ensure an appropriate level of expression of the synthetic gene within the plant and in the correct tissue, a suitable 'promotor' or gene control sequence is fused to the synthetic gene fragment to be introduced into the plant. Plant tissue can then be infected with engineered *Agrobacterium* to obtain transformed plants. This system has been used with *A. rhizogenes* containing a synthetic high quality protein to transform and regenerate sweet potato plants from hairy roots induced by the *A. rhizogenes* vector, though with less success than that employed with potato plants. Another system, employing *A. tumifaciens* with sweet potato, might be more promising (Espinoza et al., 1987).

Figure 3.2 shows in simple terms the hypothetical improvement in sweet potato protein nutritional value which might be achieved by combining plant breeding (for quantity) and recombinant DNA technology (for quality).

Similar techniques could be used to introduce genes coding for resistance to diseases. The majority of sweet potato production takes place in countries where malnutrition is prevalent and which lack resources to pay for control of pests and diseases. Genetic engineering techniques hold out a hope of improvement in quality and yield of sweet potato.

Dietary fibre

The dietary fibre content of sweet potato roots is variable, as shown in Chapter 2. In fact fibre content can be so high in some lines that they are unusable directly as human food, and have to be utilized for industrial processing. On average, however, at about 2% (fwb) of the root, fibre content is texturally satisfactory and is a valuable component of diets which include sweet potato.

There has been increasing interest of late in individual components of dietary fibre and their differing roles in physiological functioning. Though more effects have no doubt still to be discovered and those that are known are so far not fully understood, it is thought that insoluble constituents such as hemicelluloses and cellulose are largely responsible for increasing water-holding and faecal bulk, thereby increasing faecal transit times and reducing the risk of colon cancer. Soluble fibre, especially some of the pectic material, is thought to reduce cholesterol levels, and blood glucose and insulin response to meals containing carbohydrate.

The content of total dietary fibre in sweet potato roots is similar to that of other roots and tubers (Table 3.1; Holloway et al., 1985), is much higher than that of a food such as cooked white rice, but lower than that of *Phaseolus* beans (Table 3.2). The type and percentage of individual components has also been determined for sweet potato and compared with other roots and tubers and some other vegetable foods. Sweet potato has a well-balanced content of soluble and insoluble fibre in a ratio of about 1:1 (excluding lignin) (Chen and Anderson, 1981; Englyst et al., 1988; Holloway et al., 1985), unlike wheat bran, which is mainly insoluble fibre. Sweet potato in common with other roots and tubers has a relatively low hemicellulose:cellulose ratio (Holloway et al., 1985) of $\leq 2:1$ compared to the higher ratios of cereals and legumes. The water-holding capacity of dietary fibre appears to correlate positively with its hemicellulose content (Holloway and Grieg, 1984) and is thought to be an important determinant of faecal bulking and intestinal transit times which influence gastrointestinal disease. The water-holding capacity of fibre in diets high in sweet potato, or in other roots and tubers also having low hemicellulose:cellulose ratios, might therefore be less than in diets based on cereals (Holloway et al., 1985). Sweet potato fibre seems, however, to be very effective for faecal bulking. Subjects eating large amounts of sweet potato (2–3 kg/day) also produced copious quantities of soft, doughy faeces, averaging about 800 g (Orchid Island; Huang, 1982) or 1 kg (Papua New Guinea; Luyken et al., 1964) per day. The large volumes of these faeces were two to four times those of 'normal' subjects eating a more mixed diet (Oomen, 1970). Although soluble non-starch polysaccharides which are susceptible to degradation by gut bacteria account for a large proportion of total sweet potato fibre, their utilization will be partially offset by an increase in bacterial cell mass due to their availability to the intestinal flora (Holloway et al., 1985), hence increasing faecal bulk.

High levels of blood cholesterol have been implicated in the etiology of atherosclerotic coronary heart disease (Rifkind, 1987). Sweet potato dietary fibre may be an effective agent for lowering cholesterol levels. It has been suggested that pectins with a high methoxyl content are important in reducing serum cholesterol (Mokady, 1973). The methoxyl content of sweet potato pectin was high at 9.7% of a cold water extract, the highest being for onion at 11% and wheat bran having only 0.1% in a study with a series of fruit and vegetables (Holloway, 1983).

There is a popular belief in India, particularly among medical people, that consumption of high levels of roots and tubers may be related to prevalence of cardiovascular disease. Indian researchers, however, found that feeding rats with cooked roots and tubers, including sweet potato, produced lower tissue cholesterol and triglyceride levels than a comparable diet containing wheat (Prema and Kurup, 1979). Rat serum and aortic cholesterol levels were lower with sweet potato than with wheat in both cholesterol-containing and cholesterol-free diets. In comparison to wheat, sweet potato gave lower liver cholesterol values with cholesterol-free, or comparable values with cholesterol-containing, diets. Aortic, heart and liver triglycerides decreased with the sweet potato cholesterol-free diet and serum, aortic and liver triglycerides decreased with the sweet potato cholesterol-containing diet compared with wheat. There was increased lipolysis in the aorta of rats fed the sweet potato diet with cholesterol. In a Taiwanese experiment, although the hypocholesterolemic effect of four cultivars of dried, powdered sweet potato incorporated into rat diets containing 1% added cholesterol was in general not as pronounced as that of other vegetables such as legumes, one of the cultivars significantly reduced serum cholesterol and total lipids (Huang, 1982).

It has been proposed that one mechanism for the reduction of blood cholesterol levels by fruit and vegetable fibre is direct binding of cholesterol and bile acids by fibre in the large intestine. Cholesterol-binding capacity was investigated for 28 fibre samples from a variety of commonly consumed tropical fruits and vegetables including sweet potato (Lund, 1984). Binding capacity of cellulose, lignin, guar gum, pectin and cholestyramine (a commercial resin used for blood cholesterol reduction), was also determined. In view of the low binding capacity shown by most of the fruits and vegetables studied, it was concluded that direct binding in the large intestine may not be responsible for lowering of blood cholesterol. However, apart from cholestyramine with a binding value of 84%, sweet potato was by far the most effective binder at 30%. This compared with a cassava fraction at 3%, citrus pectin at 8% and the majority of samples at ≤20%.

Human blood cholesterol levels usually vary from 150 to 200 mg/100 ml. Target levels of 200 mg/100 ml for adults over age 30 years and 180 mg/100 ml for those of 30 or below have been set (National Institutes of Health, 1985), values above this leading to increased risk of coronary heart disease. It is interesting to note that villagers in Papua New Guinea living on very high intakes of sweet potato have been found to have low mean serum cholesterol levels of 108 and 128 mg/100 ml for males and females, respectively (Luyken et al., 1964). Levels rose significantly to 173 and 176 mg/100 ml respectively when subjects changed their diet as a result of hospitalization. The villager cholesterol levels were also lower than those of other subjects living in the village but having access to purchases of rice and canned fish.

High dietary levels of roots and tubers have also been implicated, particularly by medical people, in the prevalence of diabetes in India (Prema and Kurup, 1979) and Bangladesh (Poats, S., personal communication, 1983). However, this was felt to be unlikely after various roots or tubers were fed to rats and the fasting blood glucose level or the level 1 hour after a glucose load was determined (Prema and Kurup, 1979). Levels were on the whole comparable with those when glucose alone was fed. Levels 1 hour after glucose load were lower in rats fed sweet potato than in those fed glucose or wheat. The fibre components of various fruits and vegetables including sweet potato contained a high proportion of water-soluble polymers, whereas the fibre from wheat bran is mainly water-insoluble polymers (Holloway, 1983). The soluble components could be associated with gel formation in the small intestine, reducing absorption of nutrients such as glucose and thereby lowering insulin requirements in diabetic patients on high fruit and vegetable diets. Contrary to the idea of roots such as sweet potato causing diabetes, they might have beneficial effects when incorporated into diabetic diets.

Much work remains to be carried out both in quantitative and qualitative terms on the dietary fibre of tubers and roots such as sweet potato. The results of initial analyses and experiments tentatively suggest that sweet potato dietary fibre has a composition which could enable it to play a useful part in various physiological functions with known health implications. This should be investigated further. Recommendations for daily intakes of dietary fibre will remain somewhat arbitrary until analyses have been standardized and further clarification has been obtained regarding the functions of fibre. A recent WHO report (WHO, 1990) has suggested that daily intakes of non-starch polysaccharides of 25 g would be effective in preventing constipation. One hundred grams of sweet potato would thus provide about 8% of a suitable intake, although diets high in sweet potato, like those in Papua New Guinea would contain more than suggested levels.

Vitamins

Contrary to the belief of many consumers that sweet potato roots are solely an energy source, they have been shown to contain substantial amounts of ascorbic acid (vitamin C), moderate quantities of thiamin (vitamin B_1), riboflavin (B_2) and niacin, some pantothenic acid (B_5), pyridoxine and its derivatives (B_6), and folic acid. They have also been reported to contain satisfactory quantities of tocopherol (vitamin E).

One of the major contributions which sweet potatoes could make to the health and welfare of humankind, however, is that of supplying carotenoid vitamin A precursors. Vitamin A deficiency is one of the major health problems which some developing countries face at the present time. One line of attack in a multi-disciplinary approach to solving this problem is to encourage the production of food plants containing high levels of provitamin A carotenoids. Dark orange-fleshed roots are rich sources of beta-carotene, the most active provitamin A carotenoid (see Chapter 2), and yellow/orange roots supply moderate amounts. In this connection the valuable part which could be played by sweet potato leaves, which are also rich sources of active carotenoids, will be mentioned here, but is discussed in more detail under Tops, below. In view of the potential importance of the use of sweet potato as one approach to tackling the problem of vitamin A deficiency, a separate section will be devoted to this subject.

Vitamin A

Symptoms of vitamin A deficiency

An aldehyde derived from retinol (vitamin A alcohol) is an essential part of visual purple. This pigment is bleached by light, a process which stimulates the rods in the retina, thereby enabling a person to see in dim light. Mild vitamin A deficiency leads to night blindness, not in itself a serious condition, but one which warns that more dire consequences may occur if steps are not taken to remedy the deficiency.

Vitamin A is also essential for the maintenance of epithelial cells which line the body's surfaces and cavities. Deficiency causes these cells to flatten, conglomerate and undergo surface drying. This condition is most noticeable on the conjunctiva or outer lining of the eye, where it leads to a form of conjunctivitis known as xerophthalmia. If this spreads from the white of the eye to the conjunctiva covering the cornea, vision is affected and the cornea may soften – a condition known as keratomalacia. If this process is not stopped at once, the cornea perforates, the iris and lens may protrude through the gap and permanent blindness follows. The term xerophthalmia is often used to encompass the whole syndrome of severe vitamin A deficiency.

Young children are the age group most vulnerable to xerophthalmia, since vitamin A requirements are closely linked to growth rate; deficiency is often exacerbated by childhood infectious diseases and protein energy malnutrition. Infants obtain sufficient vitamin A from breast or animal milk, but are frequently weaned on to staple foods such as rice which are totally lacking in carotene. Foods such as cultivars of sweet potato with carotene-rich roots, or sweet potato leaves, could be of tremendous importance in developing countries where animal food sources are too expensive or scarce to be used for child feeding. It is a sad fact, however, that xerophthalmia is common in some environments where dark green leafy vegetables such as sweet potato leaves and tips are readily available but are not used in weaning foods. In Bangladesh, for example, where xerophthalmia is a common problem, a survey revealed that only 19 out of 496 sweet potato farmers questioned used sweet potato leaves as a vegetable for their family (Rashid, 1987). Also, sweet potato cultivars with deep yellow or orange-fleshed roots are unfortunately rejected in many developing countries in favour of white or cream-fleshed types having little or no provitamin A activity. The deep orange-fleshed lines may not only be described by consumers as having a strong, unpleasant, oily or carrot-like taste, but can also be very moist and sweet-tasting and more suitable for the preparation of desserts than for incorporation in most diets as a staple or co-staple. It is possible, however, that cultivars combining yellow or orange flesh with moderately dry mouthfeel and low sweetness would be acceptable. Nor may adult consumer prejudices extend to very young children who, if presented with carotene-rich sweet potato roots as one of their first foods, might even prefer the sweet taste and soft texture (Osei-Opare, 1987). In fact women farmers and housewives questioned in several villages of West and Central Java, Indonesia, stated that children overwhelmingly prefer the yellow- or orange-fleshed soft, watery and very sweet types of root, whereas the adult preference is for white-fleshed, 'dry' types (Watson, 1988). In some countries foods are rarely selected and cooked specially for a child in poor families. The child is more frequently fed on a portion of the family food removed from a common pot and mashed to a suitable consistency. The smaller size of sweet potato in comparison with some other roots and tubers may be advantageous in this respect; a whole sweet potato root may be cooked for a child's meal, whereas pieces must be cut off the large roots or tubers, shortening their shelf life (Osei-Opare, 1987). It is essential that an increase in production and availability of roots and leaves go hand in hand with a strenuous effort to educate people about their nutritional value, particularly for infant and child feeding. At the same time stress must be placed on the importance of including a small amount of fat in weaning diets in order to promote efficient absorption and utilization of provitamin A carotenoids (see p. 157).

World prevalence of vitamin A deficiency

In some of the rice-dependent developing countries of Asia, vitamin A deficiency is a problem of public health magnitude. It is particularly prevalent in certain regions and in the urban slums of Indonesia, India, the Philippines and throughout Bangladesh. It has been calculated that in Bangladesh alone about 12,000 rural pre-school children are surviving blind at any time and that about 45,000 have serious loss of sight in one or both eyes due to vitamin A deficiency (Cohen et al., 1986). Furthermore, approximately 750,000 pre-school children are estimated to have some degree of xerophthalmia at any time.

Prevalence rates are lower, but eye damage due to vitamin A deficiency is a recognized serious problem in Afghanistan, Nepal, Sri Lanka, Haiti, the drier northern parts of west African nations such as Nigeria and Ghana, and the north eastern states of Brazil especially in the sugar cane areas. In other countries the problem may not be extensive, but is intermittently provoked by changing seasonal, social and economic conditions. It has been estimated (FAO, 1988) that, world-wide, there are more than 500,000 cases each year of new, active corneal lesions and 6 to 7 million cases of non-corneal xerophthalmia. Countless other children are vitamin A-depleted, with consequent decreased resistance to infectious diseases and increased morbidity and mortality.

Sweet potato and prevention of vitamin A deficiency

The wide range of sweet potato root flesh carotenoid content due to genetic and other factors is discussed in Chapter 2. Total carotenoids have been found to range from 0 to >20 mg/100 g (fwb), which would be equivalent to 0 to >60 mg/100 g (dwb). It is likely that cultivars with medium to high levels contain most of their carotenoids in the form of beta-carotene. Eighty nine per cent of the total carotenoids in a cultivar with 6 mg/100 g occurred as beta-carotene (Ezell and Wilcox, 1958). Two American cultivars which contained approximately 45 mg/100 g (dwb) total carotenoids had nearly 90% of these as beta-carotene (Purcell, 1962; Purcell and Walter, 1968). Therefore, expression of total carotenoids as beta-carotene entails little error when levels are high. This may not be true at the lower end of the range of values, where a higher percentage is present as nutritionally less active carotenoids.

Sweet potato roots can be used in main meals as a staple, co-staple, vegetable or dessert, or as a savoury or sweet snack (see Chapter 6; see also Figures 3.3 and 3.4). It may be compared, therefore, in terms of vitamin contents, with other roots and tubers, with other staples and with some commonly consumed vegetables and fruits, as seen in Tables 3.9, 3.10, 3.11 and 3.12. The comparative advantage of sweet potato cultivars with medium to high beta-carotene contents can be clearly seen. The other root and tuber staples are invariably lacking or extremely low in beta-carotene with resulting negligible quantities of retinol equivalents. The only 'starchy' staple, apart from sweet potato, with a significant amount of beta-carotene is the plantain. Cereals, with the exception of some cultivars of yellow maize, have no provitamin A activity. The only commonly consumed vegetable (with the exception of the green leaves shown) which has a high carotene content, comparable to that of carotene-rich sweet potatoes is the carrot. Pumpkins have a similar content to moderate-carotene sweet potatoes. Even tomatoes, which many consumers consider to be high in provitamin A activity, are much lower in biologically active carotenoids than many cultivars of sweet potato.

The carotene content of dark yellow or orange-fleshed sweet potatoes is very high and at least comparable to that of carotenoid-rich fruit such as apricots, mangoes and peaches (Table 3.12). The analysis of one sweetened dried sweet potato product, made in the Philippines and eaten as a snack or dessert, revealed a much higher concentration of beta-carotene

Table 3.9. *Major vitamins in sweet potato roots compared with other roots and tubers (per 100 g edible portion)*

	Beta-carotene[a] equiv. (μg)	Thiamin[b] (mg)	Riboflavin[b] (mg)	Niacin[b] (mg)	Pantothenate[c] (mg)	Pyridoxine (mg)	Folic acid[c] (μg)	Ascorbic acid[d] (mg)
Sweet potato	0–>20,000[e]	0.09	0.03	0.60	0.59[f]	0.26[f]	14[f]	23.6
Cassava	0–120[g]	0.05[h]	0.04[h]	0.60[h]	0.52		24	32[h]
Plantain, ripe	390–1,035[i]	0.08[i]	0.04[i]	0.60[i]				20[i]
Potato[j]	0–Tr	0.11	0.04	1.2	0.30	0.25	24	30
Giant taro	0	0.02	0.02	0.53				16.9
Giant swamp taro	27	0.03	0.02	0.46				15.7
Taro, cocoyam	43	0.03	0.03	0.76		0.08[c]		15.1
Yam (*Dioscorea* spp.)	108	0.05	0.03	0.41	0.13			20.3
Yautia, malanga	29	0.02	0.03	0.80				13.6

Notes:
All figures on a raw basis, as figures on a cooked basis not available for some roots and tubers. equiv., equivalent; Tr, trace.
[a] Singh and Bradbury, 1988.
[b] Bradbury and Singh, 1986b; but see also figures for thiamin and riboflavin given in Table 2.7, p.72.
[c] Rao and Polacchi, 1972.
[d] Bradbury and Singh, 1986a.
[e] Range of determined values.
[f] Haytowitz and Matthews, 1984; but see also lower values for pyridoxine given in Table 2.7, p.72.
[g] The figure of 120 μg applies to 'yellow' cassava as given by Collazos et al., (1974).
[h] Average of figures from: Leung and Flores, 1961, Leung et al., 1968 and Leung et al., 1972.
[i] Leung et al., 1968. Plantain is a fruit, but is usually classed with 'starchy' root and tuber staples.
[j] Averages calculated by Woolfe (1987).
Note added in proof. Vitamin E (mg): sweet potato 4.56 (boiled 4.39); plantain 0.20; potato 0.06. Holland, B. et al., 1991. *Vegetables, herbs and spices*. Royal Society of Chemistry, Cambridge.

Table 3.10. *Vitamins in cooked sweet potato roots compared with other cooked staples (per 100 g)*

Staple	β-carotene equivalent (μg)	Thiamin (mg)	Riboflavin (mg)	Niacin (mg)	Ascorbic acid (mg)
Sweet potato (boiled in skin)	0–>20,000[a]	0.09[b]	0.06[b]	0.6[b]	17[b]
Rice (boiled, white)	0	0.02	0.01	0.4	0
Plantain (green, boiled)	345	0.04	0.06	0.6	12
Maize (boiled on cob)	240	0.12	0.10	1.4	9
(porridge)	0	0.06	0.01	0.5	0
Bread (white)	0	0.09	0.05	1.0	0
Macaroni/spaghetti (boiled)	0	0.01	0.01	0.3	0
Sorghum (porridge)	0	0.04	0.01	0.2	0
Beans (*P. vulgaris* boiled)	Tr	0.11	0.06	0.7	0

Notes:
Tr, trace.
Unless otherwise indicated, data as calculated by Woolfe (1987).
[a] Range of determined raw values assuming little loss on cooking; see Chapter 5.
[b] Watt and Merrill, 1975; riboflavin value for boiled, peeled roots is 0.14 mg/100 g according to Haytowitz and Matthews (1984).

than in dried fruit products such as apricots and peaches (Truong, 1987; and see Table 3.12). Yellow/orange-fleshed peeled roots cooked in sugar and eaten as a stewed 'fruit' dessert would contribute quantities of beta-carotene similar to many of the fresh fruits shown in Table 3.12, and are much richer sources of carotene than bananas and some of the dried fruits such as figs, dates and raisins.

As we can see from Table 3.13, 100 g of sweet potato may supply from zero to more than 100% of the daily requirement of vitamin A of all age groups, depending on its carotenoid content, which is somewhat reduced by normal cooking methods (see Chapter 6). Infants, 1–10 year-old children, 10–12 year olds and adults require 350, 400, 500 and 500–600 μg retinol equivalents, respectively, per day (FAO, 1988). The quantity of beta-carotene (or beta-carotene equivalents if other nutritionally active carotenoids are included) needed to supply these recommended safe levels of intake from wholely vegetable diets is calculated as 2100, 2400, 3000 and 3000–3600 μg per day. These quantities could be

Table 3.11. *Vitamins and minerals in sweet potato roots compared with some internationally important vegetables (per 100 g raw)*

Vegetable	β-Carotene equivalents (μg)	Thiamin (mg)	Riboflavin (mg)	Niacin (mg)	Pantothenic acid (mg)	Pyridoxine (mg)	Folic acid (μg)	Ascorbic acid (mg)	Ca (mg)	P (mg)	Fe (mg)
Sweet potato roots	0–>20,000[a]	0.09[b]	0.03[b]	0.6[b]	0.59[c]	0.26[c]	14[c]	24[b]	34	46	0.7
Carrots	12,000	0.06	0.05	0.6	0.25	0.15	27	6	48	21	0.6
Onion	0	0.03	0.05	0.2	0.14	0.10	31	10	31	30	0.3
Tomato	600	0.06	0.04	0.7	0.33	0.11	43	20	13	21	0.4
Pepper (green)	200	Tr	0.03	0.7	0.23	0.17	16	100	9	25	0.4
Pumpkin	1,500	0.04	0.04	0.4	0.40	0.06	(26)	5	39	19	0.4
Okra	90	0.10	0.10	1.0	0.26	0.08	125	25	70	60	1.0
Soybean sprouts[d]	25	0.19	0.15	0.8				10	52	58	1.1
Cabbage (white)[e]	Tr	0.06	0.05	0.6	0.21[f]	0.15[f]	26	40	44	31	0.4

Notes:
Tr, trace. Unless otherwise indicated, averages as calculated by Woolfe (1987).
[a] Range of determined values.
[b] Bradbury and Singh, 1986b (calculated); but see also figures for thiamin and riboflavin given in Table 2.7, p.72.
[c] Haytowitz and Matthews, 1984; but see also lower values for pyridoxine given in Table 2.7, p.72.
[d] Leung et al., 1972.
[e] Tan et al., 1985.
[f] Rao and Polacchi, 1972.
Note added in proof. Vitamin E (mg): sweet potato 4.56; carrots 0.56; onion 0.31; tomato 1.22; pepper (green) 0.80; pumpkin 1.06; cabbage (white) 0.20. Holland, B. et al., 1991. *Vegetables, herbs and spices*. Royal Society of Chemistry, Cambridge.

Table 3.12. *Vitamins and minerals in sweet potato roots compared with some fresh and dried fruit (per 100 g)*

Item	β-Carotene (μg)	Thiamin (mg)	Riboflavin (mg)	Niacin (mg)	Pantothenic acid (mg)	Pyridoxine (mg)	Folic acid (μg)	Ascorbic acid (mg)	Ca (mg)	P (mg)	Fe (mg)
Sweet potato roots											
fresh[a]	0–>20,000	0.09	0.03	0.6	0.59	0.26	14	24	34	46	0.7
sweet-sour, dried[b]	7,820							7			
Apricots											
fresh	1,000–2,400	0.04	0.05	0.6	0.3	0.07	(5)	7	17	21	0.4
dried	2,400–4,400	Tr	0.20	3.0	0.7	0.17	14	Tr	92	120	4.1
Banana, fresh	200	0.04	0.07	0.6	0.26	0.51	22	10	7	28	0.4
Coconut											
fresh	0	0.03	0.02	0.3	0.20	0.04	26	2			
dried	0	0.06	0.04	0.6				0	22	160	3.6
Dates, dried	50	0.07	0.04	2.0	0.80	0.15	21	0	68	64	1.6
Figs, dried	50	0.10	0.08	1.7	0.44	0.18	9	0	280	92	4.2
Guava, fresh[c]	170–330	0.06	0.04	1.3				152–633	24	31	1.3
Mango, fresh, ripe	355–13,000	0.03	0.04	0.3	0.16			10–180	10	13	0.5
Papaya, fresh[c]	205–1,500	0.03	0.03	0.4				52	21	15	0.6
Peaches											
fresh	250–1,000	0.02	0.05	1.0	0.15	0.02	3	8	5	19	0.4
dried	1,200–2,600	Tr	0.19	5.3	(0.03)	0.01	(14)	Tr	36	120	6.8
Raisins	30	0.10	0.08	0.5	(0.10)	(0.03)	(4)	0	61	33	1.6
Sultanas	30	0.10	(0.08)	(0.5)	(0.10)	(0.03)	(4)	0	52	95	1.8

Notes:

Tr., trace.

Figures in parentheses are estimated or tentative values. Unless otherwise indicated, data from: Paul and Southgate, 1978.

[a] See Table 3.11.

[b] Truong, 1987; figure for beta-carotene must be regarded as approximate as calculated from IU of vitamin A given in original text (13,033 IU/100 g).

[c] Leung et al., 1968.

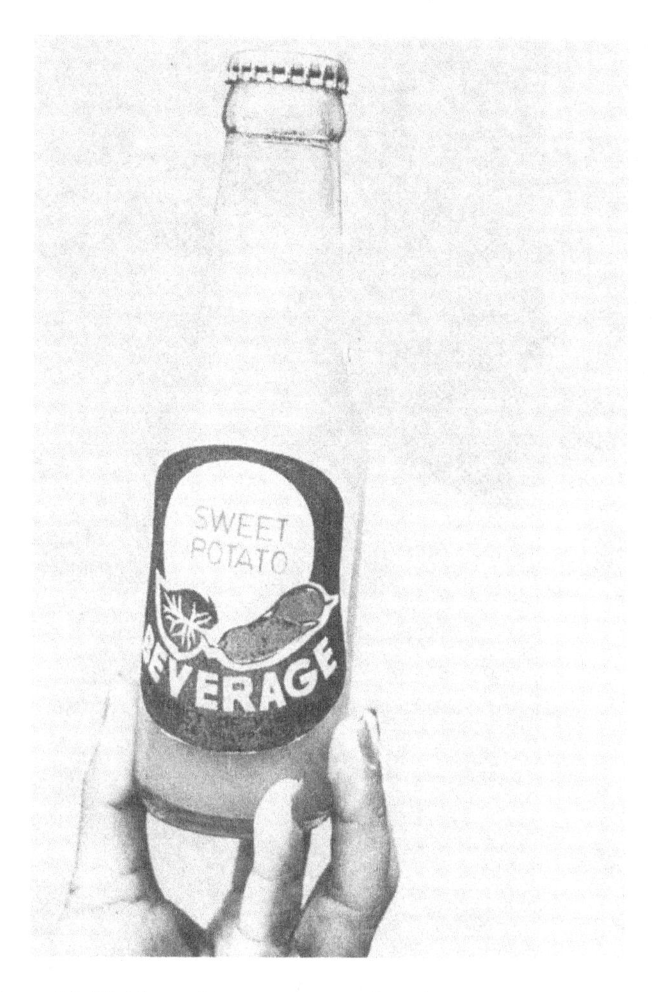

Figure 3.3. Highly nutritious products such as this non-alcoholic beverage produced in the Philippines can be made from yellow/orange fleshed cultivars of sweet potato (ViSCA).

readily supplied by approximately 100, 120, 150 and 150–180 g, respectively, of a deep yellow-fleshed sweet potato root with a beta-carotene content of approximately 2500 μg/100 g (fwb), eaten either as a vegetable or as a dessert (assuming a 20% loss on cooking). It is interesting to note that 5000 μg is the minimum beta-carotene content required for fresh root consumption of orange-fleshed cultivars in China (Zhang da Peng, personal communication). The same roots in the form of a purchased or home-processed flour could be used in smaller amounts as an addition to soups or stews. Alternatively, 100 g of the same deep yellow-fleshed root

Figure 3.4. Novel ways of promoting sweet potato and its products, such as this T-shirt from the Philippines, can help to raise public awareness of its high nutritional value (Truong Van Den).

with 2500 μg/100 g (fwb) beta-carotene could supply about 80% of a pre-school child's daily requirement, the remainder being supplied by, for example 25 g of sweet potato tops or other dark green leafy vegetables with about 2500 μg/100 g beta-carotene (again assuming a 20% cooking loss). A practical example of the utilization of beta-carotene-rich sweet potatoes to boost school children's provitamin A intake can be seen in southern Japan. A sweet potato paste (see Chapter 6), containing 12 mg/100 g beta-carotene, made by a small cooperative is

Table 3.13. *Percentages of recommended daily intakes of major nutrients provided by 100 g of boiled sweet potato roots*

Age group (years)	Vitamin A	Thiamin	Riboflavin[a]	Niacin	Folate	Pyridoxine[a]	Ascorbic acid	Fe[b]	Ca[b]
Children									
1–3	0–>100	11	18	7	22	24	85	5–14[1]	4–5
4–6	0–>100	8	13	5	22	22	85	4–11[2]	4–5
7–9	0–>100	6	11	4	11	17[1]	85	2–7	4–5
Male adolescents									
10–12	0–>100	5	9	4	11	14[2]	85	2–7	3–4
13–15	0–>100	4	8	3	7[c]		57	2–5[3]	3–4
16–19	0–>100	4	8	3	6	12	57	2–7	4–5
Female adolescents									
10–12	0–>100	6	10	4	11	17[2]	85	2–7	3–4
13–15	0–>100	5	9	4	7[c]		57	1–4[3]	3–4
16–19	0–>100	6	10	4	7	16[3]	57	1–4	4–5
Adult man (moderately active)	0–>100	4	8	3	6	12	57	2–7	4–5
Adult woman (moderately active; of child-bearing age)	0–>100	6	11	4	7	15	57	1–4	4–5

Notes:
Calculated using figures for sweet potato boiled without skin from Haytowitz and Matthews (1984) and safe levels of intake for vitamin A, folate and Fe in FAO (1988), recommended intakes of nutrients for thiamin, riboflavin, niacin, ascorbic acid and Ca in Passmore et al. (1974), and US Recommended Dietary Allowances for pyridoxine in National Research Council (1989).

[a] May be much lower; see values for riboflavin in Table 3.9 and pyridoxine in Table 2.7 (p.72):
- [1] Children 7–10 years.
- [2] Children 11–14 years.
- [3] Girls 15–18 years.

[b] Values depend on bioavailability of the mineral in the diet as a whole (factors such as quantity of foods of animal origin and presence of enhancers or inhibitors of mineral absorption):
- [1] Children 1–2 years.
- [2] Children 2–6 years.
- [3] Children 12–16 years.

[c] Children 12–16 years.

Figure 3.5. A nutritious and attractive lunch-time meal of sweet potato croquettes (foreground), and sweet potato salad (background) composed of cream, orange and purple cultivars, Japan (J.A. Woolfe).

used at the rate of 40% by weight of the dough to make bread. Primary and secondary school pupils receive 60 g and 100 g, respectively, of the bread once a week at school lunch. This quantity would supply >100% of the children's requirement of vitamin A for that day. There is increasing interest in Japan in the use of high beta-carotene sweet potato in the form of baked roots, bread or croquettes for school lunches in place of carrots which are said to be unpopular with children (see Figures 3.5 and 3.6). Strenuous efforts are therefore being made to breed cultivars with the combined traits of high beta-carotene and low sweetness (Umemura, Y., personal communication).

Most nutrients are not stored in the body and must be eaten daily if they are to be effective physiologically. Vitamin A, however, unlike the water-soluble B vitamins and ascorbic acid, is stored in the liver when more is eaten than can be utilized immediately. In this respect, sweet potatoes with high levels of beta-carotene eaten in fairly small quantities or with moderate amounts of beta-carotene eaten in larger quantities, could supply more than the daily requirement for vitamin A and hence enable consumers to build up reserves to tide them over periods when they either do not wish, or are unable, to eat carotene-rich vegetables. Thus, 200 g of sweet potato containing 7500 μg/100 g beta-carotene

Figure 3.6. Sweet potato products from Peru. Above, sweet potato bread and below (centre) sweet potato chips. Made with carotene-rich cultivars, these provide nutritious meals and snacks (International Potato Center).

would provide an adult man or woman with the recommended amount of vitamin A for 3 or 4 days, respectively, and a child for 5–6 days (again assuming a cooking loss of 20%).

These are only a few examples of how sweet potato could be used to combat vitamin A deficiency. There are many alternatives and combinations possible, depending on the cultivars of carotene-containing sweet potato roots available and the access to cultivation or purchase of sweet potato leaves. It should be mentioned, however, that there is evidence to show that an adequate fat intake is necessary to ensure proper utilization of beta-carotene (FAO/WHO, 1967). Symptoms of vitamin A deficiency were common among children in a rural area of the Philippines who were consuming large daily quantities of a yellow-fleshed sweet potato (Guzman, Guthrie and Guthrie, 1976) but who also had very low fat intakes (5–10 g/day). Many tropical diets, especially those of the poor, are low in lipids. In these cases, carotenoid intakes higher than those indicated above might be needed, but such a problem can only truly be solved by raising the economic status of the poor so that they are able to purchase the relatively expensive lipids that they need.

The preference of consumers, in some countries which could benefit most from carotenoid-rich cultivars, for white, less sweet and dry-fleshed sweet potatoes was mentioned previously. However, as described above, cultivars which could fulfil all or almost all of a person's provitamin A requirements do not necessarily have to be the very dark orange types (although the use of these could be encouraged in the form of desserts). If improved cultivars could be produced which combine moderate beta-carotene content with other desirable eating quality and agronomic characteristics and if at the same time consumers could be educated to value such cultivars, significant benefits could accrue. It is interesting to note that in a survey carried out in Asia and the Pacific region to determine, among other factors, the most important characteristics which need improvement for staple- and snack-type sweet potatoes (Lin et al., 1983) the primary concerns expressed were eating quality and nutritional content. These ranked above characteristics such as insect resistance and high yield; the chief nutritional concerns were beta-carotene and protein. Although a negative association has been found between flesh colour and dry matter (Jones, Steinbauer and Pope, 1969; Jones, 1977) promising efforts have already been made in the field of plant breeding to produce sweet potatoes with a combination of high dry matter and satisfactory beta-carotene contents (Anon., 1981).

Moreover, it is not invariably difficult to introduce orange or yellow-fleshed cultivars to an area in which they were previously unknown. At present all local cultivars grown in Bangladesh are white-fleshed (Jenkins, 1982; Rashid, 1987). However, in 1985, two improved cultivars were released to farmers, one of which has orange flesh and a high level of beta-carotene (Rashid, 1987). The farmers complained not about the colour or flavour of the flesh but about the moist texture when cooked and the fact that roots of the improved cultivars were not as sweet tasting as local cultivars. It would therefore seem that farmers would accept orange-fleshed cultivars with cooked texture and taste appropriate to local preferences.

Other vitamins

The increased consumption of carotene-containing sweet potato roots in vitamin A deficient areas or of pale-fleshed types in areas of vitamin A sufficiency would also ensure the increased intake of other essential vitamins.

Sweet potato roots are valuable sources of ascorbic acid. They may contain more than 30 mg/100 g (fwb) (Visser and Burrows, 1983; Bradbury and Singh, 1986b). If reference is once more made to Tables 3.9, 3.10, and 3.11, it can be seen that sweet potatoes have ascorbic acid contents comparable or superior to those of other roots and tubers, and to most of the commonly consumed vegetables; the only significantly richer vegetable sources, of those shown, are green peppers and cabbage.

Figure 3.7. Young child snacking on freshly cooked sweet potato root near Jinan, China (M. Iwanaga).

Cereals and legumes (unless sprouted) are totally devoid of ascorbic acid. Table 3.12 shows that whereas many dried fruits are lacking in ascorbic acid, a dried sweet potato product still maintained a concentration of 7 mg/100 g. However, some of the orange-fleshed fruits which are commonly dried are richer sources on a fresh weight basis than sweet potato.

As little as 100 g of boiled sweet potato can supply 85% of a child's recommended daily intake of ascorbic acid and over 50% of that of an adult (Table 3.13; see also Figure 3.7). A person in, for example, Peru, eating 50 g of sweet potato as a garnish or accompaniment to a traditional dish satifies >25% of their recommended daily intake solely from the sweet potato; a mixed dish of 150 g each of cooked sweet potato and rice would satisfy >100% of the recommended daily intake of, for example, a subsistence Filipino, with all the ascorbate coming from the sweet potato. In fact the sweet potato and other ascorbate containing foods are essential accompaniments to cereal-based diets otherwise lacking in ascorbic acid.

If comparisons are made for the sweet potato and other foods in terms of the B vitamins thiamin, riboflavin and niacin, it can be seen that, with a few notable exceptions, sweet potatoes have contents comparable to all the other foods shown (Tables 3.9, 3.10, 3.11 and 3.12), although data are not available which would make it possible to compare a dried 'fruit-like' product of sweet potato with the other dried fruits. Sweet potatoes are noticeably poorer in thiamin and riboflavin than soybean sprouts; they are poorer in niacin than boiled maize, wheat bread or potatoes. The United States Department of Agriculture figure (Haytowitz and Matthews, 1984) for riboflavin in boiled roots is 0.14 mg/100 g which is much higher than that usually found, but indicates that some samples may be relatively good sources of the vitamin. In contrast a sample of sweet potato analysed in Britain contained only traces of riboflavin (Kwiatkowska, Finglas and Faulks, 1989).

Few figures are available for folic acid, pantothenic acid or pyridoxine in sweet potatoes. Folic acid in sweet potato imported into the United Kingdom (Kwiatkowska et al., 1989) was 17 μg/100 g (fwb). Folic acid may have been lost during transport and storage of the roots, although the figure is a little higher than that shown in Table 3.9). The roots also had low levels of ascorbic acid. An earlier figure for folacin in food composition tables (Rao and Polacchi, 1972) gave a relatively high concentration of 52 μg/100 g (fwb). Variations in the pyridoxine content of roots have already been described in Chapter 2 (p. 81; and see Table 2.7). Only one recent source (Haytowitz and Matthews, 1984) shows pantothenate content (see Table 3.9). Figures for these three vitamins are clearly lacking and should be sought in a range of freshly harvested, stored and cooked samples using up-to-date techniques. There appear to be wide ranges in content, the reason for which is not yet known. Until more data are available it is not possible properly to assess the sweet potato as a source of these vitamins, and figures given in Tables 3.9 to 3.13 inclusive may have to be revised in the light of new knowledge.

Boiled sweet potatoes are a moderate source of thiamin; 100 g supplies 4–6% of the recommended adult daily intake. This quantity also supplies 3–4% and 8–11% of the recommended adult daily intakes of niacin and riboflavin, respectively (Table 3.13). Some roots may supply considerably lower percentages of riboflavin (see Table 2.7, p. 72). People in parts of the world where large amounts of sweet potato are eaten may obtain their entire daily requirements of these vitamins. Sweet potatoes may supply a high percentage of the recommended daily intakes of folic acid and pyridoxine (Table 3.13), although some samples may make a more significant contribution to pyridoxine requirements than others (Table 2.7, p. 72).

The levels of tocopherol found in sweet potato roots (Table 2.7, p. 72 and foot of Tables 3.9 and 3.11) indicate that sweet potato is a relatively rich source of vitamin E. In a recent analysis (Hirahara and Koike, 1989) seven out of nine cultivars investigated retained at least 1 mg/100 g (fwb) of tocopherol even after steaming (see Chapter 6). This can be compared with content of total tocopherol in some other (raw) foods: namely (mg/

100 g fwb), maize 1.7, oatmeal 2.1, butter 2.4, ox liver 1.4 and eggs 1.0. Although vitamin E deficiency states are rare except in people with fat absorption defects, United States Recommended Dietary Allowances (USRDAs; National Research Council, 1989) have been established for alpha-tocopherol equivalents; those for adults are 8–10 mg/day and those of infants and children between 3 and 7 mg/day depending upon age. One hundred grams of cooked sweet potato could provide about 20% of a small child's daily needs. The alpha- (and total) tocopherol contents of two North American baby food preparations, consisting of puréed sweet potato, were 0.5 (0.5) and 0.38 (0.74) mg/100 g (Davis, 1973). These would provide a significant proportion of the infants' daily vitamin E requirement.

Minerals and trace elements

The ash content of sweet potato roots is about 1% (fwb) and contains some important minerals and trace elements essential to various human body structures and functions. The concentrations of the major minerals and trace elements present as found by different authors are shown in Tables 2.8 and 2.9 (pp. 83 and 84). Those of three important minerals – Ca, P and Fe – are compared with contents in other staples in Tables 3.1 and 3.2.

The *iron* content is comparable to some of the other root and tuber staples (being lower than several such as arrowroot and oca) and to that of cereals cooked into porridges, noodles or pasta. It is comparable with, or superior to, on a dry weight basis as sweet potato flour, other dry or semi-dry staple foods including cassava flour, maize meal, chaphatis and bread. Even on a fresh basis it has twice the quantity of that of boiled white rice, and of some commonly consumed vegetables (Table 3.11), and a quantity similar to that of some of the yellow or orange-fleshed fresh fruits (Table 3.12). Though not an outstanding source of Fe, 100 g of boiled sweet potato roots can theoretically supply between 5% and 14% of the estimated daily Fe requirements for small children. However, a woman of child-bearing age would receive only between 1% and 4% of her Fe needs from this quantity, due to her greater demand for Fe caused by menstruation, pregnancy or lactation. A large quantity of about 500 g or more would have to be eaten by a woman to satisfy a significant proportion of her daily needs. Ranges of theoretical contributions are shown in Table 3.13 as daily requirements for iron have been estimated to vary with the bioavailability of the iron in the diet. In diets consisting almost entirely of vegetable foods, including those containing inhibitors of Fe absorption, bioavailability of dietary Fe is low and Fe requirements are consequently higher than in diets with generous supplies of animal foods and/or of the enhancer of Fe absorption, ascorbic acid. The Fe in vegetable foods is considered to be in a non-haem form, which is more poorly absorbed than the haem form present in foods of animal origin. The ascorbic acid in sweet potato roots may contribute towards the level needed to influence Fe absorption from a meal. Experiments with potatoes (*S. tuberosum*) (Fairweather-Tate, 1983) indicated that there was a positive correlation between their ascorbic acid content and the amount of Fe solubilized from them by gastric juice *in vitro*, and that a much higher proportion of the Fe from potatoes was solubilized *in vitro* than from other ascorbic acid-free foods such as kidney beans and wheat flour. It seems likely that such relationships exist for sweet potato also, but they remain to be explored. However, the sweet potato, when eaten in quantities of at least 100 g/day, can undoubtedly make a small contribution to Fe supplies.

Sweet potato is not an outstanding source of *calcium*, a characteristic shared with most other plant staples, vegetables and fruits. Giant taro and giant swamp taro have much greater quantities, as does, of course, lime-treated maize in the form of tortillas. However, sweet potato has a noticeably better Ca content than plantains, potatoes and yams, boiled rice and cereal porridge or noodle dishes. One hundred grams of sweet potato can contribute 3–5% of daily Ca requirements, depending on age (Table 3.13). Eaten in large quantities, as in some communities, sweet potato could contribute significantly to Ca intakes especially for those with little opportunity to eat dairy products. The availability of the Ca present depends on the presence or absence of substances such as phytin and oxalate, which bind and render unavailable Ca and other minerals. The small amount of oxalate occurring in sweet potato roots appears to bind about one-third of the total Ca present (Chapter 2, pp. 61–2). Phytin content does not seem to have been reported and requires investigation.

Sweet potatoes are a good source of *phosphorus*, being similar in this respect to other roots and tubers, vegetables and most cereals on a cooked basis. Cereal flours, tortillas and boiled *P. vulgaris* beans are a richer source than sweet potato even as a dried flour. One hundred grams of boiled sweet potato supplies about 6% of the USRDA for P for both adults and children.

The concentration of *potassium*, and the K:Na ratio, is high in sweet potatoes, which could therefore be used beneficially in diets designed to restrict Na intake, for example, in patients with high blood pressure.

Magnesium is another important dietary mineral. Fresh sweet potatoes supply about 6% of the adult USRDA for Mg on the basis of a 100 g serving.

The FAO has not published established daily requirements for trace elements. The RDA for *zinc* established in the United States can be compared with the average zinc content of sweet potato roots determined by recent analyses (Table 2.9). Sweet potatoes are a poor source of Zn – a 100 g serving provides less than 2% of the RDA.

Other, less extensively investigated trace elements reported to have beneficial effects in humans include *copper*, *chromium*, *manganese*, *selenium* and *molybdenum* (Passmore et al., 1974; National Research Council, 1980); for these the National Research Council (1989) provides an RDA for Se and a table of ranges for recommended intakes of Cu, Cr, Mn and Mb. Comparing these with Table 2.9 (p. 84) shows that 100 g of sweet potato can supply about 3% of the adult daily Se requirement and makes some contribution to Cu, Mn and Mb requirements.

The *fluoride* content of sweet potato has been measured as part of a study to determine the contribution of foods to endemic fluorosis in South India (Venkateswara Rao and Mahajan, 1990). At approximately 0.5 mg/kg (fwb) the fluoride content of sweet potato roots was higher than that of other roots and tubers, but low compared with most other foods analysed.

Sweet potato roots in oral rehydration therapy

In developing countries, most children have diarrhoea many times each year; those who are malnourished and have a low resistance to infection are at the highest risk. Reporting the work of others, Wilson et al. (1990) note that dehydration resulting from diarrhoea is still the world's greatest killer of children. Approximately four million die each year, accounting for 25–50% of all child mortality. Moreover, death from diarrhoea is 200 times more common among children in the poorest developing nations than among those in developed countries with high standards of living and health care.

In the late 1960s it was discovered that oral administration of a glucose and electrolyte solution, sufficient to replace diarrhoeal losses fully, could prevent or reverse dehydration, thus reducing mortality and morbidity due to diarrhoea. Oral rehydration therapy (ORT), as this was known, can reduce rates of child mortality from diarrhoea by at least half. However, packets of oral rehydration salts (ORS) produced by WHO and UNICEF are not universally available, and a simple home-prepared sugar–salt solution (SSS) can have dangerous or undesirable effects if improperly mixed or administered.

The alternative to ORS or SSS is the administration of a rehydrating fluid composed of a thick, but drinkable, mixture of a starchy food (about 50–100 g/l) with salt added to make a solution of 40–120 mEquiv./l. This is generally known as cereal-based ORT. The principle behind this preparation is that complex carbohydrates are hydrolysed during the cooking process, releasing a range of smaller molecules of dextrins and disaccharides. These are rapidly hydrolysed in the body to monosaccharides, which stimulate the sodium pump and maximize absorption of water across the intestinal tract.

In the early 1980s, research in Bangladesh and India showed that rice-based ORT significantly reduced the duration of diarrhoea, the stool volume and the quantity of rehydration fluid needed by children with acute diarrhoea. Additional clinical trials of cereal-based ORT, using other cereal grains such as wheat, maize, millet and sorghum, and tubers such as potato, all yielded results similar to those obtained when using rice. This form of ORT has immense advantages in that it can be prepared and given safely by family members, using low cost food available even in the poorest homes. Moreover, it is palatable and more culturally acceptable than standard ORS.

Although there have been many approaches to the application of cereal-based ORT in developing countries, little consideration has as yet been paid to sweet potato as an ingredient for the home preparation of an oral rehydration fluid, in spite of the fact that it would seem to be an ideal candidate for this purpose (see Chapter 5 for a description of changes in sweet potato carbohydrates on cooking). However, a pilot study in Papua New Guinea (Howard et al., 1990) has shown that sweet potato can be safe and effective as an ORT solution for infantile cases with mild to moderate dehydration. Sweet potato was chosen as a possible base for ORT fluid because in the Highlands of Papua New Guinea neither coconut water nor cereal grains are readily available, but sweet potato is universally grown and eaten as a staple food (see Chapter 8). An ORT fluid was made by boiling two average sized pieces (weight not specified) of sweet potato in 1500 ml of water for 35 min. The sweet potato was then mashed and water added to a final volume of 1 litre to make a drinkable solution. Three grams (one finger pinch) of salt were added. When analysed the sodium, potassium and glucose concentrations were, respectively, 60, 17.5 and 14 mEquiv./l.

This fluid was compared with standard WHO formula ORS in infants aged 4 to 23 months suffering from mild to moderate dehydration caused by acute diarrhoea. There was no significant difference in time to clinical rehydration between the infants given sweet potato ORT and those administered WHO ORS (36.1 and 34.8 hours, respectively). All infants in the sweet potato group seemed to like the taste of sweet potato fluid and none suffered side effects from it. Both mothers and nursing staff were enthusiastic about its use. The bacteriological quality of the fluid remained sound after as much as 2 hours of storage. Further studies are planned to confirm the efficacy and safety of the solution in the village situation.

This initial study certainly indicates that researchers in countries other than Papua New Guinea should consider the possibility of developing a sweet potato ORT fluid as an alternative to packaged ORS, especially where cereal grains are scarce or expensive. When teaching the use of, or applying, sweet potato or cereal-based ORT fluid, it must be emphasized that this fluid is given *in addition* to food and *not* as a food substitute. Moreover, cultural factors have to be considered. For example, the researchers in Papua New Guinea indicated the extreme importance of educating fathers as well as mothers in the use of ORT, where men dominate all decision making in society. See also the Report of the 1989 International Symposium on Cereal Based Oral Rehydration Therapy edited by Elliott et al. (1990).

Tops

Agriculturists and nutritionists faced with the problem of feeding the world's hungry are becoming increasingly interested in previously neglected tropical green leafy vegetables such as sweet potato greens. Though not new discoveries, these vegetables have never been properly investigated or fully exploited. Their potential as highly nutritious sources of protein, provitamin A, some of the B vitamins, iron and other minerals and trace elements is just beginning to be explored.

Indigenous tropical vegetables including sweet potato greens, have low prestige in many parts of the world, being regarded as 'poor man's salad', and as such less desirable than introduced vegetables from Western countries. This unpopularity no doubt stems from the lack of high yielding, disease and pest-resistant, tender varieties with good eating quality. The yield of sweet potato tips is low at present (about 16 tonnes/ha) compared with many other types of leafy vegetables (30–60 tonnes/ha), and harvesting is very labour intensive. Tips are considered to be tough compared to other leafy vegetables (Villareal et al., 1985). Moreover, their strong flavour makes them difficult to cook with other vegetables, whose natural flavour is diminished, and many cultivars become black and unpleasant in appearance on frying because they have a high phenolic content.

Until recently, there was no information about consumer preferences for eating different parts of the sweet potato vine such as stems, petioles or leaves, or for leaf colour, shape and size, and little has been explored from the point of view of taste, texture, marketing and preservation. Few studies to date have carried out organoleptic evaluation of leaf tips. Blanched leaf tips of 10 cultivars were evaluated for eating quality such as tenderness, flavour, stem and leaf colour and degree of hairiness by a taste panel composed of 10 Filipinos, 10 Chinese and one Indian (Villareal et al., 1979c). Interestingly, three Philippine cultivars were ranked lowest even by the Filipinos which suggests that other cultivars (for example the highest scorers in the study) could well be introduced into the Philippines for further testing. It was also found that the traits of stem colour and degree of hairiness accounted for 86% of the variation in acceptability and could be used for initial selection of lines prior to further evaluation. More studies of this nature are required, particularly in different countries to determine local preferences. In areas with a choice of alternative leafy vegetables, increased popularization of sweet potato greens may not be an easy task. In one location, Taiwan, water convolvulus (*Ipomoea aquatica*) scored more highly in all aspects during sensory testing than a sweet potato cultivar chosen for its relatively tender texture (Anon., 1985).

The disadvantages of sweet potato greens mentioned above could be overcome by research and education and are greatly outnumbered by the advantages which this vegetable bestows in terms of availability and nutritional value. They are already grown and used in many places where their productivity and popularity could be increased. These include the Philippines, Indonesia, Thailand, Taiwan, Malaysia, some rural parts of southern China, Vietnam, New Guinea, Korea, Zaïre, Sierra Leone, Tanzania and Liberia. In Liberia, the leaves are more extensively used than the roots (As-Saqui, 1982).

As sweet potato is adapted to hot, humid conditions, it is easier to grow in many tropical areas than introduced western vegetables, being less sensitive to the prevalent plant diseases and pests and to extremes of drought or heavy rainfall. Sweet potato vines are also more tolerant of high moisture than are many other leafy vegetables grown in the tropics. When the Philippines suffered serious flooding in 1972, only two green leafy vegetables could be found in urban or rural markets – sweet potato tips and water convolvulus – all other vegetables having been destroyed by about one month's continuous rain (Villareal et al., 1985). One of the few vegetables which can be grown during the monsoon season, sweet potato leaves are usually the only greens available in Taiwan markets after a typhoon (Anon., 1984).

Sweet potato tips can be harvested continuously over many weeks, not once only like many other commercially grown vegetables. Topping tends to reduce root yields, so if plants are to be grown for both roots and tops harvesting methods must be appropriate to optimizing yields of both plant parts (Dahniya, 1981). Sweet potato tips are therefore eminently suitable for production in home or school gardens and backyards as a nutritious supplement to family diets. In this role they could be particularly valuable in the diets of women and children. They have been included as one of the vegetables in a garden research programme at the AVRDC, which aims to increase nutritional status and

Figure 3.8. An experimental garden plot at AVRDC. Sweet potato leaves are suitable for inclusion in home or school gardens and backyard plots as a green vegetable which can be harvested continuously during the growing season (J.A. Woolfe).

income of people in the tropics (Tsay and Wang, 1988; see also Figure 3.8). Data on yield, and growing and harvesting periods of 146 varieties of 49 vegetables, including sweet potato tips, have been collected in order to design, for a particular area, the 4 m × 4 m garden which is most effective in providing a significant percentage of the nutritional needs of a family of five on a year-round basis. Sweet potatoes need little time for weeding, are tolerant to a wide range of environments and can be harvested for up to 1 year. In addition it has been shown that sweet potatoes are among the vegetables which can be grown for continuous harvesting of leaves in a hydroponic system suitable for an urban home garden (Tsay, J.J.S., personal communication; see Figure 3.9). Labour is required only during sowing and harvesting, the use of a nutrient solution precluding the need for watering. Sweet potatoes growing in pots and fertilized with human urine have been introduced to families in a city slum area of the Philippines (Villamayor, F., personal communication). Such a system could be nutritionally advantageous for many poor families without access to land in tropical urban situations.

The availability of sweet potato leaves during periods of heavy rainfall or drought when other crops are out of season could provide valuable supplies of protein and vitamins. Seasonality in vitamin A deficiency is

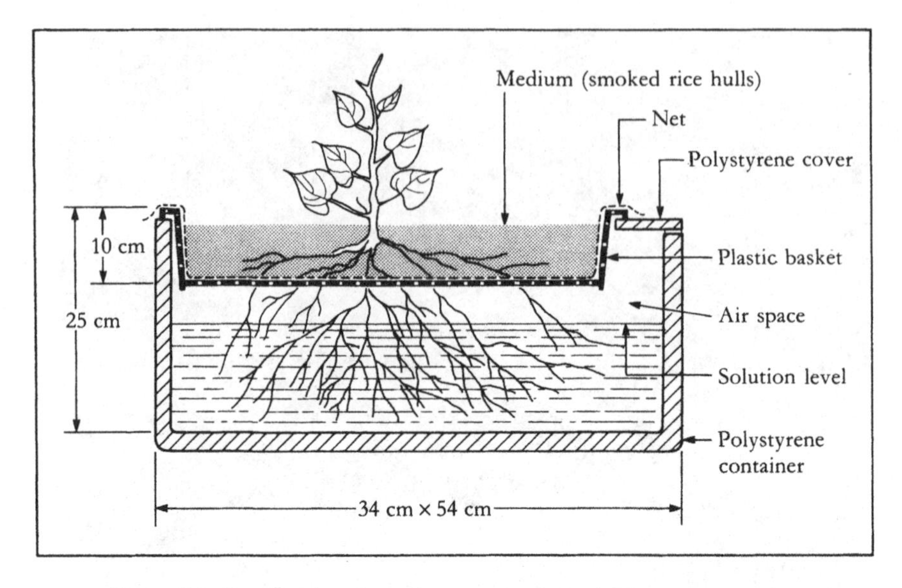

Figure 3.9. Cross-section through a non-circulating hydroponic growing unit for production of sweet potato vegetable greens in an urban backyard (AVRDC).

well known, particularly during dry periods or extended droughts when shortages of fruits and vegetables occur. The promotion of sweet potato greens as a back garden vegetable has even been suggested in the United States (Pace et al., 1985), where they could be grown during the hottest months of the year when alternatives are limited. They have been amongst the tropical vegetables explored for promotion in the Pacific region (Evensen and Standal, 1984). The predominantly prostrate vine type of sweet potato plant makes harvest on a large scale more difficult than hand topping, and selection and improvement of bush or semi-bush types could lead to higher yields and greater market potential for farm production (Anon., 1984; Villareal et al., 1985).

The possibility of extended production of sweet potato leaves in home gardens also means that they are a source of nutrients that their growers do not have to pay for. However, even when sold in the market place, sweet potato tips have been found to be among the cheapest of the leafy vegetables and to be increasingly consumed as income decreased (Villareal et al., 1979a). They were noted to be the most consumed of available leafy or yellow vegetables by the lowest income Filipino families (Urbino, Torres and Darrah, 1972). In Taiwan they were found to be the second cheapest source of provitamin A among the vegetables on sale (Popkin, B., unpublished data).

The postharvest appearance of sweet potato tips deteriorates rapidly.

Leaves wilt, dry and become darkened if displayed badly (for example in full sunlight) or if kept too long. At the same time nutrients sensitive to heat and oxidation such as ascorbic and folic acids are lost. There is a need to investigate the shelf life of different cultivars, to develop simple, cheap ways of maintaining the quality until purchase by the consumer, to formulate recipes and produce processed products which combine increased possibilities of utilization with maximum retention of nutritive value. Such work has hardly begun but, if pursued, could bring about a significant role for sweet potato leaves in feeding undernourished people of many countries. For example, instructional material in the shape of a student-developed booklet for use in the Pacific Islands describes the nutritional value of, recipes for, and methods of conservation of nutrients during cooking of, sweet potato leaves and five other vegetables (Nutrition Education and Training Program (Trust Territory of the Pacific Islands), 1980).

Nutritive value

Protein

Sweet potato leaves with about 3% protein on a fresh weight basis contain approximately twice the amount of protein of the same weight of roots. If sweet potatoes are grown specifically for their leaves, or alternatively for both roots and leaves, they produce much higher yields of protein per unit area than if only the roots are utilized. It was shown that for three cultivars, an average of only 26.25 g protein/plot (plant) was produced by the roots alone when vines were not topped (Gonzales, Cadiz and Bugawan, 1977), whereas a total of 81.4 g protein/plot (67.1 g from the tips and 14.3 g from the roots) was produced when roots and tips were harvested, tips being removed weekly.

Sweet potato leaves have a similar or inferior content of protein when compared with other tropical leafy vegetables, but they are much superior in this respect to the commonly consumed Western vegetables (Table 3.14) which, with the exception of spinach have very low protein contents.

The leaf protein of sweet potato has been little analysed for individual amino acid contents. Almost complete analyses (excluding tryptophan) for two Taiwan-grown cultivars and a sample of African-grown leaves, the contents of tryptophan, lysine, methionine and cystine in one North American sample, and the contents of all essential amino acids in a leaf protein concentrate (LPC) prepared from Taiwanese leaves are shown in Table 3.15. (LPCs are discussed further in Chapter 6). The leaves have moderate to good quantities of all the essential amino acids shown.

Table 3.14. *Protein, minerals and oxalate in sweet potato leaves and some other tropical and temperate vegetables (raw fwb)*

Vegetable	Total protein (g/100 g)	Ca (mg/100 g)	Fe (mg/100 g)	Zn (mg/100 g)	Oxalate[a] (%)
Tropical leaves[b]					
Sweet potato	2.9[c]	183[d],75[e]	1.8[d],2.4[d] 3.9[e]	0.5[d]	0.37[e]
Amaranth	2.8	176	2.8		0.88,0.77
Cassava	7.0	160	2.4		0.517
Horseradish tree	7.2	342	3.7		0.08,1.33, 0.101
Taro	3.3	96	0.95		0.426[f]
Temperate leafy and non-leafy[g]					
Cabbage	1.9	44	0.4	0.3	0.002,0.003
Carrot	0.7	48	0.6	0.4	0.023,0.005
Onion	0.9	31	0.3	0.1	0.003,0.001
Lettuce	1.0	23	0.9	0.2	
Spinach	3.2	93	3.1		0.78,0.685
Tomato	0.9	13	0.4	0.2	

Notes:

[a] Except where indicated, analyses by various authors given by Evensen and Standal (1984).

[b] Unless otherwise indicated, average figures given by Evensen and Standal (1984).

[c] Villareal et al., 1979b (average for leaf tips of three cultivars).

[d] Pace et al., 1985; lower figure is for tips, higher figure is for older leaves.

[e] Villareal et al., 1979b (average for leaf tips of ten cultivars).

[f] Holloway et al., 1989; (average of nine cultivars).

[g] Tan et al., 1985.

Cysteine/cystine, which spares methionine, whilst present only in very small amounts in the Taiwanese leaves and apparently absent in the African leaves was in a much higher concentration in the American sample and the Taiwanese LPC. The African sample whilst lower in total S-containing amino acids than the others, has an exceptionally high content of lysine. Table 3.7 demonstrates that the leaves have contents of essential amino acids similar to those in the roots, with the exception of histidine, which is twice as high in leaves as in roots, and tryptophan, which (in some samples) may be lower in leaves than roots. Though it is difficult to draw conclusions on the basis of so few analyses, the quality

Table 3.15. *Amino acids in sweet potato leaves and stems (g/16 g N)*

	HP-4[a]	Tainung 25[a]	African[b]	North American[c]	LPC[d]
Essential					
Lys	3.31	4.33	11.0	5.7	4.00
Ile	5.16	4.34	4.0	ND	5.85
Leu	7.03	9.11	8.0	ND	9.90
Met	2.75	2.69	2.2	2.15	2.04
1/2 Cystine	0.21	0.17	—	1.18	1.67
Phe	5.94	4.63	4.9	ND	5.98
Tyr	2.72	2.36	3.4	ND	3.38
Trp	ND	ND	ND	0.88	1.14
Thr	4.41	4.54	6.1	ND	3.69
Val	9.17	8.95	5.2	ND	7.08
His	3.41	3.34	3.7	ND	2.27
Non-essential					
Asp	10.46	10.53	8.0	ND	ND
Glu	12.32	13.06	10.1	ND	ND
Pro	5.83	7.34	ND	ND	ND
Gly	7.35	5.89	5.5	ND	ND
Ala	10.74	9.26	5.5	ND	ND
Arg	4.21	4.17	6.8	ND	5.30
Ser	4.84	4.68	4.7	ND	ND

Notes:
LPC, leaf protein concentrate. ND, no determination.
[a] Cheng, 1978; 2 cultivars grown in Taiwan.
[b] Maeda, 1985; Tanzanian sample.
[c] Haytowitz and Matthews, 1984.
[d] Sun et al., 1979.

of leaf protein seems to be moderately good, with a higher content of lysine than some cereals, and a higher content of S-containing amino acids than in beans. On average, sweet potato leaves appear to have less lysine and methionine than some other tropical leaves, notably cassava and horseradish tree (*Moringa* spp.), but are similar in content to others. Further analyses are needed to discover the full range of essential amino acid contents, and their relationship with genetic factors, cultivation practices etc. It would be interesting to discover whether the high lysine content of the African leaves is duplicated in other samples and whether it could be combined with more favourable S-containing amino acid and tryptophan levels.

The amino acid scores for different age groups are seen in Table 3.8, although they should be regarded as tentative at this stage. This shows that scores of roots and leaves are similar, although tryptophan may be the first limiting amino acid in leaves for the infant rather than methionine. Scores for children other than infants may be higher than are shown in those samples with a high lysine content; in those cases methionine rather than lysine may be limiting for the pre-school child. Sweet potato leaves have a fairly good amino acid score for the lower age groups and an excellent score for older children and adults. These scores combined with the fairly good quantity of protein the leaves contain (higher for example than in cooked rice) should make the leaves an attractive alternative for inclusion in weaning and infant diets being promoted by nutritionists and extension workers. Table 3.16 shows that a serving of 85 g of leaves in 1 day (85 g being an average serving determined by one researcher) can provide 10% or more of a small child's protein requirement. This serving could be eaten in two or more meals, and as indicated in the following sections provides not only protein but substantial supplies of various vitamins and minerals. Adults also benefit from the consumption of leaves, obtaining up to 9% of their daily protein requirement.

Vitamins

In common with other dark green leafy vegetables, sweet potato leaves (Figure 3.10) are rich sources of provitamin A carotenoids, chiefly beta-carotene. Contents appear to be variable (Tables 2.13 (p. 102) and 3.17), and more analyses should be carried out, but even the lowest value found in food composition tables showed leaves to have about 600 μg beta-carotene/100 g (fwb) (Haytowitz and Matthews, 1984). On average sweet potato leaves contain about 3 mg total carotenoids/100 g (fwb). As far as comparisons with other tropical leafy vegetables are concerned, sweet potato leaves generally appear to be of medium provitamin A content, being superior to Chinese cabbage and tamarind and taro leaves, but containing only about half the quantity in horseradish tree or cassava leaves. The potential for use of sweet potato roots in combating vitamin A deficiency has already been described. The role of sweet potato leaves could be even more important especially in areas where people are unwilling to accept, or cannot adapt local dishes for, cultivars with dark yellow or orange-fleshed roots. Their adoption as a back garden crop, harvested continuously, could be a valuable tool in combating seasonal vitamin A deficiency in the most vulnerable age groups. Their low cost could help to make them available even to those without land to cultivate. An 85 g serving can provide 60% of a small child's daily needs and even 50% of an adult's (Table 3.16). The

Figure 3.10. Two types of sweet potato leaf, one broad-leaved and one digitate, growing in Uganda. The leaves, which are good sources of protein, provitamin A and vitamin C, are unused or eaten infrequently in some tropical countries (M. Iwanaga).

carotenoids are less susceptible to losses during cooking than the water-soluble vitamins. This is a great advantage when it comes to cooking leaves, from which water-soluble compounds are readily leached, and which are easily overcooked with consequent heavy nutrient losses. The use of sweet potato leaves should be encouraged wherever possible in areas which suffer seasonal or constant deficiencies of vitamin A.

A potentially very important contribution which sweet potato leaves can make to nutrition lies in their relatively high level of riboflavin, which is generally deficient in Asian rice-based diets (Villareal et al., 1985). Indeed the riboflavin content of cooked cereals in general is very low (Table 3.10), as is that of roots and tubers (Table 3.9) and many non-leafy vegetables (Table 3.11). The average riboflavin content of sweet potato leaves at about 0.3 mg/100 g compares favourably with that of cow's milk (0.19), beef (0.24), chicken meat (0.16) and white fish (0.07) (figures from Tan, Wenlock and Buss, 1985). An 85 g serving of sweet potato leaves provides a substantial percentage of recommended daily riboflavin intakes, especially in the case of small children (Table 3.16). Though lower in riboflavin than horseradish tree leaves, sweet potato leaves compare very favourably on the whole with other leafy vegetables (Table 3.17). They are also moderately good sources of other B vitamins, namely thiamin, niacin and pyridoxine, further strengthening their

Table 3.16. *Percentages of recommended daily intakes of major nutrients provided by 85 g of cooked sweet potato leaves*

Age group (years)	Protein	Vit. A[a]	Thiamin[b]	Riboflavin[c]	Niacin[c]	Folic acid[d]	Ascorbic acid[d]	Fe[e]	Ca[e]
1–3	16	60	13	30	7	30	50	13–40[1]	26–33
4–6	13	60	10	22	5	30	50	11–30[2]	26–33
7–9	10	60	7	18	4	15	50	7–20	26–33
Male adol.									
10–12	9	50	6	15	4	15	50	7–20	19–22
13–15	7	40	5	14	3	9[f]	30	4–13[3]	19–22
16–19	7	40	5	13	3	8	30	7–20	22–26
Female adol.									
10–12	9	50	7	17	4	15	50	7–20	19–22
13–15	9	40	6	16	4	9[f]	30	4–12[3]	19–22
16–19	9	50	7	16	4	9	30	3–10	22–26
Adult man	7	40	5	13	3	8	30	7–20	26–33
Adult woman	9	50	7	18	4	9	30	3–10	26–33

Notes:

adol, adolescent.

Eighty five grams was suggested to be the average serving size of leaves by Vietmeyer (1978); figures calculated using recommended nutrient intakes from Passmore et al. (1974) for thiamin, riboflavin, niacin, ascorbic acid and Ca, and from FAO (1988) for vitamin A, folate and Fe.

[a] Using figure of 1745 μg/100 g for cooked leaves given by Leung et al. (1972).

[b] Assuming leaves freshly harvested and loss on cooking 25%.

[c] Assuming loss on cooking 20%.

[d] Using figures of 55 mg/100 g and 88 μg/100 g for ascorbic and folic acids, respectively shown in Table 2.13 (p.102); assuming loss on cooking 80%.

[e] Using figure of 1.8 mg/100 g for Fe given in Pace et al. (1985) for leaf tips; values depend on bioavailability of mineral in diet as a whole.

[1] Children 1–2 years.

[2] Children 2–6 years.

[3] Children 12–16 years.

[f] Children 12–16 years.

Table 3.17. *Vitamins in sweet potato leaves and tips and in other leafy vegetables (per 100 g raw)*

Leaves	β-Carotene equivalent (μg)	Thiamin (mg)	Riboflavin (mg)	Niacin (mg)	Pyridoxine (mg)	Folic acid (μg)	Ascorbic acid (mg)
Sweet potato	2700	0.13	0.35[a]	0.9	0.21[b]	88[b]	41[a],103[c]
(and tender tips)	(2290–7050)[d]		(0.29–0.41)[a]	0.9			(32–136)
Amaranth	6545	0.04	0.22	0.7		85[b]	23
Cassava	8280	0.16	0.32	1.8			82
Chinese cabbage (*Brassica chinensis*)	1200	0.04	0.14	0.5			40
Horseradish tree	8855	0.20	0.73	3.4	1.2[b]	370[b]	167
Lettuce[e]	1000	0.07	0.08	0.4	0.20[b]	34	15
Tamarind (*Tamarindus indicus*)	2510	0.10	0.11	1.5		42[b]	6
Taro	5535	0.13	0.34	1.5	0.19[b]	163[b]	63
Water convolvulus, swamp cabbage	2865	0.09	0.16	1.1		122[b]	47

Notes:

Except where indicated, data from: Leung et al., 1972.

[a] Villareal et al., 1979b.

[b] Rao and Polacchi, 1972; leaf stalk.

[c] Caldwell, 1972.

[d] Leung et al., 1968; range of 18 values.

[e] Tan et al., 1985.

supplemental value to diets based largely on polished white rice (Tables 3.10, 3.16 and 3.17).

There appears to be a very large range in ascorbic acid contents between different samples of sweet potato leaves (see Chapter 2, pp. 101–2). In general, however, the leaves have much greater quantities than do the roots. They also appear to have an ascorbic acid content superior to that of many other green leafy vegetables (Table 3.17). Moreover, their folic acid content is also favourable. Such vitamins are, however, very susceptible to cooking losses (see Chapter 6) through leaching into the cooking water, heat destruction and oxidation. Table 3.16 shows the substantial potential contribution which sweet potato leaves can make to ascorbic and folic acid supplies, even assuming a cooking loss of 80%. However, supplies could be less if cultivars with naturally low concentrations of these vitamins are used, or if concentrations are reduced through low quality of postharvest handling and excessive cooking times. During harvesting, transport and marketing leaves should be handled carefully to reduce bruising to a minimum, kept in cool and shady conditions and sold or utilized as soon as possible. Sweet potato leaves should be cooked for the minimum time possible commensurate with increasing their tenderness and acceptibility and the water in which they were cooked should also be consumed, for example as a soup, as it contains water-soluble vitamins lost by leaching.

Further investigations are required to determine vitamin contents in a greater range of samples than hitherto, particularly with respect to factors influencing concentrations. Analyses of leaves subjected to various methods of postharvest handling, preservation and cooking are also urgently needed in order to determine optimum conditions for vitamin retention.

Minerals

There have been few determinations of minerals and trace elements in sweet potato tops for the purpose of collecting nutritional data. The little information available suggests that contents of Ca and Fe vary considerably between samples grown in different places (see Table 3.14), probably due to soil concentration differences. Different parts of the vine also vary in mineral content. The contents of Ca, Fe and Zn were found to be significantly greater in the tips and older leaves than in the stem-petioles (Pace et al., 1985). However, it is clear that sweet potato tops are potentially much richer sources of Ca and Fe than the roots, and than other non-leafy vegetables (Table 3.14). They could therefore theoretically make highly significant contributions to Ca and Fe intakes, even when eaten in fairly small quantities (Table 3.16). Such calculations have to take into account the bioavailability of these minerals in the diet which may be rather low (see discussion in Roots, Minerals and trace elements, above). Sweet potato tops, in common with other leafy vegetables contain oxalic acid which can bind Ca and render it unavailable. The oxalate content is much higher than in Western vegetables with the exception of spinach. It is similar to some other leafy vegetables, but only about half as much as that of amaranth or spinach (Table 3.14; Villareal et al., 1985). The contents of Ca as calcium oxalate and Ca not combined as oxalate do not appear to have been investigated for sweet potato leaves. In taro (*Colocasia*) leaves, which have an oxalate content similar to that of sweet potato leaves, a mean of only 13% of the total Ca was found to be free rather than combined as oxalate in a number of South Pacific cultivars (Holloway et al., 1989). It remains to be seen whether a similar high percentage of total oxalates is present as calcium oxalate in sweet potato leaves.

The content of phytin, a compound which binds Ca, Fe and Zn does not appear to have been determined in sweet potato leaves. When Fe absorption from some Philippine vegetables, labelled biosynthetically with ^{59}Fe, was evaluated using rats, the Fe from sweet potato tops was only about 6% absorbed (Ortaliza et al., 1974). The mean absorption of other vegetables ranged from 4.5 to 9.1%. All vegetables, including sweet potato tops had only about 50% of the Fe absorption of an Fe salt ($FeCl_3$). The low Fe absorption from vegetable foods partly accounts for the high rate of Fe deficiency anaemia in many developing countries where calculated dietary Fe intakes apparently exceed requirements or are very close to them and where Fe is largely derived from vegetable sources. The contribution of sweet potato leaves to Fe requirements in diets based almost entirely on vegetable foods would be at the lower end of the range shown in Table 3.16. If the potential of sweet potatoes and other leafy vegetables rich in Fe is to be tapped, more research is needed to determine variations in levels of anti-nutrients such as phytin and oxalate so that cultivars with low levels can be selected. Furthermore, simple methods of cooking which improve availability of minerals should be explored.

One hundred grams of petioles or leaves supply about 1% and 3%, respectively, of the USRDA for Zn. There is no information on the availability of this element from a dietary standpoint.

The leaves, like the roots, have a high K:Na ratio and they could therefore be used in low-salt diets. It is interesting to note that one group of authors have drawn attention to the saluretic (increasing the body's elimination of Na) and the diuretic (increasing the amount of urine produced) properties of sweet potato tips and have suggested that they could be suitable vegetables for individuals with high blood pressure

(Villareal et al., 1979a), although no details of these properties have been subsequently supplied. They could be further investigated.

Little information is available with regard to other minerals and trace elements in sweet potato tops, a deficiency which could be remedied.

Dietary fibre

This is discussed in Chapter 2.

Summary

Both sweet potato roots and tops are nutritious foods which could be used to advantage in combating nutritional deficiencies in parts of the developing world if means could be found to overcome resistance to their increased consumption. The roots are not only a moderately good energy source, but also contain significant quantities of the water-soluble vitamins ascorbic acid and thiamin, besides supplying part of the daily requirement for riboflavin and niacin. Their contents of pyridoxine, folic acid and pantothenic acid may be relatively high, but up-to-date analyses for these vitamins in a wider range of samples are required. Raw leaves and tender tips are also excellent sources of ascorbic acid and some of the B-vitamins, especially riboflavin which is deficient in many Asian diets. However, high percentages of water-soluble vitamins are lost on cooking.

The deep yellow- or orange-fleshed roots and the leaves and tips are rich sources of beta-carotene and other biologically active carotenoids and would be valuable as one means in a multiple approach to prevention of vitamin A deficiency. Unfortunately, yellow/orange roots are unacceptable in many developing countries. Cultivars should be sought, therefore, which have an acceptable flavour, taste and mouthfeel combined with a carotene content which supplies a significant percentage of daily needs, particularly those of vulnerable age groups. It is possible, however, that children might be less resistant to eating cultivars high in carotenoids and these could be encouraged for planting in home gardens for both roots and leaves. Furthermore ways could be found to incorporate the roots for example in a dry form as flour in other foods, something which has already been investigated in Taiwan.

Sweet potato roots are also sources of some minerals and trace elements. They are a good source of P and though not having outstanding contents of Fe and Ca, they can make modest contributions to the recommended daily intakes of these minerals in a quantity of as little as 100 g, which also provide part of the daily allowance of Mg, Cu and Mn.

Sweet potato leaves are a potentially much richer source of Fe and Ca than the roots, but the Fe which they contain may be very poorly absorbed, and their oxalate content is such as to reduce availability of Ca. Phytate levels in roots and leaves, oxalate levels in leaves and resulting availabilities of minerals should be investigated with a view to selection of cultivars containing the minimum possible concentrations of these anti-nutrients. The role of naturally occurring ascorbic acid in both roots and leaves in promoting absorption of iron could be clarified. Methods of preparation should be devised to increase mineral availabilities.

Both parts of the sweet potato plant used for food are moderately good sources of dietary fibre, being made up of soluble and insoluble fibre, which could promote certain physiological effects such as faecal transit times and reduction of blood cholesterol levels. Other medical applications, particularly the use of roots in oral rehydration therapy, could be studied further.

Though more knowledge is needed to add to that now available with regard to the nutritional value of sweet potato roots and leaves, it is clear that they are of great nutritional benefit to those who already eat them and could be used with considerable effect in the future to nourish the increasing hungry population of the world.

References

Adolph, W.H. and Liu, H.C. 1939. The value of sweet potato in human nutrition. *Chin. Med. J.* **55**: 337–42.

Anon. 1981. *AVRDC Progress Report 1980.* AVRDC, Shanhua, T'ainan, pp. 71–2.

1984. Selection of sweet potatoes suitable for consumption as vegetable greens. *AVRDC Progress Report 1982.* AVRDC, Shanhua, T'ainan, p. 109.

1985. The effect of sweet potato starch on the digestibility of sweet potato protein. *AVRDC Progress Report 1983.* AVRDC, Shanhua, T'ainan, pp. 310–12.

Ashida, J. (Resources Council, Science and Technology Agency) (ed.). 1982. *Standard tables of food composition in Japan*, 4th revised edn, Printing Bureau, Ministry of Finance, Japan.

As-Saqui, M.A. 1982. Sweet potato and its potential impact in Liberia. In: Villareal, R.L. and Griggs, T.D. (eds.), *Sweet potato*, Proceedings of the First International Symposium. AVRDC, Shanhua, T'ainan, pp. 59–62.

Aykroyd, W.R. and Doughty, J. 1964. *Legumes in human nutrition.* FAO Nutritional Studies No. 19.

Bourke, R.M. 1985. Sweet potato (*Ipomoea batatas*) production and research in Papua New Guinea. *Papua New Guinea J. Agric. For. Fish.* **33** (3–4): 89–108.

Bradbury, J.H. 1986. Determination of energy from moisture content in foods containing small amounts of fat and dietary fiber. *J. Agric. Food Chem.* **34** (2): 358–61.

Bradbury, J.H., Baines, J., Hammer, B., Anders, M. and Millar, J.S. 1984. Analysis of sweet potato (*Ipomoea batatas*) from the highlands of Papua New Guinea: relevance to the incidence of *Enteritis necroticans*. *J. Agric. Food Chem.* **32** (3): 469–73.

Bradbury, J.H., Beatty, R.E., Bradshaw, K., Hammer, B., Holloway, W.D., Jealous, W., Lau, J., Nguyen, T. and Singh, U. 1985a. Chemistry and nutritive value of tropical root crops in the South Pacific. *Proc. Nutr. Soc. Austr.* **10**: 185–8.

Bradbury, J.H., Bradshaw, K., Jealous, W., Holloway, W.D. and Phimpisane, T. 1988. Effect of cooking on nutrient content of tropical root crops from the South Pacific. *J. Sci. Food Agric.* **43** (4): 333–42.

Bradbury, J.H., Hammer, B, Nguyen, T., Anders, M. and Millar, J.S. 1985b. Protein quantity and quality and trypsin inhibitor content of sweet potato cultivars from the highlands of Papua New Guinea. *J. Agric. Food Chem.* **33** (2): 281–5.

Bradbury, J.H. and Holloway, W.D. 1988. *Chemistry of tropical root crops: significance for nutrition and agriculture in the Pacific*, ACIAR Monograph No. 6. Australian Centre for International Agricultural Research, Canberra.

Bradbury, J.H. and Singh, U. 1986a. Ascorbic acid and dehydroascorbic acid content of tropical root crops from the South Pacific. *J. Food Sci.* **51** (4): 975–8, 987.

1986b. Thiamin, riboflavin, and nicotinic acid contents of tropical root crops from the South Pacific. *J. Food Sci.* **51** (6): 1563–4.

Bressani, R., Navarrete, D.A. and Elias, L.G. 1984. The nutritional value of diets based on starchy foods and common beans. *Qual. Plant. Plant Foods Hum. Nutr.* **34** (2): 109–15

Caldwell, M.J. 1972. Ascorbic acid content of Malaysian leaf vegetables. *Ecol. Food Nutr.* **1** (4): 313–17.

Caldwell, M.J. and Enoch, I.C. 1972. Riboflavin content of Malaysian leaf vegetables. *Ecol. Food Nutr.* **1** (4): 309–12.

Cameron, M. and Hofvander, Y. 1976. *Manual on feeding infants and young children*, 2nd edn. Protein-Calorie Advisory Group, United Nations, Geneva.

Caribbean Food and Nutrition Institute. 1974. *Food composition table for use in the English-speaking Caribbean*, Caribbean Food and Nutrition Institute, Kingston, Jamaica.

Chen, W-J. and Anderson, J.W. 1981. Soluble and insoluble plant fiber in selected cereals and vegetables. *Am. J. Clin. Nutr.* **34** (6): 1077–82.

Cheng, H-H. 1978. [Protein content and amino acid composition in tubers and stems and leaves of sweetpotato cultivars] Chinese. *J. Agric. Res. China* **27** (4): 291–5.

Church, M. 1979. Dietary factors in malnutrition: quality and quantity of diet in relation to child development. *Proc. Nutr. Soc.* **38** (1): 41–9.

Cohen, N., Jalil, A., Rahman, H., Leemhuis de Regt, E., Sprague, J. and Mitra, M. 1986. Blinding malnutrition in Bangladesh. *J. Trop. Ped.* **32** (2): 73–7.

Collazos, C. and 15 others. 1974. [*The composition of Peruvian foods*] Spanish, 4th edn, Ministry of Health, Lima.

Dahniya, M.T. 1981. Effects of leaf harvests and detopping on the yields of leaves and roots of cassava and sweet potato. In: Terry, E.R., Oduro, K.A. and Caveness, F. (eds.), *Tropical root crops: research strategies for the 1980s*, Proceedings of the First Triennial Root Crops Symposium, International Society for Tropical Root Crops, Africa branch, IDRC, Ottawa, pp. 137–42.

Davis, K.C. 1973. Vitamin E content of selected baby foods. *J. Food Sci.* **38** (3): 442–6.

Dickey, L.F., Collins, W.W., Young, C.T. and Walter, W.M. 1984. Root protein quantity and quality in a seedling population of sweet potatoes. *HortScience* **19** (5): 689–92.

Duell, B. 1990. Ways of eating sweet potatoes in Japan. In: *Tropical root and tuber crops changing role in a modern world*, Proceedings of the Eighth Symposium of the International Society for Tropical Root Crops, 30 October-5 November, 1988, Bangkok, (in press).

Elliott, K., Attawell, K., Wilson, R., Hirschhorn, N., Greenough, W.B. and Khin-Maung-U (eds.). 1990. *Cereal based oral rehydration therapy for diarrhoea*, Report of the International Symposium on Cereal Based Oral Rehydration Therapy, 12–14 November 1989, Karachi. Aga Khan Foundation, Geneva, and International Child Health Foundation, Columbia, MD.

Englyst, H.N., Bingham, S.A., Runswick, S.A., Collinson, E. and Cummings, J.H. 1988. Dietary fibre (non-starch polysaccharides) in fruit, vegetables and nuts. *J. Hum. Nutr. Dietet.* **1** (4): 247–86.

Espinoza, N.O., Yang, M.S., Jaynes, J.M. and Dodds, J.H. 1987. Regeneration of plants of sweet potato (*Ipomoea batatas* L.) transformed by *Agrobacterium rhizogenes* containing a synthetic protein gene. *Bioessays* **6** (6): 261–7.

Evensen, S.K. and Standal, B.R. 1984. *Use of tropical vegetables to improve diets in the Pacific region*, HITAHR Res. Ser. 028, HITAHR, University of Hawaii, Honolulu.

Ezell, B.D. and Wilcox, M.S. 1958. Variation in carotene content of sweet potatoes. *Agric. Food Chem.* **6** (1): 61–5.

Fairweather-Tait, S.J. 1983. Studies on the availability of iron in potatoes. *Br. J. Nutr.* **50**: 15–23.

FAO 1970 (reprinted 1972). *Amino acid content of foods and biological data on proteins*, FAO Nutr. Studies No. 24. Nutrition Division, FAO, Rome.

1988. *Requirements of vitamin A, iron, folate and vitamin* B_{12}, Report of a Joint FAO/WHO Consultation, FAO Food and Nutrition Ser. No. 23. FAO, Rome.

FAO/WHO 1967. *Report of a joint FAO/WHO expert group on requirements of vitamin A, thiamine, riboflavine and niacin.* WHO Tech. Rep. Ser. No. 362.

1973. *Energy and protein requirements.* Report of a joint FAO/WHO *Ad Hoc* Expert Committee, WHO Tech. Rep. Ser. No. 522.

FAO/WHO/UNU 1985. *Energy and protein requirements*, Report of a joint FAO/WHO/UNU Expert Consultation. WHO Tech. Rep. Ser. No. 724. WHO, Geneva.

Gonzales, F.R., Cadiz,T.G. and Bugawan, M.S. 1977. Effects of topping and fertilization on the yield and protein content of three varieties of sweet potato. *Phil. J. Crop Sci.* **2** (2): 97–102.

Goodbody, S. 1984. Nutritional characteristics of a sweet potato collection in the Papua New Guinea highlands. *Trop. Agric. (Trinidad)* **61** (1): 20–4.

Guzman, V.B., Guthrie, H.B. and Guthrie, G.M. 1976. Physical and intellectual development in Philippine children fed five different dietary staples. *Am. J. Clin. Nutr.* **29** (11): 1242–51.

Harris, M. 1963. Gold (sweet potato) in school lunch menus. *School Lunch J.* **17** (10): 18–25.

Haytowitz, D.B. and Matthews, R.H. 1984. *Composition of foods: vegetables and vegetable products*, Human nutrition information series, USDA Agric. Handbook No. 8–11, Washington, DC.

Heywood, P. and Nakikus, M. 1982. Protein, energy and nutrition in Papua New Guinea. In: Bourke, R.M. and Kesavan, V. (eds.), *Proceedings of the Second Papua New Guinea Food Crops Conference*. Department of Primary Industry, Port Moresby, pp. 303–24.

Hill, W.A. 1990. Sweet potatoes for space missions: a new approach for marketing sweet potatoes. In: *Tropical root and tuber crops changing role in a modern world*, Proceedings of the Eighth Symposium of the International Society for Tropical Root Crops, 30 October-5 November, 1988, Bangkok, (in press).

Hipsley, E.H. and Kirk, N.E. 1965. Studies of dietary intake and the expenditure of energy by New Guineans. *South Pacific Comm. Tech. Paper* No. 147.

Hirahara, F. and Koike, Y. 1989. [Tocopherol content in sweet potato tubers of different cultivars, places harvested and cooking methods] Japanese. *Jap. J. Nutr.* **47** (2): 85–91.

Holloway, W.D. 1983. Composition of fruit, vegetable and cereal dietary fibre. *J. Sci. Food Agric.* **34** (11): 1236–40.

Holloway, W.D., Argall, M.E., Jealous, W.T., Lee, J.A. and Bradbury, J.H. 1989. Organic acids and calcium oxalate in tropical root crops. *J. Agric. Food Chem.* **37** (2): 337–41.

Holloway, W.D. and Grieg, R. 1984. Water-holding capacity of hemicelluloses from fruits, vegetables and wheat bran. *J. Food Sci.* **49**: 1632–3.

Holloway, W.D., Monro, J.A., Gurnsey, J.C., Pomare, E.W. and Stace, N.H. 1985. Dietary fiber and other constituents of some Tongan foods. *J. Food Sci.* **50**: 1756–7.

Howard, P., Jenkins, C., Cerhan, J. and MacGregor, D. 1990. Proposed use of sweet potato water as a fluid for home therapy for diarrhoea. In: Elliott, K., Attawell, K., Wilson, R., Hirschhorn, N., Greenough, W.B. and Khin-Maung-U (eds.), *Cereal based oral rehydration therapy for diarrhoea*, Report of the International Symposium on Cereal Based Oral Rehydration Therapy, 12–14 November 1989, Karachi. Aga Khan Foundation, Geneva, and International Child Health Foundation, Columbia, MD, pp. 82–3.

Huang, P-C. 1982. Nutritive value of sweet potato. In: Villareal, R.L. & Griggs, T.D. (eds.). *Sweet potato*, Proceedings of the First International Symposium. AVRDC, Shanhua, T'ainan, pp. 35–6.

Huang, P-C., Lee, N-Y. and Chen, S-H. 1979. Evidences suggestive of no intestinal nitrogen fixation for improving protein nutrition status in sweet potato eaters. *Am. J. Clin. Nutr.* **32**: 1741–50.

Jaynes, J.M., Yang, M.S., Espinoza, N. and Dodds, J.H. 1986. Plant protein improvement by genetic engineering: use of synthetic genes. *Trends Biotechnol.* **4** (12): 314–20.

Jenkins, P.D. 1982. Losses in sweet potatoes (*Ipomoea batatas*) stored under traditional conditions in Bangladesh. *Trop. Sci.* **24** (1): 17–28.

Jones, A. 1977. Heritabilities of seven sweet potato root traits. *J. Am. Soc. Hort. Sci.* **102** (4): 440–2.

Jones, A., Steinbauer, C.E. and Pope, D.J. 1969. Quantitative inheritance of ten root traits in sweet potatoes. *J. Am. Soc. Hort. Sci.* **94**: 271–5.

Karafir, Y.P. 1987. The facts and problems of sweet potato in Irian Jaya. Paper presented at an International Sweet Potato Symposium, 20–26 May, ViSCA, Baybay, Leyte.

Kwiatkowska, C.A., Finglas, P.M. and Faulks, R.M. 1989. The vitamin content of retail vegetables in the UK. *J. Hum. Nutr. Dietet.* **2** (3): 159–72.

Lee, S.R. 1970. [Preparation of drum-dried weaning food based on sweet potato and soybean] Korean. *Kor. J. Food Sci. Technol.* **2** (2): 1–7.

Leung, W-T.W., Busson, F. and Jardin, C. 1968. *Food composition table for use in Africa.* US Department of Health, Education and Welfare Public Health Service, Bethesda, MD; FAO, Rome.

Leung, W-T.W., Butrum, R.R. and Chang, F.H. 1972. *Food composition table for use in East Asia.* Part I. *Proximate composition, mineral and vitamin contents of East Asian foods.* US Department of Health, Education and Welfare, Bethesda, MD; FAO, Rome.

Leung, W-T.W. and Flores, M. 1961. *Food composition table for use in Latin America.* INCAP, Guatemala; Interdepartmental Committee on Nutrition for National Defense, Bethesda, MD.

Li, L. 1982. Breeding for increased protein content in sweet potatoes. In: Villareal, R.L. and Griggs, T.D. (eds.), *Sweet potato*, Proceedings of the First International Symposium. AVRDC, Shanhua, T'ainan, pp. 345–53.

Lin, S.S.M., Peet, C.C., Chen,D-M. and Lo, H-F. 1983. Breeding goals for sweet potato in Asia and the Pacific: a survey on sweet potato production and utilization. *Proc. Am. Soc. Hort. Sci., Trop. Reg.* **27B**: 42–60.

Lopez, A., Williams, H.L. and Cooler, F.W. 1980. Essential elements in fresh and canned sweet potatoes. *J. Food Sci.* **45** (3): 675–8.

Lund, E.D. 1984. Cholesterol binding capacity of fiber from tropical fruits and vegetables. *Lipids* **19** (2): 85–90.

Luyken, M.D., Luyken-Koning, F.W.M. and Pikaar, N.A. 1964. Nutrition studies in New Guinea. *Am. J. Clin. Nutr.* **14**: 13–27.

Maeda, E.E. 1985. Effect of solar dehydration on amino acid pattern and available lysine content in four tropical leafy vegetables. *Ecol. Food Nutr.* **16** (3): 273–9.

Majumdar, B.N., Sharma, D.C. and Kehar, N.D. 1960. Effects of partial replacement of rice or wheat by tapioca or sweet potato flour on the nutritive value of poor vegetarian diets: influence of season in growth studies. *Ann. Biochem. Exptl. Med.* **20** (1): 7–14.

Malcolm, L.A. 1970. Growth retardation in a New Guinea boarding school and its response to supplementary feeding. *Br. J. Nutr.* **24**: 297–305.

Martin, F.W. and Splittstoesser, W.E. 1975. A comparison of total protein and amino acids of tropical roots and tubers. *Trop. Root Tuber Crops Newsletter* No. 8: 7–15.

Mokady, S. 1973. Effects of pectin on serum-cholesterol level in growing rats fed a cholesterol-free diet. *Nutr. Metab.* **15**: 290–4.

Nair, G.M., Ravindran, C.S., Moorthy, S.N. and Ghosh, S.P. 1987. Indigenous technologies and recent advances in sweet potato production, processing and utilization in India. Paper presented at an International Sweet Potato Symposium, 20–26 May, ViSCA, Baybay, Leyte.

National Institutes of Health. 1985. Lowering blood cholesterol to prevent heart disease. *J. Am. Med. Assoc.* **253** (14): 2080–6.

National Research Council. 1980. *Recommended dietary allowances*, 9th edn. National Academy of Sciences, Washington, DC.

1989. *Recommended dietary allowances*, 10th edn. National Academy of Sciences, National Academy Press, Washington, DC.

Norgan, N.G., Durnin, J.V.G.A. and Ferro-Luzzi, A. 1979. The composition of some New Guinea foods. *Papua New Guinea Agric. J.* **30** (1–3): 25–39.

Nutrition Education and Training Program (Trust Territory of the Pacific Islands). 1980. *Local vegetables, good nutrition plus economy.* Food and Nutrition Service, Department of Education, Commonwealth of the Northern Mariana Islands.

Ohtsuka, R., Kawabe, T., Inaoka, T., Suzuki, T., Hongo, T., Akimichi, T. and Sugahara, T. 1984. Composition of local and purchased foods consumed by the Gidra in lowland Papua. *Ecol. Food Nutr.* **15**: 159–69.

Oomen, H.A.P.C. 1970. Interrelationship of the human intestinal flora and protein utilization. *Proc. Nutr. Soc.* **29** (2): 197–206.

1971. Ecology of human nutrition in New Guinea. Evaluation of subsistence patterns. *Ecol. Food Nutr.* **1** (1): 3–18.

1972. Distribution of nitrogen and composition of nitrogen compounds in food, urine and faeces in habitual consumers of sweet potato and taro. *Nutr. Metab.* **14**: 65–82.

Oomen, H.A.P.C., Spoon, W., Heesterman, J.E., Ruinard, J., Luyken, R. and Slump, P. 1961. The sweet potato as the staff of life of the highland Papuan. *Trop. Geogr. Med.* **13** (1): 55–66.

Ortaliza, I.C., Salamat, L.A., De La Cruz, B., Trinidad, T.P. and Jacob, F.O. 1974. Iron absorption studies using biologically labelled vegetables. *Phil. J. Nutr.* **27** (3): 22–9.

Osei-Opare, A.F. 1987. Acceptability, utilization, and processing of sweet potatoes in home and small-scale industries in Ghana. In: Terry, E.R., Akoroda, M.O. and Arene, O.B. (eds.), *Tropical root crops: root crops and the African food crisis*, Proceedings of the Third Triennial Symposium of the International Society for Tropical Root Crops – Africa Branch, IDRC, Ottawa, pp. 143–5.

Pace, R.D., Sibiya, T.E., Phills, B.R. and Dull, G.G. 1985. Ca, Fe and Zn content of 'Jewel' sweet potato greens as affected by harvesting practices. *J. Food Sci.* **50** (2): 940–1.

Passmore, R., Nicol, B.M., Rao, N.M., Beaton, G.H. and DeMaeyer, E.M.

1974. *Handbook on human nutritional requirements*, WHO Monograph Ser. No. 61. FAO/WHO, Geneva.

Paul, A.A. and Southgate, D.A.T. 1978. *McCance and Widdowson's The composition of foods*, 4th edn, Medical Research Council Special Rep. Ser. No. 297. HMSO, London.

Pellett, P.L. and Young, V.R. 1988. Protein and amino acid needs for adults. *Ecol. Food Nutr.* **21** (4): 321–30.

Peters, F.E. 1958. The chemical composition of South Pacific foods. *South Pacific Comm. Tech. Paper* No. 115.

Platt, B.S. 1962. *Tables of representative values of foods commonly used in tropical countries.* Medical Research Council Special Rep. Ser. No. 302, HMSO, London (reprinted in 1985 by Department of Human Nutrition, London School of Hygiene and Tropical Medicine).

Prema, P. and Kurup, P.A. 1979. Effect of feeding cooked whole tubers on lipid metabolism in rats fed cholesterol free and cholesterol containing diet. *Indian J. Exptl. Biol.* **17** (12): 1341–5.

Purcell, A.E. 1962. Carotenoids of Goldrush sweet potato flakes. *Food Technol.* **16** (1): 99–102.

Purcell, A.E., Swaisgood, H.E. and Pope, D.T. 1972. Protein and amino acid content of sweetpotato cultivars. *J. Am. Soc. Hort. Sci.* **97** (1): 30–3.

Purcell, A.E. and Walter, W.M. 1968. Carotenoids of Centennial variety sweet potato, *Ipomoea batatas* L. *J. Agric. Food Chem.* **16** (5): 769–70.

Purcell, A.E., Walter, W.M. and Giesbrecht, F.G. 1978. Protein and amino acids of sweet potato (*Ipomoea batatas* (L.) Lam.) fractions. *J. Agric. Food Chem.* **26** (3): 699–701.

Rao, M.N. and Polacchi, W. 1972. *Food composition table for use in East Asia.* Part II *Amino acid, fatty acid, certain B-vitamin and trace mineral content of some Asian foods.* US Department of Health, Education and Welfare, Bethesda, MD; FAO, Rome.

Rashid, M.M. 1987. Indigenous technologies and recent advances in sweet potato production, processing, utilization and marketing in Bangladesh. Paper presented at International Sweet Potato Symposium, 20–26 May, 1987, ViSCA, Baybay, Leyte.

Rifkind, B.M. 1987. Diet, cholesterol and coronary heart disease: the Lipid Research Clinics Program. *Proc. Nutr. Soc.* **46** (3): 367–72.

Sakamoto, S. 1984. Dissemination to Japan and breeding of sweet potatoes. *Farming Japan* **18** (5): 14–20.

Singh, U. and Bradbury, J.H. 1988. HPLC determination of vitamin A and vitamin D_2 in South Pacific root crops. *J. Sci. Food Agric.* **45** (1): 87–94.

Sinnett, P.F. 1975. *The people of Murapin.* E.W. Classey Ltd, Faringdon, Oxon., p. 39.

Splittstoesser, W.E. and Martin, F.W. 1975. The tryptophan content of tropical roots and tubers. *HortScience* **10** (1): 23–4.

Subrahmanyan, V., Murthy, H.B.N. and Swaminathan, M. 1954. Effects of partial replacement of rice, wheat or ragi (*Eleusine coracana*) by tuber flours on the nutritive value of poor vegetarian diets. *Br. J. Nutr.* **8**: 1–10.

Sun, C.T., Tseng, S.Y., Lu, J.J. and Chang, W.H. 1979. Leaf protein concentrates from some sources available in Taiwan. *J. Chin. Agric. Chem. Soc.* **17** (1/2): 78–92.

Suzuki, K., Kanke, Y., Goto, S. and Kondo, N. 1983. The effect of root crops on nitrogen, calcium, phosphorus and magnesium utilization. *Nutr. Rep. Int.* **27** (1): 213–20.

Taira, H. and Taira, H. 1963. [Studies on amino acid contents in food crops (part 4). Amino acids in sweet potato and potato] Japanese. *J. Jap. Soc. Food Nutr.* **16** (1): 48–9.

Tan, S.P., Wenlock, R.W. and Buss, D.H. 1985. *Immigrant foods.* Second supplement to *McCance and Widdowson's The composition of foods.* HMSO, London.

Taufatofua, P. and Pole, F.S. 1987. Screening and breeding for sweet potato scab (*Elsinoe batatas*) resistance in Tonga. Paper presented at International Sweet Potato Symposium, 20–26 May, 1987, ViSCA, Baybay, Leyte.

Truong, V.D. 1987. New developments in processing sweet potato for food. Paper presented at International Sweet Potato Symposium, 20–26 May, 1987, ViSCA, Baybay, Leyte.

Tsay, J.J.S. and Wang, S.Y. 1988. Production systems program. Garden program. *AVRDC Progress Report 1988.* AVRDC, Shanhua, T'ainan.

Tsou, S.C.S. and Villareal, R.L. 1982. Resistance to eating sweet potato. In: Villareal, R.L. and Griggs, T.D. (eds.), *Sweet potato.* Proceedings of the First International Symposium. AVRDC, Shanhua, T'ainan, pp. 37–44.

Tsou, S.C.S., Kan, K-K. and Wang, S-J. 1987. Biochemical studies on sweet potato for better utilization at AVRDC. Paper presented at International Sweet Potato Symposium, 20–26 May, 1987, ViSCA, Baybay, Leyte.

University of New Hampshire. 1979. *Roots.* Video cassette, Durham, NH.

Unklesbay, N. 1978. *Vegetables.* Southeast Missouri Area Agency on Aging. Extension Division of the University of Missouri, Columbia, MO.

Urbino, M.C., Torres, E.B. and Darrah, L.B. 1972. *Consumption patterns for selected vegetables.* Dept. Agric. Econ., Univ. Phil. Coll. Agric. Staff Paper Ser. No. 124.

Venkateswara Rao, K. and Mahajan, C.L. 1990. Fluoride content of some common South Indian Foods and their contribution to fluorosis. *J. Sci. Food Agric.* **51** (2): 275–9.

Vietmeyer, N. 1978. *Poor people's crops.* Agenda. Agency for International Development, Washington, DC.

Villareal, R.L., Lin, S.K., Chang, L.S. and Lai, S.H. 1979a. Use of sweet potato (*Ipomoea batatas*) leaf tips as vegetables. I. Evaluation of morphological traits. *Exptl. Agric.* **15** (2): 113–16.

Villareal, R.L., Tsou, S.C.S., Lin, S.K. and Chiu, S.C. 1979b. Use of sweet potato (*Ipomoea batatas*) leaf tips as vegetables. II. Evaluation of yield and nutritive quality. *Exptl. Agric.* **15** (2): 117–22.

Villareal, R.L., Tsou, S.C., Chiu, S.C. and Lai, S.H. 1979c. Use of sweet potato (*Ipomoea batatas*) leaf tips as vegetables. III. Organoleptic evaluation. *Exptl. Agric.* **15** (2): 123–7.

Villareal, R.L., Tsou, S.C.S., Lo, H.F. and Chiu, S.C. 1985. Sweet potato vine tips as vegetables. In: Bouwkamp, J.C. (ed.), *Sweet potato products: a natural resource for the tropics.* CRC Press, Inc., Boca Raton, FL, pp. 175–83.

Visser, F.R. and Burrows, J.K. 1983. *Composition of New Zealand foods. 1. Characteristic fruits and vegetables.* DSIR Bull. No. 235, Wellington.

Walter, W.M. and Catignani, G.L. 1981. Biological quality and composition of sweet potato protein fractions. *J. Agric. Food Chem.* **29** (4): 797–9.

Walter, W.M., Catignani, G.L., Yow, L.L. and Porter, D.H. 1983. Protein nutritional value of sweet potato flour. *J. Agric. Food Chem.* **31** (5): 947–9.

Walter, W.M., Collins, W.W. and Purcell, A.E. 1984. Sweet potato protein: a review. *J. Agric. Food Chem.* **32** (4): 695–9.

Wang, H. 1982. The breeding of sweet potatoes for human consumption. In: Villareal, R.L. and Griggs, T.D. (eds.), *Sweet potato.* Proceedings of the First International Symposium. AVRDC, Shanhua, T'ainan, pp. 297–311.

Wang, H. and Lin, C.T. 1969. [Determination of carotene content among parental varieties and their offspring in the sweet potato] Chinese. *J. Agric. Assoc. China* **65**: 1–5.

Watson, G.A. 1988. Sweet potato in food systems. Home consumption: storage, processing and preferences. Draft summary, International Potato Center/ Lembang Horticultural Research Institute, Bandung, West Java.

Watt, B.K. and Merrill, A.L. 1975. *Handbook of the nutritional contents of foods.* Dover Publications, Inc.

WHO 1990. *Diet, nutrition and the prevention of chronic diseases.* Report of a WHO Study Group. WHO, Geneva.

Wilson, R., Greenough, W.B., Hirschhorn, N., Elliott, K., Dale, C., Sanders, D. and Attawell, K. 1990. Meeting the challenge to improve oral rehydration therapy effectiveness, safety, access, acceptance and use. In: Elliott, K., Attawell, K., Wilson, R., Hirschhorn, N., Greenough, W.B. and Khin-Maung-U (eds.), *Cereal based oral rehydration therapy for diarrhoea*, Report of the International Symposium on Cereal Based Oral Rehydration Therapy, 12–14 November 1989, Karachi. Aga Khan Foundation, Geneva; International Child Health Foundation, Columbia, MD, pp. 13–19.

Woolfe, J.A. 1987. *The potato in the human diet.* Cambridge University Press, Cambridge.

Yang, T-H. 1982. Sweet potato as a supplemental staple food. In: Villareal, R.L. and Griggs, T.D. (eds.), *Sweet potato.* Proceedings of the First International Symposium. AVRDC, Shanhua, T'ainan, pp. 31–4.

Yang, T-H. and Blackwell, R.Q. 1966. Nutritional evaluation of diets containing varying proportions of rice and sweet potato. *Proc. 11th Pacific Sci. Congr.*, Tokyo.

Yang, T-H., Tsai, Y-C., Sheu, C-T. and Chen, W-J. 1967. The influence of the partial sweet potato substitution in place of equi-caloric rice in the rice diets containing supplemental foodstuffs for young men. *J. Chin. Agric. Chem. Soc.* **5** (1–2): 47–52.

Yang, T-H., Tsai, Y-C., Sheu, C-T., Ko, H-S., Chen, S-W. and Blackwell, R.Q. 1975. [Studies on protein nutrition of the main food in Taiwan (II) Protein content and amino acid spectrum of rice and sweet potato as related to variety] Chinese. *J. Chin. Agric. Chem. Soc.* **13** (1–2): 132–8.

CHAPTER 4

Toxic and anti-nutritional factors

Toxic factors in plants include naturally occurring toxins and those which may be produced by the agency of external factors such as insect or mechanical damage, or microbial activity. External factors may be active in the field, during storage or after processing or cooking. Plant compounds having anti-nutritional effects may bind essential minerals, act as allergens or vitamin antagonists or cause bodily discomfort as a result of flatulence production. The sweet potato contains both toxins and anti-nutritional compounds, but not all of them have yet been identified as being of significance to humans. One anti-nutritional factor which binds calcium, namely oxalic acid found in sweet potato roots, and more significantly in the leaves, is discussed in Chapters 2 and 3. This chapter will describe various toxins and anti-nutritional factors occurring in sweet potato roots: the hepato-toxic and lung-toxic furanoterpenes produced as a result of fungal invasion and other injurious stimuli, the naturally occurring trypsin inhibitor, and the anti-nutritional flatulence factor(s) which have not yet been positively identified. Their significance to human and animal health will be discussed.

Toxic stress metabolites

Chemical structure

When plants are subjected to invasion by microbial pathogens, they produce antibiotic substances in the infected tissue or in very closely adjacent non-infected tissue. These substances were designated 'phytoalexins' in 1940 (Müller and Börger, 1940), and were presumed to participate in the plants' defence mechanism. Phytoalexins (also known as 'abnormal metabolites' or 'stress metabolites' can be produced by any biological, chemical or physical agent which gives the plant a continuous injurious stimulus. Such a stimulus might be as a result of microbial infection, as mentioned above, or through temperature changes, abnormal oxygen tensions, nutrient imbalance, exposure to chemical agents etc. The subject of phytoalexins has been reviewed (Kuć, 1972).

Specifically in the sweet potato, when the roots are damaged by pathogenic fungi, such as *Ceratocystis fimbriata* (Hiura, 1943) and *Fusarium solani* (Wilson, 1973) or invaded by weevil larvae of *Cylas formicarius* or *Euscepes postfasciatus* (Uritani et al., 1975), they produce a variety of sesquiterpenes, many of which have the furan ring in their structure (Schneider et al., 1984). These phytoalexins accumulate in the injured tissue and in adjacent non-injured tissue. They can also be artificially produced as a result of root treatment with toxic chemicals, such as mercuric chloride (Uritani, Uritani and Yamada, 1960). Although they are formed in the leaves and stems in response to injury, the amounts produced are small in comparison to those in the roots (Clark, Lawrence and Martin, 1981). Phytoalexins are induced in sweet potato tissue by factors produced by the fungus or other stimulus. These factors are being studied, but have not yet been identified. However, six amino acids, namely alanine, asparagine, cysteine, glutamine, glycine and proline have been demonstrated to increase terpenoid formation markedly in sweet potato roots (Kim, Oguni and Uritani, 1974). The phytoalexins themselves, however, are shown to be products of the sweet potato roots rather than of the fungus, as they only form in live, as opposed to autoclaved, root tissue (Wilson et al., 1971). A key enzyme involved in the biosynthesis of terpenoid phytoalexins in tissue responding to infection is 3-hydroxy-3-methyl-glutaryl coenzyme A (HMG-CoA) reductase (Suzuki, Oba and Uritani, 1975). This is associated in sweet potato roots with the microsomal cell membrane.

Since the beginning of the century, there have been frequent reports from Japan of cattle deaths as a result of eating mould-damaged sweet potatoes. It was also noticed that such sweet potato roots were very bitter. Efforts were made to isolate and characterize the bitter principle. Eventually an oily substance, which was called ipomoeamaron, was isolated from sweet potatoes infected with black rot (*C. fimbriata*) by a Japanese worker (Hiura, 1943). The isolate had a bitter taste and was shown to be toxic to mice and rabbits. Structural studies were performed on the isolate in Japan between 1946 and 1952. In 1952–3, its chemical structure was finally elucidated (Kubota and Matsuura, 1953) and it is now called ipomeamarone. It was the very first example of the isolation and identification of a phytoalexin from the plant kingdom (Hiura, 1943; Uritani, Suzuki and Muramatsu, 1947; Uritani and Akazawa, 1955). Since then about 30 kinds of sesquiterpene including 15-carbon,

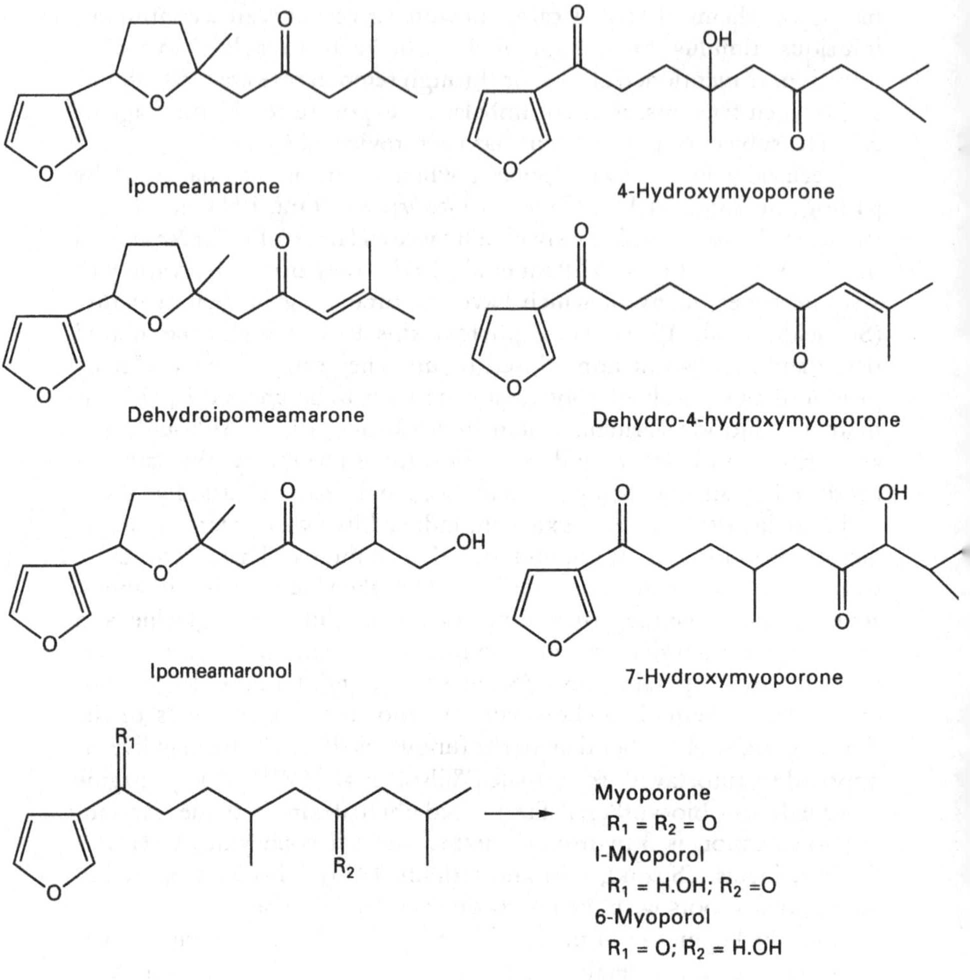

Ipomeamarone

4-Hydroxymyoporone

Dehydroipomeamarone

Dehydro-4-hydroxymyoporone

Ipomeamaronol

7-Hydroxymyoporone

Myoporone
$R_1 = R_2 = O$
l-Myoporol
$R_1 = H.OH; R_2 = O$
6-Myoporol
$R_1 = O; R_2 = H.OH$

Figure 4.1. Ipomeamarone and other hepato-toxic stress metabolites from sweet potato roots.

3-substituted furanoterpenes have been isolated from mould-damaged sweet potato roots, and their chemical structures determined. The structure of ipomeamarone and some of the other furanoterpenoid stress compounds are shown in Figure 4.1. Ipomeamarone is the most abundant sesquiterpene found in stressed sweet potatoes; indeed it is almost invariably formed in them. The other furanoterpenes are related to it, either as precursors or further conversion products in the biosynthetic pathway.

These compounds are known to be hepato-toxins; that is, their toxic effect is mainly manifested in the liver. However, no cases of liver disease due to the consumption of blemished sweet potato by humans or livestock have been reported. It has been suggested that ipomeamarone may play a part in determining host plant specificity (Kojima and Uritani, 1976). For example, furanoterpenoids including ipomeamarone are only slightly inhibitory to growth of the pathogenic strain of *C. fimbriata* on infected sweet potato tissue, whereas non-pathogenic strains from other crops, such as cacao, coffee, taro etc., are severely inhibited. Ipomeamarone production is stimulated by a variety of fungi, but these differ in the concentration they can induce (Martin, Hasling and Catalano, 1976). Some fungi have been shown to be much less sensitive than others to the concentrations of ipomeamarone which occur in sweet potato tissue, and even to be capable of degrading the terpenoids present in their vicinity in the roots (Arinze and Smith, 1980). There is also evidence that ipomeamarone accumulation varies considerably between sweet potato cultivars (Akazawa and Wada, 1961; Martin et al., 1978) when the latter are infected with the same microbial pathogens. For example, ipomeamarone synthesis (mg/g tissue) in various cultivars infected with *C. fimbriata* ranged, after 72 hours, from only 2.6 in a cultivar very susceptible to infection, to 59.5 in one which was highly resistant (Akazawa and Wada, 1961).

Reports from both the United States (Hansen, 1928) and Japan (Hiura, 1943) dating from decades ago have implicated sweet potatoes in the death of cattle from a pulmonary disease known variously as pulmonary oedema, pulmonary adenomatosis, acute bovine pulmonary emphysema and atypical interstitial pneumonia. In 1969, a devastating outbreak of bovine pulmonary disease, described as atypical interstitial pneumonia, took place in Tift County, GA, in the United States. Sixty nine cattle died out of a total herd of 275, which had been given mouldy cull sweet potatoes, previously washed and held in a barn for several days. This event provided the stimulus for much intensive research to isolate and characterize the lung-toxic factors from the remaining mouldy sweet potatoes.

Samples of these sweet potatoes exhibited varying states of decomposition and yielded more than 150 fungal isolates (Peckham et al., 1972). However, the samples showing only moderate deterioration were the most toxic and extracts from them caused death when administered to mice. These mice showed marked signs of lung disease (Wilson, Yang and Boyd, 1970), as well as toxic cellular changes in the liver, spleen and kidneys. Ipomeamarone and other hepato-toxins were found in the mouldy sweet potato roots; however, the extracted and purified ipomeamarone caused acute liver toxicity and not pulmonary disease in mice. This led to a search for a specific 'lung oedema factor'.

Of the many fungi isolated from the mouldy roots, only one type,

General formula

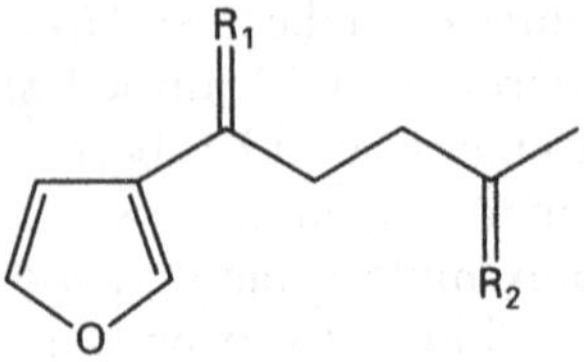

4-Ipomeanol $R_1 = O; R_2 = HOH$
1-Ipomeanol $R_1 = HOH; R_2 = O$
1,4-Ipomeadiol $R_1 = R_2 = HOH$
Ipomeanine $R_1 = R_2 = O$

Figure 4.2. Lung-toxic ipomeanols from sweet potato roots.

namely *Fusarium solani* (*javanicum*) induced lung-toxic properties to develop in viable sweet potato slices. This was demonstrated by injections of extracts into mice (Wilson et al., 1971) or by feeding artificially infected sweet potatoes to cattle (Peckham et al., 1972). Growth of *F. solani* on live sweet potato roots induced the production of ipomeamarone, ipomeamaronol and a 'lung oedema factor' (Wilson et al., 1970), one constituent of which was named 4–ipomeanol (Wilson et al., 1971).

The next step was to bioproduce the 'lung oedema factor' toxins by inoculating slices of sweet potato with *F. solani* and, after 6 days' growth of the fungus, to extract and separate the lung toxins from the infected slices using solvents and preparative gas-liquid chromatography. The lung toxins, after purification by HPLC, were identified by mass spectrometry as four closely related 1,4-dioxygenated-1-(3-furyl) pentanes (9-carbon, 3-substituted furans) (Boyd et al., 1974). These compounds, 4-ipomeanol, 1-ipomeanol, 1,4-ipomeadiol and ipomeanine are shown in Figure 4.2. Ipomeanine was present in very low concentrations in extracts containing the lung toxins. It had previously been isolated from sweet potatoes by Japanese workers (Kubota and Ichikawa, 1954), but its toxicity had not been reported. The compound 4-ipomeanol was reported as being the most abundant of the four and probably most responsible for the lung toxicity (Boyd et al., 1974). Later work (Catalano et al., 1979) found 1- and 4-ipomeanol to be present in approximately equal quantities.

It was also discovered that *F. solani* converts the 15-carbon furanoterpenoid hepato-toxin, 4-hydroxymyoporone, to all four 9-carbon furanoterpenoids (Burka et al., 1977). Hence the fungus serves both as a stress initiator and a converter of at least one resulting compound to the potent lung toxins. This *in vitro* conversion could not be demonstrated with *C. fimbriata* (Burka et al., 1977), although *C. fimbriata* was found to be capable of reducing ipomeanine to the other three lung toxins. Thus it has been suggested that the production of ipomeanols is a specific response to infection with *F. solani*, *F. oxysporum* or certain other species of *Fusarium* (Wilson and Burka, 1983). It takes place in raw, but not autoclaved, sweet potato and is therefore a response by the root and not a product of the fungus per se. However, it has been shown that *C. fimbriata* and other pathogens, apart from *Fusarium* spp. can induce formation of ipomeanols *in vivo* (Clark et al., 1981). Two outbreaks of respiratory disease among 102 cattle in Taiwan which resulted in 17 deaths were attributed to the formation of a lung oedema toxin in mouldy sweet potatoes fed to the cattle (Liu, 1982). *Ceratocystis* spp., but not *Fusarium* spp., were isolated from the contaminated sweet potato. The finding that various pathogens can induce ipomeanols, plus the fact that they can also be induced by mercuric acetate, led to the suggestion that ipomeanols form as a result of general furanoterpene biosynthesis by the host, or as a result of degradation of ipomeamarone. However, the low levels of ipomeamarone induced by some pathogens is not simply due to its degradation to other compounds, since the same pathogens also induced correspondingly low levels of ipomeanols and total furanoterpenes (Clark et al., 1981). On the basis of their capacity to cause furanoterpenoid accumulation, various pathogens of sweet potato were classed as follows: non-inducers (*Streptomyces ipomoea*, *Monilochaetes infuscans* and internal cork virus); low-level inducers (*Rhizopus stolonifer*, *Erwinia carotovora*) or high-level inducers (*C. fimbriata*, *F. solani*, *Plenodomus destruens*, *Diaporthe batatatis*, *Diplodia tubericola*, *Macrophomina phaseoli* and *Sclerotium rolfsii*). The interrelationship of the many furanoterpenoids involved during biogenesis of sweet potato stress metabolites has been shown (Schneider et al., 1984).

Toxicity

The toxic effects of stress metabolites with an undegraded sesquiterpene structure (the hepato-toxins) have received little attention compared to those of the lung toxins (Wilson and Burka, 1983). The hepato-toxins have never been demonstrated to cause liver disease in humans or livestock. Ipomeamarone has been shown to be toxic to mice and rabbits (Hiura, 1943) and the lethalities of ipomeamarone and five other furanosesquiterpenes have been compared in mice. With the exception of 6-myoporol, the LD_{50} values (dose lethal to 50% of mice) of ipomeamarone (Seawright and Mattocks, 1973), ipomeamaronol, 4-hydroxymyoporone, 7-hydroxymyoporone and dihydro-7-hydroxymyoporone (Wilson and Burka, 1979) were about 200 mg/kg body weight by intraperitoneal injection. The most lethal was 6-myoporol with an LD_{50} value of 84 mg/kg. Death of the mice usually occurred in less than 24 hours, and not more than 48 hours. Post-mortem examination revealed an enlarged and congested liver with extensive necrosis of the hepatocytes. There was no evidence of pulmonary oedema or damage to the kidneys.

Earlier, a Japanese researcher (Kondo, 1971) had investigated the toxic effect of an ether-soluble substance containing ipomeamarone and other furanosesquiterpenes extracted from sweet potato roots infected with *C. fimbriata* (see Figure 4.3) on mice, rabbits and goats. LD_{50} value of the crude extract for both mice and rabbits was about 1 g/kg body weight. Pathological examination of all the animals involved in poisoning revealed the disappearance of glycogen from, and degeneration and necrosis in, the liver and abnormal changes in the kidneys, pancreas, lungs and nervous system.

Veterinary reports on cattle suffering from pulmonary disease as a result of eating mouldy sweet potatoes indicate that signs of disease may manifest themselves as early as 1 day after feeding commences, with death following a few days later. The main symptoms of, and findings in relation to, the disease have been described (Wilson and Burka, 1983). The most obvious sign is dyspnea with rapid breathing which becomes more laboured until death, which apparently results from asphyxiation. Certain visceral organs, apart from those involved in respiration, have been found to be damaged in cattle as a result of some sweet potato poisoning outbreaks. However, the principal pathology seems to be confined to the lower respiratory tract. No particular therapeutic measures have been successful in preventing death of animals once the disease is well advanced.

The bovine is among the species most susceptible to 4-ipomeanol. Eighteen-month-old heifers died within 3 to 4 days after receiving 9–14 mg toxin/kg body weight (Doster et al., 1978). Their bodies showed many of the symptoms characteristic of natural outbreaks of sweet potato poisoning causing atypical interstitial pneumonia in cattle. All four of the lung-toxic furanoterpenes (ipomeanols) produced identical reactions in the lungs of mice by whichever means they were administered. Their relative toxicities for mice as shown by LD_{50} values are ipomeanine > 4-ipomeanol > 1-ipomeanol > 1,4-ipomeadiol (Boyd et al., 1974), although the contribution of ipomeanine, the most potent of the four, may be small due to its low concentration. Adult male mice surviving near-lethal levels of lung-toxins often later succumbed to non-pulmonary diseases which particularly involved the kidneys. This renal toxicity was especially associated with 1- and 1,4-ipomeanols.

a

b

Figure 4.3. Roots with (a) black rot disease *Ceratocystis fimbriata* (M. Iwanaga) or (b) after attack by the sweet potato weevil *Cylas formicarius* (P. Ewell) may contain high levels of toxic sesquiterpenes and could be unsuitable for feeding to livestock.

Other species vary in their susceptibility to 4-ipomeanol. Rats, rabbits and guinea pigs have been found to be quite susceptible, with lethal doses in the range 10–60 mg/kg (Dutcher and Boyd, 1979). The hamster was more resistant to the pulmonary effects of 4-ipomeanol, with an LD_{50} value of about 150 mg/kg. At high dose levels, liver lesions were also seen in the hamster. In birds, including the chicken and the Japanese quail, 4-ipomeanol was found in the liver, rather than the lungs or kidneys. This may have been related to the lack, in these avian species, of the ciliated and non-ciliated bronchiolar lining cells found in other vertebrates (Buckpitt and Boyd, 1978) – see below.

Distribution of ^{14}C-4-ipomeanol has been determined in the rat (Boyd, Burka and Wilson, 1975). A characteristic pattern emerged regardless of the dose given, its means of administration or the type of radioactive compound used. Tissue radioactivities peaked within 1 to 2 hours after injection and thereafter reached relatively high plateaus representing residual activity. The greatest concentration of radioactivity occurred in the lungs. Seventy to ninety per cent of the residual radioactivity in the lungs, liver and kidneys was apparently covalently bound to tissue macromolecules. The lung contained the highest bound activity followed by the liver and kidneys. Subsequent studies, demonstrating activation of 4-ipomeanol by mixed function oxygenase enzymes in target organ tissues and consequent covalent binding of the activated compound to organ microsomes have been reviewed (Wilson and Burka, 1983). Certain compounds, shown to decrease covalent binding also decrease toxicity of the toxin. Enzyme systems such as the lung mixed-function oxygenase system have xenobiotic activity; that is, they are capable of converting and removing foreign chemical compounds entering the lungs. However, this activity may sometimes result in serious consequences for the organism through the production of highly reactive intermediate metabolites responsible for cytotoxicity and carcinogenesis (Doster, Farrell and Wilson, 1983). Use of radio-labelled 4-ipomeanol (Boyd, 1977) has enabled localization of the mixed-function oxygenase system in the respiratory tract of rats. The prime cellular targets for covalent binding of 4-ipomeanol were non-ciliated bronchiolar (Clara) cells of the terminal airways. The *in situ* binding of toxin results in cell necrosis. There has been a suggestion that this finding may indicate potential vulnerability of these particular lung cells to other xenobiotic agents, including inhaled carcinogens (Wilson and Burka, 1983). Identification of specific cells involved in pulmonary diseases, their enzyme systems and biochemical properties, by use of agents such as 4-ipomeanol may also aid advances in therapeutic measures for these diseases (Doster et al., 1983).

Table 4.1. *Levels of ipomeamarone in retail sweet potatoes in two cities of the USA*

Sample source	Ipomeamarone content (mg/g sp)	Ipomeamarone content (mg in whole SP)
Lexington, KN	7.6	950
Nashville, TN	1.1	138
Nashville, TN	0.24	20.4
Nashville, TN	0.12	14.1
Lexington, KN	0.10	9.4

Notes:
SP, sweet potato.
From: Boyd and Wilson, 1971.

Furanoterpenes

Content in sweet potatoes sold for human consumption or animal feed

There has been little research carried out to determine furanoterpene levels in sweet potato roots sold for human consumption, or on their stability to cooking or processing. This probably reflects the lack of reports of any acute human poisoning due to eating blemished sweet potatoes, although concern has been expressed about possible toxic effects.

Using gas chromatography as an analytical tool for quantitative estimation of ipomeamarone in sweet potato extracts, a number of samples of sweet potato from local food stores and then at various retail and wholesale markets around Nashville, Tennessee, and Lexington, Kentucky, in the United States, were examined (Boyd and Wilson, 1971). A typical sweet potato, from the food store, described by the authors as showing only minor blemishes, contained 1.1 mg ipomeamarone/g sweet potato tissue. This relatively high concentration suggests that the root may have been more than slightly blemished. The retail and wholesale markets revealed numerous samples which contained ipomeamarone and other toxic compounds. The ipomeamarone contents of representative samples were as shown in Table 4.1.

In addition, ipomeamarone levels have been measured in whole samples of roots imported into the United Kingdom and purchased at random in retail shops and markets in four urban centres (Coxon, Curtis and Howard, 1975). Roots described as rotten, or partially rotten, and soft or sprouted contained between 0.1 and 0.9 mg ipomeamarone/g fresh weight. Some samples also contained small quantities (not measured) of ipomeanine and 4-ipomeanol. Roots in good condition generally contained low levels of ipomeamarone (<0.04 mg/g), but two samples apparently free of any external or internal damage contained 0.068 and 0.328 mg ipomeamarone/g fresh tissue.

These findings show that there can be little room for complacency about levels of toxic furanoterpenes in sweet potato roots on sale to the public if at any time in the future these compounds are shown to be a public health hazard. They were higher and more widespread than previously thought. Toxins were reported to be present in older roots of the American samples that showed some evidence (however slight) of decomposition or discoloration. A narrow black ring just below the peeling in many cases is often associated with abnormal metabolite formation in the adjacent cells (Wilson, 1973).

Levels of ipomeamarone have also been studied by gas chromatography in bruised or otherwise damaged roots from retail grocery stores (Wood and Huang, 1975) and in sound and healthy, blemished, or severely blemished and diseased sweet potatoes (Catalano et al., 1977). Blemished and diseased roots in the latter study were obtained from sweet potatoes discarded at the grading line of packing sheds or from washed roots which had become diseased during storage. No ipomeamarone was detected in healthy roots; levels in bruised and damaged roots from grocery stores (Wood and Huang, 1975) were 0.0026 to 0.046 mg/g; blemished root levels (Catalano et al., 1977) ranged from 0.01 to 1.8 mg/g and in severely diseased roots from 5.4 to 25.5 mg/g.

Lung toxins 4-ipomeanol and 1,4-ipomeadiol have been measured in necrotic tissue of sweet potato inoculated with various pathogens (Clark et al., 1981). They were always present in a much lower concentration than was ipomeamarone. For example, a sample inoculated with *F. solani* contained 4580 μg/g ipomeamarone, 124 μg/g 4-ipomeanol and 272 μg/g 1,4-ipomeadiol. The potency of the lung toxins is greater than that of the hepato-toxins as shown by mice LD_{50} values (Burka et al., 1974, 1977). However, there has been little attempt to quantify the ipomeanols in blemished sweet potatoes offered for sale.

There do not appear to have been any analyses for toxic furanoterpenes carried out on sweet potato roots sold in developing-country markets.

Effects of cooking and processing on furanoterpenes

There have been contradictory findings on the effect of cooking on toxic stress metabolites in sweet potato roots. Preliminary studies showed that they were not destroyed by the normal cooking procedures of boiling and baking (Boyd and Wilson, 1971). Later experiments by other workers show that the most effective way to eliminate toxic factors is by careful removal of diseased parts by peeling and trimming.

The effect of heat during cooking has been investigated (Catalano et al., 1977). In the majority of samples of either moderately or severely blemished and diseased roots baking reduced the ipomeamarone concentration of peelings and trimmings, though it did not eliminate it entirely. Peeling and trimming blemished and diseased roots 3–10 mm beyond the affected areas, depending on the degree of infection, either before or after boiling or baking, produced residual healthy tissue with little or no ipomeamarone present. The flesh remaining after peeling and trimming following cooking was healthy, showing that cooking does not cause a migration of ipomeamarone from diseased to healthy tissue. The absence of ipomeamarone in apparently healthy tissue immediately surrounding diseased tissue could account for the lack of reports of human poisoning if visibly diseased portions are discarded from sweet potato roots during preparation for cooking (Martin et al., 1976).

Other workers (Cody and Haard, 1976) found that baking (for 45 min at 204°C) and microwave cooking destroyed 90% of ipomeamarone in sweet potato roots. The lung toxin 4-ipomeanol was more heat-stable than ipomeamarone, but also decreased as a result of normal cooking.

The effect of processing on ipomeamarone was determined with roots severely affected by soil rot fungi (Catalano et al., 1977). They were divided into three parts and either left unpeeled, lye-peeled in hot sodium hydroxide or lye-peeled and trimmed to remove infected tissue. Each portion was then canned in syrup and the contents analysed for ipomeamarone. No ipomeamarone could be detected in the canned sweet potato prepared by the normal commercial processing operation which involved lye-peeling and trimming. Lye-peeling alone was not as effective.

In some parts of the world, sweet potato roots are commonly cut into chips and dried before storage for later use as an animal feedstuff. The drying process may not be carried out efficiently so that moisture levels are not reduced below those available to microorganisms, or may be pursued in unhygienic conditions. In these cases the chips are susceptible to microbial contamination either during processing or in storage. A variety of pathogenic and saprophytic fungi were found on sweet potato chips collected from 46 main production areas in Taiwan (Yang and Yu, 1976). The furanoterpenoid content of most samples was between 0 and 4 mg/g chips, but four samples from one area ranged in content from 9.3 to 17.4 mg/g chips and were thought to be the cause there of toxicosis of swine. The authors attributed the severity of contamination of the chips by microorganisms and consequent production of furanoterpenoids to the production process and the conditions under which the chips were subsequently stored. The furanoterpenoids in the sweet potato chip tissue were found to be quite stable to various treatments including up to 15 hours in sunlight at 25°C, heating at 121°C for 15 min or exposure to ultraviolet radiation.

Significance to humans

The chief apparent effect which human beings encounter from the formation of furanoterpenes in sweet potatoes is economic loss. A loss to farmers occurs when domestic livestock are fed toxic roots and either die or become unhealthy. The feeding of animals with roots considered to be unfit for human consumption means that outbreaks of lung-toxic poisoning will continue to occur from time to time. It has also been pointed out that cattle may often eat sweet potatoes containing small amounts of furanoterpenes accumulated by roots in injured small spots in the field where they are attacked by a range of microorganisms (Uritani et al., 1981). It may be important therefore to elucidate the chronic effects of furanoterpenes on animal physiology.

Formation of furanoterpenes leading to a bitter flavour is a secondary factor in loss of root quality through weevil or fungal attack in the field. For example, in the Camotes Islands, Philippines, roots are attacked by weevil in a large proportion of fields. Farmers remove damaged, bitter parts of the roots with knives, but the remaining undamaged parts are unfit for sale and are retained for home use (Bartolini, Hirose and Sawayama, 1984).

The occurrence of ipomeamarone in damaged or defective roots used for processing may lead to the production of a bitter off-flavour in a processed product. For example, when all defective roots are not removed before the production of *shochu*, a Japanese alcoholic liquor made from sweet potatoes, bitter components (ipomeamarone) are sometimes detected in the drink. *Shochu* production involves many hundreds of tonnes of raw roots. Defective roots are at present removed by hand selection (Figure 4.4), a tedious process which involves large numbers of workers and therefore adds considerably to overall production costs. A suggested process for removal of 94% of the ipomeamarone from *shochu* by stirring with activated carbon has been described (Kudo and Hidaka, 1984).

There have been no authenticated reported cases of acute human poisoning from eating sweet potatoes containing either hepato-toxins or lung-toxins. The possibility that sweet potato might be involved in endemic respiratory disease among Papua New Guinea people was

Figure 4.4. Hand-trimming damaged and diseased sweet potato roots at a *shochu* (alcoholic drink) factory in Japan. Diseased parts may contain bitter compounds which would adversely affect the taste quality of the drink (S. Sugama).

investigated in 1978 (Wilson, 1982). However, examination of methods of cultivation, storage and preparation for eating of sweet potato roots failed to suggest a role for this food as a causative agent. Analyses of sweet potato obtained from prepared human food samples also provided no clear-cut evidence that lung-toxins were involved in the pulmonary disease.

The need for more knowledge through research on the possible chronic long-term effects of naturally occurring toxicants such as the lung-toxins of sweet potato has been emphasized (Curtis, 1986). There has been speculation as to whether mouldy foods such as sweet potatoes found to be toxic to other animal species would be eaten in localities where extreme hunger exists, and whether undernourished humans might not be more susceptible to the toxins than would individuals having adequate diets (Davis et al., 1975). Although these speculations involved foods containing mycotoxins, they could also be applied to those containing other types of hepato- or lung-toxins. There have certainly been reports that the quality of sweet potatoes sold in some developing countries is poor due to deficient postharvest handling methods. Such roots could contain high levels of phytoalexins, including hepato- and lung-toxins, induced as a result of damage and disease. Until such a time when it is unequivocally shown that furanoterpenoid toxins do not constitute any form of public health hazard either in the short or long term, every effort should be made to maintain conditions under which these abnormal metabolites are prevented from forming.

Research utilizing sweet potato lung toxins in laboratory animals could lead to the discovery of new lines of attack by scientists in the study of disease processes of pulmonary tissue in humans.

Inhibition of formation

It has already been emphasized elsewhere that efforts to improve significantly the quality of roots sold to consumers (which in some areas at present leaves much to be desired) may raise the status of sweet potato and encourage increased consumption. Such an improvement in quality would at the same time ensure that consumers purchase healthy roots free from hepato-toxins and lung-toxins.

Measures to diminish postharvest spoilage include careful harvesting methods to minimize stress to the roots and the application of plant fungicides. Roots should be properly cured to promote wound healing and thus prevent microbial attack. They can also be washed with hypochlorite solution, dried and then packed in boxes using paper to separate layers of roots during storage or transport. Unfortunately these measures are expensive and may be impractical in some developing country situations.

It has been suggested (Martin et al., 1978) that differences found between cultivars in their potential to produce furanoterpenes lead to the possibility of selecting and releasing cultivars with low furanoterpene-inducing potential. For example cultivars with little or no concentration of HMG-CoA reductase (see p. 189) could be produced by plant breeding (Uritani, I., personal communication). However, a drawback to this line of attack is the involvement of HMG-CoA reductase in the synthesis of carotenoids. A cultivar lacking HMG-CoA reductase might not therefore form beta-carotene simultaneously. Though the plant's secondary defence mechanism employing furanoterpenes would be lost, the strengthening of alternative primary defence factors, which inhibit attack by microbes, such as the high molecular weight spore agglutination and germ tube growth inhibiting factors found in sweet potato roots (Kojima and Uritani 1978a,b; Kojima, Kawakita and Uritani, 1982) could be explored. A two-pronged attack of the problem could entail breeding of sweet potatoes without the genetic potential to produce furanoterpenes, but which still produce other non-toxic pathogenic inhibitors, and at the same time plant seedlings free from contamination by living organisms which stimulate furanoterpene production (Uritani et al., 1981). A further alternative strategy might be to breed sweet potatoes with much thicker, tougher skins capable of resisting damage and fungal penetration (Uritani, 1990).

Note:
In view of the conditions under which sweet potatoes are dried into chips and then stored in some parts of the world (see above), the presence of aflatoxins in these chips should be investigated. Aflatoxins, produced by moulds such as *Aspergillus flavus*, have been found in a variety of foods, although not a single report of their presence in sweet potatoes was found in the literature. They have been shown to cause illness or death in animals. No acute effects of these mycotoxins have been found in humans. Their possible chronic effects are still a subject of considerable debate and research on this topic continues. Until their significance for humans is clarified, however, production and storage of foods should be undertaken in ways to discourage the growth of moulds.

Trypsin inhibitors

A broad class of polypeptides and proteins occurring in plants and other life forms, which inhibit the action of proteolytic enzymes and are known therefore as proteinase inhibitors, includes those which specifically inhibit the important digestive enzyme trypsin. Others inhibit a related enzyme chymotrypsin. The subject of proteinase inhibitors has been reviewed (Ryan, 1981). The occurrence of these inhibitors has nutritional implications for human and animal feeding as they are found in many important food plants eaten both raw and cooked. The first known plant proteinase inhibitor was the trypsin inhibitor crystallized from soybeans (Kunitz, 1945). Other proteinase inhibitors have since been described in a variety of plants (Ryan, 1981), especially the Graminae, Leguminosae and Solanaceae families. The exact physiological function of proteinase inhibitors in plants is still in question, but it is thought that they may play a role in plant protection by inhibiting the digestive enzymes of invading insect pests or pathogens (Ryan, 1981).

The first non-leguminous plant reported to contain a trypsin inhibitor was the sweet potato (Sohonie and Bhandarkar, 1954). The isolated impure inhibitor fraction was very thermolabile. Since this initial report, several researchers have investigated the nature of the sweet potato trypsin inhibitor, all reporting that it consists of more than one molecular species, but differing in the number of individual fractions found and in some cases the characteristics of these fractions.

The strong inhibition of trypsin which has been demonstrated to occur *in vitro* with sweet potato trypsin inhibitor could indicate an interference with protein digestion *in vivo* thus having nutritional implications in humans, especially those whose protein intake is already marginal, and who have the habit of snacking on raw sweet potato. There has been some suggestion that trypsin inhibitors may be implicated in the incidence of *Enteritis necroticans* ('pigbel') in Highland Papua New Guinea (see below). Furthermore, trypsin inhibitors in roots fed raw, either fresh or dried, to animals may reduce feed efficiency (although some researchers have indicated that the role of starch may be as significant as that of trypsin inhibitors in reducing protein digestibility of feeds – see Chapter 7). This section describes the present knowledge about the characteristics of, factors affecting, and nutritional implications of, the sweet potato trypsin inhibitors.

Characteristics

In 1973, three different trypsin inhibitors were isolated from Japanese sweet potato roots (Sugiura et al., 1973). Inhibitors II and III were calculated to have molecular weights of 23,000 and 24,000. They were acidic proteins, a finding confirmed by others (Lin, Cheng and Fu, 1983), who also, in a Taiwanese sample, found three to four different molecular species of inhibitor with such similar physical and chemical properties that they were difficult to separate by conventional protein purification techniques. The inhibitor activities of the Japanese sample were quite stable over a pH range of 2–11. Nigerian workers isolated 10 trypsin inhibitors by gel filtration and ion-exchange chromatography (Obidairo and Akpochafo, 1984) from roots grown locally. The molecular weights of the three most active inhibitors were estimated to be 12,000, 10,000 and 9300, and they were found to show maximum activity at pH 7.5–8.5. Seven trypsin inhibitor fractions were demonstrated in American sweet potatoes by disc-gel electrophoresis (Dickey and Collins, 1984). The sweet potato obviously contains more than one protein exhibiting trypsin inhibitor activity (TIA); the final number of inhibitor fractions found may await development of even more refined methods of protein isolation and separation. The extent of trypsin inhibition varies between fractions (Obidairo and Akpochafo, 1984). Inhibitors were found to lose their activity when combined amino groups containing arginyl residues were chemically modified (Sugiura et al., 1973), suggesting that arginine groups may be important sites for interaction of the inhibitor with the enzyme. Fractions which strongly inhibited trypsin only weakly inhibited other proteinases such as plasmin and kallikrein and had no effect on chymotrypsin (Sugiura et al., 1973), nor was a chymotrypsin inhibitor present in nine Papua New Guinea cultivars which contained trypsin inhibitor activity (Bradbury et al., 1985b). However, a small amount of chymotrypsin inhibitor activity was detected in cultivars from the Solomon Islands (Bradbury et al., 1985a).

Measurement of activity

Details of methods vary, but all are based on the same principle. The proteolytic activity of trypsin when incubated with a pure protein such as casein (Lin and Chen, 1980) or other suitable substrate (Bradbury et al., 1984) is compared with its activity on the same protein in the presence of a sweet potato extract exhibiting TIA or of a trypsin inhibitor fraction. Alternatively the enzyme and inhibitor may be pre-incubated for a specific length of time before being added to the protein solution. The difference between control and test solutions is assessed in some way, for example by measuring the difference in absorbance of the solutions at a specific wavelength. The degree of inhibitor activity found is expressed in arbitrary units which vary according to the way the method is carried out and the particular researcher involved. Results of findings cannot therefore be compared quantitatively between different laboratories.

Within one laboratory (Bradbury et al., 1985a), measurement of the average TIA of a number of different root crops from the South Pacific in terms of trypsin inhibitor units per g root tissue, gave the following results: 25.4, 3.5, 27, 267, 3.0, 0.3, 0.56, 0.0 for sweet potato (Solomon Islands), sweet potato (Papua New Guinea), taro, giant taro, giant swamp taro, yautia, winged yam and Chinese yam, respectively (see Table 3.1, p. 122, for botanical names). Although standard deviations were high for all crops, it is apparent that sweet potato is generally higher in TIA than the other root crops with the exception of giant taro, which has a very active trypsin inhibitor. In a comparison of the TIA of various fruits and vegetables, sweet potato was found to be moderately inhibitory compared to the higher activity of potatoes and sweet corn, the lower activity of vegetables such as cucumber, cabbage, lettuce and cauliflower and the negligible activity shown by various fruits (Chen and Mitchell, 1973).

Factors affecting activity

The degree of TIA demonstrated by sweet potato samples depends on both cultivar and environment. The differences found, and their relationship with root total protein content, may be exploited to identify samples with consistently low levels of TIA.

Distribution of TIA has even been determined within individual roots. In four American cultivars examined there was a longitudinal gradient of TIA with highest activity in the proximal or stem end of the root (Dickey and Collins, 1984), but the gradients were not large enough to enable reliable elimination of TIA by the removal of one section. A cross-sectional gradient, the extent of which depended on cultivar, was also found, with TIA concentrated in the cortical region of all roots. Over 50% of TIA was found in the cortical region of 'Jewel' and 'Caromex' cultivars, a finding which the authors concluded could be useful when considering heat deactivation of the whole roots.

Trypsin inhibitor content was found to vary in a similar way between roots within one plant, between plants of the same cultivar and across environments for the same cultivar (Bradbury et al., 1985b). This variability was much larger than that due to experimental errors of measurement. Some cultivars showing either consistently low or consistently high TIA were identified.

Significant genetic variability in TIA has been reported (Lin and Chen, 1980; Bradbury et al., 1984; Dickey et al., 1984). Cultivar differences were suggested to be quantitative rather than qualitative in nature after the same seven trypsin inhibitor protein bands were found, by gel electrophoresis, in four different cultivars (Dickey and Collins, 1984). The existence and distribution of these seven fractions was constant in the four cultivars between roots harvested at different times, analysed before and after storage, or sampled in different years. The only difference between the protein bands appeared to be quantitative, with two of the bands particularly prominent in the two cultivars with the highest TIA levels.

The relationship between TIA and protein concentration has been explored in view of the fact that trypsin inhibitors are proteins and therefore might be expected to increase if efforts are made to breed for higher root protein levels. There have been conflicting findings by different authors with regard to this relationship.

A significant positive correlation between protein content and TIA was found for cultivars planted in Taiwan under similar conditions in a single season (Lin and Chen, 1980), or during four different seasons (Bouwkamp and Tsou, 1983). In contrast, other workers found no correlation between TIA and protein content in either North American (Dickey et al., 1984) or Papua New Guinea (Bradbury et al., 1984) samples. Extending the study with Papua New Guinea sweet potatoes to over 60 samples representing many cultivars and several environments confirmed the overall lack of a correlation between TIA and protein (Bradbury et al., 1985b). However, there was a positive correlation ($r=0.46$ to 0.96) within a particular cultivar and between cultivars grown in the same environment ($r=0.5$). No correlation was observed between TIA and skin or flesh colour.

A significant positive correlation was also observed between dry matter content and TIA in three out of the four seasons' plantings for Taiwanese samples (Bouwkamp and Tsou, 1983). In order to assess the potential for selecting low TIA–high protein types, multiple regression lines based on protein and dry matter contents and TIA, were drawn for each of the four seasons' plantings. Low TIA types were taken to be those which showed the greatest negative deviation from predicted TIA values. The selected lines were generally higher than average in dry matter, near the mean values for protein and much lower than the population mean for TIA. This procedure could represent a strategy for identifying consistently low TIA genotypes. However, an alternative could be to accept known levels of trypsin inhibitors and try to increase protein and dry matter levels, given that sweet potato trypsin inhibitors represent only a small fraction of total protein (Bradbury et al., 1984) and are present at levels only 1/50 to 1/100 that of the soybean trypsin inhibitor (Bouwkamp and Tsou, 1983).

Deactivation

It is important to identify the minimum conditions for destroying sweet potato trypsin inhibitors or reducing them to very low levels for both human and animal feeding. Heating is the usual method employed, but it is desirable to expend as little energy as possible in the case of sweet potato for animal feed, otherwise its use becomes uneconomic or limits the size of an animal farm. For both animal and human feeding, excessive use of heat may reduce availability of amino acids, notably lysine. Generally speaking normal methods of cooking for human consumption effectively destroy all or most of the TIA.

Investigations of time–temperature combinations which reduce or destroy inhibitors have been carried out on sweet potato extracts or fractions displaying inhibitor activity, and on chopped or whole roots. Isolated inhibitors from a Japanese sweet potato cultivar were reported to be thermostable, losing only 5% of their activity when heated at 70°C and only 30–35% even at 90°C for 30 min (Sugiura et al., 1973). However, thermostability at 70°C for 10 min was found to vary with cultivar (Lin and Chen, 1980) in crude extracts of inhibitors. This cultivar difference was also found for chopped root tissue (Dickey and Collins, 1984). Trypsin inhibitor activity was reduced by 50% between 75°C and 85°C in 'Pope' and 'Centennial', at 75°C in 'Caromex' and at only 65°C in 'Jewel'. The trypsin inhibitors did not appear to be heat stable in chopped tissue. Moreover, TIA was reduced below 10% of its original activity in all four cultivars when whole roots were boiled for 15 min (Dickey and Collins, 1984). Inhibitor activity was also reported by

Australian researchers to be largely destroyed by boiling or baking roots until they are cooked (Bradbury et al., 1985b). Inhibitor activity in the most active individual fractions was completely destroyed by boiling for 40 min (Obidairo and Akpochafo, 1984). A crude extract needed 30 min at 130°C for almost complete destruction (Chien and Lee, 1980). The inhibitors therefore seem more sensitive to heat in the whole root than when isolated; perhaps the presence of other root constituents increases their thermolability.

Trypsin inhibitor activity was found to be little affected by sun-drying of roots (Chien and Lee, 1980). Preparation of dried chips for animal feed by this method as practised in some areas would not be effective alone in destroying TIA. Pressing of juice from sweet potato roots only partially succeeded in expelling TIA, 67% of which remained in the residue (Chien and Lee, 1980).

Addition of partially purified sweet potato trypsin inhibitor to rat diets containing 10% protein significantly reduced the protein efficiency ratios (see p. 135) of the diets (Tsou, Kan and Wang, 1987). Pre-heating to destroy TIA enhanced the nutritive value of sweet potato as demonstrated with rats (Yang, 1982). A feed containing pre-heated sweet potato gave higher feed efficiency, nitrogen retention and growth rate than one containing raw sweet potato.

Relevance to human feeding

Although no harmful effects have been reported in communities which eat raw roots, for example as a between-meal snack, the presence of trypsin inhibitors could decrease protein digestion and utilization in people whose protein intake may already be low. Roots should preferably be properly cooked before consumption.

The disease *Enteritis necroticans* (EN) is endemic in the Papua New Guinea Highlands, where until recently it was the main cause of death in children over the age of 1 year. EN has also been reported in Africa, Southeast Asia and America, and in Europe before 1949, particularly during World War II (Lawrence and Walker, 1976). It develops as a result of damage to the gut wall caused by a protein beta-toxin produced by *Clostridium perfringens* Type C (originally known as *C. welchii* Type C) in the gut. The Highland Papua New Guinea diet, being based on sweet potato as the staple food, is very low in protein (see Chapter 3). An occasional meal high in protein, usually pig meat, eaten at, for example, a feast, can induce EN, hence the local name for the disease 'pigbel'. It is thought that the sudden intake of a large quantity of protein causes rapid growth of *C. perfringens* Type C, whose origin may be the meat or which may already be present in the gut, with the subsequent production of beta-toxin (Bradbury et al., 1984). Studies with monkeys have shown that severely protein-deficient diets cause a lowering of protease production in the intestine (Gyr et al., 1975). This could result in a reduction in the attack on, and destruction of, beta-toxin by trypsin and chymotrypsin. Experimental pigbel was produced in guinea pigs on a low protein diet in conjunction with intake of raw sweet potatoes (Lawrence and Cooke, 1980). It has been postulated that the presence of heat-stable trypsin inhibitors in sweet potatoes eaten by Highland Papuans, combined with their low protein diet and consequent low intestinal protease activity, might retard the tryptic breakdown of the beta-toxin produced by *C. perfringens* Type C hence causing pigbel (Lawrence and Walker, 1976). Anti-tryptic activity in sweet potato cooked by Papua New Guineans was reported at Goroka in the Highlands (Lawrence and Walker, 1976).

Given the difficulties of administering a vaccine available against EN to more than 50% of the population at risk, the possibility that sweet potato TIA was involved in pigbel was investigated (Bradbury et al., 1984). This was done indirectly by determining the trypsin inhibitor contents of cultivars from low and high incidence EN regions of the Papua New Guinea Highlands (Bradbury et al., 1984). However, there was no significant difference in mean TIA of raw cultivars from the two regions. The authors concluded that if the incidence of EN had simply been related to high TIA, the TIA should have been greater in the region with a high incidence of EN. However, the effect of the difference in altitude at which the two communities lived on the boiling point of water and extent of TI destruction during cooking was not investigated. The community with a high incidence of EN lived at twice the altitude of, and on a more monotonous diet than, that with a low incidence of EN.

The presence of trypsin inhibitors in sweet potato roots appears to present little problem to human consumers provided roots are thoroughly cooked to reduce TIA to a low level or destroy it. The effects of boiling at temperatures below 100°C on TIA in sweet potatoes cooked by communities living at high altitudes should be investigated.

Flatulence factors

Formation of flatus as a result of physiological digestive processes occurs when the microflora of the colon or large intestine ferment certain substrates, not previously absorbed in the upper digestive tract, to produce gases. These gases, namely carbon dioxide, hydrogen and sometimes methane, can accumulate until pressure forces the sphincter muscle to relax and allows them to leave the body via the anus. The increase in gas pressure in the intestine can cause varying amounts of discomfort and even pain, although complaints in this respect are necessarily subjective and may depend on the individuals pain threshold as well as the actual magnitude of gas production. The association of intake of a particular food with the consequent production of flatus may decrease the popularity of that food to such an extent that it is omitted from the diet, thus depriving that person of an otherwise valuable source of nutrients. In that case flatulence factors are acting as anti-nutritional agents. Investigations into their nature, concentration and means of elimination or destruction are hence important to remove the grounds for at least one source of discrimination against the food.

Flatulence caused by sweet potatoes

It seems to be quite generally recognized that consumption of sweet potatoes causes flatulence which can be severe (Palmer, 1982). Authors mention that flatulence production by sweet potato is well known to the Asian consumer, for example (Tsou and Yang, 1984). Breaking of wind as a result of eating sweet potatoes has featured in a number of anecdotes and stories from previous centuries in Japan (Duell, 1984). The occurrence of flatulence has been suggested as a possible factor contributing to the low acceptability of sweet potato and the resistance shown by consumers to increasing sweet potato intake (Tsou and Villareal, 1982).

However, there has been little attempt to investigate the extent to which flatus formation following consumption of sweet potato leads to its rejection as a food. During a survey designed to identify social and cultural factors leading to acceptance or rejection of sweet potatoes in a community in North Carolina, United States, the most common reason given, by 14 out of 100 respondents, for disliking sweet potatoes was 'indigestion', interpreted by the investigator (Fitzgerald, 1976) as flatulence.

Evidence for flatulence factors in sweet potatoes

There are no reports of clinical tests with human subjects to determine the truth of the flatulence-producing effects ascribed to sweet potatoes. However, sweet potato has been shown to induce flatus in rats (Tsou and Yang, 1984). The exhaled hydrogen produced by rats kept in a metabolic chamber and fed on raw or cooked dried sweet potato chips was determined. Volumes of hydrogen varied from 26 to 71 ml/g dry sweet potato (raw) and 3.2–58 ml/g dry sweet potato (cooked) for various different cultivars. Respiratory hydrogen volume produced by rats fed various sweet potato carbohydrate fractions was so much greater than that produced on a corn diet that sweet potato was concluded to be a flatus inducer.

Components which may act as flatulence factors

Flatulence is a common phenomenon associated with the ingestion of legumes. The oligosaccharides raffinose, stachyose and verbascose have been heavily implicated in flatus production from legumes (Rackis, 1975). Soybeans contain (by weight) about 1% raffinose and 2.5% stachyose (Rackis, 1975) and winged bean 1–2% raffinose, 2–4% stachyose and 0.2–1% verbascose (Garcia and Palmer, 1980). Other legumes contain verbasose as the major oligosaccharide. Thus the concentrations of these compounds have been investigated in sweet potatoes as a first step in the search for a flatulence factor. However, an oligosaccharide-free residue from beans was also shown to induce flatus in rats (Wagner et al., 1976). In addition, it has been reported that a variety of polysaccharides from the plant cell wall can induce flatus (Salyers, Palmer and Balascio, 1979). A further group of workers showed that among unabsorbed carbohydrates reaching the colon from dietary sources is as much as 20% of undigested starch (Stephen, Haddah and Phillips, 1983). This could also be a fermentable source of carbohydrate for intestinal microorganisms. In other words, any food polysaccharide which finds its way into the colon is a potential source of flatus (Palmer, 1982).

Oligosaccharides have been eliminated as major factors leading to flatulence with sweet potatoes. They were found to be absent (Roxas, Fukuba and Mendoza, 1985) or present in very low concentrations (Palmer, 1982; Truong, Biermann and Marlett, 1986) by most authors. Raw and cooked Philippine sweet potato roots contained 0.23–0.4% cellobiose, negligible raffinose, verbascose in only trace amounts and no detectable stachyose (Truong et al., 1986). Raffinose was reported at 0.5% of the fresh weight in baked sweet potatoes (Palmer, 1982). Nine cultivars grown in Taiwan contained raffinose at levels ranging from undetectable to 1.08% (dwb) and, whilst seven of the nine did not contain detectable amounts of stachyose, two cultivars contained 0.2% and 0.99% (dwb) (Tsou and Yang, 1984). However, concentrations are low compared with those in legumes such as soybeans. A soluble sugar fraction, extracted from the sweet potatoes and containing the oligosaccharides, when fed as part of a diet to rats kept in a metabolic chamber failed to induce more than an insignificant quantity of respiratory hydrogen (Tsou and Yang, 1984).

Starch and dietary fibre fractions, however, produced comparably large respiratory hydrogen volumes from the rats. Since starch accounted for 50% of sweet potato dry matter, the authors concluded that about 85% of the hydrogen gas produced resulted from the ingestion of starch and the remaining 15% from dietary fibre. Isolated starches of sweet potato, banana and potato fed to rats produced similar flatulence-sized volumes of exhaled hydrogen, significantly greater than those from wheat, cassava, mungbean, rice or corn starches (Tsou et al., 1987). For a series of whole sweet potato samples, hydrogen production was significantly correlated with starch content ($r=0.622$) (Tsou and Yang, 1984). Hydrogen production rate reached its maximum level 10 hours, and ended 20 hours, after feeding.

The same investigation found that cooking whole sweet potato or sweet potato starch reduced, but did not entirely eliminate, hydrogen production. Fully gelatinizing starch greatly reduces its flatus production. However, the residual flatulence after cooking could be associated with a fraction of the starch remaining resistant to digestive enzymes and hence entering the colon; in addition, substances in the dietary fibre could also be responsible for flatus gases. The cooking effect on gas-producing properties of sweet potato has been found to vary with cultivar (Tsou and Yang, 1984). In addition the quantities of starch from test meals passing unabsorbed through the ileum vary among different human subjects (Stephen et al., 1983). The chemical form of this resistant starch was not investigated.

Measures to reduce flatulence formation

More study is needed of the impression that flatus necessarily accompanies human ingestion of sweet potato. Variations in both root characteristics and human responses to them should be determined. Although the study described above indicated that both starch and constituents of dietary fibre may play a part in flatus induction, further investigations into the exact chemical nature of the flatus factors are required.

Improved techniques of flatus gas measurement giving more reproducible results which could be used to compare flatulence properties of different sweet potato cultivars are needed (Tsou and Yang, 1984). Knowledge of the characteristics of flatulence factors and genetic variations in concentrations could lead to breeding or selection for low-flatulence lines.

Storage and processing methods which can eliminate flatulence should be explored (Tsou and Villareal, 1982). The production of sweet potatoes without distasteful characteristics could be one avenue of approach to the problem of popularizing this root and fully exploiting its nutritional potential in a hungry world.

References

Akazawa, T. and Wada, K. 1961. Analytical study of ipomeamarone and chlorogenic acid alterations in sweet potato roots infected by *Ceratocystis fimbriata*. *Plant Physiol.* **36** (2): 139–44.

Arinze, A.E. and Smith, I.M. 1980. Antifungal furanoterpenoids of sweet potato in relation to pathogenic and non-pathogenic fungi. *Physiol. Plant Pathol.* **17** (2): 145–55.

Bartolini, P.U., Hirose, S. and Sawayama, S. 1984. Root crop survey in Visayas (Leyte and the Camotes Islands, Cebu) B. The Camotes Islands survey. In: Uritani, I. and Reyes, E.D. (eds.), *Tropical root crops – postharvest physiology and processing*. Japan Scientific Societies Press, Tokyo, p. 23.

Bouwkamp, J.C. and Tsou, S.C.S. 1983. Trypsin inhibitors in sweet potatoes: genotype and environment effects. *Proc. Am. Soc. Hort. Sci. Trop. Reg.* **27B**: 126–35.

Boyd, M.R. 1977. Evidence for the Clara cell as a site of cytochrome P450-dependent mixed-function oxidase activity in lung. *Nature (Lond.)* **269**: 713–15.

Boyd, M.R., Burka, L.T., Harris, T.N. and Wilson, B.J. 1974. Lung-toxic furanoterpenoids produced by sweet potatoes (*Ipomoea batatas*) following microbial infection. *Biochim. Biophys. Acta* **337** (2): 184–95.

Boyd, M.R., Burka, L.T. and Wilson, B.J. 1975. Distribution, excretion and binding of radioactivity in the rat after intraperitoneal administration of the lung-toxic furan, [^{14}C]4-ipomeanol. *Toxicol. Appl. Pharmacol.* **32** (1): 147–57.

Boyd, M.R. and Wilson, B.J. 1971. Preparation and analytical gas chromatography of ipomeamarone, a toxic metabolite of sweet potatoes (*Ipomoea batatas*). *J. Agric. Food Chem.* **19** (3): 547–50.

Bradbury, J.H., Baines, J., Hammer, B., Anders, M. and Millar, J.S. 1984. Analysis of sweet potato (*Ipomoea batatas*) from the highlands of Papua New Guinea: relevance to the incidence of *Enteritis necroticans*. *J. Agric. Food Chem.* **32** (3): 469–73.

Bradbury, J.H., Beatty, R.E., Bradshaw, K., Hammer, B., Holloway, W.D., Jealous, W., Lau, J., Nguyen, T. and Singh, U. 1985a. Chemistry and nutritive value of tropical root crops in the South Pacific. *Proc. Nutr. Soc. Austr.* **10**: 185–8.

Bradbury, J.H., Hammer, B., Nguyen, T., Anders, M. and Millar, J.S. 1985b. Protein quantity and quality and trypsin inhibitor content of sweet potato cultivars from the highlands of Papua New Guinea. *J. Agric. Food Chem.* **33** (2): 281–5.

Buckpitt, A.R. and Boyd, M.R. 1978. Xenobiotic metabolism in birds, species lacking pulmonary Clara cells. *Pharmacologist* **20** (3): 181.

Burka, L.T., Kuhnert, L., Wilson, B.J. and Harris, T.M. 1974. 4-Hydroxy-myoporone, a key intermediate in the biosynthesis of pulmonary toxins produced by *Fusarium solani* infected sweet potatoes. *Tetrahedron Lett.* **46**: 4017–20.

Burka, L.T., Kuhnert, L., Wilson, B.J. and Harris, T.M. 1977. Biogenesis of lung-toxic furans produced during microbial infection of sweet potatoes (*Ipomoea batatas*). *J. Am. Chem. Soc.* **99** (7): 2302–5.

Catalano, E.A., Hasling, V.C., Dupuy, H.P. and Constantin, R.J. 1977. Ipomeamarone in blemished and diseased sweet potatoes (*Ipomoea batatas*). *J. Agric. Food Chem.* **25** (1): 94–6.

Catalano, E.A., Hasling, V.C., Pons, W.A. and Schuller, W.H. 1979. Analysis of sweet potato products for lung edema toxins (Abstr.). *HortScience* **14**: 124–5.

Chen, I. and Mitchell, H.L. 1973. Trypsin inhibitors in plants. *Phytochemistry* **12** (2): 327–30.

Chien, S.L. and Lee, P.K. 1980. [The effect of physical treatment on the available lysine and trypsin inhibitor of sweet potatoes] Chinese. *Taiwan Livestock Res.* **13** (1): 75–84.

Clark, C.A., Lawrence, A. and Martin, F.A. 1981. Accumulation of furanoterpenoids in sweet potato tissue following inoculation with different pathogens. *Phytopathology* **71** (7): 708–11.

Cody, M. and Haard, N.F. 1976. Influence of cooking on toxic stress metabolites in sweet potato root. *J. Food Sci.* **41** (2): 469–70.

Coxon, D.T., Curtis, R.F. and Howard, E. 1975. Ipomeamarone, a toxic furanoterpenoid in sweet potatoes (*Ipomoea batatas*) in the United Kingdom. *Food Cosmet. Toxicol.* **13** (1): 87–90.

Curtis, R.F. 1986. Plant toxicants in the human diet. In: Battaglia, R. (ed.), *Proceedings of the European Food Toxicologists II. Interdisciplinary Conference on Natural Toxicants in Food*, University of Zurich, pp. 1–14.

Davis, N.D., Wagener, R.E., Dalby, D.K., Morgan-Jones, G. and Dibner, U.L. 1975. Toxigenic fungi in food. *Appl. Microbiol.* **30** (1): 159–61.

Dickey, L.F. and Collins, W.W. 1984. Cultivar differences in trypsin inhibitors of sweet potato roots. *J. Am. Soc. Hort. Sci.* **109** (5): 750–4.

Dickey, L.F., Collins, W.W., Young, C.T. and Walter, W.M. 1984. Root protein quantity and quality in a seedling population of sweet potatoes. *HortScience* **19** (5): 689–92.

Doster, A.R., Farrell, R.L. and Wilson, B.J. 1983. An ultrastructural study of bronchiolar lesions in rats induced by 4–ipomeanol, a product from mold-damaged sweet potatoes. *Am. J. Pathol.* **111** (1): 56–61.

Doster, A.R., Mitchell, F.E., Farrell, R.L. and Wilson, B.J. 1978. Effects of 4–ipomeanol, a product from mold-damaged sweet potatoes, on the bovine lung. *Vet. Pathol.* **15** (3): 367–75.

Duell, B. 1984. Anthropological problems connected with the introduction and diffusion of the sweet potato into Japan (II). *J. Int. College Comm. Econ.* **29**: 51–73.

Dutcher, J.S. and Boyd, M.R. 1979. Species and strain differences in target organ alkylation and toxicity by 4–ipomeanol. Predictive value of covalent binding in studies of target organ toxicities by reactive metabolites. *Biochem. Pharmacol.* **28** (23): 3367–72.

Fitzgerald, T.K. 1976. *Ipomoea batatas*: the sweet potato revisited. *Ecol. Food Nutr.* **5**: 107–14.

Garcia, V.V. and Palmer, J.K. 1980. Carbohydrates of winged beans, *Psophocarpus tetragonologus* (L). *J. Food Technol.* **15**: 477–84.

Gyr, K., Wolf, R.H., Imondi, A.R. and Felsenfeld, O. 1975. Exocrine pancreatic function in protein-deficient patas monkeys studied by means of a test meal and an indirect pancreatic function test. *Gastroenterology* **68** (3): 488–94.

Hansen, A.A. 1928. Potato poisoning. *N. Am. Vet.* **9**: 31–4.

Hiura, M. 1943. [Studies on storage and rot of sweet potato (2)] Japanese. *Rep. Gifu Agric. Coll.* **50**: 1–5.

Kim, W.K., Oguni, I. and Uritani, I. 1974. Phytoalexin induction in sweet potato roots by amino acids. *Agric. Biol. Chem.* **38** (12): 2567–8.

Kojima, M., Kawakita, K. and Uritani, I. 1982. Studies on a factor in sweet potato root which agglutinates spores of *Ceratocystis fimbriata*, black rot fungus. *Plant Physiol.* **69**: 474–8.

Kojima, M. and Uritani, I. 1976. Possible involvement of furanoterpenoid phytoalexins in establishing host-parasite specificity between sweet potato and various strains of *Ceratocystis fimbriata*. *Physiol. Plant Pathol.* **8**: 97–111.

1978a. Studies on factor(s) in sweet potato which specifically inhibits germ tube growth of incompatible isolates of *Ceratocystis fimbriata*. *Plant Cell Physiol.* **19**: 71–81.

1978b. Isolation and characterization of factors in sweet potato root which agglutinate germinated spores of *Ceratocystis fimbriata*, black rot fungus. *Plant Physiol.* **62**: 751–3.

Kondo, S. 1971. [Studies on poisoning caused (by) black-rotten sweet potatoes in domestic animals] Japanese. *Bull. Fac. Agric., Tokyo Univ. Agric. Technol.* No. 15.

Kubota, T. and Ichikawa, N. 1954. On the chemical constitution of ipomeanine, a new ketone from the black-rotten sweet potatoes. *Chem. Ind.*, pp. 902–3.

Kubota, T. and Matsuura, T. 1953. [Chemical studies on black rot disease of sweet potato] Japanese. *J. Chem. Soc. Japan* **74**: 101–9, 197–9, 248–51, 668–70.

Kuć, J. 1972. Compounds accumulating in plants after infection. In: Kadis, S., Ciegler, A. and Ajl, S. (eds.), *Microbial toxins: a comprehensive treatise*. VIII *Fungal toxins*. Academic Press, New York, pp. 211–48.

Kudo, T. and Hidaka, T. 1984. [Removal of bitter components from shochu produced of defected sweet potatoes by carbon treatment] Japanese. *J. Brewing Soc. Japan* **79** (12): 900–1.

Kunitz, M. 1945. Crystallization of a trypsin inhibitor from the soyabean. *Science* **101** (2635): 668–9.

Lawrence, G. and Cooke, R. 1980. Experimental pigbel: the production and pathology of necrotizing enteritis due to *Clostridium welchii* Type C in the guinea-pig. *Br. J. Exp. Pathol.* **61** (3): 261–71.

Lawrence, G. and Walker, P.D. 1976. Pathogenesis of *Enteritis necroticans* in Papua New Guinea. *Lancet* **1** (7951): 125–6.

Lin, Y-H. and Chen, H-L. 1980. Level and heat stability of trypsin inhibitor activity among sweet potato (*Ipomoea batatas* L.) varieties. *Bot. Bull. Acad. Sinica* **21** (1): 1–13.

Lin, Y-H., Cheng, J-F. and Fu, H-Y. 1983. Partial purification and some properties of trypsin inhibitors of sweet potato (*Ipomoea batatas* Lam.) roots. *Bot. Bull. Sinica* **24** (2): 103–13.

Liu, C-I. 1982. Moldy sweet potato related respiratory distress in cattle. *J. Chin. Soc. Vet. Sci.* **8** (2): 155–9.

Martin, W.J., Hasling, V.C. and Catalano, E.A. 1976. Ipomeamarone content in diseased and nondiseased tissues of sweet potatoes infected with different pathogens. *Phytopathology* **66** (5): 678–9.

Martin, W.J., Hasling, V.C., Catalano, E.A. and Dupuy, H.P. 1978. Effect of sweet potato cultivars and pathogens on ipomeamarone content of diseased tissue. *Phytopathology* **68**: 863–5.

Müller, K.O. and Börger, H. 1940. [Experimental investigation into the *Phytophthora*-resistance of the potato] German. *Arb. Biol. Reichanstalt. Landw. Forstw. Berlin* **23**: 189–231.

Obidairo, T.K. and Akpochafo, O.M. 1984. Isolation and characterization of some proteolytic enzyme inhibitors in sweet potato (*Ipomoea batatas*). *Enzyme Microb. Technol.* **6**: 132–4.

Palmer, J.K. 1982. Carbohydrates in sweet potato. In: Villareal, R.L. and Griggs, T.D. (eds.), *Sweet potato*, Proceedings of the First International Symposium. AVRDC, Shanhua, T'ainan, pp. 137–8.

Peckham, J.C., Mitchell, F.E., Jones, O.H. and Doupnik, B. 1972. Atypical interstitial pneumonia in cattle fed moldy sweet potatoes. *J. Am. Vet. Med. Assoc.* **160**: 169–72.

Rackis, J.J. 1975. Oligosaccharides of food legumes: alpha-galactosidase activity and the flatus problem. In: Jeanes, A. and Hodge, J. (eds.), *Physiological effects of food carbohydrates*. Am. Chem. Soc., Washington, DC, pp. 207–22.

Roxas, M.J.T., Fukuba, H. and Mendoza, E.M.T. 1985. The absence of oligosaccharides in storage roots of sweet potato (*Ipomoea batatas* L.). *Phil. J. Crop Sci.* **10** (3): 161–3.

Ryan, C.A. 1981. Proteinase inhibitors. In: Stumpf, P.K. and Conn, E.E. (eds.), *The biochemistry of plants*, Vol. 6 *Proteins and nucleic acids*. Academic Press, Inc., New York, pp. 351–70.

Salyers, A., Palmer, J.K. and Balascio, J. 1979. Digestion of plant cell wall polysaccharides by bacteria from the human colon. In: Inglett, G. and Falkehag, S.T. (eds.), *Dietary fibers: chemistry and nutrition*. Academic Press, New York, pp. 193–201.

Schneider, J.A., Lee, J., Naya, Y., Nakanishi, K., Oba, K. and Uritani, I. 1984. The fate of the phytoalexin ipomeamarone: furanoterpenes and butenolides from *Ceratocystis fimbriata* infected sweet potatoes. *Phytochemistry* **23** (4): 759–64.

Seawright, A.A. and Mattocks, A.R. 1973. The toxicity of two synthetic 3-substituted furan carbamates. *Experientia* **29** (10): 1197–200.

Sohonie, K. and Bhandarkar, A.P. 1954. Trypsin inhibitors in Indian food-stuffs. Part 1. Inhibitors in vegetables. *J. Sci. Ind. Res.* **13B**: 500–3.

Stephen, A.M., Haddah, A.C. and Phillips, S.F. 1983. Passage of carbohydrate into the colon. Direct measurements in humans. *Gastroenterology* **85** (3): 589–95.

Sugiura, M., Ogiso, T., Takeuti, K., Tamura, S. and Ito, A. 1973. Studies on trypsin inhibitors in sweet potato. I. Purification and some properties.

Biochim. Biophys. Acta **328** (2): 407–17.

Suzuki, H., Oba, K. and Uritani, I. 1975. The occurrence and some properties of 3-hydroxy-3-methyl-glutaryl coenzyme A reductase in sweet potato roots infected by *Ceratocystis fimbriata. Physiol. Plant Pathol.* **7** (3): 265–76.

Truong, V.D., Biermann, C.J. and Marlett, J.A. 1986. Simple sugars, oligosaccharides and starch concentrations in raw and cooked sweet potato. *J. Agric. Food Chem.* **34** (3): 421–5.

Tsou, S.C.S., Kan, K-K. and Wang, S-J. 1987. Biochemical studies on sweet potato for better utilization at AVRDC. Paper presented at International Sweet Potato Symposium, 20–26 May, ViSCA, Baybay, Leyte.

Tsou, S.C.S. and Villareal, R.L. 1982. Resistance to eating sweet potato. In: Villareal, R.L. and Griggs, T.D. (eds.), *Sweet potato*, Proceedings of the First International Symposium. AVRDC, Shanhua, T'ainan, pp. 39, 42, 43.

Tsou, S.C.S. and Yang, M.H. 1984. Flatulence factors in sweet potato. *Acta Horticult.* **163**: 179–86.

Uritani, I. 1990. Biochemical aspects of phytoalexins in sweet potato. In: *Tropical root and tuber crops changing role in a modern world.* Proceedings of the Eighth Symposium of the International Society for Tropical Root Crops, 30 October–5 November, 1988, Bangkok (in press).

Uritani, I. and Akazawa, T. 1955. Antibiotic effect on *Ceratostomella fimbriata* of ipomeamarone, an abnormal metabolite in black rot. *Science* **121**: 216–17.

Uritani, I., Oba, K., Takeuchi, A., Sato, K., Inoue, H., Ito, R. and Ito, I. 1981. Biochemistry of furano-terpenes produced in mold-damaged sweet potatoes. In: Ory, R.L. (ed.), *Antinutrients and natural toxicants in foods.* Food & Nutrition Press, Inc., Westport, CT, pp. 1–16.

Uritani, I., Saito, T., Hondo, H. and Kim, W.K. 1975. Induction of furanoterpenoids in sweet potato roots by the larval components of the sweet potato weevils. *Agric. Biol. Chem.* **39** (9): 1857–62.

Uritani, I, Suzuki, S. and Muramatsu, K. 1947. Toxic effects on microorganisms of bitter substances in black-rotted sweet potato. *Agric. Hortic.* **22**: 515–17, 554–6.

Uritani, I., Uritani, M. and Yamada, H. 1960. Similar metabolic alterations induced in sweet potato by poisonous chemicals and by *Ceratocystis fimbriata. Phytopathology* **50**: 30–4.

Wagner, J.R., Becker, M.R., Gumbmann, M.R. and Olson, A.C. 1976. Hydrogen production in the rat following ingestion of raffinose, stachyose and oligosaccharide-free bean residue. *J. Nutr.* **106** (4): 466–70.

Wilson, B.J. 1973. Toxicity of mold-damaged sweet potatoes. *Nutr. Rev.* **31**: 73–8.

1982. Mycotoxins and toxic stress metabolites of fungus-infected sweet potatoes (*Ipomoea batatas*). In: Hathcock, J.N. (ed.), *Nutritional toxicology.* Academic Press, New York, pp. 239–302.

Wilson, B.J., Boyd, M.R., Harris, T.M. and Yang, D.T.C. 1971. A lung oedema factor from mouldy sweet potatoes (*Ipomoea batatas*). *Nature (Lond.)* **231**: 52–3.

Wilson, B.J. and Burka, L.T. 1979. Toxicity of novel sesquiterpenoids from the stressed sweet potato (*Ipomoea batatas*). *Food Cosmet. Toxicol.* **17** (4): 353–5.

1983. Sweet potato toxins and related toxic furans. In: Keeler, R.F. and Tu, A.T. (eds.), *Handbook of natural toxins*, Vol. I *Plant and fungal toxins*. Marcel Dekker Inc., New York, pp. 3–41.
Wilson, B.J., Yang, D.T.C. and Boyd, M.R. 1970. Toxicity of mold-damaged sweet potatoes (*Ipomoea batatas*). *Nature (Lond.)* **227**: 521–2.
Wood, G. and Huang, A. 1975. Detection and quantitative determination of ipomeamarone in damaged sweet potatoes (*Ipomoea batatas*). *J. Agric. Food Chem.* **23** (2): 239–41.
Yang, T.H. 1982. Sweet potato as a supplemental staple food. In: Villareal, R.L. and Griggs, T.D. (eds.), *Sweet potato*, Proceedings of the First International Symposium. AVRDC, Shanhua, T'ainan, pp. 31–4.
Yang, Y.S. and Yu, P.H. 1976. [Studies on the contamination of furanoterpenoids in sweet potato in Taiwan] Chinese. *Plant Protect. Bull.* **18** (1): 1–12.

CHAPTER 5

Postharvest procedures: I. Storage and cooking

Postharvest handling of sweet potatoes begins at the moment of harvest and therefore is inevitable in some form. It ranges in degree from simple lifting of roots, carrying them from field to house and immediate consumption after cooking, to sophisticated methods of curing and storage under controlled conditions followed by processing into a variety of high quality food products. Postharvest saleability, quality and nutritional value of roots or leaves and the presence or absence in roots of bitter, toxic furanoterpenoid phytoalexins or mycotoxins depends greatly on the degree and types of treatment to which produce is subjected. Climatic and soil conditions before harvest and contamination or attack by microorganisms or insect pests in the field may initiate or enhance subsequent postharvest deterioration. Careless postharvest handling can lead to both quantitative and qualitative losses which may be extremely high in some circumstances. Research has concentrated on the improvement of preharvest conditions to increase yield and lower disease rates. However, such efforts are wasted unless they go hand in hand with others designed to reduce the high degree of loss associated with careless postharvest handling.

Sweet potatoes have a high moisture content, and a relatively thin and delicate skin. They remain metabolically active after harvest and are an easily damaged, highly perishable, commodity, which makes their postharvest handling and storage more difficult than that of, for example, the dry grain crops. In some situations, therefore, they may be processed into a dry form before storage as an alternative to the more difficult or expensive procedure of storing the fresh produce.

Because of the difficulties of storing sweet potato roots, storage avoidance is practised in many parts of the tropics. In other words, roots are harvested when and as they are needed (sometimes being left in the ground for several months after attaining adequate size), and used directly for consumption as soon as they are harvested. In Papua New Guinea, roots of a desirable size are removed from individual plants leaving smaller roots to increase in size for later harvest. In Uganda, the largest roots are removed first, the remainder being harvested 3 or 4 months later. Alternatively subsistence farmers may plant a range of sweet potato cultivars suitable for harvesting at different times, or sweet potato may form part of a series of crops which can be harvested successively, thus ensuring a constant supply of food.

After harvest in the tropics, roots may be left for a short period in the sun to dry. If they are to be sold they are usually transported to market almost immediately. However, lack of proper curing, packing and storage leads to skinning, wounding, dessication and attack by pathogens, resulting in losses and such reduction in quality that many roots are unmarketable, or command only a very low price.

Storage is necessary in some circumstances, for example to extend availability of fresh roots throughout the year where production is essentially seasonal. In temperate regions production is impossible in the winter (Morris, 1981) and in Bangladesh during the wet monsoon season (Jenkins, 1982), because the crop is sensitive to chilling and waterlogging, respectively. Storage can also be used to avoid gluts (occurring, for example, in parts of India where sweet potato production is confined to one season) which cause a sharp drop in price, depriving growers of a satisfactory return (Prasad, Srinivasan and Shanta, 1981). A limited amount of storage is employed in parts of the Papua New Guinea Highlands to overcome shortages expected as a result of frosts (Keleny, 1965).

Recipes for cooked sweet potato dishes from various parts of the world can be found in Appendix 1. Cooking increases palatability by producing changes in texture and flavour as well as improving the digestibility of starch and increasing the availability of certain nutrients. Heat treatment of roots may be necessary to reduce levels of antinutrients such as trypsin inhibitors (see Chapter 4).

Postharvest handling operations produce changes in the quality and nutritional value of sweet potato roots and leaves. Such changes have hardly been studied in leaves, which for human consumption are either eaten directly by producers or marketed mainly in a fresh form. Rough handling during transport to market can result in damage and bruising of leaves; exposure to high temperatures and direct sunlight during sale can produce drying and wilting, resulting in loss of quality and nutrients. Prolonged cooking can reduce levels of nutrients such as water-soluble vitamins to low levels. These are lost unless the cooking water is utilized for soup. Studies on nutritional changes during postharvest handling of leaves are needed in order better to devise ways of minimizing quality and nutrient losses for human feeding.

General compositional changes during curing and storage of roots have been extensively studied, but largely in the temperate regions, under controlled conditions of temperature and humidity. As will be described below, roots are subject to a variety of practices which, especially in the tropics, do little to enhance keeping quality or nutritional value of the stored produce. Even under controlled environment storage in the United States significant losses occur. These were reported to be 19% during storage and wholesaling, 4% during retailing and 12% within the household (that is a total of about 35%) (Friedman, 1960). A later study in New York showed similar retail and household losses (Ceponis and Butterfield, 1974). Higher percentage losses are bound to occur in the technologically less advanced areas of the world where simpler and less effective storage methods are used. For example, for sweet potatoes stored by traditional methods in Bangladesh losses in root fresh weight were estimated to be 30–35% after 3–4 months of storage (Jenkins, 1982).

This chapter briefly discusses methods of storage and cooking used in various parts of the world in relation to sweet potato in the human diet. It reviews the information at present available regarding both qualitative and nutritional changes taking place during these operations, and suggests areas of interest for future research.

A. Curing and storage

Methods and improvements

Reasons for storage losses

Quantitative or qualitative losses or a combination of the two arising from postharvest storage result from physical, physiological or pathological factors or various combinations of these (Booth, 1974). Physical factors include mechanical damage, much of which is sustained during harvest itself. Harvesting in the tropics is usually manual, employing a variety of implements such as digging sticks, spades, hoes and machetes or knives. Mechanical harvesting, using tractor- or animal-drawn ploughs, or specially designed machines, is confined to areas of large-scale production. Sweet potato roots are often cut, grazed, skinned and bruised by the harvesting implement. In India 24% of roots harvested using a spade were found to be damaged after lifting and handling (Prasad et al., 1981). A similar estimate for the percentage of damaged roots has been made for some parts of China where hoes are used for harvesting (Zhang, D.P., personal communication). In countries where roots are transported loose or in sacks by a variety of means such as lorry, motor cycle, cart or mule back over rough road surfaces to the storage area, increased root damage and loss of quality frequently results.

When roots are left in the hot sun after harvest, or during uncontrolled conditions of storage are exposed to high temperatures, moisture losses and the susceptibility to decay increase. Some cultivars develop pithiness, a texture defect resulting from an increase in the volume of intercellular spaces in root tissue. Cold wet soils before harvest or subsequent exposure to temperatures below 10°C cause chilling damage resulting in tissue breakdown, development of off-flavours, and/or 'hardcore', an internal disorder manifested as permanently hard areas of flesh after cooking. Chilling also renders roots more susceptible to attack by microorganisms such as certain fungi, perhaps through an increased sensitivity of the tissue to pathogenic pectic enzymes and to a reduced capacity for synthesizing phytoalexins (Arinze and Smith, 1982).

There is a natural respiratory loss of dry matter during storage as well as a transpiratory (or wilting) loss of water. Sweet potatoes stored under controlled conditions have been shown to lose water and CO_2 in such a way that the dry matter:water ratio changes little (Kushman and Wright, 1969), but respiratory losses are much greater under tropical temperatures (Jenkins, 1982). High temperatures also encourage sprouting with consequent increases in water and respiratory losses.

Pre- or postharvest attack by pathogenic microorganisms (fungi, bacteria and to a lesser extent viruses) is a very serious cause of postharvest losses of storage roots. The relative importance of the major pathogens can differ considerably between localities and with time of year. For example, *Ceratocystis fimbriata*, which causes black rot of sweet potatoes in the United States, is not found in Australia. Conversely *Pythium ultimum*, a major rot in Australia, is not found in the United States (Morris, 1981). Physical damage and physiological changes in the roots increase susceptibility of produce to pathogenic attack, as microorganisms invade root tissue more readily through wounds. Pathogens can produce quantitative losses by causing rots involving tissue breakdown, or qualitative losses through surface blemishes and diseases making roots unattractive and unmarketable. Roots are stimulated by certain fungi to produce phytoalexins which are lung- and hepato-toxic to some animals (see Chapter 4) and cause bitter off-flavours. Internal blemishes, which may be caused by viruses, may also reduce the final quality.

In addition losses may be increased by attack from pests, the most important of which is the sweet potato weevil *Cylas formicarius* and other species. Roots may be initially attacked during storage or may be contaminated with eggs or larvae from the field. Such contamination may not be readily visible to the naked eye and apparently healthy roots may be stored only to be attacked when eggs hatch and larvae begin to feed. Previously uninfested roots are then also open to attack. The weevil may go through several life cycles during a prolonged storage period. Weevil damage produces quantitative losses and aesthetically unappealing roots which may be discoloured and bitter tasting. Like the fungi, weevils stimulate the production of phenolic compounds, leading to brown discoloration of the flesh (see Chapter 2) and also the formation of phytoalexins such as ipomeamarone (see Chapter 4).

Sweet potatoes stored by traditional methods are also lost through attack by other pests such as rats (Jenkins, 1982). Such losses may be significant in some places, but small, in comparison to those already detailed, in others.

The main requirements for minimizing the losses described above are gentle harvesting and careful handling at all times to avoid wounding, followed by curing at high temperatures and humidities (see below) and then storage at temperatures above 12°C.

Curing

The process of curing results in the healing of wounds incurred during initial postharvest handling. One of the first noticeable changes to occur is desiccation of several layers of the outermost parenchyma cells exposed to the air on wounding. It has been shown (Walter and Schadel, 1983) that beneath the desiccated cells there is a subsequent deposition of a polymeric material in the parenchyma cells. This is normally referred to as 'suberization', although 'lignification' is probably a more accurate term. This process provides a barrier to further moisture loss and impedes microbial invasion of the tissue. The polymer deposited has an aromatic aldehyde, lignin-like character and is very little like suberin in nature. The final stage of wound healing is the formation of a wound periderm beneath the 'suberized' cells. The wound periderm is similar to normal periderm in terms of its aromatic compounds, but contains much less suberin. Suberin itself is a saponifiable lipid consisting mainly of hydroxy fatty acids. During wound periderm formation a general skin strengthening, which is genetically determined, may occur in uninjured areas of the periderm. A rapid method of determining curing progress in sweet potato roots has been described (Walter and Schadel, 1982). The intensity of colour developing, when a saturated solution of phloroglucinol in strong acid is applied to the underside of detached wound tissue, is assessed. The most intense colour coincides with termination of wound healing.

The formation of the wound periderm over areas of broken skin not only acts as a barrier to pathogen penetration but also reduces periderm

Figure 5.1. Sweet potato roots covered with banana leaves await collection in West Java, Indonesia, during which time curing may occur incidentally (M. Potts).

permeability at the wound site, resulting in lower weight loss during storage. The percentage weight loss during 113 days of storage of cured and uncured roots in the West Indies was 17% and 42%, respectively (Thompson, 1972). The storage life of uncured sweet potatoes is very short, but that of cured roots can be as long as a year under optimum conditions. Ideally, curing is carried out at 30–33°C and 85–95% relative humidity for 5–7 days (Kushman and Pope, 1972). In temperate regions such as the United States, curing is carefully controlled in special purpose-built storage houses. Promoting storability using procedures such as dipping roots in wax and exposing them to ethylene as alternatives to curing (Buescher, 1980) were unsuccessful. In tropical regions curing is infrequently pursued as a deliberate course of action. However, in the prevailing high ambient temperatures and humidities, curing of roots may take place incidentally (see Figure 5.1). For example, in roots stored under traditional farm conditions in India, losses due to rots showed no increase after 4 weeks of storage (Prasad et al., 1981), a fact the authors attributed to natural curing under ambient temperatures and humidities which resembled those ideally used for curing. Nor did artificial curing of sweet potatoes bring any additional benefits during storage of roots in Bangladesh (Jenkins, 1982), natural wound healing taking place at a fairly rapid rate under ambient conditions.

However, more purposeful curing could be beneficial in some areas of the tropics. Suitably high temperatures are not difficult to achieve, but humidities may often fall to levels which encourage desiccation by evaporation of moisture through the skin. In many places curing can be achieved with simple modifications of existing practices using inexpensive materials mainly to maintain humidities at suitable levels. For example, lining crates or boxes with polyethylene maintained a high humidity environment around roots, while allowing some ventilation. Six days in these modified containers with ambient conditions of 27–30°C and 40–60% relative humidity increased the percentage of marketable roots after 90 days of storage from 48% to 85% (Gull and Duarte, 1974). In Jamaica, covering roots with dried grass, straw or moist coir or placing such materials between the roots and a tarpaulin or polyethylene cover while allowing adequate ventilation is advised for curing (Lawrence, 1985). Curing may be allowed to take place from 4 to 20 days in the tropics depending on environmental conditions.

Genetic variations in response to curing have been reported (Gull and Duarte, 1974). Some cultivars sustain much greater losses than others during storage without curing; a shorter curing period is needed for some cultivars than for others in the same curing environment. In addition, preharvest soil conditions can influence the degree of rot which roots undergo during curing and storage (Ahn, Collins and Pharr, 1980). Roots harvested from cold (4°C) dry soil did not rot during curing, contrasting with those harvested from cold flooded or especially warm (24–32°C) flooded soil, which rotted during curing and subsequent storage.

Traditional methods of storage

Storage of sweet potato roots is not a recent innovation. Ancient methods, practised by the Maoris of New Zealand for hundreds of years, have been described (Cooley, 1951; Keleny, 1965). These consisted of underground storage houses, with timber roofs, dug into the side of a hill. The roots were placed on the floor, which was previously covered with gravel and a dunnage of dried *manukau* (*Leptospermum* spp. or fern brush). The seed stock was placed in first, at the back with the food roots at the front. Any cut or bruised roots were placed nearest the entrance to be used first. The whole store was then sealed and left for some time, presumably to allow curing under the influence of respiratory heating. Alternatively, subterranean well-like pits were dug which, after filling, were sealed. However, these methods entailed heavy losses due to decay and rat damage (Keleny, 1965).

Pit storage of sweet potatoes has also been reported in Zimbabwe and

Figure 5.2. External view of semi-underground sweet potato store used by a farmer near Kumamoto, Japan (J.A. Woolfe).

Malawi, where the roots are placed in pits with alternate layers of wood ash (Lancaster and Coursey, 1984), and in Papua New Guinea where roots are alternated with layers of grass in grass-lined pits (Siki, 1979). Similarly in Cameroon roots placed in a dry leaf-lined hole and sprinkled with wood ash are finally covered with dry leaves or grass (Numfor and Lyonga, 1987). In rare cases a mound of earth is built on top of the heap. The Kakoli people of the Kaugel Valley in Papua New Guinea dig groups of three to four tunnels into natural banks, and place about 11 kg of roots in each tunnel before sealing the tunnel with turf (Yen, 1974). For semi-underground storage in fields in the south of Japan, farmers dig a shallow pit about 0.5 m in depth and 2 m in diameter and line it with rice straw. Roots are piled in and covered with a conical structure of rice straw which is then partially covered in earth. Rat damage is said to be the major cause of spoilage. A covering of sharp pine needles may be used to deter rats. In another part of Japan, on-farm stores take the form of a semi-underground storage room dug into the earth, with (nowadays) concrete-lined walls (Figure 5.2). The above-ground portion is covered with a 1 m thick layer of earth. Ventilation is by means of a hole in the ceiling which communicates with the air by means of a chimney-like shaft. Various types of pit store are in use in China. These are described in some detail below.

Sweet potatoes are commonly stored in farmers' dwellings rather than in special pits or structures. For example, in Bangladesh roots are stored on the floor of houses in shallow piles or sometimes piled on a bench-like structure made of bamboo and called a *macha* (Jenkins, 1982). Occasionally bundles of the larger roots are suspended by jute ropes from the roof of the house. Only a very few farmers construct, outside the dwelling house, a *dool* or special storing structure made of bamboo and plastered with mud (Rashid, 1987). Seventy five per cent of the harvest is stored in the home, either on the floor or suspended from the ceiling, for 1–3 month periods in Vietnam (Truong and Vander Zaag, 1987). This storage period is reduced if weevils are present. In Nigeria, sweet potatoes are traditionally stored in the house, heaped on the floor or laid on a rock or shelf. Fires may be lit once or twice weekly to fumigate the roots, which are often covered with ashes having fungistatic properties. Losses of up to 95% were reportedly encountered by these methods (Olorunda, 1979). Similarly roots may be placed on platforms in houses in Papua New Guinea in a dark, well-ventilated area where the smoke from cooking fires seems to aid in curing (Keleny, 1965; Siki, 1979). Roots are stored only for a short time. A few days is said to improve the eating quality due to loss of moisture raising the dry matter and hence the energy content, and also due to conversion of starch to sugars (Kimber, 1972). Similarly, farmers in Indonesia may expose moist mouthfeel roots to the sun for up to 1 day before storage to reduce moisture content and improve the texture (Watson, 1988). A similar practice by Peruvian farmers exposes roots for several days. In Cameroon, exposure of roots to the sun is followed by storage of roots in a dry place or in a basket covered with banana leaves or grass (Numfor and Lyonga, 1987).

The Bantoc of Luzon in the Philippines often adapt their *'alangs* (wooden rice store houses) to store sweet potato roots, after the rice is almost used up. Additional space is created for the greater bulk of the root crop by the use of plank bins (Yen, 1974). Storage is needed because the entire sweet potato crop is harvested at one time to make space in the fields for a second crop. In the Koraput district of Orissa State, India, roots are heaped up, covered with a thin layer of rice straw, and plastered over with soil mixed with cow dung (Nair et al., 1987). The mud covering is pierced with ventilation holes. As the huts are dark and cool during the day and the climate is relatively cool, farmers claim a storage life of 6 months. However, simulated farm conditions of storage in Bihar and Orissa, India, as a single layer of roots either in a room or a ventilated yard, gave 70% and 50% total losses, respectively, of sweet potatoes stored for 2 months.

In Korea, sweet potatoes are stored throughout the winter under relatively low temperatures which range from −5°C to 15°C. The temperature in traditional stores is not controlled by artificial heating, but by utilizing natural residual soil heat (Hong, 1982). For example, roots may be stored in underground pits covered with a building to protect it from cold winds, or in underground tunnels. In one part of Korea, sweet potatoes are kept in rooms warmed by an *ondol* or traditional floor heated from below.

In China, during the period of peasant organization into communes, roots were communally stored in large coal-heated store houses which considerably reduced the incidence of black and soft rots. Since the dissolution of the communes in recent years, individual farmers have reverted to traditional methods of storage in tunnels or simple holes dug into the ground (see Figure 5.3). Fungal attack is controlled by soaking roots in a fungicide solution prior to storage. Heavy over-use of extremely toxic fungicide is a common problem (Zhang, D.P., personal communication). The construction of pit or cellar stores, for on-farm use in China, where 80% of the world's sweet potato crop is grown, has been discussed in detail (Zhao and Jia, 1985) and will be described briefly. Fuller details are given in an English translation (International Potato Center archives).

Storage in China

Two categories of storage – cellar or pit, and indoor – are used in China.

Cellar storage

Pit or cellar stores are those which are dug into the ground. Certain fundamental principles should be observed for their construction. A suitable site, free from wind, on high terrain with a compact soil and good drainage should be selected. The ground water table should be at least 35 cm below the bottom of the pit. In the case of permanent pits the bottom should be above the maximum level of rise in ground water during the rainy season so that the pit is not damaged and its operational life shortened.

When old pits are reused, they must first be cleaned by scraping a 2 cm layer of soil off the walls to expose new soil and removing the old soil. Disease-infected pits should be disinfected by burning wood or straw or by fumigation with sulphur.

The size of the pit is determined on the basis of the cultivated area of sweet potato to be stored and the estimated production and consumption of roots. Capacity of a pit is recommended to be 1 m^3 of space per 500–600 kg of roots, allowing 30–50% of spare space in order to maintain aerobic conditions. Construction should be complete at least 15 days prior to harvest, or earlier in areas where heavy rains might cause flooding of pits and consequent poor storage.

Figure 5.3. The entrance to a farmer's pit store in China (M. Iwanaga).

Well cellar

The construction and dimensions of the well cellar, which needs a high location, solid soil and a high level of underground water, are shown in Figure 5.4. After the well neck has been dug, a platform is left in the bottom centre of the neck and a door (0.75 m wide by 1.2 m high) dug for a storage room on each side of the neck. The door is expanded into a storage room, the size of which can be varied according to the quantity of sweet potato to be stored. In the diagram, each storage room is approximately 1.7 m high by 1.5 m wide by 2.6 m long and holds 1500–2000 kg of roots. The floor of the storage room should be covered in a layer of sand 10–13 cm deep. The mouth of the well should be about 0.3 m higher than the ground surface to prevent rain entering the cellar.

Well cellars in Shandong Province may have three to four storage rooms. In Henan and Hebei Provinces twin well cellars with multiple storage rooms can be seen. A storage efficiency of 99% is claimed for this method, with root weight reduction as low as 2–3%. Well cellars of a similar construction with four storage rooms are also used in some parts of Japan. One seen by the author was 35 years old and needed no attention. It was claimed that the temperature and relative humidity stayed constant at 15°C and 80%, respectively, the year round and that spoilage rates are very low.

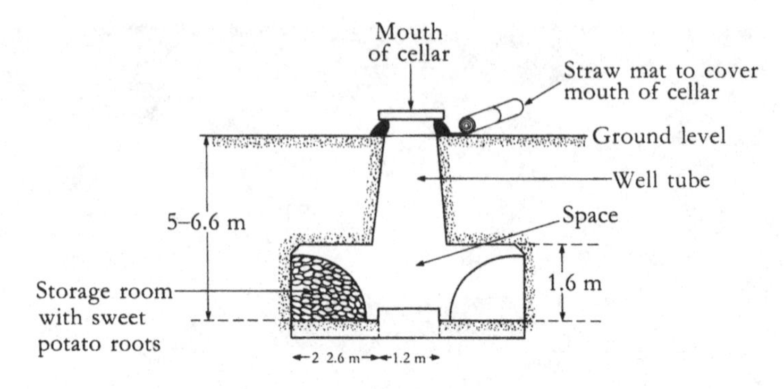

Figure 5.4. A well cellar.

Cave cellar or mountain cave storage

Cave cellar construction is shown in Figure 5.5. A channel 1–1.3 m in diameter is dug downwards for about 1.5–2 m, then a passageway is dug horizontally. This passageway is 1.2 m high by 1–1.2 m wide by 2.5–4.5 m long. In the case of large-quantity storage, the passageway can be extended in length. Storage rooms are dug alternately on both sides of the passageway. These storage rooms or 'caves' are roundish in shape. The ground and walls of the storage rooms are lined with dry straw. Sweet potato roots are piled up on the straw leaving a space of about 33 cm between the top of the pile and the roof. The pile is then covered with straw. The door of the store can be sealed with straw, earth or bricks. A bamboo pole inserted through the door allows ventilation.

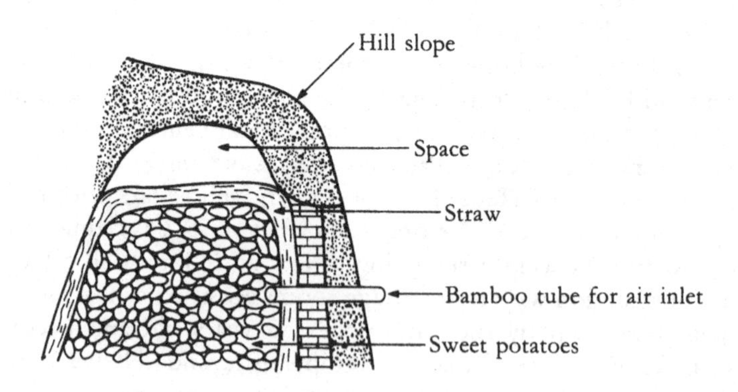

Figure 5.5. A cave cellar.

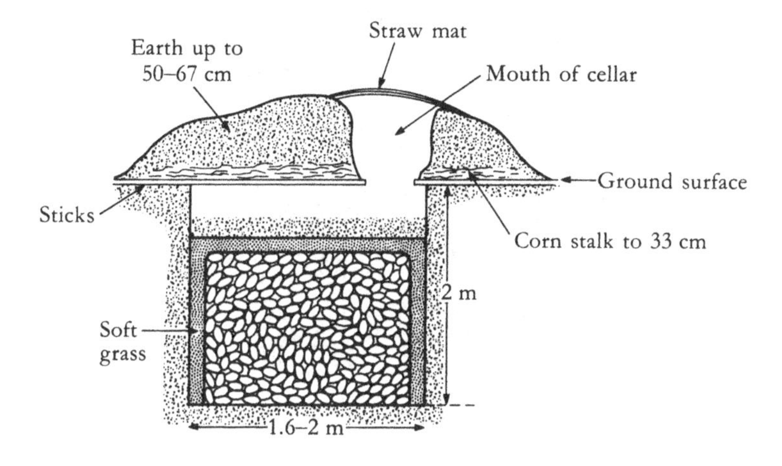

Figure 5.6. A canopy cellar.

Canopy cellar

This has a relatively simple structure as shown in Figure 5.6. It is similar to vegetable storage cellars of north China, with a rectangular shape but rounded corners. Initially a rectangular pit 1.5–1.8 m wide by 3.5–4 m long by 1.9 m deep is dug. If the underground water level is high the cellar can be dug half underground and the upper half constructed with earth. It can also be dug as a sort of cylinder 1.8 m in diameter. The inside is lined with soft grass, for insulation, and the sweet potato roots piled up solidly from the bottom. The final height of the pile should be such as to leave at least 1 m of space at the top of the pile to allow for thermal dissipation in the initial stages of storage. If external temperatures are low, the pile should be covered with 10–13 cm of fine bran or dry bean leaves when the inside temperature is 14–15°C, to enhance thermal insulation.

The mouth of the cellar is covered with sticks, on top of which is laid straw to a thickness of 0.3 m, and this is covered with earth to a depth of 0.5–0.7 m, leaving an entrance at the southeast corner. The entrance should be covered with a straw mat or a basket full of soft straw, which acts like the stopper of a bottle.

To control both thermal dissipation during early storage and storage temperature, criss-crossing air-inlet channels, which connect to the surface via the walls, can be dug in the bottom of the cellar. These are covered with sticks to prevent their blockage by the roots.

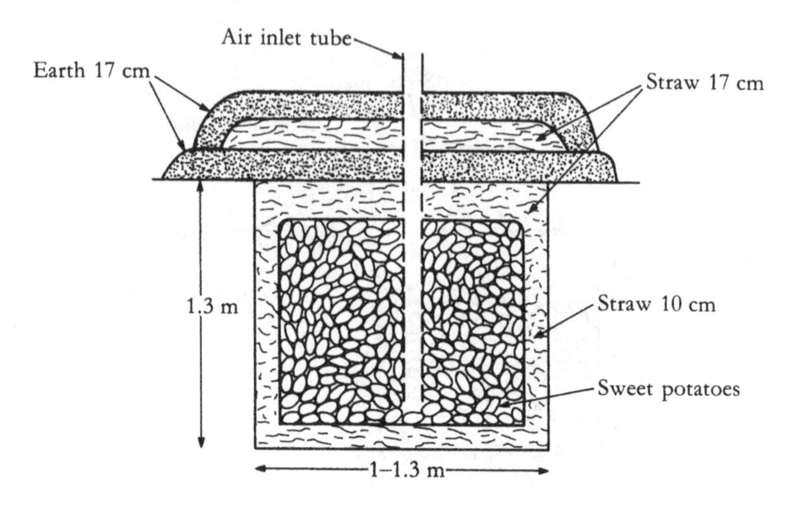

Figure 5.7. A ditch cellar.

Ditch cellar

The structure of the ditch cellar is simple, as shown in Figure 5.7. It has the disadvantage of being a sealed structure so that checking the condition of the contents is not easy. To counteract this, thermometers should be placed at the bottom, middle and top of the sweet potato pile. A ditch, 1–1.3 m wide by 1.3 m deep and of variable length depending upon the storage capacity required is dug. Criss-crossing air-inlet slots are dug in the bottom of the ditch and connected to the surface. These can be covered with sticks to prevent blockage by roots.

A layer of straw 10 cm thick is used to line the walls and bottom of the ditch and sweet potatoes are piled up to a height of about 1.2 m. Soft straw to a depth of 17 cm is used to cover the heap. Alternate layers of earth and straw can be used to insulate thermally the top of the ditch, depending on the external temperatures.

Indoor storage

Sweet potatoes may also be stored indoors in a heated house structure (Figure 5.8). This may be wholely above ground, or partly underground, where the water table is high. Such a heated structure is beneficial for curing and reducing root losses due to rots, but may promote sprouting and formation of root cavities.

The house is constructed with thick earth walls, a low roof made of alternate layers of earth and straw (six in all) and pairs of small air-inlet

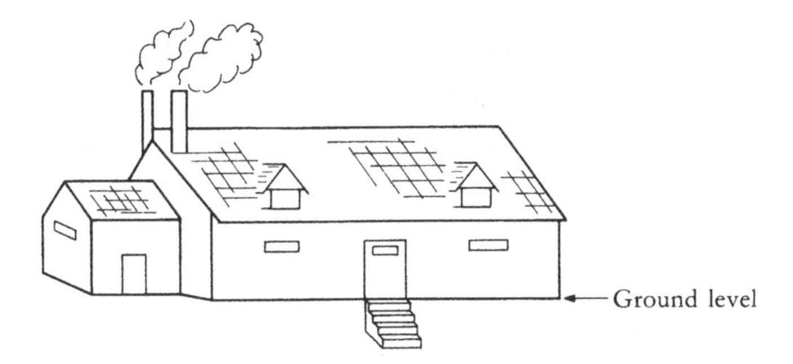

Figure 5.8. A semi-underground heated store house.

windows opening on opposite walls. A smaller house may be built on one end for installation of the heating stove and heat is transmitted from the stove to the main store via a funnel made of bricks or earth bricks. Smoke is emitted via two chimneys, one on each side of the stove, built into the wall. The store is lined, apart from the windows, with straw 33–40 cm thick. A passageway 1 m wide leading from the entrance is left clear and sweet potato roots are piled on either side of it at least 33 cm away from the heating funnel. Sweet potatoes are not placed directly on the ground, but on a straw mat covering on top of hollow bricks. Each side of the house may be divided up into three equal areas in which are stored heaps of sweet potatoes with air-inlet devices placed in each heap.

Management of stores

The management of sweet potato roots in store varies with the method of storage and with external climatic conditions. During the early part of storage and vigorous respiration of the roots, care must be taken to ensure heat dissipation and efficient air exchange. Store entrances must therefore be opened and closed for the correct lengths of time required to maintain temperatures as near as possible to 12–14°C and a relative humidity of 80–90%. In many sweet potato producing areas of China temperatures drop significantly during the winter months, which requires sweet potato storage structures to be thermally insulated to prevent cold damage to the stored roots. During mid- and late-storage, therefore, when the respiration rate of the roots slows down, structures are usually sealed, leaving a small inlet for air exchange. Canopy cellars used as stores for sweet potatoes which will be removed at intervals for food are not sealed, but the thickness of the straw mat covering the entrance is enhanced.

Further attention should be paid to rat-proofing of cellars and providing adequate drainage around stores where rainfall is high.

Large-scale storage

Ideally sweet potato roots should be stored at temperatures between 13°C and 16°C and 85% relative humidity. In countries such as the United States where roots are stored for many months these conditions are achieved by the operation of large-scale storage houses where temperature and humidity are carefully controlled by the use of heating and humidifiers. However, even in the United States less expensive alternatives which can be used by the smaller producer are being investigated. Cured roots stored inside an inflated, double-walled, black plastic, green-house-like structure with a fresh-air exchange system were 80% marketable after 3 months of storage (Paterson et al., 1980; Wagner, Burns and Paterson, 1983). This system was not only relatively inexpensive compared to large-scale stores, but could be situated in the area of production and tailored to suit the size of the sweet potato producer's operation.

Investigations have also been initiated into the possibility of curing and storing roots in large continuous volume piles to improve the efficiency of space utilization and handling of the many thousands of tonnes of sweet potatoes harvested by bulk mechanical means in the United States every year. Initial trials have determined the air flow characteristics (Abrams and Fish, 1982) and maximum possible depths (Abrams, 1984) of such large bulk piles, although a system for placing roots into these piles and removing them for market has not yet been devised. In addition, an infrared image method for the detection of surface bruises in sweet potatoes and other commodities has been developed in Japan (Miyazato, Ishiguro and Dannao, 1981) for grading of produce after removal from storage, although grading is at present done by eye and hand.

Improvements in small-scale storage

In most parts of the developing world the market value of the sweet potato crop is insufficiently high to justify the expense involved in constructing sophisticated store houses. Alternatives utilizing simple and inexpensive methods and materials are in use or being sought in several areas.

In Papua New Guinea, for example, mounds have been constructed in the form of a conical structure made of sticks, lined with kunai (*Imperata cylindrica*) grass (Aldous, 1976; and see Figure 5.9). Sound roots were placed in the structure, which was then covered completely in a 10 cm thick layer of kunai grass. A hurricane lamp was placed in a trench dug beneath the base of the mound, so that the heat given off could promote

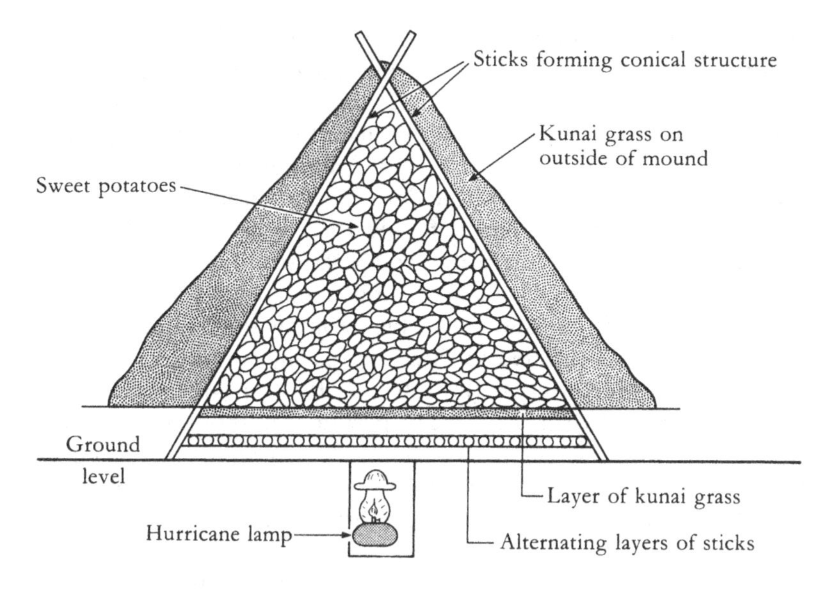

Figure 5.9. An experimental mound store (Papua New Guinea) (after Aldous, 1976).

curing. Roots were stored fairly successfully for 40–50 days. Recommended improvements to the structure included increasing the layer of kunai to a thickness of 20–25 cm, leaving the apex of the stick shell a little open to allow adequate ventilation, and sheltering the mounds to minimize fluctuating temperatures.

A further experiment involved roots stored on shelves in a house in which a fire was kept constantly burning as a heat source, and from the rafters of which hung copra bags moistened regularly to elevate the relative humidity. With this method, roots could be kept only for 2–3 weeks. Covering roots with dry grass and then soil to form a clamp maintained them for 30 days at one site in Papua New Guinea (Bourke, 1985).

Roots were stored successfully for up to 4 months in Barbados, West Indies, by using clamps (Anon., 1960). The latter consisted of rectangular pits 8–10 cm deep, the floor of which was covered with a 'trash' of dried grass, weeds, coconut fronds, banana leaves etc. Sound sweet potatoes were piled on this to a height of about 1 m. The heap was covered in 'trash', then a layer of soil and finally more 'trash'.

In the Philippines, roots stored in a trench 50 cm deep covered in sand and sheltered by a roof kept for 55 days. However, 35% were decayed and 45% sprouted (Cadiz and Bautista, 1967).

Traditional storage of roots in Malawi entails the placement of roots in pits directly after harvest or after sun-curing for about a day (Kwapata, 1983/84). The pits are dug under maize or groundnut silos. The bottom and sides of the pits and subsequent layers of roots are sprinkled with ash collected from the fires used for cooking family meals. In these conditions roots are stored for 3–4 months. Extending storage for longer periods, which would be desirable for ensuring a supply of roots during months when most family food reserves are depleted, entails heavy losses. Though curing temperatures of 25–30°C are not difficult to achieve, maintenance of high relative humidity and a storage temperature of 12–15°C is usually a problem for the small farmer. The pit storage life of roots subjected to three sun-curing times (0, 12 and 24 hours) and two ash treatments (no ash and 500 g of ash from *Gmelina* spp.) was therefore investigated over a period of 6½ months (Kwapata, 1983/84). After pits 60 cm by 60 cm by 30 cm were dug, 5 kg of undamaged roots were placed in each. Following application or otherwise of ash, roots were covered by a 5 cm layer of dried grass and then 5 cm of soil. A grass thatched shade was constructed above the pits to prevent contact with rain. After storage for 6½ months, roots were dug out, graded and the weights of undamaged and rotten roots were obtained. The degree of sprouting (number of roots sprouted divided by total number of undamaged roots, multiplied by 100) and storability (final weight of undamaged roots divided by the initial root weight multiplied by 100) were assessed. The results are shown in Table 5.1. Increasing the sun-curing time from 0 to 24 hours significantly decreased storability and increased the weight of rotten roots, but the increase in degree of sprouting was not significant. The application of ash from *Gmelina* spp. had no significant effect on the measured parameters. There was no significant interaction between curing time and ash application. Sound roots were rated organoleptically equivalent to recently harvested roots in terms of flesh firmness and taste. Rot of roots was primarily due to *Rhizopus stolonifer* soft rot. In this study the storage of recently harvested roots, which had not lost excessive moisture through evaporation, in pits covered with layers of insulating dry grass and soil, provided conditions to maintain high relative humidity in the pits. Furthermore, construction of a thatched shade over the pits prevented rain entering and at the same time modified external high temperatures. The prolonged exposure of roots to direct sunlight before storage, however, appeared to do more harm than good, perhaps causing excessive loss of moisture leading to tissue breakdown. Alternatively, elevation of root temperature with consequent increased respiration rate could have raised temperatures in the pits favouring sprouting and promoting microbial activity. The researcher concluded that traditional pit storage without initial prolonged sun-curing has potential, but

Table 5.1. *The effect of sun-curing and ash application during pit storage in Malawi*

Treatment	Storability (%)	Weight of sound roots (kg)	Weight of rotten roots (kg)	Degree of sprouting (%)
Curing time (h)				
0	83.2 x[a]	4.16 x	0.17 x	18.0
12	81.0 x	4.05 x	0.35 y	31.0
24	55.6 y	2.78 y	0.79 z	32.0
Ash (kg)[b]				
0	74.3	3.71	0.43	27.3
0.5	72.3	3.61	0.43	26.7

Notes:
Based on: Kwapata, 1983/4.
[a] xyz: means in same column followed by different letters significantly different at $p < 0.05$.
[b] From *Gmelina* spp.

further investigation is needed to assess the effectiveness of ashes collected from different plant species and also the storability of different local sweet potato clones.

The storage of freshly harvested Indonesian roots in sand, soil, ash, rice husks or gunny bags (in an attempt to reduce temperatures and stabilize relative humidity) exhibited no advantage over the simple maintenance of control roots in open baskets (Antarlina, 1990). After 6 weeks, more than 50% of all samples were mouldy, weight loss in general was more than 25% and test samples had significantly higher degrees of sprouting than the control sample.

Sprouting occurs frequently during prolonged storage in conditions of high temperature and humidity. In the tropics, sprouts are generally broken off as they appear. Sprout suppressants could be used but storage facilities for such chemicals are frequently not available and their use is limited by economic and managerial factors (Booth, 1974). In some countries the use of gamma-irradiation for suppression of sprouting has been explored. Japanese workers found that sweet potato needs higher doses of gamma-irradiation, in the range 50–200 krad (0.05–2.00 kGy), than potato to suppress sprouting (Uritani, I., personal communication), but that such doses accelerate the production of enzymes including polyphenolase and peroxidase (Ogawa, Hyodo and Uritani, 1969; Ogawa and Uritani, 1969). This was disadvantageous when the roots were used for processing. The application of a lower dose, 3–5 krad (0.03–0.05 kGy), markedly suppressed sprouting in Taiwan sweet potatoes (Wang et al., 1982). In an unirradiated control sample, 100% sprouted after 1 month of storage, whereas those treated with 0.05 kGy gamma radiation were only 2–2.5% and 9% sprouted after 1 and 5 months, respectively. Sprouting was reported to be controlled for 2 months in two North American cultivars by a radiation dose of 0.1 kGy (Lu et al., 1987) and sweet potato weevils to be sterilized by 0.25 kGy. Although effective, gamma-irradiation is very expensive and, as indicated above, also results in various biochemical changes in the roots (Ajlouni and Hamdy, 1988; Lu et al., 1987; Wang et al., 1982; and see Nutritional and quality changes, below) which above certain dose levels have an adverse effect on sensory quality.

Sprouting was reduced in roots stored for 4–8 weeks when the crop was sprayed 2 weeks before harvest with a 0.5% maleic hydrazide solution (Gooding and Campbell, 1964). Roots from an unsprayed crop were also successfully stored in containers in a confetti impregnated with a methyl ester of naphthalene acetic acid in acetone.

Research in the Philippines has reduced sprouting by 99% when roots are stored in diffused light in a hut with slatted bamboo walls and closed bamboo flooring (Icamina, 1985). Bamboo water troughs have also been introduced into storage huts to increase humidity and improve storage time. As described above, a large percentage of roots is frequently lost through the development of storage rots. A wide variety of microorganisms is responsible and no treatment has so far been found effective against all. Among the more commonly encountered microorganisms causing storage rots in various parts of the world are *Rhizopus stolonifer* (soft rot), *Erwinia chrysanthemi*, *Ceratocystis fimbriata* (black rot), *Fusarium oxysporum* (surface rot), *Diaporthe phaseolorum* var. *batatatis* (dry rot), *Macrophomina phaseolina* (charcoal rot) and *Botryodiplodia theobromae* (Java black rot). R. *stolonifer* is of considerable economic importance, since under favourable conditions it can destroy the entire root in a few days. *B. theobromae* is often a serious problem in the tropics, as its optimum growth temperature is about 28°C. It will be noted that among these organisms are those described in Chapter 4 as high level inducers of ipomeamarone and other furanoterpenoid phytoalexins.

Roots may be washed before storage to remove soil bearing microorganisms, but washing may result in bruising, and reduced rather than improved storability (Tereshkovitch and Newsom, 1965). A disinfectant may be used in the wash water to inhibit growth of microorganisms. Soil may also be removed by gentle polishing. The application of fungicides which inhibit spore germination, growth and production of extracellular cellulytic and pectic enzymes by fungi (Arinze, Naqvi and Ekundayo,

Table 5.2. *Hot water dip time–temperature combinations for delaying pathological decay during storage of sweet potato*

Temperature	Dip time (s)	Time to initial rot (days)[a]
High temperature – short time		
Untreated	0	58
90°C	2	78*
80°C	2	102**
80°C	4	86**
80°C	10	92**
Medium temperature – medium time		
Untreated	0	77
80°C	10	109*
70°C	10	112*
Low temperature – long time		
Untreated	0	38
40°C	120	51*

Notes:
Based on: Scriven et al., 1988.
[a] Significantly different from untreated at: $*p < 0.05$, $**p < 0.01$.

1975), helps to prevent storage rots. In the West Indies roots are cured at 31–36°C for 3–8 days and then dipped in 500 p.p.m. of dichloran, before being packed in crates to await export (Anon., 1988).

An alternative to the application of fungicides, in the form of hot water dipping of roots before storage, has been investigated (Scriven, Ndunguru and Wills, 1988). Roots inoculated with *Fusarium* and *Rhizopus* were dipped in water for various time–temperature combinations, ranging from 40°C to 100°C and 2 to 240 s, high temperatures being associated with short dip times. Roots were subsequently stored loosely in polyethylene at 25°C and about 55% ambient relative humidity. The time to rotting of 10% of the sweet potato root surface was recorded. Roots were weighed during storage and their respiration rate assessed by measuring their CO_2 emission. Time–temperature combinations found significantly to prolong time to the development of rotting compared to untreated controls, without affecting weight loss or respiration, are shown in Table 5.2. All dips at 100°C and 80°C for more than 10 s accelerated decay, respiration and weight loss. The potential usefulness of hot water dipping to prolong sweet potato storage life could be restricted in some developing countries by the need for fairly careful temperature control and the requirement of scarce fuel for heating the water.

In some parts of the tropics, one of the most serious problems limiting storage is the sweet potato weevil. Unfortunately, genetically resistant cultivars are not yet available, nor is spraying the crop with insecticide a very effective means of control. One way of avoiding losses is not to store roots from a crop suffering from weevil infestation in the field. However, roots carrying weevil eggs may not be detected. Of apparently weevil-free roots stored under traditional conditions in India, 32% were lost to weevil damage after 2 months of storage (Prasad et al., 1981). Studies at IITA, Nigeria (Hahn and Anota, 1982, p. 73), showed that weevils could be killed if the storage temperature was lowered to 20°C, roots were immersed in water at 52°C or 62°C for 10 min or 42°C for 30 min. Weevils also died when near the soil surface in underground (burial) storage, probably due to high day-time temperatures. In India, where sweet potato roots are left exposed in heaps on the godown (warehouse) floor, damage from weevils may rise to 60% within 1 month (Rajamma, 1984). Efforts to control weevil infestation during 2 months by covering the roots with dried red earth, sand, wood ash, sawdust or mixtures of sand and red earth with dusts of malathion or carbaryl were successful except in the case of carbaryl. However, the roots under sand or sawdust completely dried up. Storage in gunny bags sprayed with insecticide controlled infestation for 1 month, after which the insecticide lost its toxicity.

The use of heaps or sacks for storage may lead to physical damage which could be prevented if roots were carefully packed in boxes, cartons or crates. A number of simple measures have been recommended for improvement of sweet potato root storage in the tropics (Jenkins, 1982):

1. *Careful handling* at all times to minimize damage.
2. Removal of soil from roots prior to storage.
3. Quality control followed by separate storage of damaged roots and their rapid utilization.
4. Utilization of small roots first.
5. Covering storage layers or piles of roots with suitable materials to build up and maintain high humidity.

The need for further research into improved simple methods of sweet potato storage is very important. Where a significant percentage loss is experienced (see Figure 5.10), any technique producing even a small loss

Figure 5.10. Poor storage can lead to a high level of wastage in roots dehydrated or diseased. A merchant in Indonesia pares old roots for sale (G.A. Watson).

reduction would be of immense benefit to farmers (Rashid, 1987). In addition, selection of cultivars which are particularly suitable for storage, for example those with a higher level of genetic resistance to pathogens (Morris, 1981) or weevils (Wilson, 1979), or with genetic variation in susceptibility to weight loss and sprouting (Rashid, 1987), or with resistance to mishandling would be an important contribution to improved storage. The Argentinian cultivar 'Inta Morada', for example, is said to be very resistant to damage during handling and hence later deterioration, though no research has yet been carried out to establish the reasons for this. Screening of genotypes for storability has started in the Philippines (Data, 1985). Of 20 genotypes stored for 3 months, five decayed completely before the storage period ended. Of the remainder, weight loss, degree of shrivelling, degree of decay and incidence of sprouting varied considerably between genotypes. Five cultivars did not develop any signs of sprouting even after 3 months of storage.

Storage during retailing

Further losses of roots are encountered as a result of conditions during retailing, which may also be considered to be a form of storage if roots are displayed for some time before being purchased. Mechanical injuries sustained by retailing in bulk in New York supermarkets produced about 5% of unsaleable roots through *Rhizopus* soft rot decay and moisture loss (Ceponis and Butterfield, 1974). Rough handling sustained during retailing also caused subsequent decay due to soft rot in roots held in conditions simulating those practised by consumers in the home. Total losses from soft rot and moisture loss during 1 week's display in bulk in a United States retail store amounted to about 13% (Ceponis, Kaufman and Tietjen, 1973). Packaging of roots on trays in a plastic film sleeve (open at both ends) or complete overwrapping, as a means to prevent injuries due to handling, decreased moisture losses, but not soft rot. Total losses were reduced to 2.8% by impregnating the complete overwrap with a fungicide (2,6-dichloro-4-nitroaniline). Losses of roots piled in heaps in tropical markets or being sold by the side of the road (Figures 5.11 and 5.12) are likely to be much higher than those in a North American supermarket, but the extent of such losses has not been explored.

Storage of dried roots

An alternative to storage of roots in the fresh state which is employed in some areas is to dehydrate them and store them as a dry product. This has the advantages of prolonging the interval in which they can be maintained in an acceptable condition, reducing storage area and minimizing losses. However, difficulties may be experienced in rehydration of the dried product and the rehydrated form may not be suitable for incorporation into local dishes.

In parts of East and West Africa where there is a pronounced dry season, a proportion of roots is peeled, sliced and sun-dried for storage. When needed, the dried roots are ground into a flour and used in making local dishes (Jana, 1982; Lancaster and Coursey, 1984; Numfor and Lyonga, 1987). Dried chips, made by peeling, boiling, slicing and drying over the hearth, are a famine security product in several provinces of

Figure 5.11. Heaped roots waiting to be sold in a market in Kampala, Uganda (M.Iwanaga).

Figure 5.12. Roadside sweet potato sales, West Java, Indonesia (G.A. Watson).

Cameroon (Numfor and Lyonga, 1987). The chips, which can be stored in a basket, calabash or other utensil for up to a year are eventually consumed directly. The production of solar-dried chips is the principle technique for preserving roots for storage in central Vietnam (Truong and Vander Zaag, 1987). The chips are stored in clay pots for an indefinite period. Subsistence farmers in some parts of East Java process surplus sweet potatoes into sun-dried cubes (Widodo, 1990). After steaming, the roots are peeled, cut into small cubes and sun-dried for 5 days to a moisture content of about 12%. In this form the product can be stored for 6 months. When needed it is reconstituted in warm water, steamed and eaten with coconut. In China, large quantities of sweet potato chips are sun-dried in the field (Figure 5.13), and then stored in the open air at starch or alcohol processing plants to allow the plants to operate all year round (Wiersema, Hesen and Song, 1989). It is recommended (Zhao and Jia, 1985) that chips be dried to a moisture content of less than 17%, 12.5% and 11% if they are to be stored in temperatures of 6–10°C, 15–20°C and 27–29°C, respectively. Chips may be stored in piles sealed with sand, earth bricks or wheat husks to prevent moisture uptake. Where chips are infested with insects, drying in strong sunshine to a moisture content of 8.5% followed by heating the chips before storing, to kill insects, is recommended.

In the Philippines, dried chips are prepared for storage and then pounded into a flour as required for use in the preparation of a gruel (Lancaster and Coursey, 1984). Roots are cleaned, sliced and dehydrated in the sun in open yards in India (Nair et al., 1987). They are then ground into flour. In Taiwan farmers normally produce dried sweet potato chips suitable for storage and use as an animal feed or for starch production. Recently drying has been introduced into some areas where it was previously not practised. Sweet potatoes washed, cut up into chips, sun-dried on a piece of plastic for 2–3 days and placed in bags were stored in Papua New Guinea for up to 1 year (Aldous, 1976). It was suggested that the dried product should be fed to pigs and poultry, making more fresh roots available for human consumption in times of scarcity. It is reported (Jenkins, 1982) that drying of root slices, which can subsequently be sealed in polyethylene bags and kept for 6 months in storage without loss, is practised to a small extent in Bangladesh. However, many farmers do not have access to even the minimum facilities required to produce dried chips of good keeping quality (Rashid, 1987). In many places sweet potato is harvested just as the rainy season sets in and sun-drying is impossible. There is scope for a great deal of research into simple methods of preservation which would provide alternatives to the difficult storage of fresh roots. One possibility is the production of dry products prepared by traditional processes which grate the roots,

Figure 5.13. Sun-drying of root slices to dried chips, such as these in baskets awaiting collection near Jinan, China, can be an alternative to storage of fresh roots (M. Iwanaga).

squeeze out juice and then roast the pieces over a fire to produce a dry flour, such as is used with cassava processing in West Africa for *gari* making and in Brazil for *farinha de mandioca* (see Chapter 6).

Quality and nutritional changes

The eating characteristics of sweet potatoes after cooking are often a result of changes in constituents due to a combination of postharvest procedures. In the case of some compounds, particularly the carbohydrates, it is therefore unsuitable to discuss the changes taking place during curing or storage in isolation from those taking place during cooking or processing. For this reason, some of the findings related to cooking and processing are in fact reviewed in this section.

The quality of roots after cooking or processing is a complex character comprising a combination of aroma, taste, mouthfeel or texture, colour and fibre content. The way quality is perceived by the consumer depends on personal, local, or regional preferences and eating habits (Martin and Roderiguez-Sosa, 1985; Villareal et al., 1979). Quality attributes are influenced by postharvest handling procedures, including curing and storage, and may be quite different if the roots are used directly after harvest. Even when controlled, curing and storage conditions may produce undesirable characteristics in processed products, for example dark coloured chips or loss of firmness in canned sweet potatoes. In addition, the final concentration of nutrients in cooked or processed sweet potato depends partly on conditions and length of curing and storage. Investigations of nutritional and quality changes in sweet potato during curing and storage have been carried out largely under carefully controlled conditions in the United States and are presented here only as a guide to the changes which may be expected in roots held in uncontrolled conditions in tropical countries. More research is required into changes in roots stored under simple conditions with a view to minimizing undesirable nutritional alterations and producing roots of a quality commensurate with local expectations. Furthermore, where nutritional changes cannot be controlled they should be documented so that nutritionists, dieticians and extension workers can make appropriate assessments of the dietary contributions of stored sweet potatoes. It will be obvious from reading the following sections that the activities of amylase enzymes and associated changes in carbohydrates leading to specific characteristics of texture and taste in cooked roots still require further investigation and clarification. Research carried out so far has been confined mainly to North American cultivars restricted in their range of characteristics (most are 'moist'-fleshed and sweet tasting) and should be extended to include a much greater variety of distinctive clones.

Carbohydrates

The carbohydrates in sweet potato roots play an important part not only in the degree of sweetness of cooked roots, but also in their mouthfeel described as 'moistness' or 'dryness' (which is independent of moisture content), in their texture through fibrous constituents and in the colour and firmness of processed products such as chips and canned roots. The organoleptic sensations of moistness or dryness especially are extremely complex and are the sum of a number of factors, not all of which are by any means fully understood at present. The combined effects of moisture and dry matter losses, respiration, and activities of the amylase enzymes (see Chapter 2, Enzymes) during curing and storage as well as changes during cooking, all differ in extent between cultivars. These effects are responsible for variations in the concentrations of individual carbohydrates and in their proportions in the cooked roots. Freshly dug roots are relatively high in starch and non-starch polysaccharides (alcohol-insoluble solids) and relatively low in sugars. Though cultivars differ in mouthfeel and taste when cooked directly after harvest, in general they become moister in mouthfeel and sweeter in taste after curing and storage. The changes in carbohydrates taking place during optimum curing and storage conditions have been reviewed (S-101 Technical Committee, 1980). Results for carbohydrate conversions in raw roots during curing were said to be contradictory. Some studies found a decrease in starch and a corresponding increase in reducing (glucose and fructose) and total (glucose, fructose, sucrose and other) sugars. Other workers showed negligible or no changes in carbohydrate fractions. Net accumulation of sugar depends on a more rapid hydrolysis of starch to sugar than of respiratory decomposition of sugar to carbon dioxide and water (Edmond and Ammerman, 1971, p. 238). The relative rates of these transformations no doubt vary between cultivars, producing no net change in sugar concentration in some, and increases or decreases in others. Neither maltose (Picha, 1986a) nor dextrins (Hammett and Barrantine, 1961) were found in raw roots. The storage of one Filipino cultivar in ambient conditions of 28–32°C and 66–83% relative humidity for 17 days resulted in increases of total sugars, sucrose, glucose and fructose as percentages of root fresh weight (Kawabata et al., 1984). Very small amounts of maltose (59–116 mg/100 g root tissue) were produced only between days 7 and 12. The ratio of sucrose to total sugar increased while that of glucose and fructose decreased; in other words the degree of increase in sucrose was greater than that of the other two sugars.

When sweet potato roots are cooked, activity of the amylase enzymes is initiated causing a hydrolysis of starch to maltose, and dextrins of varying molecular weight. The extent of this conversion, including the final molecular sizes of the dextrins, which depends on the level of enzyme activity, may be one important factor influencing the mouthfeel of cooked roots (Walter, Purcell and Nelson, 1975) and the viscosity of sweet potato purée (Hamann, Miller and Purcell, 1980; Rao, Hamann and Humphries, 1975b). That is, the more dextrins produced, and the lower their molecular weight, the more 'moist' the mouthfeel and the lower the viscosity, respectively. Freshly harvested sweet potatoes contain relatively low levels of alpha-amylase. However, when alpha-amylase activity was measured after 1 week of curing plus 2 weeks of storage, it had increased nearly four-fold in six cultivars (Walter et al., 1975). This increase continues as storage progresses (Ikemiya and Deobald, 1966; Walter et al., 1975). It is not clear whether the potentiation of alpha-amylase and initiation of carbohydrate conversions are important consequences of curing or merely reflect more rapid reactions at the higher temperature during curing than during storage (S-101 Technical Committee, 1980). The change in beta-amylase during curing is unclear. For six cultivars only small changes in beta-amylase activity were observed and the direction of change varied between cultivars (Walter et al., 1975). However, the beta-amylase activity of an uncured cultivar stored for 15 days at 24°C did not change, (Ajlouni and Hamdy, 1988). Cultivars low in beta-amylase activity have been developed in Japan. In one such cultivar the level of beta-amylase activity remained low throughout storage for 6 months (Baba et al., 1987). In spite of this, reducing and non-reducing sugar contents increased and starch content decreased significantly, which suggests that alpha-amylase, rather than beta-amylase, was responsible for the changes. This finding indicates that cultivars which are perceived as non-sweet immediately after harvest may not maintain this characteristic after a period of storage.

Large differences in carbohydrate fractions between cured and uncured roots may be evident after cooking. Baked cured roots have been shown to have higher levels of reducing sugars, total sugars and dextrins and less starch than roots baked immediately after harvest (Hammett and Barrantine, 1961), resulting in increased sweetness and 'moistness'. These changes were also shown in a recent study, but, with the exception of total sugars, were not significant (Purcell and Walter, 1988). The level of maltose produced in baked roots which had been cured was found to be less than that in roots baked immediately after harvest in the case of six North American cultivars (Picha, 1986a), perhaps because of altered beta-amylase activity. It was reported (Jenkins and Gieger, 1957) that of cured and uncured raw and baked sweet potatoes, only cured baked roots contained measurable quantities of dextrins and that these influenced the texture of cured roots. Dextrins have been found in uncured baked roots, but at lower levels than in cured baked roots (Hammett and Barrantine, 1961; Walter et al., 1975; Purcell and Walter, 1988). Evaluation of baked cured or uncured roots by a trained flavour and texture panel (Hamann et al., 1980) indicated that cured roots were sweeter and less starchy than uncured roots, and had a stronger caramel flavour. Cured roots were also perceived as 'moister' in mouthfeel and less 'chalky' than uncured roots. However, these differences were found to disappear if cured and uncured roots were stored for several weeks after harvest (Walter, 1987). No statistical differences could be distinguished between the carbohydrates of roots placed directly into storage without curing and those of cured roots even after as little as 9 days after harvest. In canned sweet potatoes, starch and alcohol-insoluble solids were lower in roots cured and stored for 2 weeks than in roots processed immediately after harvest, but this decrease was even greater in uncured roots held at room temperature for the same time (Kattan and Littrell, 1963). The accumulation of sugars during storage is independent of storage temperature (Picha, 1986b). This is in contrast to potatoes (*Solanum tuberosum*) which accumulate sugars most rapidly at temperatures below 4°C and which can be reconditioned at 21°C to reduce sugar concentration. These findings suggest that changes in carbohydrates noted in roots cured and stored under controlled conditions are likely to take place in a similar way in roots stored in ambient conditions in the tropics. However, no information is available about the rapidity of such changes at varying temperatures and humidities.

The measurements of alpha-amylase in the raw root, the amount and molecular sizes of dextrins formed in cooked roots and the quantity of starch remaining after cooking have been suggested as a means of predicting degree of mouthfeel 'moistness' or 'dryness' (S-101 Technical Committee, 1980). However, significant levels of dextrins were not found in three baked cultivars of varying mouthfeel after curing and storage (Kays and Horvat, 1984), leading the authors to suggest that, in these cultivars at least, beta-amylase was the enzyme primarily active in the hydrolysis of starch during baking. Alpha-amylase, however, is important in increasing the effectiveness of beta-amylase through the formation of additional end groups open to attack.

The average changes of root carbohydrates in four cured and stored cultivars before and after baking as determined by one group of workers are shown in Table 5.3. The general decrease in starch and increase in total sugars which takes place in raw sweet potatoes on curing continues during storage, as a result of metabolic activity. The concentrations of sucrose, glucose and fructose generally rise during storage (Picha, 1986b; Hoover, Walter and Giesbrecht, 1983), but are little affected by cooking. The rates and extent of concentration increases in the various sugars differ between cultivars (Deobald et al., 1971; Hoover et al., 1983; Martin, 1986). Sucrose was, moreover, found to decrease in baked roots as storage of dry white-fleshed cultivars progressed, and to increase in orange-fleshed cultivars (Picha, 1986a), indicating that sucrose metabolizing pathways, or enzyme activities differed in the two types of root. The quantity of maltose produced as a result of starch hydrolysis during

Table 5.3. *Changes in carbohydrates of sweet potato roots during curing and storage*

	% total sugar		% reducing sugar[a]		% Sucrose		% Starch		% Dextrins	
	Raw	Baked	Raw	Baked	Raw	Baked	Raw	Baked	Raw	Baked
Harvest	1.82	9.16	0.43	7.28	1.39	1.88	13.61	7.43	0	1.01
Curing[b]	4.60	11.04	0.65	7.12	3.95	3.92	11.92	4.18	0.24	2.09
Cured + 4 weeks' store[c]	4.60	11.64	0.70	7.39	3.90	4.25	12.07	2.63	0.26	2.20
Cured + 13 weeks	5.19	11.77	1.00	7.32	4.19	4.45	11.78	2.14	0.24	2.03
Cured + 21 weeks	5.50	11.56	0.94	6.99	4.56	4.57	10.97	2.70	0.23	1.65

Notes:

Based on: Ali and Jones, 1967. Average of four cultivars.

[a] The large increase in reducing sugar after baking is due to maltose production.

[b] Cured at 30°C and 85% relative humidity for 2 weeks.

[c] Stored at 15.5°C and 72% relative humidity.

cooking declines during the later stages of storage (Hoover et al., 1983; Picha, 1986a); probably less starch is available as substrate for beta-amylase activity. There are significant differences in the way cultivars respond to storage which affect the extent of starch conversion to sugars and dextrins. The percentage of starch converted to dextrins in several North American cultivars after 71 days of storage was about 27% in cultivars classed as 'moist' after cooking and only 9% in a 'drier' cultivar (Walter et al., 1975). Similarly, alpha-amylase activity increased most during storage in the 'moist' and least in the 'dry' cultivars.

Chips prepared from several North American cultivars were lightest coloured when made just after harvest, but darkened on storage of the raw roots, due to increased reducing sugar production (Picha, 1986b). Reducing sugars react with amino acids in the Maillard reaction to produce brown compounds known as melanoidins. Darkening was independent of whether curing was carried out and also of storage temperature. There were significant genetic differences in the extent of chip darkening as a result of storage, reflecting varying levels of reducing sugar increase in different cultivars.

Carbohydrate concentration changes, particularly a decrease in starch and an increase in sucrose, have been noted in irradiated and stored sweet potatoes (Lu et al., 1987; Ajlouni and Hamdy, 1988). However, there is some indication (Wang et al., 1982) that starch is unaffected at low dose rates of up to 0.07 kGy, and that the extent of changes in enzymes and sugars may vary with cultivar. Adverse textural changes in the form of tissue softening, probably associated with disruption of carbohydrates such as starch and pectin, were observed in sweet potatoes irradiated with 1.5 kGy or above (Lu et al., 1987).

Changes in non-starch polysaccharides, registered in roots during curing and storage may also play a part in determining the texture of cooked or processed roots. Cured and stored roots are generally softer and less firm than those cooked or canned immediately after harvest. The percentage of water-soluble pectin significantly increased and the percentages of calgon-soluble pectin, hemicellulose and cellulose all decreased during storage of two North American cultivars (Sistrunk, 1977). Greater changes in hemicellulose and cellulose occurred during the storage of 'Centennial' than that of 'Georgia Jet'. Changes in pectic fractions were found to be closely associated with the firmness of canned sweet potato (Baumgardner and Scott, 1963). Holding roots for 8 days at 30°C before processing resulted in a decrease in protopectin, accompanied by an increase in oxalate-soluble pectin. Total pectin did not change in raw roots during these 8 days, but declined in the processed product. Other work showed a decrease in total pectin in raw roots during curing (Ahmed and Scott, 1958). Intrinsic viscosity values and molecular weights of water-soluble, oxalate-soluble and acid-soluble pectic fractions were greatly reduced by curing. Apparent decreases in pectic materials (Baumgardner and Scott, 1963) and cellulose and hemicellulose (Sistrunk, 1977) in stored roots after cooking have been related to changes in solubility rather than losses in these compounds.

Textural properties of cooked sweet potatoes may also be influenced by interactions between pectic compounds and sugars or dextrins (Walter et al., 1975). The addition of sugars or chemically prepared dextrins altered the flow of pectin solution by reducing its viscosity (Chen and Joslyn, 1967a,b). Such interactions may be perceived by humans in cooked sweet potato as increased 'moistness'. The effect of cooking on carbohydrate constituents will be discussed further under Domestic cooking, below.

In summary, metabolic changes in carbohydrates and amylolytic enzymes which take place during curing and/or storage of sweet potato are manifested in cooked roots as reduction in the levels of starch and solubilities of non-starch polysaccharides, increases in the concentrations of sugars and dextrins and reductions in the molecular weights of dextrins, and interactions between pectic components, sugars and dextrins. Such changes result in increased sweetness, 'moistness' and softness of cooked roots, decreased viscosity of puréed roots and loss of firmness in canned roots. The extent of carbohydrate transformations varies significantly between cultivars. This fact offers the possibility of selecting cultivars which undergo minimal carbohydrate changes during storage and cooking for use in tropical areas where consumers express a preference for less sweet and 'drier' textured sweet potatoes.

Losses of dry matter (most of which is carbohydrate) during storage, can result in significant reductions in the energy value of roots. Dry matter storage losses are significant though small under optimum conditions of temperature and humidity (Hammett and Miller, 1982). They may be extremely pronounced at high ambient temperatures which encourage maximum root respiration rates and the occurrence of sprouting (Jenkins, 1982). Total root weight loss from a cultivar stored under ambient conditions in the Philippines (see above) was 45.2% after 17 days (Kawabata et al., 1984). Calculations reveal that approximately one third of this was dry matter loss, which is similar to an earlier finding for roots stored under controlled conditions for 4 months (Speirs et al., 1953). Lowering of root energy values may be particularly significant in the nutrition of subsistence families, relying on stored roots during seasons of scarcity of fresh roots or other staples. As there is little in the way of extended storage of sweet potatoes in tropical areas at present, this is not likely to be a widespread problem.

Lipids

The lipids of cured and stored sweet potatoes were described in detail in Chapter 2 (pp. 62–4). Studies on the changes taking place in the lipids of two North American cultivars as a result of curing, and storage at 15.5°C for 3 months indicated that there was a tendency for shorter chain saturated fatty acids, including palmitic, to decrease, while unsaturated fatty acids including linoleic and linolenic acids increased (Boggess et al., 1967). There was an increase in total lipids during storage, but not curing. However, this could have been due either to an increase in extractability of lipids with respirational composition changes of roots or to an actual synthesis of lipids from non-lipid components. Organoleptic values of baked roots, especially in relation to flavour, deteriorated with storage time, but this may have been related to increased oxidation of fatty acids. Changes in total lipids and lipid fractions were more pronounced at the low temperatures, which cause chilling injury in sweet potato roots during storage, than at 15.5°C. Increases in total lipid and changes in fatty acid concentrations during storage of several cultivars at 10°C or 15.5°C have also been noted by Charoenpong (1984). There do not appear to have been any investigations into lipid changes at tropical storage temperatures and humidities.

Nitrogenous compounds

Sweet potato roots appear to undergo active nitrogen metabolism while in storage as evidenced by decreases in total protein and changes in non-protein nitrogen (NPN). Several authors report losses in total protein (N × 6.25) in cured roots during storage, both during pit storage in straw at 8–10°C and 80–85% relative humidity for 4 months (Sharfuddin and Voican, 1984) and during controlled temperature storage at 13°C and 50–80% relative humidity for about 6 months (Purcell, Walter and Giesbrecht, 1978). Others (Li and Oba, 1985) noted the complete disappearance of the major soluble storage protein from roots stored at 10–12°C for 1 year. Cutting, or infection of the roots by *Ceratocystis fimbriata*, caused a degradation of this high molecular weight protein to proteins of lower molecular weight. However, the much greater metabolic loss of carbohydrates than of nitrogen during storage can result, for some cultivars, in an apparent increase in total protein with storage time (Purcell et al., 1978; Collins and Walter, 1985).

Changes in the amino acids of total protein during storage appear hardly to have been studied. Histidine decreased by 39% after 69 days storage at 13°C (Dickey et al., 1984) whereas the concentration of tryptophan remained the same.

The percentage of NPN has been found to decrease to a minimum during storage and then to increase once more (Purcell et al., 1978). Part of the NPN consists of purine nitrogen. Recent investigations suggest that nitrogen may be lost from stored tubers and roots, including sweet potato, by purine degradation resulting in the release of ammonia (Osuji and Ory, 1986, 1987), and that a feedback mechanism also operates to conserve this purine nitrogen.

Amounts of the various amino acids in NPN changed during storage at 13°C and uncontrolled humidity for 280 days (Purcell and Walter, 1980). Aspartic acid, threonine and serine increased during the first part of storage and then decreased. Lysine increased slightly towards the end of the storage period. Glycine, cysteine and ammonia did not change significantly, and all other amino acids decreased during the initial part of storage and then increased toward the end. NPN was suggested to be part of a metabolically active nitrogen pool, its relatively large amounts of asparagine being available for root synthesis of amino acids on demand, and aspartic and glutamic acids taking part in transamination reactions. The fall in NPN to a minimum during storage might represent a resting period before the root prepares itself for sprouting. The suggestion that once NPN has fallen to its minimum, storage conditions might be modified to maintain this minimum level and hence increase storage life is not practical in tropical situations. However, the finding, if proved to be true of roots under differing conditions of storage, could be nutritionally useful. As described in Chapter 3, the nutritional value of sweet potato protein decreases as the percentage of NPN increases. Hence roots removed from storage at the time of known minimum NPN levels, might contain higher concentrations of essential amino acids.

The nutritional implications of changes in nitrogenous compounds during storage will not be clear until such changes have been investigated in more detail and under varying storage conditions.

Vitamins

Carotenoids

Changes in the carotenoid pigments during curing and storage could affect their quality as perceived in terms of colour, but more importantly could alter the nutritional value of roots. In general, workers have found either increases in carotene or total carotenoids during storage, or little actual change. These results depended not only on whether root weight losses were taken into account for purposes of calculation, but also on genetic differences and temperature of storage.

Early work on this topic has been reviewed (Speirs et al., 1953).

Results from different investigators varied, carotenoid content being found to increase greatly during several months of storage when root weight losses were not taken into consideration. Two major groups of workers in this field found either retention of the original carotene content with no significant gain or loss (Speirs et al., 1945, 1953) or a nutritionally significant increase (Ezell and Wilcox, 1948, 1952) the extent of which depends on cultivar and storage temperature. The former results took root weight losses into account, carotene contents being calculated back to weights of roots at harvest. Little change in carotene or total carotenoids took place in any cultivar during storage at 10–11°C (Ezell and Wilcox, 1952; Speirs et al., 1953). Carotene and total carotenoids increased at 20°C, but not as rapidly as at 15.5°C, the temperature usually recommended for storage, which was therefore optimum for carotene retention (Ezell and Wilcox, 1952), at least for the cultivars under test. However, one cultivar lost carotene and total carotenoids at all storage temperatures. Another worker (McNair, 1956) found a 10% increase (calculated back to harvest basis) in one season's roots stored at 21°C for 6 months. A second season's roots showed no loss or gain in the first 3 months of storage, but lost about 10% in the succeeding 3 months.

High carotene content at harvest does not impede carotene synthesis on storage. During two seasons, in two North American cultivars cured and then stored at 15.5°C for several months, the actual increase in beta-carotene and total carotenoids was greater for each cultivar in the season when it had a relatively higher concentration of carotene (Ezell and Wilcox, 1958). Similarly, when three uncured Japanese-grown cultivars (representing high, medium and low levels of flesh carotenoids) were stored at about 13.6°C, the greatest real increase in beta-carotene and total carotenoids took place in the cultivar with the highest initial concentrations (Yamamoto and Tomita, 1958). The percentage increase in beta-carotene and carotenoids in the North American cultivars during storage was significantly greater in the low- than in the high-carotenoid cultivar in both seasons, whereas the pale-fleshed Japanese cultivar with a low initial carotenoid content actually lost beta-carotene and total carotenoids during storage. Cultivars with the highest concentrations of carotene at harvest also have the highest concentrations at any stage during storage (Ezell and Wilcox, 1958; Yamamoto and Tomita, 1958). The increase in carotenoid pigments during storage has been shown to be primarily due to an increase in beta-carotene (Yamamoto and Tomita, 1958).

More recent research generally confirms that carotenoids are not lost during storage and that they may even increase. One worker noted a marked decrease in beta-carotene in four cultivars at 10°C and 75%

relative humidity, but a rapid increase followed by a slight decline in one other cultivar cured, and stored at 15.5°C and 85% relative humidity (Charoenpong, 1984). However, total carotenoids were reported to increase significantly from an average of 13.5 to 18.3 mg/100 g (fwb) in four cultivars cured, and stored for 7 months at 16°C (Reddy and Sistrunk, 1980). There is no indication as to whether weight losses were taken into account in the calculations. There was an overall tendency for Taiwanese cultivars to increase in carotenoids when stored at room temperature (15°C to 20°C) during the winter season (Wang, 1974). Total carotenoids were higher (though significantly so in only one cultivar) after curing and short-term (4–5 weeks) storage at 7°C, 15.6°C or 26.6°C than at harvest in four North American cultivars (Picha, 1985). Data were corrected for weight loss during storage. The carotenoid content of two seasons' harvests of 'Jewel' cultivar cured and then stored at 12.8°C and 85–90% relative humidity for several months did not change in the first season, but increased significantly in the second (Hammett and Miller, 1982).

Beta-carotene content of sweet potatoes was reportedly unaffected by gamma-irradiation with doses of up to 0.07 kGy and subsequent storage for 5 months (Wang et al., 1982). Nor was there any apparent effect on total carotenoids with doses of up to 2 kGy and 2 weeks of storage (Lu et al., 1987).

In general therefore sweet potato roots containing nutritionally significant levels of beta-carotene and other carotenoids appear to retain their nutritional value during storage especially at temperatures above 10°C. However, there are indications that this may not be true of all cultivars. It has been suggested that carotene is both synthesized and degraded in the roots during storage, and that a true increase depends upon the rate of enzymically controlled synthesis being greater than that of degradation (Yamamoto and Tomita, 1958). Hence when retention of carotenoids during storage is nutritionally vital, carotenoid changes in individual cultivars under varying conditions of storage should be studied. Moreover, the decrease in carotenoid levels in sweet potatoes attacked by moulds during storage (see Chapter 2, p. 79) should be measured.

Ascorbic acid

Ascorbic acid has been found to decrease significantly during curing and storage (Ezell, Wilcox and Hutchins, 1948; Hollinger, 1944; Speirs et al., 1945; Yamamoto and Tomita, 1960; Hammett and Miller, 1982; Izumi, Tatsumi and Murata, 1984; Reddy and Sistrunk, 1980). The extent of ascorbic acid losses found varied according to curing or storage and between cultivars. When considering the losses reported by different workers it is also necessary to note whether total (reduced + dehydroascorbic acids) or only the reduced form was measured and whether root weight losses were taken into account. The measurement of only reduced ascorbic acid overestimates losses (Bradbury and Singh, 1986a; McCombs and Pope, 1958), whereas ignoring weight changes may underestimate them (Speirs et al., 1953). Most authors have measured only reduced ascorbic acid. Losses of up to 50% of the reduced form (Speirs et al., 1945, 1953; Ezell et al., 1948) or of total ascorbic acid (Yamamoto and Tomita, 1960) have been reported to occur as a result of several months of storage. In general, losses are most rapid during curing and the first part of the storage period, followed by little change or a more gradual decrease during the rest of storage (Izumi et al., 1984; McCombs and Pope, 1958; Speirs et al., 1945, 1953). During approximately 1 week of curing time, 16 cultivars lost an average of 6 mg reduced ascorbic acid/100 g (calculated taking root shrinkage losses into account), but only a further 4 mg/100 g were lost during an ensuing 4 months of storage (Speirs et al., 1953). The total true loss of reduced ascorbic acid was on average 43% of that present at harvest, but the concentration in roots in their actual condition at the end of storage was only 25% less than that in the freshly harvested roots, due to moisture loss.

Figure 5.14 shows the changes in reduced and total ascorbic acid taking place in one cultivar over 7 months of storage at three different temperatures. The increase in total ascorbic acid during the final part of storage was due to an increase in the dehydro form and may have been due to increased metabolic activity associated with sprouting. At each stage of storage, the percentage loss of total ascorbic acid was somewhat less than that of the reduced form, presumably because part of the reduced form was oxidized to the dehydro form.

Genetic differences in ascorbic acid losses during curing and storage have been reported. One group (Ezell et al., 1948) reported losses ranging from 12% to 41% in different cultivars during curing, but another group (Speirs et al., 1953) found genetic differences to be small and insignificant. However, both groups found significant genetic differences in ascorbic acid losses during storage. There was an indication that the cultivars with the highest initial ascorbic acid content experienced the greatest percentage losses during storage (Speirs et al., 1953).

Within the range 10–21°C, temperature of storage was not found to influence ascorbic acid losses greatly (Ezell and Wilcox, 1952). However, in experiments in which humidity was uncontrolled and during which roots experienced very high levels of shrinkage, true losses of reduced

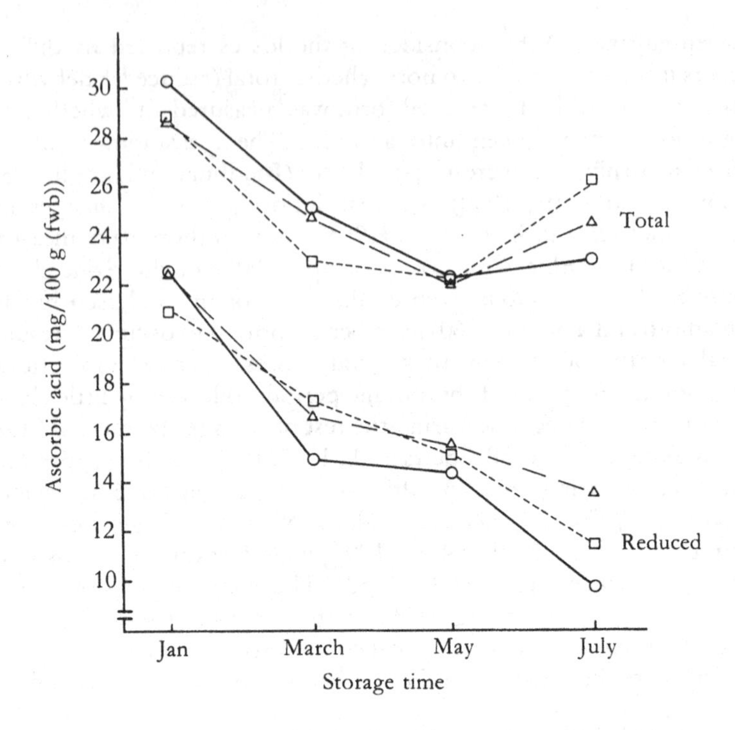

Figure 5.14. Changes in total and reduced forms of ascorbic acid during sweet potato root storage at three temperatures: (○) 13°C, (△) 15.5°C, (□) 18°C. The samples were cured for 10 days and the points shown are the average of 3 years' results. The retention of ascorbic acid was 73% after 4 months and (○) 76%, (△) 80% and (□) 86% after 6 months. (From McCombs and Pope, 1958.)

ascorbic acid were also extremely high, being 70% of the original value at one location (Speirs et al., 1953). Therefore it would appear that conditions which favour shrinkage also favour ascorbic acid losses, a finding with repercussions for storage under tropical conditions.

More recent results also show highly significant losses of reduced ascorbic acid during storage. The concentration of ascorbic acid in roots cured and then stored for 7 months at 16°C (relative humidity not given) was reduced from 20.75 mg/100 g to 15.64 mg/100 g (fwb), an apparent loss of about 25% (Reddy and Sistrunk, 1980). From figures given for cured roots with a much lower initial ascorbic acid content stored for 4 months in a pit between layers of straw at 8–10°C and 80–85% relative humidity, it can be calculated that the actual loss of original reduced ascorbic acid was 47% (Sharfuddin and Voican, 1984). The same cultivar ('Jewel') harvested, and then cured and stored under apparently the same conditions in two successive seasons lost ascorbic acid more rapidly in the second season than in the first (Hammett and Miller, 1982), even though the initial concentration of ascorbic acid was the same.

A more recent study (Bradbury and Singh, 1986a) determined the loss of both reduced and total ascorbic acid during short-term storage for 28 days at 25°C in a Solomon Islands' hybrid. This sample had previously experienced 1 week in transit at about 30°C (during which time natural curing had presumably taken place), and apparently did not undergo spoilage during the 28 days in store. The loss in total ascorbic acid was only 17% compared to 26% of the reduced form. However, obviously the decrease in reduced ascorbic acid was greater than the increase in the dehydro form of the vitamin.

Total ascorbic acid decreased significantly in one North American cultivar ('Jewel') treated with gamma-irradiation doses of 1.5 and 2 kGy and stored for 2 weeks, compared to untreated or lower dosage treated samples of the cultivar (Lu et al., 1987). Another cultivar ('Georgia Jet') was unaffected.

Further and more meaningful investigations of ascorbic acid losses during storage at elevated temperatures and uncontrolled humidities are required, taking into account transformations of reduced ascorbic acid into the other nutritionally active dehydro form as well as shrinkage (dry matter and moisture losses).

It should be remembered, however, that even where sweet potato roots are not stored as such for long periods in the tropics, they are often transported and marketed under conditions which may lead to a rapid loss of ascorbic acid. Wounding of roots by careless handling and transportation, and even short-term delays between harvest and sale, in conditions which encourage the growth of fungi and other pathogens, leads to initiation of tissue darkening by polyphenolic oxidation. Ascorbic acid, which acts as a reducing agent, may be lost during inhibition of this reaction (see Chapter 2). The equivalent of home storage for 24 hours at room temperature (figure not given) in India was found to reduce total sweet potato ascorbic acid by 35% (Pasricha, 1967), compared to only 14% during refrigeration at 4°C.

The finding that roots lose ascorbic acid most rapidly during high temperature curing and initial periods of storage would also appear to indicate that there is little that can be done in tropical conditions to minimize ascorbic acid losses other than careful handling to prevent wounding and infection. In cases where there is a delay between harvest and consumption therefore, pre-cooking as well as cooking losses must be considered when calculating the contribution which roots make to dietary ascorbic acid intakes.

There do not appear to have been studies carried out on the storage changes in ascorbic acid levels in sweet potato dried by traditional methods and stored as chips or flour. Losses during storage, however, must be unimportant considering that the drying process itself reduces ascorbic acid to very low levels (as little as 3% of the original value) in the dried product (see Chapter 6). Even samples of sweet potato oven-dried at 40°C for 2–3 days lost 20% of their total ascorbic acid (Bradbury and Singh, 1986a). Samples dried in traditional conditions are likely to undergo browning and attack by moulds as well as heat degradation and oxidation, which combine to produce a high ascorbic acid loss. It would be invaluable to have further information on the levels of ascorbic acid present in roots when they are actually utilized in various tropical diets.

There does not appear to have been any investigation into the diminution of ascorbic acid in sweet potato leaves which are not consumed immediately after harvest. It has been noted, however, that in at least one part of the world (Malaysia) marketed leaves may sometimes be displayed on cloths on the ground exposed to the sun; some if not sold may be held for at least 24 hours and are seen for sale after the leaves have darkened (Caldwell, 1972). The average ascorbic acid loss in a number of green leafy vegetables (not including sweet potato leaves) held covered for 24 hours at room temperature (not stated exactly) in India (Pasricha, 1967) was about 35% (range 7% in cabbage to 69% in soy leaves). Amaranthus leaves were reported to lose 23% and 37% of their ascorbic acid when kept for 4 and 8 hours, respectively, in partial sunlight (Oñate et al., 1970). Similar losses might be expected in sweet potato leaves subjected to delays between harvest and sale and kept in exposed conditions.

Other vitamins

There is virtually no information on the changes in other sweet potato root vitamins during storage. It seems likely that folic acid, which is susceptible to oxidation, will be lost in a manner similar to that of ascorbic acid. However, this will remain mere speculation until investigations into vitamin changes during storage are carried out. Mention is made in the literature (Junek and Sistrunk, 1978) that pantothenic acid increased and niacin decreased in 'Centennial' cultivar during curing, but no figures were given. Decreases were found in the levels of thiamin and riboflavin in three Japanese cultivars during 3 months of storage on rice straw in a wooden box at 13.6°C (Yamamoto and Tomita, 1960). These decreases are shown in Figure 5.15. Approximately 36% of riboflavin and between 20% and 30% of thiamin (depending on cultivar) were lost. Losses in thiamin, riboflavin, niacin and beta-carotene as a result of drying roots, which could be relevant to the nutritional value of traditionally dried products for storage, are discussed in Chapter 6.

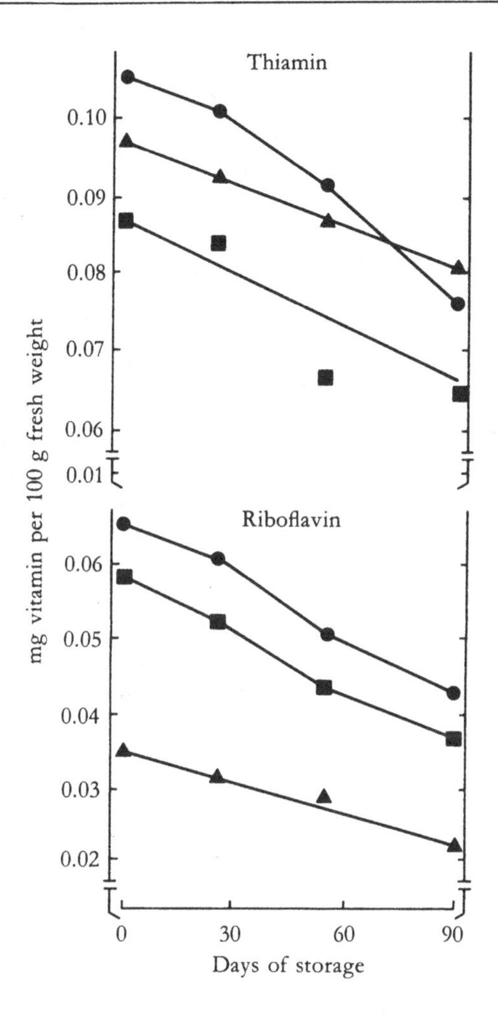

Figure 5.15. Changes in thiamin and riboflavin during sweet potato root storage at constant temperature (approximately 14°C). (▲) 'Nōrin no.2', (■) 'Hayato', (●) 'Kagoshima 7–1061'. After 90 days, the retention of thiamin was (▲) 81%, (■) 71% and (●) 71%, and that of riboflavin was (▲) 63%, (■) 65% and (●) 64%. (From Yamamoto and Tomita, 1960.)

The behaviour of tocopherol (vitamin E) during storage may vary with cultivar. In a recent study (Hirahara and Koike, 1989), the tocopherol content of 'Satsumahikari', a low-sweet cultivar, decreased by 50% after 8 months of storage, whereas that of 'Benihyato', a high beta-carotene type, did not change significantly during the same period.

Again, there is as yet no information on the levels of vitamins in sweet potato leaves kept for some time between harvest and consumption.

Minerals

There is no information on changes in mineral constituents of sweet potato roots during storage. However, losses in minerals are likely to be insignificant as minerals are normally lost through physical removal, for example by trimming of unwanted plant parts, migration into parts such as sprouts which are discarded, or leaching into wash- or cooking water.

B. Domestic cooking

Methods

Sweet potato roots and leaves are consumed mainly in the home in tropical areas, although they may also be eaten in restaurants as part of a meal or on the street as a snack. The basic methods of cooking roots used in all areas are boiling, steaming, baking or roasting, and frying. Variations on these basic methods may be used to produce a variety of dishes with characteristics to suit the tastes of local consumers (see Appendix 1). Most methods of preparation are simple, however, consisting of boiling or baking the sweet potato in its skin, after which the root is peeled or the flesh is scooped out and eaten. Alternatively roots may be peeled and cut into pieces before being boiled. Boiling is the most common method among many people who find the use of oil for frying or the addition of sugar or other ingredients too costly, for example farmers in some rural areas of West and Central Java, Indonesia (Watson, G., unpublished material). Sweet potato boiled in its skin is a popular part of evening tea in parts of India (Balagopalan, C., personal communication). In Peru and Taiwan, boiled sweet potato is often eaten mixed with boiled white rice. Small pieces of sweet potato are mixed and cooked with rice to make a form of heavy porridge eaten for breakfast in some areas of China (Zhang D.P., personal communication). In Uganda, a dish called *acok* (Iteso) or *umugoyo* (Baganda) is prepared by boiling peeled sweet potato and beans separately and then mashing them together in the ratio 2:1 (McDowell, 1970).

Baking the roots in country ovens until the skin is charred is very common among Indian villagers (Ghosh, S.P., personal communication). Sweet potatoes are baked together with other food items in stone-lined underground ovens on the islands of Micronesia (Dayrit, 1987). Where ovens are not available, whole roots may be roasted in the embers of a fire as in East and Central Kenya (Jana, 1982) or in Papua New Guinea (Kimber, 1972). In Tokyo and the surrounding areas of Japan street vendors carry around a wood-burning stove which heats a bed of small smooth stones into which sweet potatoes are placed for roasting

Figure 5.16. Seller of roasted sweet potato roots in Kawagoe, Japan (J.A. Woolfe).

and sold as a snack during cold seasons (Duell, 1990). *Tsuboyaki* sweet potato shops are also found, although less frequently. These have a large earthenware urn, heated from the bottom by some burning fuel, in which sweet potatoes, strung together on wire hooks, are suspended (see Figure 5.16). Baked yellow-orange fleshed sweet potatoes are also popular as a street snack in China (Sheng and Wang, 1988; and see Figure 5.17). Daily consumption of baked sweet potato in Jinan, Shandong Province, is estimated at 4 tons! (Wiersema et al., 1989). A frequent method of cooking roots in Japan is to prepare *tempura* by dipping peeled slices or pieces in light batter and frying. Sweet potato roots are steamed in a *mumu* in the Highlands of Papua New Guinea, that is they are wrapped in banana leaves and placed between white-hot stones in a pit (Earland, J., personal communication). As a vegetable, they are incorporated into curries in Bangladesh (Rashid, 1987) and Sri Lanka (Jayawardena and De Silva, 1987), where they are also spiced and used as a filling for pastries. In Korea considerable quantities of sweet potato are fried (Hong, 1982; see also Figure 5.18). The roots are usually sliced after peeling and then fried, for example in Peru, or India. Oil-fried chips are used as a sweet snack in the Yellow River Valley of China during the Chinese traditional spring festival (Sheng and Wang, 1988). Sweet potatoes are frequently made into desserts in some countries. In Thailand they may be boiled with sugar and ginger or coconut milk

Figure 5.17. Baked sweet potato sold as a roadside snack near Jinan, China (M.Iwanaga).

Figure 5.18. Different types of sweet potato can be used to prepare (clockwise) savoury potato-like sticks, sweet crunchy balls or deep-fried cakes, West Java, Indonesia (G.A. Watson).

(Poolperm, 1987). Local delicacies such as *lidgid* and *guinata'an* using sweet potato, sugar, coconut milk and flavourings are prepared by many households in the Philippines (Villamayor, 1987). *Ondei ondei*, made by boiling balls of mashed sweet potato stuffed with brown sugar and rolled in grated coconut, is a snack or dessert in Malaysia (Tan, S.L., personal communication). A variety of desserts made with varying proportions of sweet potato are served in some coffee shops in Tokyo (Figure 5.19). Indeed in some parts of Japan, there exist restaurants which specialize in complete meals incorporating sweet potato into soups, salads, main course and even icecream and crème caramel!

In the United States, baking is the most usual method of preparation, being carried out in a convection oven at a carefully controlled temperature. Microwave baking is also becoming increasingly popular. In some countries, notably Japan, attractive leaflets and booklets containing a variety of recipes and suggestions for domestic preparation are used to popularize sweet potato (including cultivars high in beta-carotene or anthocyanins), often through the medium of the press.

The heat treatments used for cooking are relatively mild compared to those which the roots undergo during processing and should therefore cause less reduction in nutritional value. The most popular traditional

Figure 5.19. Tokyo sweet potato coffee shop serving and selling a variety of sweet potato desserts (J.A. Woolfe).

methods cook the sweet potato intact, the skin acting as a barrier to loss of nutrients to the environment or into the cooking water.

Sweet potato leaves or tips are normally boiled or steamed, either alone or mixed with other foods. They may be incorporated into soups. In the Philippines, a dish called *sinapaw* is prepared by cooking the leaves with rice (Oliveros and Sumabat, 1968). In Tonga and Fiji a favourite way of eating the leaves is to to cook them briefly with some onion, then marinade them in coconut cream and eat them cold as a salad (Foss, P., personal communication). Specialized dishes such as petioles boiled in soy sauce and vine tip tempura (vine tips dipped in batter and deep-fried) are prepared in Kawagoe, Japan (Duell, 1990). Petioles are also consumed in Korea as a substitute for other vegetables undergoing seasonal shortages (Chin, M.S., personal communication). Sweet potato tops are rarely used for human food in China; this has been attributed to browning of the leaves during cooking. *Ipomoea aquatica* is said to be preferred as a vegetable in China as, during boiling, it retains an attractive green colour (Zhang, D.P., personal communication).

Sweet potato roots are generally not eaten raw, although yellow-fleshed cultivars are consumed by both adults and children in the raw state in Bangladesh (Rashid, 1987). They are sometimes eaten raw in a 'fruit salad' in Indonesia (Watson, G., unpublished material). Raw orange-fleshed sweet potato is occasionally used as a salad ingredient (in a way similar to carrot) in parts of the United States (Collins, W.W., personal communication). Strips of blanched (partially cooked) white, orange- or purple-fleshed root may be included in salads served in some Japanese restaurants.

Cooking changes the texture and flavour of the roots as well as increases the digestibility of certain nutrients especially starch. It may also serve to reduce the levels of toxic terpenoid phytoalexins (Cody and Haard, 1976) and anti-nutritional trypsin inhibitors (see Chapter 4). However, prolonged cooking causes losses of nutrients through heat degradation, oxidation and reduction of bioavailability. The majority of investigations into changes on cooking have taken place in the United States, with an emphasis on the taste and textural changes during conventional oven baking. These investigations aim to determine the most important biochemical changes associated with changes in organoleptic quality and to assess those characteristics most preferred by American sweet potato consumers. Less emphasis has been placed on determining nutritional losses, although some information is available. Very little research has been carried out into nutritional changes in sweet potato leaves and tips in association with human feeding.

Quality and nutritional changes

Carbohydrates

Roots

The changes in carbohydrates during cooking have been mentioned above (see Curing and storage). Most people in the tropics at the time of writing do not store their sweet potatoes but cook and eat them directly after harvesting or after a very short interval between harvest and purchase. As indicated previously, these roots will generally be less sweet in taste, and firmer and 'drier' in texture than equivalents which have been cured and/or stored for some time.

The mouthfeel and sweetness of the cooked roots depends largely on the activities of the amylase enzymes, which vary among cultivars or clones (Almazan, 1987; Martin and Deshpande, 1985; Walter et al., 1975). The relative 'dryness' of freshly harvested cooked roots is no doubt partially attributable to the low levels of alpha-amylase with consequently small production of dextrins and hence low dextrin-pectin interaction (see Curing and storage, above). The amylases in 'moist' cultivars may also be active over a greater range of temperature than those of 'dry' cultivars (Shen and Sterling, 1981). There are many possible sources of variation in enzyme activity which may affect sugar concentrations and hence sweetness in cooked roots, for example differences in enzyme concentration, presence (or absence) of enzyme inhibitors and degree of molecular branching in starch (Martin and Deshpande, 1985; Walter et al., 1975). The major change associated with cooking is the amylolytic hydrolysis of starch and the consequent production of maltose (see Table 2.2, p. 44) and dextrins. However, there are also changes associated with non-starch polysaccharides such as pectins, hemicelluloses and cellulose (see Table 5.4).

Starch apparently has to be gelatinized before enzymic hydrolysis can occur. Maltose was found in baked sweet potatoes only above the gelatinization temperature of 70–75°C (Lee and Ahn, 1981). The amylase enzymes are relatively heat stable and remain active for several minutes at the temperatures which disrupt starch granules. The sequence of events thought to occur during baking has been described (S-101 Technical Committee, 1980) as follows:

1. When the temperature of the flesh reaches about 77°C, starch is degraded at a rate proportional to the quantity of alpha-amylase present. Beta-amylase produces maltose from the fragments of starch produced by alpha-amylase.
2. As the temperature of the flesh rises, beta-amylase is inactivated but alpha-amylase proceeds to break down starch until it is finally inactivated at about 95°C. Without beta-amylase to degrade starch fragments to maltose, dextrins accumulate until the inhibition of alpha-amylase takes place.

Correlations between sensory panel scores and objective measurements suggest (S-101 Technical Committee, 1980) that moistness of cooked roots may be predicted by measurement of (a) alpha-amylase in the raw root, (b) the quantity and sizes of dextrins formed, (c) the amount of starch remaining after baking, (d) the viscosity of a purée made from cooked tissue or (e) determination of pectin in the cooked roots. The simplest procedure is the determination of viscosity, and the higher the viscosity the 'drier' the texture of the cooked root. Rheological measurements and their relationship to cooked sweet potato mouthfeel have been investigated (Rao et al., 1974, 1975a,b).

The differences in 'dry' and 'moist' cultivars after cooking are shown by the data in Table 5.4 (Shen and Sterling, 1981). Although the raw roots of both types of cultivar are essentially similar in carbohydrate composition, on baking only the non-reducing sugars and cellulose remained the same. In the 'moist' cultivar the amount of reducing sugars increased much more, the starch content decreased more rapidly and to a lower level, the water-soluble pectin increased and the content of hemicelluloses markedly decreased. The authors speculated that hemi-

Table 5.4. *Carbohydrate changes in 'dry' and 'moist' roots during baking (% dwb)*

	Raw		Baked	
Variable	Jersey ('Dry')	Garnet ('Moist')	Jersey ('Dry')	Garnet ('Moist')
Starch	49.1	46.2	22.9	2.61
Amylose	13.4	13.6	5.19	0.70
Reducing sugars	3.44	4.44	12.4	20.0
Total sugars	19.4	22.4	29.2	37.6
Pectin (water-sol.)[a]	2.39	1.95	2.68	3.32
Pectin (Calgon-sol.)[b]	0.53	0.47	0.67	0.31
Hemicellulose	4.95	3.82	5.73	1.00
Cellulose	3.26	2.66	2.39	2.54
(Moisture	70.6	70.2	66.9	65.8)

Notes:
sol, soluble
Based on: Shen and Sterling, 1981.
[a] High methoxyl pectin (forms gel with sugar and acid).
[b] Low methoxyl pectin and pectic acid (forms gel with calcium salts).

cellulose fibrils in the cell wall break down more on baking in the 'moist' cultivars. Similarly the increase of water-soluble pectin in baked 'moist' types may indicate a breakdown of part of the pectin in the middle lamella between cells as well as in the cell wall itself. These changes could result in cell rupture, creating large intercellular spaces and release of cell sap giving a 'syrupy' texture.

There are indications that the method of cooking itself can affect the degree of sweetness and 'moistness' of roots particularly by its effect on the length of time for which the amylases can remain active. Roots baked at 200°C or 230°C were softer, sweeter and more 'moist' and contained a greater percentage of reducing sugars than those baked at 150°C or 180°C (Losh et al., 1981). The internal temperature of the former roots rose more rapidly than the latter and hence reached the optimum temperature for amylase activity more quickly, enabling it to work for a longer time. Less starch was converted to sugars in roots cut up into strips and then rapidly cooked by steaming than in baked roots (Walter and Hoover, 1984), presumably because the amylases were denatured more quickly in the strips. Boiled roots of any given cultivar are generally less sweet and 'moist' than those which have been baked, as boiling more rapidly raises the temperature of the flesh to that at which the enzymes are inactivated, and less starch is hydrolysed. Significant carbohydrate composition differences were found between baked and boiled roots (Sistrunk, 1977) including a lower percentage of starch, and higher percentages of hemicellulose and cellulose in the former than in the latter. Baked or microwave-cooked roots were generally higher in reducing sugars, total sugars and pectins and lower in starch than were boiled or steamed roots, which were not significantly different in carbohydrate composition (Reddy and Sistrunk, 1980). The microwave-cooked roots were apparently higher in hemicellulose and cellulose than the other roots. These increases in hemicelluloses and cellulose probably reflect changes in solubility and extractability rather than increases per se.

For any given cooking method, the activity of the amylases and hence the quantity of starch converted to reducing sugars varies with cultivar. For several North American cultivars, the percentage of starch converted was lowest at 54% in a relatively 'dry' cultivar after baking, compared to 63–69% conversion in 'moist' cultivars. Ninety nine per cent of the hydrolysed starch was converted to maltose. This high level of starch conversion to maltose on cooking, even immediately after harvest, was thought to be characteristic of all sweet potato cultivars. However, in recent years a number of low-sweet or non-sweet clones have been distinguished in Nigeria, Japan and Puerto Rico (Almazan, 1987; Kukimura et al., 1989; Martin and Deshpande, 1985; Martin, 1986). A very low-sweet cultivar, 'Satsumahikari', was released in Japan in 1987. In such cultivars, reducing sugars either do not increase on cooking or increase only slightly, a finding interpreted as the absence, or a very low concentration, of beta-amylase (Kumagai et al., 1990; Martin, F.W. and Deshpande, D.S., unpublished data). Their degree of sweetness then depends on their total sugar content after cooking. On the basis of the sweetness rating and total soluble sugar content after baking of a clone consistently rated as non-sweet by a taste panel, a total soluble sugar content after baking of 20% or less on a dry weight basis has been suggested as an indicator of non-sweetness (Almazan, 1987). On this basis, 60 of 380 clones in the germ plasm collection of the IITA were classed as non-sweet (Almazan, 1987). The reducing, non-reducing and total sugars and the mouthfeel of a number of clones ranging from very sweet to non-sweet after baking is shown in Table 5.5. Only one non-sweet clone, 'Ninety nine', was found to be generally unacceptable for culinary purposes by taste-testing (Martin and Beauchamp de Caloni, 1987), as its texture was extremely dry and its flesh a grey unpleasant colour on cooking. Some of the selections were not suitable for baking, being too dry and hard, but were acceptable either boiled and mashed, or fried (Martin and Beauchamp de Caloni, 1987). The non-sweet clones

Table 5.5. *Sugars in cultivars of varying sweetness before and after baking*

		Reducing sugars		Total sugars		
Cultivar	Type	Before baking	After baking	Before baking	After baking	Mouthfeel[a]
'Gem'	Dessert	7.7	25.6	23.6	39.7	5
'Frita'	Tropical	2.8	15.6	13.2	25.8	3
'Nutty'	Tropical	8.5	22.1	13.8	28.9	3
'Spud'	Tropical	4.6	18.8	12.5	26.4	3
'Francia'	Sub-staple	7.6	12.8	19.0	22.0	3
'Limonette'	Sub-staple	3.7	3.8	17.0	16.8	2
'Papota'	Sub-staple	2.4	6.8	9.2	13.4	2
'Mojave'	Staple	3.2	5.7	8.3	11.1	2
'Ninety nine'	Staple	0.8	0.8	6.2	6.5	1
'Sahara'	Staple	3.2	3.1	15.2	12.7	1

Notes:
Based on: Martin, 1986.
[a] Rated 1 to 6, from very dry to very moist.

have been designated staple, and the low-sweet clones sub-staple, types. They could fulfil a role in the diet similar to that of rice, potatoes and other tropical roots and tubers, and are much easier to produce in the hot humid tropics than are potatoes (Martin, 1987). The Japanese cultivar 'Satsumahikari' can be processed (see Chapter 6) into granules, flakes and snacks with taste and colour very similar to those of corresponding potato (*S. tuberosum*) products (Baba, T., personal communication). While most of the selections have white flesh, a few contain carotenoids and have a yellow or orange flesh colour (Martin and Beauchamp de Caloni, 1987) and could therefore be useful in areas where a drier-fleshed, less sweet sweet potato is preferred, but where the presence of provitamin A carotenoids is nutritionally desirable or essential. Moreover, a vigorous programme of crossing to breed new cultivars which combine low sweetness with high levels of beta-carotene or anthocyanin pigments is being pursued in Japan (Umemura, Y., personal communication).

One of the major nutritional changes in sweet potato carbohydrate on cooking is the increased digestibility of starch. This has been demonstrated *in vitro* by the action of bacterial alpha-amylase and *in vivo* by the growth performance of rats (Cerning-Beroard and Le Dividich, 1976) compared to those fed maize starch, by comparison of the starch from sweet potatoes either raw or boiled in water for 30 min. Cooked sweet potato starch was much more susceptible to enzymic hydrolysis than the raw starch and produced a higher growth performance in rats, not significantly different from that of maize starch. This increased digestibilty on cooking is essential if sweet potato is to act as an efficient energy source for human feeding and also to ensure the maximum digestibility of other organic constituents, such as protein, which may be reduced by carbohydrate of low digestibility.

Boiling and steaming have been found to increase the moisture content of sweet potato roots by about 4% and 2%, respectively, but baking decreases it by about 7% (Bradbury et al., 1988). Dry baking was therefore calculated to confer advantages, compared with boiling or steaming, of 40% or 30% increases respectively in energy per unit weight of root, hence decreasing the bulkiness of the food. It was pointed out that this would be a particular advantage for the young child whose intake capacity is limited. Baking (especially with the skin intact) would therefore appear to be the method of choice for preparing roots for child feeding, as it also minimizes loss of vitamins and minerals (see below).

A significant increase in dietary fibre was found to take place in boiled, steamed or baked Tongan sweet potatoes (Bradbury et al., 1988). This was attributed by the authors to a modification of part of the starch on cooking so that it became 'resistant starch' (see Chapter 2) which was measured as a fraction of the dietary fibre. The nutritional significance of resistant starch is not known at present, but it is possible that it plays a role in flatulence production (see Chapter 4). Cooking also initiates changes in non-starch polysaccharides which constitute the dietary fibre fraction of sweet potatoes. For one cultivar it has been shown that hemicellulose significantly decreases and cellulose slightly decreases on baking (Shen and Sterling, 1981). The percentage of water-soluble pectin increases and that of acid-soluble pectin decreases during baking (Lee, Shin and Ahn, 1985). Total pectins decrease. The physiological effects of such changes in sweet potato as a source of dietary fibre have not been investigated.

Tops

Food composition tables (Leung, Butrum and Chang, 1972; Haytowitz and Matthews, 1984) show no significant change in the moisture content of leaves on cooking. The slight increase in concentration of carbohydrates which is shown to take place is therefore no doubt due to a concentration effect brought about by loss of soluble nutrients, especially protein. An experimental study (Oñate et al., 1970) showed a slight increase in moisture content on cooking of young sweet potato shoots and a decrease in total carbohydrate from 9% to 6% (fwb).

Nitrogen

Roots

The normal cooking methods of boiling, steaming and baking have little effect on either the quantity or quality of sweet potato root total protein. Experimental boiling, steaming and baking of peeled Tongan sweet potato roots cut into chunks (Bradbury et al., 1988) or boiling of similarly prepared roots imported into the United Kingdom (Faulks, R.M. and Kwiatkowska, C.A., unpublished data) did not significantly alter the total protein content. Food composition tables indicate similar results (Haytowitz and Matthews, 1984; Ashida, 1982; Leung et al., 1972). Candying of roots as carried out in the United States can decrease the total protein on a fresh weight basis by dilution through addition of ingredients such as sugar and butter. The total protein of fresh or boiled roots was shown to decrease from 1.7% to 1.3% (fwb) (Watt and Merrill, 1975) or 0.9% (Haytowitz and Matthews, 1984) in candied roots.

The quality of a protein may be reduced by heat treatment which is severe enough to initiate amino acid–sugar reactions leading to the production of brown pigments (melanoidins). This is known as non-enzymic or Maillard browning and may render part or all of the content of a susceptible amino acid such as lysine nutritionally unavailable. Protein damage increases with increase in severity of heat treatment, but is usually minimal during mild methods of heating such as are employed during domestic cooking. Though little work has been done to investigate the effects of cooking on sweet potato root protein quality, one group of workers (Purcell and Walter, 1982) showed that convection oven baking in the skin for 90 min at 190°C produced little change in amino acids compared to the processes of canning or drum-drying (see Table 6.11, p. 362). Furthermore, the amino acid composition of the baked sweet potatoes was similar to that calculated for the composite of nitrogenous fractions extracted from raw sweet potatoes. The lysine content appeared to be little affected by baking. These results make it seem unlikely that the quality of sweet potato protein is affected by boiling or steaming.

Tops

Though the effect of cooking on the quality of sweet potato leaf protein does not seem to have been investigated as yet, the quantity of total protein is significantly reduced by cooking. This is no doubt due to leaching of soluble protein and free amino acids into the cooking water. The total protein content of young shoots decreased from 3.7% to 2.5% (fwb) when cooked in water for 4 min (Oñate et al., 1970). Food composition tables show the total protein content (fwb) of North American leaves to have decreased from 4% to 2.3% (Haytowitz and Matthews, 1984) and that of Asian leaves from 3.2% to 2.6% (Leung et al., 1972) on steaming or boiling. The nitrogenous compounds lost into the cooking water can be utilized if the water is retained and used in a soup or stew.

Vitamins

Roots

There have been few determinations of the changes in vitamin content of sweet potato roots on cooking and these have been confined mainly to the determination of the reduced form of ascorbic acid. Some information can be obtained from food composition tables, but with the disadvantage that comparisons of foods before and after cooking are based on the assumption (which may be false) that raw and cooked samples were drawn from the same original batch. Food composition tables have the drawback that details of cooking treatments are not supplied, being limited to basic descriptions such as boiled, baked etc., a fault also common among authors who sometimes fail to note details such as whether the roots were peeled or not before cooking. Such details are important in that the destruction or preservation of vitamins can vary greatly with the method of cooking employed. The final vitamin content is influenced by the presence or absence of the sweet potato root skin, which helps to prevent leaching of water-soluble constituents into the cooking-water, the degree of heat employed and the presence of oxygen and metal ion catalysts in the case of vitamins such as carotenoids or ascorbic acid susceptible to oxidation. Mild cooking can enhance absorption of carotenoids by increasing digestibility, or releasing carotenoids from their complexes (in raw vegetables) with other constituents (Erdman, Poor and Dietz, 1988), whereas overcooking reduces their bioavailability. Hence it was discovered, when examining the literature or calculating losses of vitamins from food composition tables, that losses found by different authors vary widely. Methods of cooking differ from country to country. Most information is at present confined to the most common methods of boiling and baking, there being little for roasting and frying. Much more information is required, both for a greater range of vitamins and also for local, popular methods of cooking sweet potato roots.

Present knowledge of vitamin losses on domestic cooking is summarized in Table 5.6. These have been taken directly from authors or calculated from food composition tables.

Losses of total carotenoids, which contain many double bonds, are

Table 5.6. *Percentage losses of vitamins from sweet potato roots during domestic cooking*

Treatment	Carotenoids	Ascorbic acid	Thiamin	Riboflavin	Niacin	Pantothenic acid	Pyridoxine	Folic acid	Tocopherol
Boiled in skin[a]	20–25[b]	20	10	0	0				
Boiled, peeled[1]	20–25[b]	18–78[c]	20[d]–60[e]	5[d]–20[f]–40[f]	0[e]–20[f]–40[f]	10[d]	5[d]–40[e]	20[d]–50[e]	
Baked in skin	30–40[b]	15–20[a]	0[d]–10[a]	0[a]–14[d]	0[a]–10[d]	0[d]	6[d]		60–75[g]
Baked, peeled[2f]		45–68[c]	20	20	20[f]				
Steamed[b]		44	20	20	38				25–45[g]
Roasted[b]		41	8	16	6				
Fried[3]		13[i]–56[j]							

Notes:

[1] Lowest losses = boiled whole; medium losses = boiled, cut up, water retained; highest losses = boiled cut up, water discarded.

[2] Lowest loss = baked 30 min; highest loss = baked 45 min.

[3] Lowest loss = fried simply in oil; highest loss = stir-fried in cast iron vessel with added salt and condiments.

Calculated or derived from:

[a] Watt and Merrill, 1975.

[b] Losses due to degradation and isomerization, see Table 5.7.

[c] Bradbury and Singh, 1986a.

[d] Haytowitz and Mattews, 1984.

[e] Kwiatkowska et al., 1989.

[f] Bradbury and Singh, 1986b.

[g] Hirahara and Koike, 1989.

[h] Ashida, 1982.

[i] Watson, 1976.

[j] Pasricha, 1967.

mainly due to oxidative degradation promoted by light, heat, metals and peroxides. Losses on boiling are apparently fairly small. A Brazilian cultivar containing 9 mg total carotenoids/100 g (fwb) (80% of which were beta-carotene) lost less than 10% of carotenoids and beta-carotene when placed in water already boiling and cooked for 20 min (Cascon et al., 1988). Losses on oven baking have been found to vary with cultivar (Wentworth and Dempsey, 1961). Baking in the skin reduced total carotenoid content by about 25%, significantly more than the loss of about 10% by boiling in the skin (McNair, 1956). A recent study (Chandler and Schwartz, 1988) found a 31.4% loss of total beta-carotene from 'Jewel' as a result of oven baking the whole root wrapped in aluminium foil for 80 min at 191°C. However, others found that, for an average of four cultivars, there was no difference in the carotenoid content whether roots were oven-baked, boiled, steamed or microwave-cooked (Lanier and Sistrunk, 1979).

Determinations of carotenoid losses alone, however, do not present the whole picture in terms of the provitamin A activity to be expected in the heated food. In fresh vegetables the all-*trans* isomeric forms of the carotenoids, particularly all-*trans* beta-carotene which has the highest biological activity of the various carotenoid isomers, predominate. However, thermal processing, including cooking, can cause conversion of part of the all-*trans* carotenes to *cis* isomers which have lower provitamin A activity than the all-*trans* forms. Determination of the total provitamin A potential of sweet potato therefore requires a knowledge not only of total carotenoid losses, but also of the extent of isomer formation during cooking or processing and a calculation of the total activities of the various isomers finally present in the heated food. Several investigators have found the percentage of *cis* isomers in raw sweet potato roots to be nil (Chandler and Schwartz, 1987) or very low (Chandler and Schwartz, 1988; Panalaks and Murray, 1970; Sweeney and Marsh, 1971) as shown in Table 5.7. The presence of small amounts of *cis* isomers in raw roots has been attributed to prolonged storage before analysis (Chandler and Schwartz, 1988). After boiling (Sweeney and Marsh, 1971) or baking (Chandler and Schwartz, 1988), however, the percentage of *cis* isomers in the total beta-carotene, remaining after losses due to degradation, is significantly increased (Table 5.7). The biological value, or percentage equivalent activity of all isomers expressed as all-*trans* beta-carotene (biological value taken as 100%), shown in column 5 of Table 5.7 can be used to calculate the effective provitamin A activity of the cooked sweet potato. If the loss of activity through isomerization is accounted for as well as that by degradation, the true loss of carotenoid activity is somewhat greater than that shown solely by determination of total carotene before and after cooking (see notes to Table 5.7).

Table 5.7. *Isomeric changes in beta-carotene as a result of cooking*

Sample of sweet potato	All-*trans* β-carotene (%)	13-*cis* isomer[a] (%)	9-*cis* isomer[b] (%)	Biological value (%)
Fresh[c]	96.7	0.1	3.2	97.9
Boiled (20 min)[c]	69.7	24.9	5.4	85.6
Fresh[d]	94.9	5.1		97.6
Baked (80 min)[d]	77.0	23.0	Trace	89.2

Notes:
If boiled root loses 10% (data from authors not listed) of total beta-carotene by degradation, and that remaining has biological value of 85.6% then total loss of activity by degradation and isomerization = 23%.
If baked root loses 31% (Chandler and Schwartz, 1988) of total beta-carotene by degradation and that remaining has biological value of 89% then total loss of activity by degradation and isomerization = 38.5%.
[a] Assuming activity 53% of all-*trans*.
[b] Assuming activity 38% of all-*trans*. (Zechmeister, 1962).
[c] Sweeney and Marsh, 1971.
[d] Chandler and Schwartz, 1988.

As might be expected, losses of other vitamins are in general lowest when the sweet potato is boiled or baked with its skin intact, the skin acting particularly as a barrier against leaching of nutrients into the cooking-water. Some investigators have found higher levels of vitamins in baked than in raw roots, but this is undoubtedly a concentration effect resulting from loss of moisture at high temperatures.

Losses of ascorbic acid appear on the whole to be the highest for any vitamin (although most have not yet been investigated thoroughly), but vary greatly depending on the presence or absence of the skin during cooking, the length of cooking time and the conservation or otherwise of the cooking-water. Losses can, however, be overestimated if only the reduced form of the vitamin is determined. Peeled and cut sweet potato pieces boiled for 20 or 30 min, and analysed with the cooking-water, apparently lost 42% and 69% of ascorbic acid, respectively, when the reduced form only was determined (Bradbury and Singh, 1986a). When both reduced and dehydroascorbic acids were determined these losses were found to be 18% and 50%, respectively. Moreover, discarding the cooking-water increased respective total ascorbic acid losses to 74% and 84%. Whenever possible, therefore, sweet potato cooking-water should be used for preparing soups or stews. This is a frequent practice in some tropical countries. The losses described above are somewhat greater than those of reduced ascorbic acid found by other authors. Boiled intact roots lost 10–20% (McNair, 1956) and boiled peeled roots about 20–30% (Haytowitz and Matthews, 1984; Kwiatkowska, Finglas and Faulks, 1989; Pasricha, 1967; Watson, 1976) when cooked for 30 min or less. Pressure cooking reduced ascorbic acid by only 14% according to one author (Pasricha, 1967), and by about 45% according to others (Odachi et al., 1980).

Baking of intact roots caused a loss of 12–20% reduced ascorbic acid (McNair, 1956). Oven baking of peeled roots, said to simulate cooking in the coals of a fire as used in the South Pacific (Bradbury et al., 1988) reduced total ascorbic acid by 11% with 15 min of baking and up to 68% with 45 min of baking (Bradbury and Singh, 1986a). Simple frying in oil caused only a small loss in ascorbic acid (Watson, 1976; and see Table 5.6). However, frying in an iron vessel (which could have served to catalyse oxidation) with added salt and other condiments produced a much greater loss (Pasricha, 1967; and see Table 5.6). One group of workers found no significant difference in the average reduced ascorbic acid content of four cultivars with boiling, baking or steaming, but microwave-cooked roots had a significantly lower concentration (Lanier and Sistrunk, 1979). It is not clear, however, whether roots were peeled or left intact before cooking.

One author (McNair, 1956) determined the total loss of reduced ascorbic acid sustained by roots subjected to the combined effects of storage and cooking. Roots stored for 3 months and then boiled or baked unpeeled lost respectively 30% or 25–37% of ascorbic acid. Those stored for 6 months lost 43–61% and 45–67%, respectively. The ranges indicate cultivar differences.

Changes in sweet potato root folic acid on cooking have hardly been investigated. However, folic acid appears to be susceptible to losses of about 50% on boiling (Kwiatkowska et al., 1989). It suffers from heat instability and susceptibility to leaching similar to those of ascorbic acid.

Losses of B vitamins on cooking have not been thoroughly investigated. Recent analyses of seven types of root and tuber crops including sweet potato revealed no differences between crops with respect to B vitamin losses on cooking (Bradbury and Singh, 1986b). Losses of thiamin, riboflavin and niacin averaged 20% after boiling peeled roots for 20 min and retaining the cooking-water, 40% after boiling and discarding the water and 23% after 30 min of baking. The obvious leaching of these water-soluble nutrients again emphasizes the need to retain and utilize the water in which sweet potato is cooked. Pressure cooking of roots was found to reduce thiamin by 25%, compared to 15% for normal boiling for 25 min (Odachi et al., 1980). Baking retained significantly more niacin and riboflavin than did boiling or steaming in

North American roots (Lanier and Sistrunk, 1979). Riboflavin was also reduced by microwave cooking, but niacin was not affected.

Little is known about the behaviour of pantothenic acid and vitamin B_6 in cooked sweet potato roots. However, they appear to be little affected, at least during boiling and baking (see Table 5.6), although one group of authors (Kwiatkowska et al., 1989) detected 40% loss of vitamin B_6 on boiling peeled cut roots. There was no difference in pantothenic acid content between baked, boiled or microwave-cooked roots, but steaming reduced pantothenic acid significantly relative to the other treatments (Lanier and Sistrunk, 1979).

Losses of alpha-tocopherol (vitamin E) during steaming, baking and microwave cooking have recently been studied in a number of Japanese cultivars (Hirahara and Koike, 1989). Losses were variable both between cultivars and cooking methods. Baking caused greater losses (about 60–75%) than steaming (about 25–45%), whereas the microwave cooker hardly affected tocopherol content, with the exception of one cultivar, 'Simon 1', which suffered losses of more than 60% by all methods of cooking. In spite of such high tocopherol losses, cooked sweet potatoes eaten in quantity remain an important source of vitamin E.

Tops

Sweet potato tops, in common with other tropical green leafy vegetables, have been almost completely ignored by investigators seeking information about changes in vitamin content on cooking. Only one study showing analyses for five vitamins in both raw and cooked sweet potato tops was found in the literature (Oñate et al., 1970). In order to obtain most of the information given here it was necessary to calculate losses from figures given for raw and cooked leaves in food composition tables. Even this information is sparse. This situation must be remedied if efforts to increase the consumption of tropical vegetables are to have any meaning and support.

The carotenoid content of Philippine sweet potato tips apparently increased on boiling in water for 4 min, when calculated back to a fresh weight basis (Oñate et al., 1970). Perhaps the extractability of the carotenoids increased during cooking. Other sources found losses varying from 10% (Haytowitz and Matthews, 1984) to about 35% (Leung et al., 1972). Perhaps losses are greater from leaves than from roots due to the large surface area of leaf exposed to the cooking-water. Alternatively, carotenoids may be less tightly bound to other compounds or be present in sites more easily exposed to oxidation in leaves than in roots. In addition to losses there may be some conversion of all-*trans* beta-carotene to *cis* isomers as described above for the roots.

Although such a change has not been measured directly for sweet potato leaves, an indication of its extent can be gained from an investigation of other types of green leafy vegetables (Sweeney and Marsh, 1971). The biological value (as defined above) for several types of fresh greens was about 90%. This decreased to about 87% after 5–10 min of boiling. Thus the total loss of beta-carotene in green leaves that apparently lose 10% on cooking would be about 22% if isomerization is taken into account. Such changes should be explored for sweet potato greens cooked in different ways so that their final provitamin A activity can be more accurately assessed.

Losses of 80% of reduced ascorbic acid have been reported both on boiling Ghanaian leaves (Watson, 1976) and in preparing a Filipino dish called *sinapaw* where leaves were cooked for 8–9 min with rice (Oliveros and Sumabat, 1968). A smaller loss of 44% was found for *sinigang*, in which leaves were cooked for 4 min with citric acid added (Oliveros and Sumabat, 1968). A portion of freshly prepared *sinigang* was determined to provide approximately 75% of the Filipino daily requirement for ascorbic acid. Another Filipino study found that reduced ascorbic acid decreased from 98 to 73 mg/100 g (fwb), which represents only a 25% loss, when sweet potato tips were cooked for 4 min (Oñate et al., 1970). The cooking-water contained 3.2 mg/100 g of ascorbic acid. Much of the loss was therefore due to degradation of the vitamin. Reduction of ascorbic acid in cooked leaves to very low levels of only about 1 mg/100 g (fwb) has been reported by several sources (Leung et al., 1972; Haytowitz and Matthews, 1984; Evensen and Standal, 1984). It would be useful if dieticians could devise recipes adapted to local needs which minimized the loss of water-soluble and heat-sensitive constituents. Alternatively consumers should be advised to utilize the water used for cooking leaves to prepare soups and stews. This may only be possible if cultivars with a mild flavour are developed.

Loss of thiamin through cooking sweet potato tips in water for 4 min was negligible (Oñate et al., 1970). Losses of riboflavin and niacin appeared to be about 24% and 8%, respectively, but significant quantities of the vitamins were found in the cooking-water. Calculations using food composition tables indicate that thiamin losses can range from about 25% to 50%. Similarly riboflavin losses are about 20–30% and niacin losses 10–20%. Leaves can be a good source of vitamins if used when fresh and cooked for the minimum time, and followed by subsequent utilization of the cooking-water.

Minerals

There are only two means by which minerals can be lost during cooking of roots or tubers, namely during peeling, which removes the outer skin and variable percentages of the underlying flesh, and through leaching of soluble minerals such as K and Na into the surrounding cooking-water. Minerals may also be lost from leaves through leaching.

Roots

It has already been indicated in Chapter 2 (pp. 82–3) that the peel of sweet potato has been found to contain a higher percentage of total ash and Ca, Mg, Fe, Zn and Mn than the remaining pulp. Peeling of the root is therefore likely to reduce the concentration of these minerals and trace elements before cooking.

The effect of cooking itself can be deduced by comparing mineral contents of raw and cooked roots after peeling. Most food composition tables indicate that the effects of boiling or steaming peeled roots on most mineral contents is negligible (Faulks, R.M. and Kwiatkowska, C.A., unpublished data; Haytowitz and Matthews, 1984; Leung et al., 1972) with the exception of K which decreases (Faulks, R.M. and Kwiatkowska, C.A., unpublished data; Haytowitz and Matthews, 1984). An apparent slight increase in some minerals on roasting (Ashida, 1982) or baking (Haytowitz and Matthews, 1984) can be accounted for by loss of moisture in the former, but not in the latter, which had the same moisture content as the raw roots. The skin of the baked sweet potatoes, presumably removed after baking, accounted for 22% of the total weight, whereas the peelings and trimmings from the raw roots, with which they were compared, accounted for 28% of the weight (Haytowitz and Matthews, 1984). It is possible therefore that the higher concentrations of Ca, Mg, P, K, Cu and Mn in the edible portion of the baked roots was due to the lower percentage of peel removed.

Boiling peeled cut root pieces in distilled water, steaming them or oven baking them, significantly reduced ash content in the first two methods (Bradbury et al., 1988). Results for individual minerals are shown in Table 5.8. In boiled roots there was no significant change in any mineral except Zn, which decreased. Insignificant decreases in Na and K in boiled, Ca, Mg, Na, Mn, Al and B in steamed and Ca, Mg, Mn, Al and B in baked roots were recorded. Decreases were attributed to solubilization in water. The smallest changes took place in baked roots, a further reason for encouraging their use for child feeding. Roots should preferably be baked intact and the skin removed with the greatest care after baking to minimize mineral loss.

Tops

Losses of Ca and Fe from sweet potato tips due to boiling in an equal weight of water for 4 min were small (8% and 6%, respectively), but the

Table 5.8. *Percentage changes in minerals in sweet potato roots during domestic cooking*

Minerals	Boiled	Steamed	Baked
Ca	0.5	−6.7	−2.0
P	1.0	1.4	1.0*
Mg	2.8	−3.7	−0.6
Fe	NS[a]	NS	NS
Na	−12.7	−10.4	−2.7
K	−36	47*	37
S	1.1	1.1	0.8
Cu	NS	NS	NS
Zn	−0.05**	0.01	0.06
Mn	0.01	−0.03	−0.01
Al	0.18	−0.01	−0.03
B	0	−0.01	−0.01

Note:
From: Bradbury et al., 1988. Roots peeled and cut up. Differences from raw control samples calculated on basis of fresh weight. Asterisks indicate a significant change: $*p<0.05$, $**p<0.01$.
[a] NS, no significant change, but percentages not given by authors.

P content was reduced by 50% (Oñate et al., 1970). Comparisons (which should be treated with caution as indicated above) of leaves before and after cooking from food composition tables indicates that losses can be quite high. It can be calculated that leaves lost Ca (30%), Fe (40%), P (30%), K (8%) and no Mg or Na on steaming (Haytowitz and Matthews, 1984). Losses of Ca, Fe, P and K in cooked Asian leaves can be calculated as 70, 85, 25 and 18%, respectively (Leung et al., 1972). Losses may be influenced by the mineral content of the water or the vessel in which they are cooked, but this remains to be investigated.

References

Abrams, C.F. 1984. Depth limitations on bulk piling of sweet potatoes. *Trans. Am. Soc. Agric. Eng.* **27** (6): 1848–53.

Abrams, C.F. and Fish, J.D. 1982. Air flow resistance characteristics of bulk piled sweet potatoes. *Trans. Am. Soc. Agric. Eng.* **25** (4): 1103–6.

Ahmed, E.M. and Scott, L.E. 1958. Pectic constituents of the fleshy roots of the sweet potato. *Proc. Am. Soc. Hort. Sci.* **71**: 376–87.

Ahn, J.K., Collins, W.W. and Pharr, D.M. 1980. Influence of preharvest temperature and flooding on sweet potato roots in storage. *HortScience* **15** (3): 261–3.

Ajlouni, S. and Hamdy, M.K. 1988. Effect of combined gamma-irradiation and storage on biochemical changes in sweet potato. *J. Food Sci.* **53** (2): 477–81.

Aldous, T. 1976. Storage of sweet potato tubers. In: Wilson, K. and Bourke, R.M. (eds.), *Papua New Guinea Food Crops Conference Proceedings 1975*, Department of Primary Industry, Port Moresby, pp. 229–36.

Ali, M.K. and Jones, L.G. 1967. The effect of variety and length of storage on the carbohydrate contents and table quality of sweet potatoes. *Pak. J. Sci. Ind. Res.* **10** (2): 121–6.

Almazan, A.M. 1987. Selecting nonsweet clones of sweet potato from the IITA germ plasm collection. In: Terry, E.R., Akoroda, M.O. and Arene, O.B. (eds.), *Tropical root crops: root crops and the African food crisis*, Proceedings of the Third Triennial Symposium of the International Society for Tropical Root Crops, IDRC, Ottawa, pp. 76–8.

Anon. 1960. Storing sweet potatoes. *World Crops* **12** (2): 73.

1988. Good storage makes root crops go further. *Agric. Int.* **40** (7/8): 171.

Antarlina, S.S. 1990. Post harvest processing of sweet potato for food diversification in Indonesia. Paper presented at the User's Perspective with Agricultural Research and Development (UPWARD) Workshop, 2–7 April, Los Baños.

Arinze, A.E., Naqvi, S.H.Z. and Ekundayo, J.A. 1975. Storage rot of sweet potato (*Ipomoea batatas*) and the effect of fungicides on extracellular cellulytic and pectolytic enzymes of the causal organism. *Int. Biodeterior. Bull.* **11** (2): 41–7.

Avinze, A.E. and Smith, I.M. 1982. Effect of storage conditions on the resistance of sweet potato tissues to rotting by *Botryodiplodia theobromae* (Pat.) and other fungi. *J. Stored Prod. Res.* **18** (2): 37–41.

Ashida, J. (Resources Council, Science and Technology Agency) (ed.). 1982. *Standard tables of food composition in Japan*, 4th revised edn. Printing Bureau, Ministry of Finance, Japan, pp. 47–8.

Baba, T., Nakama, H., Tamaru, Y. and Kono, T. 1987. [Development of snack foods produced from sweet potatoes. V. Changes in sugar and starch contents during storage of new type sweet potato (low beta-amylase activity in roots)] Japanese. *J. Jap. Soc. Food Sci. Technol.* **34** (4): 249–53.

Baumgardner, R.A. and Scott, L.E. 1963. The relation of pectic substances to firmness of processed sweet potatoes. *Proc. Am. Soc. Hort. Sci.* **83**: 629–40.

Boggess, T.S., Marion, J.E., Woodroof, J.G. and Dempsey, A.H. 1967. Changes in lipid composition of sweet potatoes as affected by controlled storage. *J. Food Sci.* **32**: 554–8.

Booth, R.H. 1974. Post-harvest deterioration of tropical root crops: losses and their control. *Trop. Sci.* **16** (2): 49–63.

Bourke, R.M. 1985. Sweet potato (*Ipomoea batatas*) production and research in Papua New Guinea. *Papua New Guinea J. Agric. Forest. Fish.* **33** (3–4): 89–108.

Bradbury, J.H. and Singh, U. 1986a. Ascorbic acid and dehydroascorbic acid content of tropical root crops from the South Pacific. *J. Food Sci.* **51** (4): 975–8, 987.

1986b. Thiamin, riboflavin, and nicotinic acid contents of tropical root crops from the South Pacific. *J. Food Sci.* **51** (6): 1563–4.

Bradbury, J.H., Bradshaw, K., Jealous, W., Holloway, W.D. and Phimpisane, T. 1988. Effect of cooking on nutrient content of tropical root crops from the South Pacific. *J. Sci. Food Agric.* **43** (4): 333–42.

Buescher, R.W. 1980. Storability of wounded sweet potato roots as affected by curing and noncuring treatments. *Ark. Farm Res.* **29** (6): 6.

Cadiz, T.G. and Bautista, O.D.K. 1967. Sweet potato. In: Knott, J.E. and Deanon, J.R. (eds.), *Vegetable production in Southeast Asia.* College of Agriculture, University of the Philippines, Los Baños, Laguna, pp. 48–65.

Caldwell, M.J. 1972. Ascorbic acid content of Malaysian leaf vegetables. *Ecol. Food Nutr.* **1** (4): 313–17.

Cascon, S.C., Alves, I.T.G., Bittencourt, A.M. and Leal, N.R. 1988. [Carotenoids of Jerimu sweet potato: stability on domestic cooking and industrial processing] Portuguese. *Congr. Brasil. Cien. Tecnol. Aliment., Recife, Abstr.* **11**: 126.

Ceponis, M.J. and Butterfield, J.E. 1974. Retail and consumer losses in sweet potatoes marketed in metropolitan New York. *HortScience* **9** (4): 393–4.

Ceponis, M.J., Kaufman, J. and Tietjen, W.H. 1973. Effects of DCNA and prepackaging on the retail quality of sweetpotatoes. *HortScience* **8** (1): 41–2.

Cerning-Beroard, J. and Le Dividich, J. 1976. [Feeding value of some starchy tropical products: in vitro and in vivo study of sweet potato, yam, malanga, breadfruit and banana] French. *Ann. Zootech.* **25** (2): 155–68.

Chandler, L.A. and Schwartz, S.J. 1987. HPLC separation of *cis–trans* carotene isomers in fresh and processed fruits and vegetables. *J. Food Sci.* **52** (3): 669–72.

1988. Isomerization and losses of *trans-β*-carotene in sweet potatoes as affected by processing treatments. *J. Agric. Food Chem.* **36** (1): 129–33.

Charoenpong, C. 1984. A study on beta-carotene and lipid composition of sweet potatoes and the effect of low oxygen during storage. *Diss. Abstr. Int. B* **45** (6): 1721.

Chen, T-S. and Joslyn, M.A. 1967a. The effect of sugars on viscosity of pectin solution. 1. Comparison of corn syrup with sucrose solutions. *J. Colloid. Interface Sci.* **23**: 399–406.

1967b. The effect of sugars on viscosity of pectin solution. 2. Comparison of dextrose, maltose and dextrins. *J. Colloid. Interface Sci.* **25**: 346–52.

Cody, M. and Haard, N.F. 1976. Influence of cooking on toxic stress metabolites in sweet potato root. *J. Food Sci.* **41** (2): 469–70.

Collins, W.W. and Walter, W.M. 1985. Fresh roots for human consumption. In: Bouwkamp, J.C. (ed.), *Sweet potato products: a natural resource for the tropics.* CRC Press, Inc., Boca Raton, FL, pp. 153–73.

Cooley, J.S. 1951. The sweet potato, its origin and primitive storage practices. *Econ. Bot.* **5**: 378–86.

Data, E.S. 1985. Screening of different sweet potato genotypes for longer storability. In: *Root crops.* ViSCA, Baybay, Leyte, Ann. Rep. Phil. Root Crop Res. Training Center, pp. 68–9.

Dayrit, R.S. 1987. Status of sweet potato (*Ipomoea batatas*) cultivation and utilization in Micronesia. Paper presented at an International Sweet Potato Symposium, 20–26 May, ViSCA, Baybay, Leyte.

Deobald, H.J., Hasling, V.C. and Catalano, E. 1971. Variability of increases in alpha-amylase and sugars during storage of Goldrush and Centennial sweet potatoes. *J. Food Sci.* **36** (2): 413–5.

Dickey, L.F., Collins, W.W., Young, C.T. and Walter, W.M. 1984. Root protein quantity and quality in a seedling population of sweet potatoes. *HortScience* **19** (5): 689–92.

Duell, B. 1990. Ways of eating sweet potatoes in Japan. In: *Tropical root and tuber crops changing role in a modern world*, Eighth Symposium of the International Society for Tropical Root Crops, 30 October–5 November, Bangkok, (in press).

Edmond, J.B. and Ammerman, G.R. 1971. *Sweet potatoes: production, processing, marketing*, AVI Publ. Co., Inc., Westport, CT.

Erdman, J.W., Poor, C.L. and Dietz, J.M. 1988. Factors affecting the bioavailability of vitamin A, carotenoids, and vitamin E. *Food Technol.* **42** (10): 214–21.

Evensen, S.K. and Standal, B.R. 1984. *Use of tropical vegetables to improve diets in the Pacific region.* HITAHR Res. Ser. 028, University of Hawaii.

Ezell, B.D. and Wilcox, M.S. 1948. Effect of variety and storage on carotene and total carotenoid pigments in sweetpotatoes. *Food Res.* **13**: 203–12.

1952. Influence of storage temperature on carotene, total carotenoids and ascorbic acid content of sweetpotatoes. *Plant Physiol.* **27**: 81–91.

1958. Variation in carotene content of sweet potatoes. *Agric. Food Chem.* **6** (1): 61–5.

Ezell, B.D., Wilcox, M.S. and Hutchins, M.C. 1948. Effect of variety and storage on ascorbic acid content of sweetpotatoes. *Food Res.* **13**: 116–22.

Friedman, B.A. 1960. Market diseases of fresh fruit and vegetables. *Econ. Bot.* **14**: 145–56.

Gooding, H.J. and Campbell, J.S. 1964. Improvement of sweet potato storage by cultural and chemical means. *Emp. J. Exp. Agric.* **32**: 65–75.

Gull, D.D. and Duarte, O. 1974. Curing sweetpotatoes (*Ipomoea batatas*) under tropical conditions. *Am. Soc. Hort. Sci. Proc. Trop. Region 22nd* **18**: 166–72.

Hahn, S.K. and Anota, T. 1982. 1. The effect of temperature on the development of the sweet potato weevil. 2. The effect of water treatment on the survival of the sweet potato weevil. 3. Mortality pattern of weevils buried in soil at different depths. *IITA Annu. Rep. 1981*, Ibadan, Nigeria.

Hamann, D.D., Miller, N.C. and Purcell, A.E. 1980. Effects of curing on the flavor and texture of baked sweet potatoes. *J. Food Sci.* **45** (4): 992–4.

Hammett, H.L. and Barrantine, B.F. 1961. Some effects of variety, curing and baking upon the carbohydrate content of sweetpotatoes. *Proc. Am. Soc. Hort. Sci.* **78**: 421–6.

Hammett, L.K. and Miller, C.H. 1982. Influence of mineral nutrition and storage on quality factors of 'Jewel' sweet potatoes. *J. Am. Soc. Hort. Sci.* **107** (6): 972–5.

Haytowitz, D.B. and Matthews, R.H. 1984. *Composition of foods: vegetables and vegetable products.* Human Nutrition Information Ser., USDA Agric. Handbook No. 8–11, pp. 428–39.

Hirahara, F. and Koike, Y. 1989. [Tocopherol content in sweet potato tubers of different cultivars, places harvested and cooking methods] Japanese. *Jap. J. Nutr.* **47** (2): 85–91.

Hollinger, M.E. 1944. Ascorbic acid value of the sweetpotato as affected by variety, storage, and cooking. *Food Res.* **9**: 76–82.

Hong, E.H. 1982. The storage, marketing and utilization of sweet potatoes in Korea. In: Villareal, R.L. and Griggs, T.D. (eds.), *Sweet potato*, Proceedings of the First International Symposium. AVRDC, Shanhua, T'ainan, pp. 405–11.

Hoover, M.W., Walter, W.M. and Giesbrecht, F.G. 1983. Preparation and sensory evaluation of sweet potato patties. *J. Food Sci.* **48** (5): 1568–9.

Icamina, P.M. 1985. From subsistence to supermarket: sweet potatoes go commercial. *The IDRC Reports*, October, pp. 35–7.

Ikemiya, M. and Deobald, H.J. 1966. New characteristic alpha-amylase in sweet potatoes. *J. Agric. Food Chem.* **14** (3): 237–41.

Izumi, H., Tatsumi, Y. and Murata, T. 1984. [Effect of storage temperature on changes of ascorbic acid content of cucumber, winter squash, sweet potato and potato] Japanese. *J. Jap. Soc. Food Sci. Technol.* **31** (1): 47–9.

Jana, R.K. 1982. Status of sweet potato cultivation in East Africa and its future. In: Villareal, R.L. and Griggs, T.D. (eds.), *Sweet potato*, Proceedings of the First International Symposium. AVRDC, Shanhua, T'ainan, pp. 63–72.

Jayawardena, S.D.G. and De Silva, K.P.U. 1987. Indigenous technologies and recent advances in sweet potato production, processing, utilization and marketing in Sri Lanka. Paper presented at an International Sweet Potato Symposium, 20–26 May, ViSCA, Baybay, Leyte.

Jenkins, P.D. 1982. Losses in sweet potatoes (*Ipomoea batatas*) stored under traditional conditions in Bangladesh. *Trop. Sci.* **24** (1): 17–28.

Jenkins, W.F. and Gieger, M. 1957. Curing, baking time and temperatures affecting carbohydrates in sweet potatoes. *Proc. Am. Soc. Hort. Sci.* **70**: 419–24.

Junek, J. and Sistrunk, W.A. 1978. Sweet potatoes high in vitamin content but content is affected by variety and cooking method. *Ark. Farm Res.* **27** (5): 7–8.

Kattan, A.A. and Littrell, D.L. 1963. Pre- and post-harvest factors affecting firmness of canned sweet potatoes. *Proc. Am. Soc. Hort. Sci.* **83**: 641–50.

Kawabata, A., Sawayama, S., del Rosario, R.R. and Noel, M.G. 1984. Effect of storage and heat treatment on the sugar constituents of tropical root crops. In: Uritani, I. and Reyes, E.D. (eds.), *Tropical root crops: postharvest physiology and processing.* Japan Scientific, Tokyo, pp. 243–58.

Kays, S.J. and Horvat, R.J. 1984. A comparison of the volatile constituents and sugars of representative Asian, Central American, and North American sweet potatoes. In: *Proceedings of the Sixth Symposium of the International Society for Tropical Root Crops.* International Potato Center, Lima, pp. 577–86.

Keleny, G.P. 1965. Sweet potato storage. *Papua New Guinea Agric. J.* **17** (3): 102–8.

Kimber, A.J. 1972. The sweet potato in subsistence agriculture. *Papua New Guinea Agric. J.* **23** (3 & 4): 80–95.

Kukimura, H., Yoshida, T., Komaki, K., Sakamoto, S., Tabuchi, S., Ide, Y. and Yamakawa, O. 1989. 'Satsumahikari': a new sweet potato cultivar. *Bull. Kyushu Nat. Agric. Expt. Stn* **25** (3): 250.

Kumagai, T., Umemura, Y., Baba, T. and Iwanaga, M. 1990. The inheritance of β-amylase null in storage roots of sweet potato, *Ipomoea batatas* (L.) Lam. *Theor. Appl. Genet.* **79** (1): 1–8.

Kushman, L.J. and Pope, D.T. 1972. Causes of pithiness in sweet potatoes. *N. Carolina Agric. Exp. Stn Tech. Bull.* No. 207.

Kushman, L.J. and Wright, F.S. 1969. *Sweet potato storage*. USDA Agric. Handbook No. 358.

Kwapata, M.B. 1983/84. Effect of sun curing time and application of ash on storability of sweet potato (*Ipomoea batatas*). *Bunda Res. Bull.* **12**: 171–80.

Kwiatkowska, C.A., Finglas, P.M. and Faulks, R.M. 1989. The vitamin content of retail vegetables in the UK. *J. Hum. Nutr. Dietet.* **2** (3): 159–72.

Lancaster, P.A. and Coursey, D.G. 1984. Traditional post-harvest technology of perishable tropical staples. *FAO Agric. Serv. Bull.* **59**: 38–40.

Lanier, J.J. and Sistrunk, W.A. 1979. Influence of cooking method on quality attributes and vitamin content of sweet potatoes. *J. Food Sci.* **44** (2): 374–6, 380.

Lawrence, J. 1985. Post-harvest storage of yam and sweet potato. *Carib. Farming*, July, p. 32.

Lee, E.H. and Ahn, S.Y. 1981. Studies on the change of sugars in sweet potatoes on heating. *J. Korean Agric. Chem. Soc.* **24** (4): 245–50.

Lee, K.A., Shin, M.S. and Ahn, S.Y. 1985. The changes of pectic substances in sweet potato cultivars during baking. *Korean J. Food Sci. Technol.* **17** (6): 421–5.

Leung, W-T.W., Butrum, R.R. and Chang, F.H. 1972. *Food composition table for use in East Asia*. Part I. *Proximate composition, mineral and vitamin contents of East Asian foods*. US Department of Health, Education and Welfare, Bethesda, MD; FAO, Rome.

Li, H.-S. and Oba, K. 1985. Major soluble proteins of sweet potato roots and changes in proteins after cutting, infection, or storage. *Agric. Biol. Chem.* **49** (3): 737–44.

Losh, J.M., Phillips, J.A., Axelson, J.M. and Schulman, R.S. 1981. Sweet potato quality after baking. *J. Food Sci.* **46**: 283–6, 290.

Lu, J.Y., White, S., Yakubu, P. and Loretan, P.A. 1987. Effects of gamma irradiation on nutritive and sensory qualities of sweet potato storage roots. *J. Food Qual.* **9** (16): 425–35.

Martin, F.W. 1986. Sugars in staple type sweet potatoes as affected by cooking and storage. *J. Agric. Univ. Puerto Rico* **70** (2): 121–6.

1987. Introducing staple-type sweet potatoes. A potential new food crop for the tropics. *Agric. Int.* **39** (4): 114–18.

Martin, F.W. and Beauchamp de Caloni, I. 1987. Culinary characteristics of new selections of sweet potato. *J. Agric. Univ. Puerto Rico* **71** (4): 365–72.

Martin, F.W. and Deshpande, S.N. 1985. Sugars and starches in a non-sweet sweet potato compared to those of conventional cultivars. *J. Agric. Univ. Puerto Rico* **69** (3): 401–6.

Martin, F.W. and Roderiguez-Sosa, E.J. 1985. Preference for color, sweetness, and mouthfeel of sweet potato in Puerto Rico. *J. Agric. Univ. Puerto Rico* **69** (1): 99–106.

McCombs, C.L. and Pope, D.T. 1958. The effect of length of cure and storage temperature upon certain quality factors of sweet potatoes. *Proc. Am. Soc. Hort. Sci.* **72**: 426–34.

McDowell, J. 1970. Food utilization in Uganda. Report, Makerere University, Kampala. [Mimeo]

McNair, V. 1956. Effect of storage and cooking on carotene and ascorbic acid content of some sweet potatoes grown in northwest Arkansas. *Ark. Agric. Exp. Stn Bull.* No. 574.

Miyazato, M., Ishiguro, E. and Dannao, A. 1981. [Quality evaluation of agricultural products by infrared imaging method. IV. Grading of fruits and root vegetables for bruise and other surface defects] Japanese. *Bull. Fac. Agric. Kagoshima Univ.* **31**: 149–56.

Morris, S.C. 1981. Postharvest storage and handling of sweet potatoes. *CSIRO Food Res. Quart.* **41** (3–4): 63–7.

Nair, G.M., Ravindran, C.S., Moorthy, S.N. and Ghosh, S.P. 1987. Indigenous technologies and recent advances in sweet potato production, processing and utilization in India. Paper presented at an International Sweet Potato Symposium, 20–26 May, ViSCA, Baybay, Leyte.

Numfor, F.A. and Lyonga, S.N. 1987. Traditional postharvest technologies of root and tuber crops in Cameroon: status and prospects for improvement. In: Terry, E.R., Akoroda, M.O. and Arene, O.B. (eds.), *Tropical root crops: root crops and the African food crisis*, Proceedings of the Third Triennial Symposium of the International Society for Tropical Root Crops – Africa branch. IDRC, Ottawa, pp. 135–9.

Odachi, J., Fujita, T., Kanbe, T. and Oshiba, K. 1980. [Changes of vitamins and amino acids in foods after pressure cooking] Japanese. *Jap. J. Nutr.* **38** (5): 267–73.

Ogawa, M., Hyodo, H. and Uritani, I. 1969. Biochemical effects of gamma radiation on potato and sweet potato tissues. *Agric. Biol. Chem.* **33**: 1220–2.

Ogawa, M. and Uritani, I. 1969. Metabolic changes in sweet potato roots induced by gamma radiation in response to cutting. *Radiation Res.* **39**: 117–25.

Oliveros, M.S. and Sumabat, L.M. 1968. Ascorbic acid losses in some cooked vegetables. I. Petsay and kamote leaves. *Phil. J. Nutr.* **21** (4): 241–51.

Olorunda, A. 1979. Storage and processing of some Nigerian root crops. In: Plucknett, D. (ed.), *Small-scale processing and storage of tropical root crops.* Westview Press, Boulder, CO, pp. 90–9.

Oñate, L.U., Arago, L.L., Garcia, P.C. and Abdon, I.C. 1970. Nutrient composition of some raw and cooked Philippine vegetables. *Phil. J. Nutr.* **23** (3): 33–44.

Osuji, G.O. and Ory, R.L. 1986. Purine degradative pathway of the yam and sweet potato. *J. Agric. Food Chem.* **34** (4): 599–602.

1987. Regulation of allantoin and allantoic acid degradation in the yam and sweet potato. *J. Agric. Food Chem.* **35** (2): 219–23.

Panalaks, T. and Murray, T.K. 1970. The effect of processing on the content of carotene isomers in vegetables and peaches. *Can. Inst. Food Technol. J.* **3** (4): 145–51.

Pasricha, S. 1967. Effect of different methods of cooking and storage on the ascorbic acid content of vegetable. *Ind. J. Med. Res.* **55** (7): 779–84.

Paterson, D.R., Jones, A., Wagner, A.B., Earhart, D.R., Walker, D.W. and Fuqua, M.C. 1980. Sweet potato storage under plastic. (Abstr.). *HortScience* **15** (3, sect. 2): 378

Picha, D.H. 1985. Crude protein, minerals, and total carotenoids in sweet potatoes. *J. Food Sci.* **50** (6): 1768–9.

1986a. Sugar content of baked sweet potatoes from different cultivars and lengths of storage. *J. Food Sci.* **51** (3): 845–6, 848.

1986b. Influence of storage duration and temperature on sweet potato sugar content and chip colour. *J. Food Sci.* **51** (1): 239–40.

Poolperm, N. 1987. Indigenous technologies and recent advances in sweet potato production, processing, utilization and marketing in Thailand. Paper presented at an International Sweet Potato Symposium, 20–26 May, ViSCA, Baybay, Leyte.

Prasad, S.M., Srinivasan, G. and Shanta, P. 1981. Post harvest loss in sweet potato in relation to common method of harvest and storage. *J. Root Crops, India* **7** (1): 69–73.

Purcell, A.E. and Walter, W.M. 1980. Changes in composition of the nonprotein-nitrogen fraction of 'Jewel' sweet potatoes (*Ipomoea batatas* (Lam.)) during storage. *J. Agric. Food Chem.* **28** (4): 842–4.

1982. Stability of amino acids during cooking and processing of sweet potatoes. *J. Agric. Food Chem.* **30** (3): 443–4.

1988. Comparison of carbohydrate components in sweet potatoes baked by convection heating and microwave heating. *J. Agric. Food Chem.* **36** (2): 360–2.

Purcell, A.E., Walter, W.M. and Geisbrecht, F.G. 1978. Changes in dry matter, protein and non-protein nitrogen during storage of sweet potatoes. *J. Am. Soc. Hort. Sci.* **103**: 190–2.

Rajamma, P. 1984. Control of *Cylas formicarius* during storage of sweet potato (*Ipomoea batatas*) tubers. *J. Food Sci. Technol. India* **21** (3): 185–7.

Rao, V.N.M., Hamann, D.D. and Humphries, E.G. 1974. Mechanical testing as a measure of kinesthetic quality of raw and baked sweet potatoes. *Trans. Am. Soc. Agric. Eng.* **17** (6): 1187–90.

1975a. Apparent viscosity as a measure of moist mouthfeel of sweet potatoes. *J. Food Sci.* **40** (1): 97–100.

1975b. Flow behaviour of sweet potato purée and its relation to mouthfeel quality. *J. Texture Studies* **6** (2): 197–209.

Rashid, M.M. 1987. Indigenous technologies and recent advances in sweet potato production, processing, utilization and marketing in Bangladesh. Paper presented at an International Sweet Potato Symposium, 20–26 May, ViSCA, Baybay, Leyte.

Reddy, N.N. and Sistrunk, W.A. 1980. Effect of cultivar, size, storage, and cooking method on carbohydrates and some nutrients of sweet potatoes. *J. Food Sci.* **45** (3): 682–4.

S-101 Technical Committee. 1980. *Sweet potato quality*. Southern Coop. Ser. Bull. No. 249.

Scriven, F.M., Ndunguru, G.T. and Wills, R.B.H. 1988. Hot water dips for the

control of pathological decay in sweet potatoes. *Sci. Hortic.* **35**: 1–5.
Sharfuddin, A.F.M. and Voican, V. 1984. Effect of plant density and NPK dose on the chemical composition of fresh and stored tubers of sweet-potato. *Indian J. Agric. Sci.* **54** (12): 1094–6.
Shen, M.C. and Sterling, C. 1981. Changes in starch and other carbohydrates in baking *Ipomoea batatas. Starch/Staerke* **33** (8): 261–8.
Sheng, J. and Wang, S. 1988. The status of sweet potato in China from 1949 to the present. Xuzhou Institute of sweet potato, Jiangsu, China. [Mimeo]
Siki, B.F. 1979. Processing and storage of root crops in Papua New Guinea. In: Plucknett, D. (ed.), *Small-scale processing and storage of tropical root crops*, Westview Press, Boulder, CO, pp. 64–82.
Sistrunk, W.A. 1977. Relationship of storage, handling and cooking method to color, hardcore tissue, and carbohydrate composition in sweet potatoes. *J. Am. Soc. Hort. Sci.* **102**: 381–4.
Speirs, M., Cochran, H.L., Peterson, W.J., Sherwood, F.W. and Weaver, J.G. 1945. *The effects of fertilizer treatments, curing, storage and cooking, on the carotene and ascorbic acid content, of sweetpotatoes.* Southern Coop. Ser. Bull. No. 3.
Speirs, M. and 18 others. 1953. *The effect of variety, curing, storage, and time of planting and harvesting on the carotene, ascorbic acid, and moisture content of sweetpotatoes grown in six southern states.* Southern Coop. Ser. Bull. No. 30.
Sweeney, J.P. and Marsh, A.C. 1971. Effect of processing on provitamin A in vegetables. *J. Am. Dietet. Assoc.* **59**: 238–45.
Tereshkovich, G. and Newsom, D.W. 1965. Some effects of date of washing and grading on keeping quality of sweet potatoes. *Proc. Am. Soc. Hort. Sci.* **86**: 538–41.
Thompson, A.K. 1972. Storage and transport of fruit and vegetables in the West Indies. In: Tai, E.A., Phelps, R.H. and Rankine, L.B. (eds.), *Proceedings of the Seminar/Workshop on Horticultural Development in the Caribbean*, 12–15 March, Matarin, Venezuela. Department of Crop Science, University of the West Indies, St Augustine, Trinidad, pp. 170–6.
Truong, V.H. and Vander Zaag, P. 1987. Sweet potato in Vietnam. *Agric. Int.* **39** (7/8): 221–3.
Villamayor, F.B. 1987. Indigenous technologies in sweet potato production and utilization. Paper presented at an International Sweet Potato Symposium, 20–26 May, ViSCA, Baybay, Leyte.
Villareal, R.L., Tsou, S.C., Lai, S.H. and Chiu, S.L. 1979. Selection criteria for eating quality in steamed sweet potato roots. *J. Am. Soc. Hort. Sci.* **104** (1): 31–3.
Wagner, A.B., Burns, E.E. and Paterson, D.R. 1983. The effects of storage systems on sweet potato quality. *HortScience* **18** (3): 336–8.
Walter, W.M. 1987. Effect of curing on sensory properties and carbohydrate composition of baked sweet potatoes. *J. Food Sci.* **52** (4): 1026–9.
Walter, W.M. and Hoover, M.W. 1984. Effect of pre-processing storage conditions on the composition, microstructure, and acceptance of sweet potato patties. *J. Food Sci.* **49** (5): 1258–61.
Walter, W.M., Purcell, A.E. and Nelson, A.M. 1975. Effects of amylolytic enzymes on "moistness" and carbohydrate changes of baked sweet potato

cultivars. *J. Food Sci.* **40** (4): 793–6.

Walter, W.M. and Schadel, W.E. 1982. A rapid method for evaluating curing progress in sweet potatoes. *J. Am. Soc. Hort. Sci.* **107** (6): 1129–32.

1983. Structure and composition of normal skin (periderm) and wound tissue from cured sweet potatoes. *J. Am. Soc. Hort. Sci.* **108** (6): 909–14.

Wang, H. 1974. Study on the carotene content of sweet potato – Effect of storage on the carotene content of sweet potato varieties. *J. Agric. Assoc. China* **87**: 50–6.

Wang, U.P., Lee, C.Y., Chang, J.Y. and Yet, C.L. 1982. [Gamma-radiation effects on Taiwan sweet potatoes] Chinese. *J. Chin. Agric. Chem. Soc.* **20** (3/4): 133–8.

Watson, G. 1988. Home consumption: storage, processing and preferences. Unpublished review (Indonesia), International Potato Center, Lima. [Mimeo]

Watson, J.D. 1976. Ascorbic acid content of plant foods in Ghana and the effects of cooking and storage on vitamin content. *Ecol. Food Nutr.* **4** (4): 207–13.

Watt, B.K. and Merrill, A.L. 1975. *Handbook of the nutritional contents of foods*, Dover Publications, Inc., New York.

Wentworth, J. and Dempsey, A.H. 1961. The carotene content of sweet potatoes. *Georgia Agric. Res.* **3** (2): 12.

Widodo, Y. 1990. Incorporating sweet potato into food crops agricultural development in Indonesia. Paper presented at the User's Perspective with Agricultural Research and Development (UPWARD) Workshop, 2–7 April, Los Baños.

Wiersema, S.G., Hesen, J.C. and Song, B.F. 1989. Report on a sweet potato postharvest advisory visit to the People's Republic of China, 12–27 January, International Potato Center, Lima. [Mimeo]

Wilson, J.E. 1979. The potential for genetic improvement in storage quality of root crops. In: Plucknett, D. (ed.), *Small-scale processing and storage of tropical root crops*, Westview Press, Boulder, CO, pp. 166–8.

Yamamoto, Y. and Tomita, Y. 1958. Studies on the bio-pigments and vitamins: IV. Correlative changes in the carotene, total carotenoids and the other constituents of sweet potatoes during storage (1). *Mem. Fac. Agric., Kagoshima Univ.* **3** (2): 63–8.

1960. Studies on the bio-pigments and vitamins: V. Correlative changes among the carotene, total carotenoids and the other constituents of sweet potatoes in relation to the variety and storage (2). *Mem. Fac. Agric. Kagoshima Univ.* **4** (1): 13–19.

Yen, D.E. 1974. *The sweet potato and Oceania.* Bishop Museum Bull. 126, Honolulu.

Zechmeister, L. 1962. *Cis–trans isomeric carotenoids, vitamin A and arylpolyenes.* Springer-Verlag, Vienna.

Zhao, Z. and Jia, F. 1985. [*Safe storage and indigenous processing of sweet potato*] Chinese. The Publishing House of Agriculture, People's Republic of China.

CHAPTER 6

Postharvest procedures: II. Processing

Processing methods vary in sophistication from the simple slicing and field sun-drying of roots as practised in some developing countries to the large-scale, multi-stage production of, for example, frozen, canned or flaked products tailored to consumer expectations, which is performed in developed countries such as the United States and Japan. The former simple processing of sweet potato into dried 'chips' may be only an intermediate stage in the final production of other products such as flour, snacks, starch or alcohol. On a home or village level, however, even such intermediate processing can add considerably to the value of a farmer's crop. In addition, the home production of simple traditional processed sweet potato foods, such as is practised by women and children in parts of the Philippines (Alcober and Parrilla, 1987) can usefully increase family income. Although the main emphasis is on processing of roots at the present time, methods of processing leaves are being explored in some areas. In contrast to the processing of whole roots or leaves for direct use as human food, there are considerable possibilities for processing to produce items which can be incorporated indirectly into the diet. Such items include starch, sugar syrups, alcoholic beverages, food colorants, enzymes and leaf protein among others.

The main advantages of producing processed products from sweet potatoes for human feeding are various:

Decreasing losses of food and hence increasing the quantity available (for example by utilization of lower-grade produce not suitable for the fresh market).

Promoting year-round, as opposed to seasonal, consumption.

Providing a greater variety and convenience of uses, especially for the roots, by making products with characteristics distinct

from those of the raw material or which can be used in other product formulations.
Increasing the economic value of the crop to producers.
Increasing the efficiency of the food delivery system thus freeing time for other occupations.

In spite of these advantages, a sweet potato product will not succeed commercially, however relatively favourable its nutritional value in comparison to similar products from other crops, unless it is economically viable. At the same time it must be available at a price that the consumer can afford and be presented in an acceptable form.

The literature provides numerous examples of the variety of products attainable from sweet potatoes, but very few cases where commercialization of these has been achieved. Consumer oriented surveys to determine potential markets for proposed products, and project feasibility studies to predict processing economics correctly have rarely been carried out. The assessment of locally grown cultivar suitability for industrial purposes, in terms of characteristics such as dry matter, starch, fibre and latex contents, nutritional value, pectin and amylase enzyme content, extent of polyphenolic discoloration etc. has been pursued in only a few countries, for example Brazil and Thailand (Cereda et al., 1982; Prabhuddham et al., 1987). Studies such as one carried out in Taiwan to assess the suitability of various clones for production of specific products such as flakes, fillings and chips (Chen and Chiang, 1985a) are even rarer. Such surveys, feasibility studies and analyses are vital if an increase in the quantity and variety of processed products, which will act as a stimulus to increased crop production, is to be achieved.

The first part of Section A deals briefly with the technology involved in methods of processing at present employed to manufacture products used for direct human feeding. The section describes explorations into the utilization of sweet potato for processing in various countries, including innovations in products derived from primary processes described, difficulties experienced in research or processing, and success or failure of attempts at marketing these products where known.

Section B describes quality and nutritional changes known to occur in processed products. Unfortunately research on such changes has not been extensive and so far has been confined mainly to those in the products of the developed world. Where possible, implications for products made by simple, small-scale technology will be discussed.

Alternative or indirect uses of sweet potato as a food will be discussed in Section C, with the exception of animal feeding, which is discussed separately in Chapter 7. It is not the purpose of this chapter to give extensive details of individual processing steps and where possible processes will be summarized in the form of flow diagrams. Useful addresses for further details of processes and products referred to here, where available, may be obtained from the author. Finally, in Section D, the major constraints to increased utilization of sweet potato through processing are summarized and discussed.

A. Processing for direct use as human food

Technology

Dehydration

Roots

Dehydration of sweet potato roots has been traditionally practised in many developing countries for generations and in improved forms has great potential for increasing the quantity which could be preserved. In traditional practice, the roots, which may or may not be peeled, are sometimes cooked but are more often directly cut up into pieces and spread out in the sun to dry. They yield chips or slices which can be stored as such or ground in a mortar to a flour which is then sieved. In Indonesia, fresh roots are sometimes soaked in 8–10% salt solution for about an hour before cutting and drying, a practice which is reported to inhibit microbial growth during drying (Winaro, 1982). On a laboratory or commercial scale, roots are often treated with a solution of sodium metabisulphite to inhibit enzymic browning, and hence discoloration of the finished product. In traditional processing, however, the occurrence of browning may not necessarily be considered a disadvantage. In parts of rural Japan, for example, steamed dumplings made from raw, untreated, sun-dried slices ground into flour have a characteristic and desirable dark brown colour and strong flavour.

Dehydration presents various advantages over canning and freezing which, while in common use in temperate zones, may be too expensive in some tropical situations. Drying yields a light, compact, relatively cheaply packaged, easily stored and transported material which can be used in a great variety of dishes and further food formulations. Some genotypes may, however, be more appropriate than others for drying. The percentage of flour, for example, which is recovered by drying can vary greatly between genotypes. In the Philippines, 'Georgia Red' and 'Ilocos Sur' cultivars were reported to yield 12 and 37 kg, respectively, of flour from 100 kg of fresh roots (Truong, 1984). As traditionally practised, sun-drying has various drawbacks including poor control of energy input and product quality, as well as frequent contamination of the food by microorganisms, dust and insects.

Figure 6.1. Slicing sweet potato to be sun-dried in the field, Jinan, China (M. Iwanaga).

Small-scale dehydration

Cabinet-, tunnel-, drum- or spray-drying as used in large Western commercial enterprises are highly technical processes using large amounts of energy, which add greatly to the cost of the final product. They are not suitable to the socio-economic realities of many tropical areas (Martin, 1984a). Innovations in simple techniques and mechanics of dehydration suitable for home, village or commercial scale operations are at present being explored in various countries. It is not possible in the scope of this book to detail all of these. Hence only a few examples will be given.

China

In China many thousands of tonnes of sweet potato are dried every year in the form of chips or slivers by traditional sun-drying on the farm (see Figures 6.1, 6.2, 6.3). The major part of this dried product is then sent on to starch or alcohol factories for further processing. Methodology for the improvement of traditional drying to produce a higher quality product which can be stored without spoilage have been developed (Zhao and Jia, 1985). These include: paying attention to the weather forecast before cleaning or slicing of roots to avoid mildew brought on by damp weather; if rain does start during drying, protection of slices

Figure 6.2. Farmers collecting dried sweet potato slices after sun-drying, Jinan, China (M. Iwanaga).

Figure 6.3. Dried sweet potato chips at an alcohol factory in Xuzhou, China (M. Iwanaga).

from moisture and care to prevent over-piling which could cause heating and lead to rotting; on cessation of rain, frequent turning of the slices to speed up drying; suspension of slices on threads or use of a drier in areas where it rains on successive days during the drying season; preparation of store houses before harvest in case roots have to be stored temporarily during rain, and for rapid storage of dried slices to prevent contamination with dust, sand etc.; possible fumigation of slices with sulphur, which prevents incompletely dried slices from rotting or being attacked by moulds whilst maintaining a pleasing white appearance in the dried product, and rapid transfer to sunshine as soon as possible to complete drying. If dried chips are required to be stored they have to be sealed into some sort of container, under dry conditions and at as low a temperature as possible, very rapidly to avoid moisture uptake and contamination by insects or microorganisms. If the dried product is to be stored for a long period, the moisture content should be reduced below 10%; it can then be piled up and sealed with sand, earth bricks or wheat husks. If the product is infested with insects, it should be heated by exposure to strong sunlight before storage.

Taiwan

It has been suggested that pressing of sweet potato chips to extract a high percentage of the juice before drying the chips would save fuel costs (Hong, Su and Sung, 1977). An added benefit is the high amylase content of the expressed juice which could be utilized in processing. A double pressing of the wet chips extracted 80% of the total amylase.

Peru

Solar dehydration, a process combining drying and temperature control, air flow regulation and technical innovations designed to minimize the effects of climatic changes and protect food from contamination, can increase sun-drying efficiency and improve product quality. A solar drier for potatoes, which could be adapted for use with sweet potatoes, has been developed by the International Potato Center, Lima, Peru. Cooked, shredded potatoes are placed on a shaded screen oriented towards the wind, and the moisture reduced by about 45% in 18 hours. Reorientation to direct sunlight the following day dries the product to about 10% moisture.

Hawaii

The University of Hawaii, Manoa, has also designed a solar dehydrator which can function directly with sunlight, indirectly with solar-heated air, or in a combined direct and indirect mode (which proved to be the most efficient). The construction of this drier has been described and illustrated (Moy and Chi, 1982). It was suggested that supplementary heating using biogas could be used during temporary unfavourable climatic conditions. The dried slices produced might be marketed as a snack in the same category as dried fruit. Alternatively they could be ground into a flour and used to formulate noodle or pasta-type products. The authors also describe an 'osmosol' process in which the slices of sweet potato were partially dried by immersion in a recirculating concentrated sugar solution (60–65°Brix) for several hours after which they were solar dried as before. Further studies are needed to improve these processes and achieve ways of marketing the products.

The Philippines

A series of tools and equipment for use by small-scale rural industries drying root crops, including sweet potato, has been developed in the Philippines (Truong and Guarte, 1985). These include a pedal-operated peeler and a slicer (Figure 6.4), the power transmission units of which consist of a bicycle chain and sprockets, a manually operated slicer/cuber (Figure 6.5), and a steamer (for a pre-drying blanch) heated by burning wastes such as coconut husks or shells. Drying, in areas where an electric power source is unavailable, can be accomplished with a natural convection drier (Figure 6.6) burning wastes as above to heat a cylinder which in turn heats the air surrounding the root chips. It is claimed that a minimum of 50 kg of root chips can be dried in 7–8 hours to a high quality product.

An excellent example of a dried sweet potato product which can be incorporated directly into a traditional dietary item has been produced in the Philippines using the above-described equipment (Truong, 1990). This is the small-scale manufacture of dried sweet potato cubes. Cultivars of white, orange or even purple flesh can be used to produce cubes of varied and attractive appearance. The dried sweet potato is mixed with dried cubes of cassava, cocoyam, jack fruit and banana and packaged in plastic. On reconstitution the cubes are cooked with rice, coconut milk, brown sugar and vanilla to make a 'fruit soup' called *guinata'an* which is a traditional Filipino dish. During processing the sweet potato roots are peeled, washed and sliced 1 cm thick. The slices are then manually cut into cubes with a lever-operated cuber. The cubes are steamed and then dried. The waste from cubing is steamed, dried, and ground into a highly nutritious flour which could be used in baby food formulations. It is suggested that the product might have considerable export potential among ex-patriate Filipinos. Besides *guinata'an*,

Figure 6.4. A pedal-operated sweet potato slicer developed at ViSCA in the Philippines (ViSCA).

there are many alternative uses for the cubes, which are claimed to rehydrate easily and be very similar, when reconstituted, to fresh cubes. They may be used as a substitute for carrots in soups or in *empanada* (pastry case) fillings and can also be fried.

A product resembling dried fruit has been prepared in the Philippines (Truong, 1987) from orange-fleshed roots. The sweet potatoes are peeled, sliced lengthwise, soaked in 2% (w/v) metabisulphite solution and cooked in a 60°Brix syrup containing 0.8–1.0% citric acid. They are then mechanically dried and packaged in plastic bags. The technology

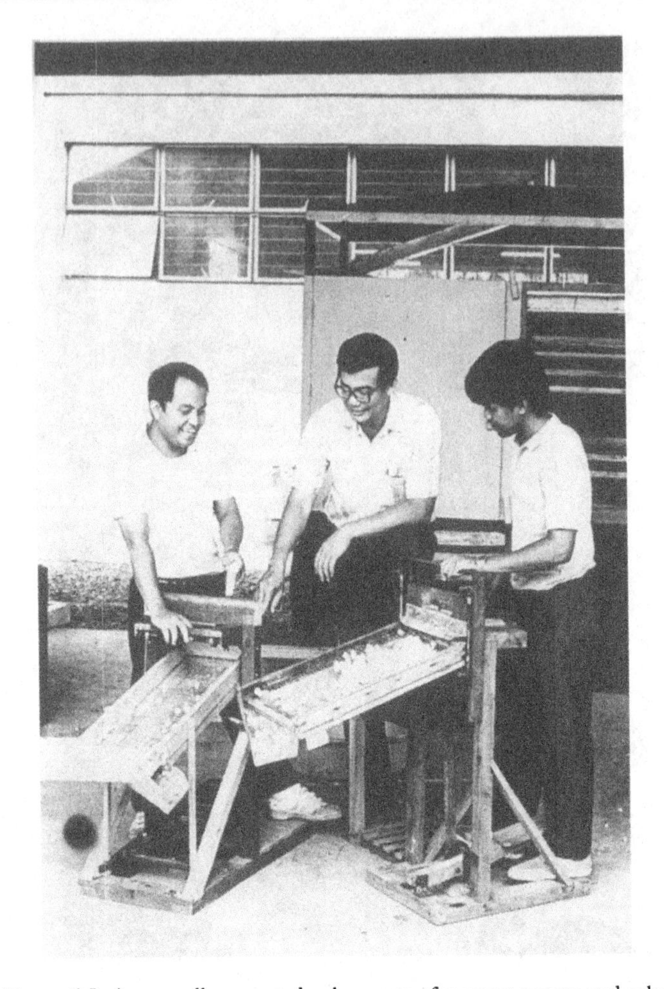

Figure 6.5. A manually operated cuber-sorter for sweet potato and other root crops at ViSCA in the Philippines (ViSCA).

Figure 6.6. (a) A natural convection drier, fuelled by agro-wastes, developed at ViSCA in the Philippines and (b) dried sweet potato cubes being removed from the natural convection drier (ViSCA).

a

b

Table 6.1. *Composition of a dried 'fruit-like' sweet potato product, dried mango and dried apricot*

Composition	Dried sweet potato product	Dried mango	Dried apricot
Moisture (%)	17.5	9.4	16.6
Soluble solids (%)	70	68.5	—
pH	2.8	3.6	—
Carbohydrates (%)	78.3	87.3	66.6
Protein (%)	1.2	1.6	5.3
Lipid (%)	0.4	1.0	0.6
Crude fibre (%)	1.6	1.8	—
Ash (%)	0.9	0.7	—
Ascorbic acid (mg/100 g)	7.2	28.9	12.6
Carotene[a] (μg/100 g)	7820	25	6540

Notes:
From: Truong, 1987.
[a] Converted from IU vitamin A by 1 IU = 0.6 μg beta-carotene.
A dash indicates data not determined.

has been transferred to a private company for commercialization, and the product is being marketed under the trade name Tropical Delight. It can be used in the same ways as dried fruits or eaten as a snack. The composition of this product, as known at present, is compared with that of two dried fruit products in Table 6.1.

Malaysia and the United States

Another form in which fruit is dried and eaten as a dessert or snack is as an edible 'leather', also known as fruit roll (South Africa) or crush (Australia). The possibility of presenting dehydrated orange-fleshed sweet potato either as the main ingredient, or together with a fruit, as a 'leather', has been explored in Malaysia (Yaacob and Raya, 1983) and the United States (Collins and Washam-Hutsell, 1987) respectively. The flesh of a Malaysian cultivar was cooked, mashed and sieved, mixed with 0.5% (w/w) carboxymethyl cellulose (a binder), 200 p.p.m. sodium bisulphite and 7% (w/w) sugar and formed into a sheet 1 mm thick which was oven-dried to 10–17% moisture and then packed in plastic film. Samples dried at 75°C were chewier than those dried at 55°C or 65°C, which were crisper or crunchier. Deep frying the leather increased its sweetness and improved its taste. The 'leather' was well received by a sensory panel.

Figure 6.7. Handicrafting sweet potato products behind a shop in Kawagoe, Japan (J.A. Woolfe).

Formulations of 'leather' were made with two North American cultivars each of which was baked, puréed, mixed with honey or high fructose corn syrup, or with apple purée and one of the aforementioned sweeteners, spread into sheets of 1.5 mm thickness, oven-dried at 55°C and rolled into a scroll. Only baking (as opposed to boiling or canning), to cook the roots before processing, produced an unbroken leathery product. The authors attribute this to the extensive production of dextrins by amylolytic hydrolysis of starch during baking. A sensory panel evaluation for texture, flavour and general acceptability indicated that only texture was affected by cultivar and type of sweetener added. The water activities of all the formulations were below the minimum required for growth of microorganisms, and could be maintained as such during storage if adequately packaged.

Japan

Sweet potato has been grown in Japan for about 300 years. Although Japan is now a leading world economic power and has large-scale up-to-date processing, it still retains many traditional products including those made from sweet potato (Figure 6.7). The preparation of such products is relatively simple and could be adopted elsewhere. An intermediate moisture (25%) product called *mushikiriboshi*, popular in Ibaraki and Chiba Prefectures, is made by steaming peeled roots which are then left whole if very small, or sliced lengthwise and dried at 15°C. During drying, maltose (Nakajima, 1970) produced as a result of starch saccharification during cooking oozes to the surface and dries there, forming an attractive white film resembling a powdered sugar dusting. The product is eaten as a snack generally after roasting or baking. It has a chewy texture. The preparation of *mushikiriboshi* with high beta-carotene roots, though not a popular practice in Japan, gives a more nutritious product and one which has a flavour reminiscent of dried apricots.

Another traditional product found in the Kawagoe area is *sembei* consisting of dry-fleshed cultivars which are left unpeeled, thinly sliced, baked until crisp and dry between two hot metal plates and coated in sugar syrup; it is hard and translucent when dry. These sweet 'crackers' may be decorated and flavoured with ginger or sesame seeds before drying. The use of red-skinned cultivars ensures a thin strip of colour round the edge, adding a touch of elegance to a product which is often purchased as a gift for others (Figure 6.8; Duell, 1990).

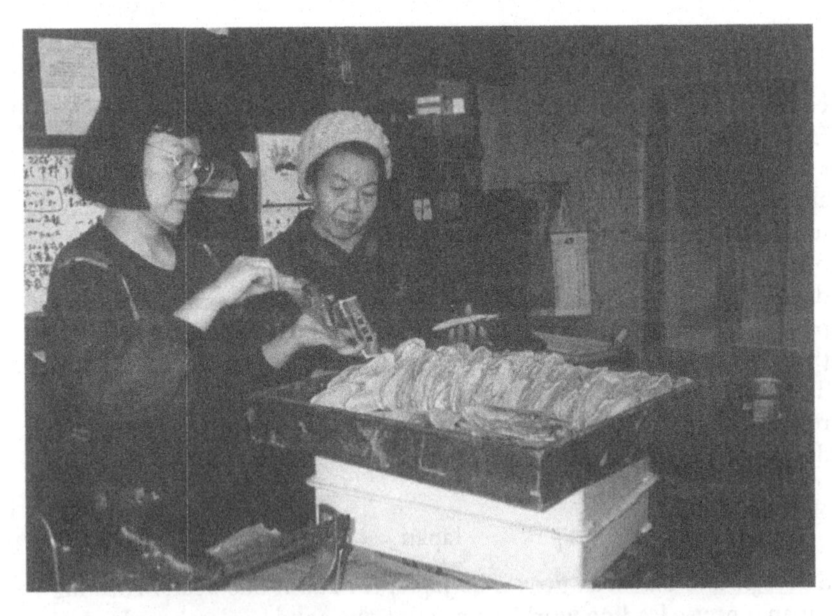

Figure 6.8. Packing *sembei* crackers – crisp, sugared and flavoured dried slices of sweet potato – at a small business establishment in Kawagoe, Japan (J.A. Woolfe).

India

A three-tier system for the production of dried chips and flour from potatoes (*Solanum tuberosum*) in India is to be adapted to using sweet potatoes for the same purpose (Nave, R., personal communication). The first tier consists of small farmers and low-income families who carry out processing of the potatoes/sweet potatoes as far as the preparation of sun-dried chips. The second tier consists of a centre where the chips are ground into flour, and the chips/flour packaged and despatched. The third tier carries out marketing and sales of the products. Thus each tier works within its capabilities and specializes in one set of operations.

The equipment used is appropriate to the village level of processing. A washer/peeler is made from an oil drum lined with coarse abrasive carbide grit. The axle on which it is turned is a hollow, water-fed pipe. A crank handle is turned by one or two people. The drum has an inspection door so that completion of peeling can be viewed. Two types of slicer are available. Sweet potatoes are fed into a rotating disc which is hand-cranked, in which case the capacity is 70–80 kg/hour. The other slicer, which is more expensive, consists of a rotating head operated by a cycle or motor. The roots are held against the cutting edge by centrifugal force, and the capacity using the motor is 400–500 kg/hour. The rack for sun-drying is constructed of light-weight electric conduit piping with removable legs so that it is easy to store. This skeleton is stretched with a coarse wire mesh, which in turn is covered with nylon mosquito netting in such a way that the netting does not touch the wire. For sun-drying, the sliced roots are placed on the netting, which can easily be removed for washing. Drying takes from 4 hours to 2 days depending on climatic conditions. Though the above-described equipment is appropriate for sweet potatoes, a number of factors remain to be explored in relation to sweet potato processing. It will be necessary to determine how locally available cultivars behave in relation, for example, to discoloration and the need for blanching and application of sulphite, and to the retention of beta-carotene from orange-fleshed cultivars.

Sweet potato grated, sun-dried and milled into flour has been produced by the Central Tuber Crops Research Institute, Kerala, India, and is now being marketed. It can be used to substitute part of the wheat flour used to make chaphatis and other baked goods.

A laboratory-scale high temperature short time (HTST) pneumatic drier has been designed and fabricated in India for the pre-treatment prior to conventional drying of various vegetables including sweet potato roots (Jayaraman et al., 1982). Dried cubes produced by this technique were expanded and porous, resulting in shorter drying and rehydration times and improved texture and rehydration characteristics.

Peeled, cubed and blanched sweet potato roots were subjected to HTST pneumatic drying followed by conventional cabinet drying to 5% moisture. When suitably packaged, the dried cubes retained their acceptability and improved rehydration characteristics for up to a year at ambient temperatures (25–30°C). Such cubes could be used as quick-cooking convenience vegetables or in soup mixes etc.

Brazil

In Brazil experiments have been carried out to produce from sweet potato roots a dried product resembling *farinha de mandioca* (dried cassava flour), which is a traditional product made particularly in rural areas. It is an integral part of the every day rural and urban diet in many parts of Brazil, notably Amazonas and the northeast. The process as carried out in a traditional Amazon production location and some of the equipment used has been described and illustrated (Guedes, 1986). After peeling, the roots are grated and then pressed to remove some of the moisture. The shreds are then dried (toasted) on a large dried and hardened clay surface heated from below by fire. The dried 'flour' is then sieved to produce fine and coarse grades. The yield of sweet potato flour was found to be superior to that of cassava whilst its energy content was similar and its total protein content more than twice that of cassava flour. Other Brazilian researchers (Carvalho, Moura and Pape, 1981) found that by this traditional method the process of peeling was unnecessary except in the case of roots with coloured skins. This produced an advantage over cassava, which has a tough skin, the obligatory removal of which caused a 20% wastage. However, under uncontrolled conditions in a rural cassava processing plant, pressing the grated flesh of sweet potato reduced the content of total protein, some of the sugar and the soluble minerals, which were lost in the extracted juice, so that the final flour had no qualitative advantage over cassava flour. An alternative process in which the sweet potato was cut into slices or cubes and dried in a circulating air oven at 50°C before milling and sieving produced a flour of high nutritional value with about 3% total protein.

Puerto Rico

Techniques and problems associated with the small-scale production of sweet potato flour have been investigated in Puerto Rico using a series of sweet potato cultivars (Martin, 1984a). Hand-peeled and trimmed roots were sliced or shredded using a household device. The slices or shreds were placed on screen trays and allowed to partially dehydrate, under shelter, by natural airflow for 16–18 hours. The tray was then exposed to direct sunlight for up to 5 hours. The dried material was ground and packed in polyethylene bags. Various problems were encountered during processing. Small, irregular-shaped or diseased roots were difficult to handle. When cut, some cultivars released latex which interfered with the operation of the shredder. The shreds from some cultivars readily underwent polyphenolic oxidation, becoming a dirty, unpleasant grey colour. Yields of flour varied greatly between cultivars not only because of differences in dry matter content, but also through peeling and trimming losses. Flours from different cultivars varied widely in nutritional composition. When flours from samples uncooked before drying were boiled in water they resulted in a cooked food with a brown colour and a disagreeable taste and odour. On the other hand pre-cooking before drying to inactivate polyphenolases, eroded shreds and rendered them difficult to handle.

The reactants producing the undesirable changes which could have been due to non-enzymic Maillard, as well as enzymic, browning were reduced by diffusion processing (Martin, 1984b). This entailed placing the cut pieces of sweet potato in five times their weight of water for three consecutive 30 min intervals (changing the water after each interval) to remove soluble reactants such as polyphenolics. Starch could be recovered from the water by settling, decanting and drying. Diffusion thus obviated the need for the use of additives (to inhibit darkening), such as metabisulphite or ascorbic acid, which may be unsuitable for home or small-scale production due to lack of control or abuse. A technique involving peeling, shredding, diffusion processing, draining, air-drying, sun-drying, final oven-drying at 60°C, grinding and packaging (Martin, 1985) produced a product with improved appearance, odour and taste which could be incorporated into dishes such as pancakes, fritters, small cakes and a type of 'porridge' (Martin, 1986).

Industrial dehydration in developing countries

The literature provides two examples of industrial schemes for the production of dried sweet potato in developing countries. The first provides details in the form of a case history of the manufacture, and reasons for commercial failure, of *kaukau* (sweet potato) rice in Papua New Guinea (Thomas, 1982). *Kaukau* rice – dehydrated, diced sweet potato thought to be suitable as a local substitute for imported rice – was successfully produced on a laboratory and pilot plant level. Projected costs of production on a large scale estimated from these trials showed the product to be uncompetitive with imported rice at that stage. However, before further trials could be conducted, a glut of sweet potatoes in 1976 promoted a hasty decision to build a large-scale unit capable of handling 3 tonnes of sweet potatoes per day. Unfortunately the surplus of sweet potatoes was not repeated in succeeding seasons and the plant ceased to operate due to insufficient raw material, low sales due to high cost and unacceptability to consumers (probably through lack of appropriate cultivars for processing) and poor marketing and promotion.

The other example discusses a project feasability study for a proposed scheme for drying sweet potatoes in the Philippines. This will involve the growing of improved sweet potato cultivars in a pilot area and their primary processing into a food grade flour from high quality roots, and into a secondary animal feed product from lower quality roots, peelings and vine tops (Taylor, 1982). The proposed flour production plant has a starting capacity of 5 tonnes/day of fresh roots.

Research is continuing into improvements in dehydration processes of sweet potato and into the subsequent utilization of resulting products, some of which will be described further in the next section.

Large-scale dehydration

The production of high quality flakes and other dried sweet potato products may entail many unit operations (individual processing steps) which are outlined in Figures 6.9 and 6.10. The pre-processing operations shown in Figure 6.9 are also common to other processes such as canning and freezing. Benefits of pre-heating include reduction of enzymic discoloration (polyphenolic oxidation) and of peeling time (Bouwkamp, 1985). Deep peeling to remove the high percentage of polyphenolics located in the outer root tissues (see Chapter 2, pp. 66–7) is also avoided.

Steam peeling is often used for removal of the skin as conditions are fairly readily controlled and waste disposal is easier than with the use of lye. Peeling with lye (sodium hydroxide solution) entails a 5–6 min exposure to 20% lye solution for cured roots, and 3–6 min exposure to a 10% solution for freshly harvested roots, at 104°C (Kays, 1985). Peeling losses increase with decrease in size of roots (greater surface area to volume exposure) and increase in peeling time (greater depth of peel removed) (Bouwkamp, 1985). Peeling with super-heated steam followed by flash cooling by direct injection of cold water into the peeler chamber has been found to increase yields of peeled roots (Smith et al., 1982). The effects of mechanical, steam and lye peeling have been investigated and compared by various authors (Burkhardt, Merkel and Scott, 1973; Lee and Lee, 1984; Walter and Giesbrecht, 1982; Walter and Schadel, 1982). Thorough washing is necessary after peeling particularly after the use of lye. Trimming to remove surface imperfections due to mechanical damage, disease or insects and fibrous ends then follows.

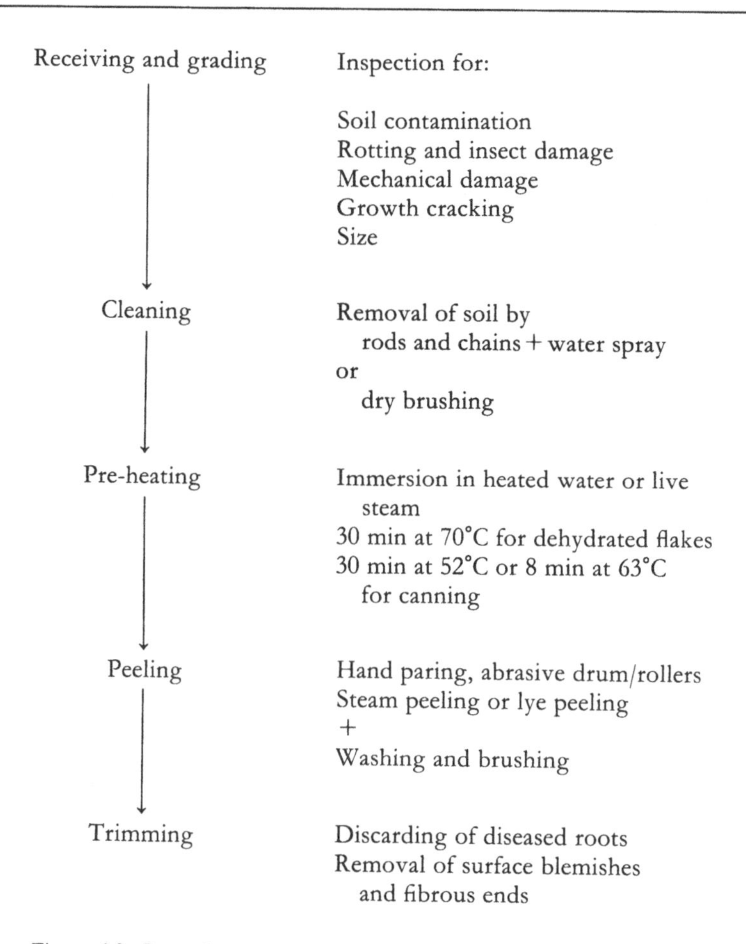

Figure 6.9. General preprocessing procedures.

If slices, dice (cubes) or strips of dried sweet potato are to be manufactured, cutting into the requisite shapes and sizes is next carried out. The cut pieces are then blanched, as a result of an early discovery that uncooked dried sweet potatoes gave a poor product which did not rehydrate satisfactorily. After cabinet- or tunnel-drying, the dehydrated pieces should be suitably packaged to exclude oxygen and moisture for maximum keeping quality. Maximum drying efficiency has been studied using different shapes of cut sweet potato, temperatures varying from 40°C to 70°C and both co- and counter-current air circulation in the drying chamber (Taharazako, 1984).

More usually pre-processed sweet potatoes are dried and reduced to the form of flakes, which can be reconstituted into mashed sweet potato or incorporated into a variety of other products such as pies, pasties, cakes etc. In Japan, flakes are produced commercially both by private

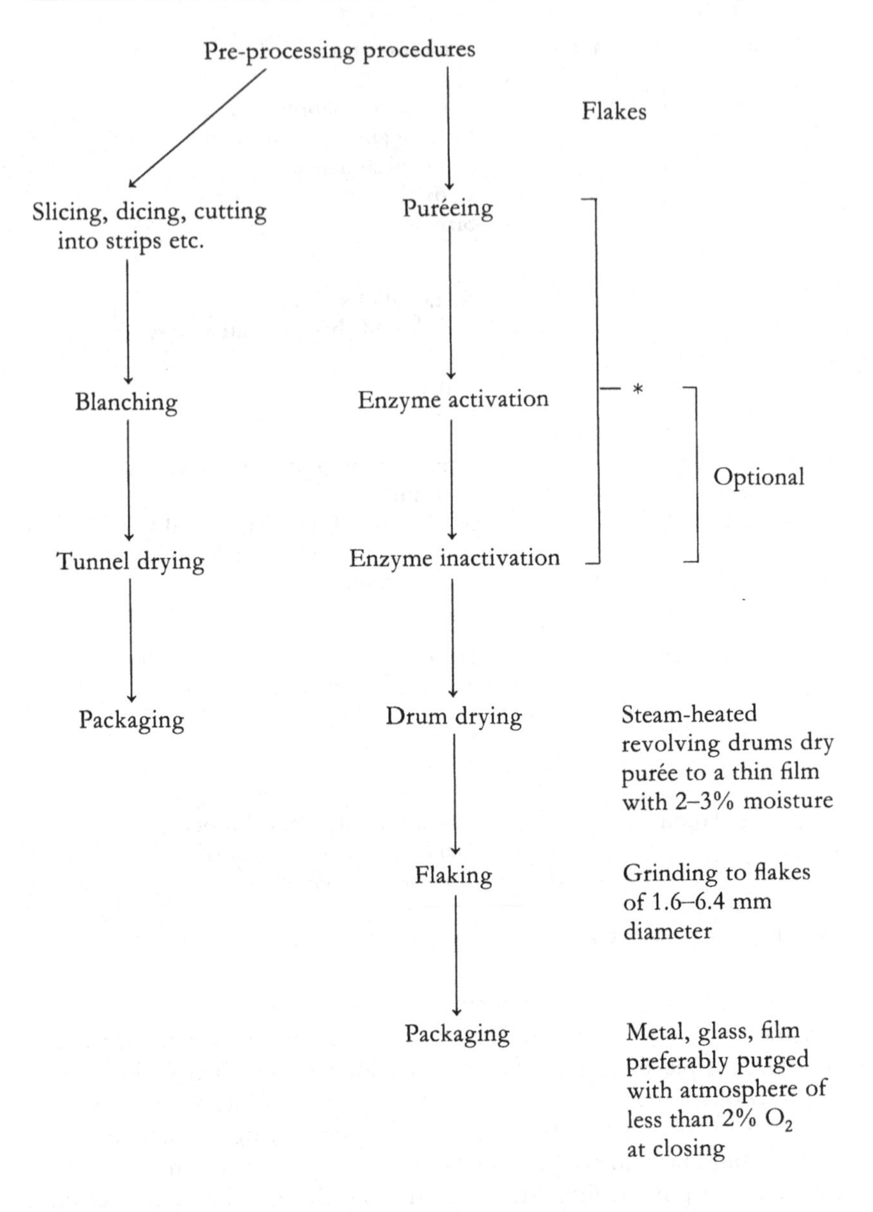

Figure 6.10. Dehydration process. *, see Figure 6.12.

firms and by large agricultural cooperatives. Although most production at present involves white-fleshed sweet potato cultivars, there is also some production of flakes from cultivars high in beta-carotene, and of purple flakes from high anthocyanin types. The pigmented flakes have an excellent colour and can be incorporated both for increased nutritional value and attractive appearance in a variety of cakes, desserts, icecreams etc. (see below). After pre-processing, the roots are steamed, crushed, and fed directly to a single drum drier heated to 120°C; the dry film is broken up into flakes and the flakes are packed in 10 kg sacks. High beta-carotene flakes have also been made successfully in Taiwan (Tsai, W., personal communication).

In the United States, the pre-processed roots are puréed in a pulper where blades force the roots through a 0.8 mm screen, removing much of the fibrous material. Differences in raw material due to cultivar, curing, storing and handling which used to result in a finished product of very variable quality have been minimized by the employment, after puréeing, of an enzyme activation process. This gelatinizes the starch and activates amylolytic enzymes naturally present in the roots by heating the purée to 74–85°C using steam injection. The timing of the partial degradation of starch which follows can be adjusted for variations in raw material properties to produce a consistently textured final product. Flash heating subsequently inactivates the enzymes and completes cooking. The purée is then dried to 2–3% moisture on steam-heated drums, flaked and immediately packaged in metal, glass or film containers which exclude moisture and oxygen. An alternative process employing a commercial alpha-amylase to hydrolyse only a part of the pre-gelatinized purée, which is then blended with the untreated portion, has been developed (Szyperski, Hamann and Walter, 1986) to produce a consistent product independent of raw material variations. In addition, sodium acid pyrophosphate or citric acid may be added to the purée before drying (Manlan et al., 1985) to control non-enzymic browning which causes discoloration of the reconstituted flakes.

In order to promote an increased institutional use of sweet potato flakes with a high provitamin A content, investigations into the possibility of fortifying flakes with high protein supplements of soy or cottonseed flours have been conducted (Walter and Purcell, 1978). By this means it was hoped to increase the proportion of total energy provided by protein. Sweet potato flakes supplemented with soy flour, soy flour plus methionine, or defatted cottonseed flour had protein energy to total energy ratios of 0.185, 0.217 and 0.257, respectively, compared with unfortified flakes at 0.093 or the national average for all foods of 0.118. However, the water-binding capacity of fortified flakes was lower than that of unsupplemented flakes and decreased still further on storage. There were also changes in some amino acids and the development of an undesirable off-flavour at a high storage temperature. In Thailand, sweet potato flakes prepared with 53% mashed sweet potato, 7.6% rice flour, 15% soy flour, 7.6% milk powder, 15% sugar and 1.5% salt blended together into a paste and drum-dried are planned as a supplementary food for children (Prabhuddham et al., 1987), for example as a breakfast food mixed into milk.

A marketing study on sweet potato flakes indicated that consumers normally visualize the flakes only in the form of mashed sweet potatoes (USDA, 1964) and that recipes and demonstrations help them to become aware of alternative uses in casseroles, pies and candies. However, although sweet potato flakes have been produced for many years in the United States, they are still a product of quantitatively minor importance when compared with canned sweet potatoes (Kays, S., personal communication).

A further development which has taken place in recent years in Japan is the production of sweet potato granules by the add-back process similar to that for potato (*S. tuberosum*) granules. The sweet potatoes are peeled, steamed, crushed (moisture content 68%), dry granules added back (moisture content reduced to 35–45%), conditioned, granulated and flash dried (Baba, T., personal communication). An industry proposing to manufacture 100 tonnes of granules per year has been established in Kagoshima. The granules can be used in, for example, croquettes or snack foods. Sweet potato granules closely resembling those of the white potato have been produced on an experimental basis using the very low-sweet cultivar 'Satsumahikari'.

Quality changes in stored dehydrated flakes

One important constraint to the widespread consumer acceptance of sweet potato flakes has been their limited shelf-life due to the development of a strong hay-like off-odour. This could be prevented by storing the flakes in a low oxygen atmosphere or in nitrogen. The addition of antioxidants has been found to be less effective (Deobald and McLemore, 1964).

The highly unsaturated nature of sweet potato root lipids has already been discussed in Chapter 2. The unsaturated fatty acids and carotenoids (which form part of the lipid fraction) contain double bonds very susceptible to oxidation, with the consequent production of compounds with off-flavours. The rapid oxidation of carotenoids in flakes stored in air leads not only to off-flavours, but also to loss of colour and an undesirable decline in nutritional value.

A histochemical study of sweet potato flakes (Purcell, A.E., unpublished data) was reported (Walter, Purcell and Hansen, 1972) to have found a portion of the carotenoids and other lipids to be spread on the surface of the carbohydrate matrix and therefore in contact with oxygen, while the rest were tightly bound inside the matrix and therefore less susceptible to oxidative attack. When sweet potato flakes were stored in air at 31°C, carotenoids and unsaturated fatty acids, especially linolenic acid, were destroyed by oxidation much more rapidly in the surface than in the bound lipid fractions (Walter and Purcell, 1974). The flakes were found to develop a strong off-flavour, which was postulated to be the result of surface lipid oxidation, after 29 days of storage.

During processing and storage of sweet potato flakes canned in a reduced oxygen atmosphere and stored at 23°C for 6 or 12 months, unsaturated linolenic acid with three double bonds decreased the most in all types of lipid, whereas the saturated fatty acids palmitic and stearic increased slightly (Alexandridis and Lopez, 1979). An off-odour, characterized as hay-like, was detectable after 6 months of storage and coincided with a marked decrease in unsaturated fatty acids and in the ratio of unsaturated:saturated fatty acids, leading to the conclusion that these occurrences were linked. The authors concluded also that the extended storage period had allowed oxidation of lipids throughout the product; that is, both surface and bound lipids.

It is not known whether the more simple methods of production of sweet potato flour in developing countries have the same effect on sweet potato lipids, producing free and bound fractions, nor have any studies been reported about the storage properties of such flours in terms of flavour stability. However, it seems likely that these flours if they are to be stored for long periods will require packaging in material which excludes oxygen as well as moisture. The effects of dehydration on beta-carotene will be discussed more fully in Section B.

Tops

The drying of tops is not a common occurrence, although it is practised to some extent, notably in Africa, where leaves are dehydrated by exposure to direct sunlight, for later use as a food ingredient. Improvement in dehydration technology combined with an attempt to maximize leaf nutrient retention has been carried out experimentally with the construction of an enclosed solar drier provided with shade by painting black the inside of its transparent plastic cover. The design of this drier has been illustrated (Maeda and Salunkhe, 1981). Leaf nutrient changes during drying with this method are detailed in Section B.

Mashing, pasting or puréeing

Puréeing of sweet potatoes in its simplest form as practised in some countries entails merely the boiling or steaming of roots followed by mashing. This mash may then be used either as the outer casing for, or as the stuffing of, other products. In Japan, sweet potato paste made from either white- or orange-fleshed cultivars is produced commercially by manufacturers and cooperatives. A small cooperative enterprise in Kiire, southern Japan, produces a commercialized high (12 mg/100 g (fwb)) beta-carotene paste for incorporation into bread or icecream. The bread is supplied once a week to about 8000 children as part of the school lunch programme. The process used, which yields about 60–70% by weight of raw material, is shown in Figure 6.11. A deep purple paste, made with high anthocyanin cultivar 'Yamakara Murasaki', has also been produced on an experimental basis and could be used as an ingredient of icecream, tarts, yoghurts etc. A luxury paste, prepared by cooking the sweet potato roots with an infra-red or near infra-red energy source (to mimic the flavour of a traditional product cooked in hot stones), cutting, scooping the flesh out by hand and pressing, is made by one factory in Japan. Sweet potato paste is frequently used, in desserts, to substitute for more expensive ingredients (for example sweet chestnut or mung bean stuffing in Japan and Taiwan, respectively).

The method of producing a purée as used on a large scale in the United States has already been described (see p. 311 and Figure 6.12). Purées may be packaged in cans or jars, sealed and then heat sterilized, flash heated to a very high temperature and aseptically packed, or frozen. Freezing produces a product of higher quality than that made by conventional retorting. However, the Auburn process, using flash heating by direct injection of super-heated steam followed by aseptic packing is claimed to produce a purée with quality equal to that of a frozen purée and can be stored at room temperature (Smith, Harris and Rymal, 1983). This was found to be particularly useful for the preparation of large institutional packs of sweet potato purée (Smith et al., 1982), as excessive retort times and inconveniently long thaw times could be avoided.

Purées may be used for incorporation into pie fillings, frozen patties, soufflés, baby foods and many other products. Its major use at present in the United states is as baby food. As such it is packaged in glass jars of 133 or 222 ml capacity, either alone or mixed with a butter sauce. Two consistencies, finely strained and slightly coarser, are available (Kays, 1985).

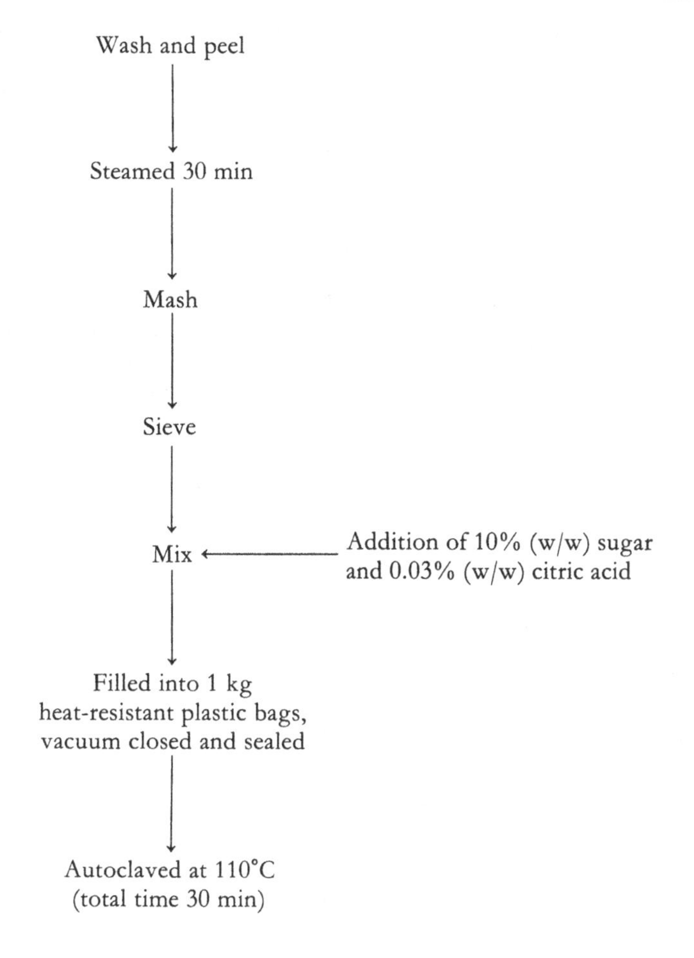

Figure 6.11. Production of a thermally processed paste produced by Kiire Cooperative, Japan.

Canning

Technology

Considerable quantities of sweet potato roots are canned in the United States. Of the limited number of sweet potato products at present available to the consumer in the United States, only canned sweet potatoes are widely distributed and consumed (Walter and Hoover, 1986). The establishment of a sweet potato canning industry has also been considered in other developed countries, for example Australia (Mason, 1982), and canning investigations have been carried out in Taiwan (Chew, 1972). Although canning is too expensive a process to be

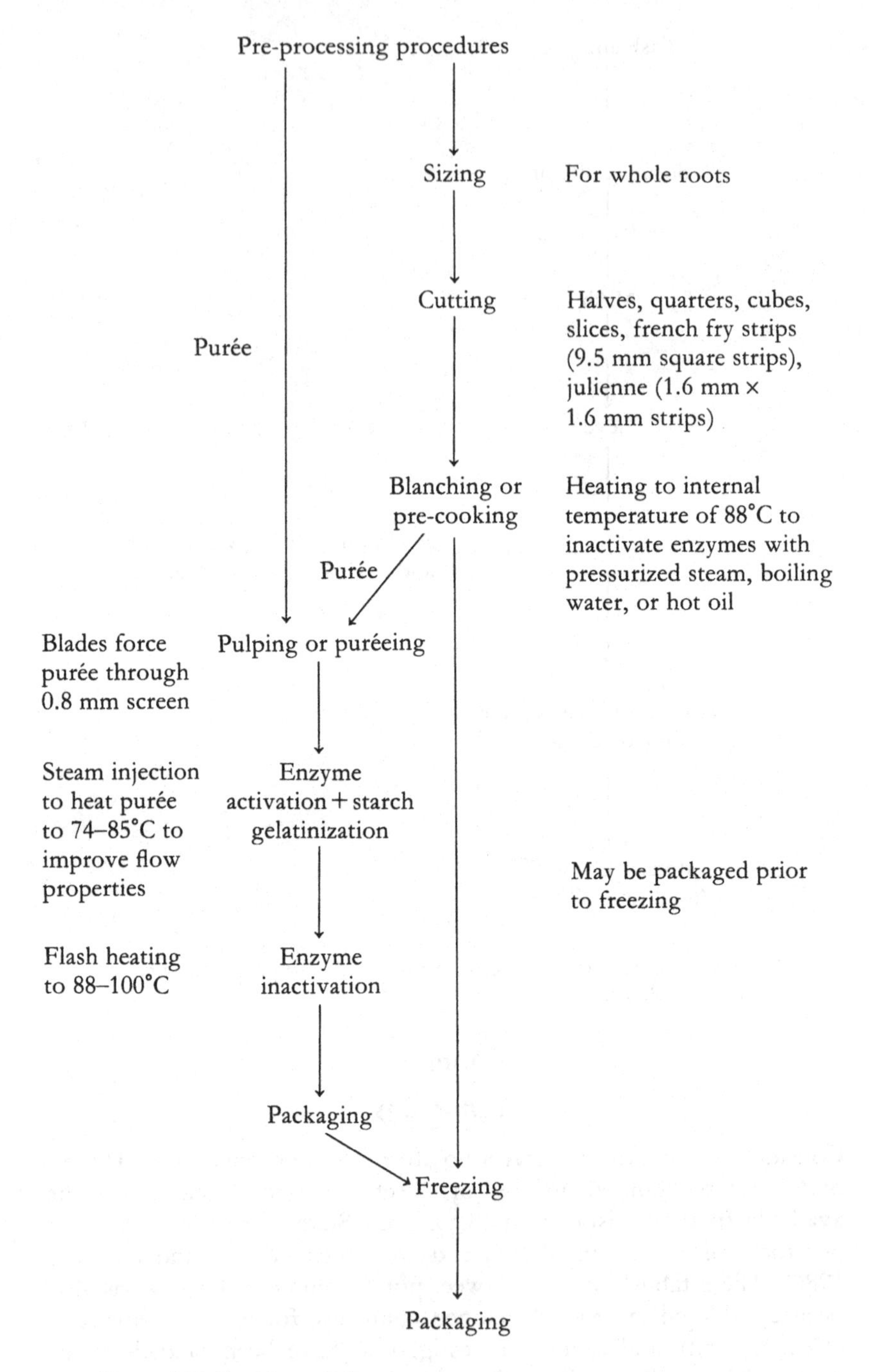

Figure 6.12. Puréeing and freezing processes.

considered economical for most developing countries, it could in future be used to counteract seasonal sweet potato gluts in those countries with increasing middle class populations such as India and Brazil.

Sweet potatoes may be canned whole, halved or otherwise cut into chunks, either in a syrup or under vacuum without a surrounding liquid. Alternatively they can be puréed and canned as a solid pack. Recently flash sterilization followed by aseptic packaging has been found to increase the storability and quality of solid packs (Smith et al., 1982). In addition to the canning of roots alone, syrup packs containing 85% sweet potato, and 15% pineapple titbits in 40°Brix syrup with 20% orange juice, 0.2% orange peel and 0.2% citric acid for extra flavour, have been investigated (Chew, 1972). Canning of sweet potato greens in a 3% NaCl solution has also been explored in the United States (Pace, Dull and Phills, 1985).

The unit operations, or sequence of steps, leading to the production of canned sweet potato roots are shown in Figure 6.13. The initial operations are common to other forms of processing. In the case of canning, pre-heating drives off intercellular gases, helping to maintain good can vacuums and reduce stress on cans during processing.

If sweet potato roots are to be cut up before canning, a uniform size and shape of pieces (for ease of can filling) suitable to local tastes and preferences should be used. Grading the peeled roots to remove substandard poorly peeled or under sized pieces improves final quality (Nanz, 1953). Blanching roots before filling the cans helps to expel gases and raise the initial temperature of can contents thus improving vacuum maintenance. Retort schedules for solid packs, syrup packs of 20°Brix or less, syrup packs of 20–40°Brix and vacuum packs are given in Tables 6.2 to 6.5 inclusive.

An alternative to the use of cans in retort processing is the use of flexible pouches (Rizvi and Acton, 1982). The pouch shape provides a much higher surface-to-volume ratio than metal or glass containers of equal volume. Retort processing time can be greatly reduced and in the case of sweet potato purée, for example, has significantly enhanced retention of thiamin and riboflavin.

Quality problems of firmness and colour

Firmness is one of the most important attributes determining the quality and marketability of canned sweet potato roots. It is influenced by a number of factors including cultivar, geographical location of growth, cultivation practices, curing and storage, and processing techniques. These have been fully reviewed elsewhere (S-101 Technical Committee, 1980) and will be discussed only briefly here.

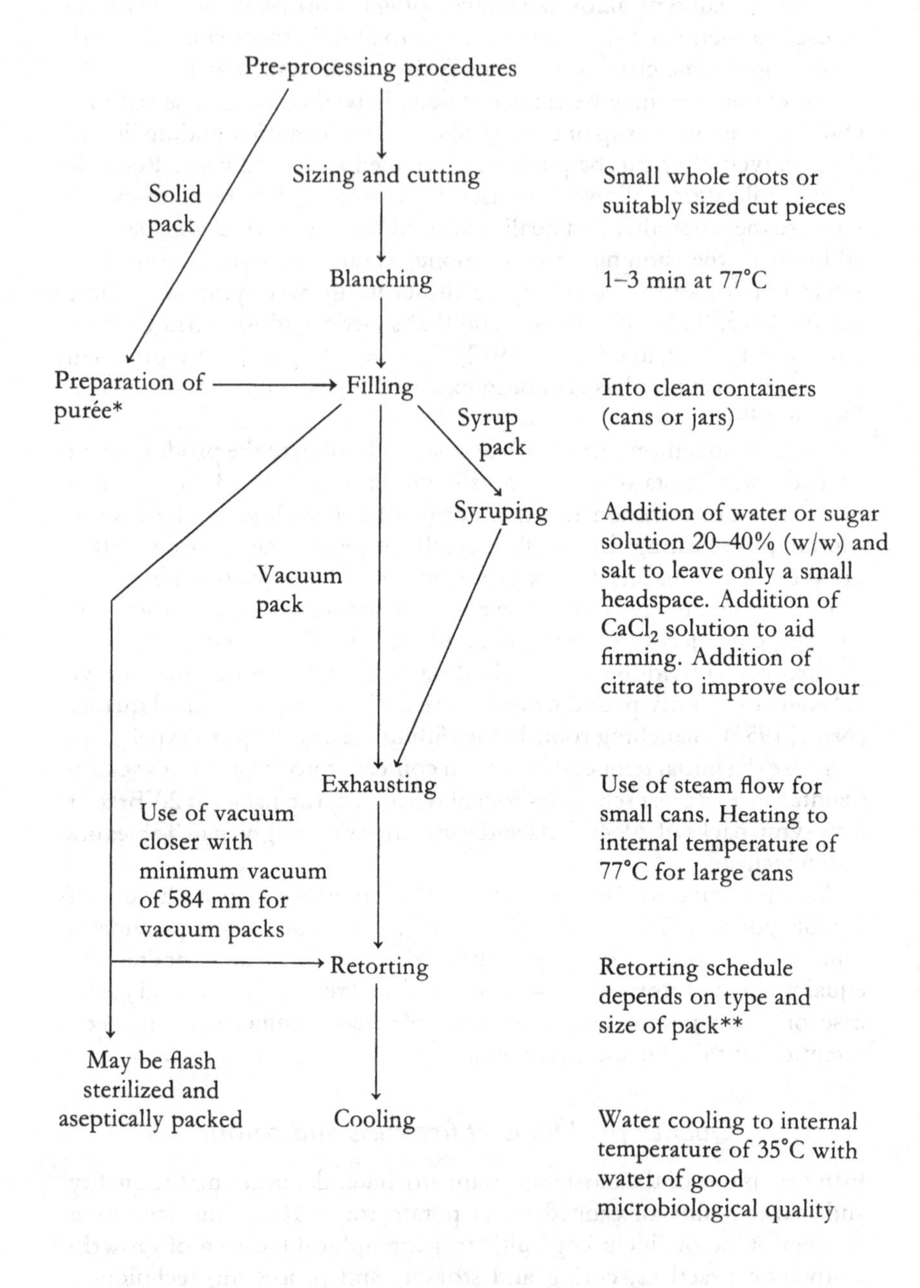

Figure 6.13. Canning of roots. *, see Figure 6.12; **, see Tables 6.2 to 6.5.

Table 6.2. *Retort schedules for solid pack sweet potatoes*

Can size		Minimum initial temperature	Time at retort temperature 116°C
US	Metric (mm)	(°C)	(min)
211 × 400	68.3 × 101.6	49	73
		66	68
		72	61
307 × 409	87.3 × 115.9	49	105
		66	100
		82	84
401 × 411	103.2 × 119.1	49	130
		66	120
		82	105

Notes:
Reprinted with permission from Bouwkamp, J.C. 1985. Processing of sweet potatoes – canning, freezing, dehydrating. In: Bouwkamp, J.C. (ed.), *Sweet potato products: a natural resource for the tropics*, pp. 185–203. Copyright CRC Press, Inc., Boca Raton, FL.

Table 6.3. *Retort schedules for syrup pack (up to 20° Brix), freshly dug or stored, whole, cut or sliced sweet potatoes*

Can size		Minimum initial temperature	Time at retort temperature (min)		
US	Metric (mm)	(°C)	116°C	118°C	121°C
303 × 406	81 × 111.1	21	36	28	23
		38	35	27	22
		60	32	24	20
401 × 411	103.2 × 119.1	21	52	42	37
404 × 307	108.0 × 87.3	38	49	40	35
401 × 602	103.2 × 155.6	60	45	37	32
603 × 700	157.2 × 177.8	21	57	46	40
		38	54	44	38
		60	50	40	35

Notes:
Reprinted with permission from: Bouwkamp, J.C. 1985. Processing of sweet potatoes – canning, freezing, dehydrating. In: Bouwkamp, J.C. (ed.), *Sweet potato products: a natural resource for the tropics*, pp. 185–203. Copyright CRC Press, Inc., Boca Raton, FL.

Table 6.4. *Retort schedules for syrup pack (20° to 40°Brix), freshly dug or stored, whole, cut or sliced sweet potatoes*

Can size		Minimum initial temperature	Time at retort temperature (min)		
US	Metric (mm)	(°C)	116°C	118°C	121°C
303 × 406	81.0 × 111.1	21	42	33	26
		38	49	30	24
		60	35	27	21
401 × 411	103.2 × 119.1	21	57	47	39
404 × 307	108.0 × 87.3	38	53	43	35
401 × 602	103.2 × 155.6	60	47	37	30
603 × 600	157.2 × 177.8	21	77	59	46
		38	69	52	41
		60	57	43	34

Notes:
Reprinted with permission from: Bouwkamp, J.C. 1985. Processing of sweet potatoes – canning, freezing, dehydrating. In: Bouwkamp, J.C. (ed.), *Sweet potato products: a natural resource for the tropics*, pp. 185–203. Copyright CRC Press, Inc., Boca Raton, FL.

Apart from intrinsic genetic differences which affect firmness, and variations between the same cultivars grown in different locations (Jenkins, Anderson and Watson, 1956), the application of fertilizers may increase firmness (for example nitrogen) or decrease firmness (phosphorus or potassium) (Constantin, Jones and Hernandez, 1977). Irrigation was reported to increase firmness (Kattan and Littrell, 1963). Planting and harvest dates may influence firmness. It was noted that when some roots were planted and harvested earlier than others the early roots produced firmer canned sweet potatoes. However, increase in softness was found to be related to chronological age of roots rather than planting date per se (Scott and Bouwkamp, 1975).

Many investigators have observed the fact that sweet potatoes canned immediately after harvest are firmer than those previously cured or stored (Magoon and Culpepper, 1922; Blackwell, 1955; Baumgardner and Scott, 1962; Constantin and McDonald, 1968; Abdulla, 1970; Yang, 1979). Most reported that even a few days' delay between harvest and processing could cause a significant decrease in firmness. Furthermore the temperature of holding affects firmness. Storage temperatures of 15°C to 30°C caused greater softening of canned roots than did 0°C (Baumgardner and Scott, 1962). At the higher temperatures, however,

Table 6.5. *Retort schedules for vacuum pack sweet potatoes*

Can size		Minimum initial temperature (°C)	Time at retort temperature 116°C (min)
US	Metric (mm)		
404 × 307	108.0 × 87.3	21	45

Notes:
Reprinted with permission from: Bouwkamp, J.C. 1985. Processing of sweet potatoes – canning, freezing, dehydrating. In: Bouwkamp, J.C. (ed.), *Sweet potato products: a natural resource for the tropics*, pp. 185–203. Copyright CRC Press, Inc., Boca Raton, FL.

cured roots stored for 1 month at 27°C were firmer than those stored at 16°C (Constantin and McDonald, 1968). Changes in pectic fractions (see Chapter 5, pp. 251–2), are thought to be responsible for the decreased firmness of previously stored, canned roots (Baumgardner and Scott, 1963). As firmness values decreased in the canned product, insoluble 'protopectin' also decreased, while oxalate soluble pectin increased. Starch changes, in contrast, were small and not constantly related to firmness.

An efficient and economic sweet potato canning industry requires an extended supply of roots. In those areas where harvesting is seasonal, therefore, curing and storage of some roots before canning is unavoidable. Methods of improving firmness of canned cured and stored roots have therefore been developed. Based on the association between firmness and pectic substances, a calcium treatment in the form of a pre-processing soak has been found to firm-up canned roots. Peeled roots have been soaked, for varying lengths of time, in either 1% or 2% (w/v) calcium chloride solution alone (Lutz, 1949) or with the addition of citric acid (Scott and Twigg, 1969) or buffers (Sistrunk, 1971) to reduce the pH. The addition of acid produced a more uniformly firm product than calcium alone. Increasing the temperature of the soak increases firmness, and pre-process soaking is more effective than addition of calcium chloride or citric acid to the can liquid (Sistrunk, W.A., unpublished data). Soaking roots in calcium hydroxide solutions can also be used for increasing firmness after storage (Rao and Ammerman, 1974). Such a soak can be followed by canning the roots in a syrup with 2% added pectin (Williams and Ammerman, 1968). The addition of amylopectin to the syrup was also suggested to be a reliable method of firming small, whole, canned roots (Abdulla, 1970).

The conditions under which the roots are processed can affect their texture. Canning in heavy syrups with high sugar concentrations produced firmer roots than did canning in light syrup, and firmness was slightly greater with sucrose than with corn syrup (Woodroof, Dupree and Cecil, 1955). Pre-peeling heat treatment of roots to prevent discoloration and/or to aid in peeling has been found to decrease firmness (Woodroof et al., 1955). It has been suggested that utilizing a combination of appropriate cultivars and suitable storage temperatures can be used to control firmness of canned roots by improvement of raw material quality (S-101 Technical Committee, 1980).

The addition of citric acid to the can syrup to improve the colour of the canned product is illustrated in Figure 6.13. Research (Sistrunk, 1948) has found that holding peeled sweet potatoes in citric acid solution for 24 hours, or application of a combination of citric acid and calcium chloride before canning improves the colour of canned yellow/orange-fleshed sweet potatoes. Lowering the pH of can contents to 3.0 by the use of citric acid buffers was found to be the most effective treatment (Sistrunk, 1971). Presumably this pH is the most inhibitory to enzymic browning reactions.

Freezing

The use of freezing to preserve sweet potatoes is unlikely to be adopted on a large scale by any but a limited number of developed countries, due to prohibitive cost of plant maintenance and a lack of suitably equipped retail outlets outside the major cities. It will therefore be described very briefly here.

Sweet potatoes may be frozen in many different forms: as whole roots, halves, quarters, slices, cubes, french fries, mash/paste, or as purée (see p. 314 and Figure 6.12). The outline of a procedure for production of frozen cubes, slices and paste as used in Japan is shown in Figure 6.14.

Blanching, to inactivate enzymes otherwise responsible for off-colours or flavours in the frozen product, may be carried out in pressurized steam. Alternatively, in the case of french fries, chips, strips or cubes, hot oil may be used and the frozen fried product is merely reheated by the consumer.

Freezing should take place rapidly under strictly hygienic conditions, as no further heating of the product takes place after blanching. Packaging may precede freezing, or follows it in the case of whole roots or pieces individually quick frozen.

Studies on a North American convenience food in the form of a frozen mash produced from sweet potatoes lye-peeled, trimmed, steam-cooked, mashed, stuffed into casings and frozen at −19°C, showed that low grade culls or extra large jumbo roots could be used (McMillen, 1981), as

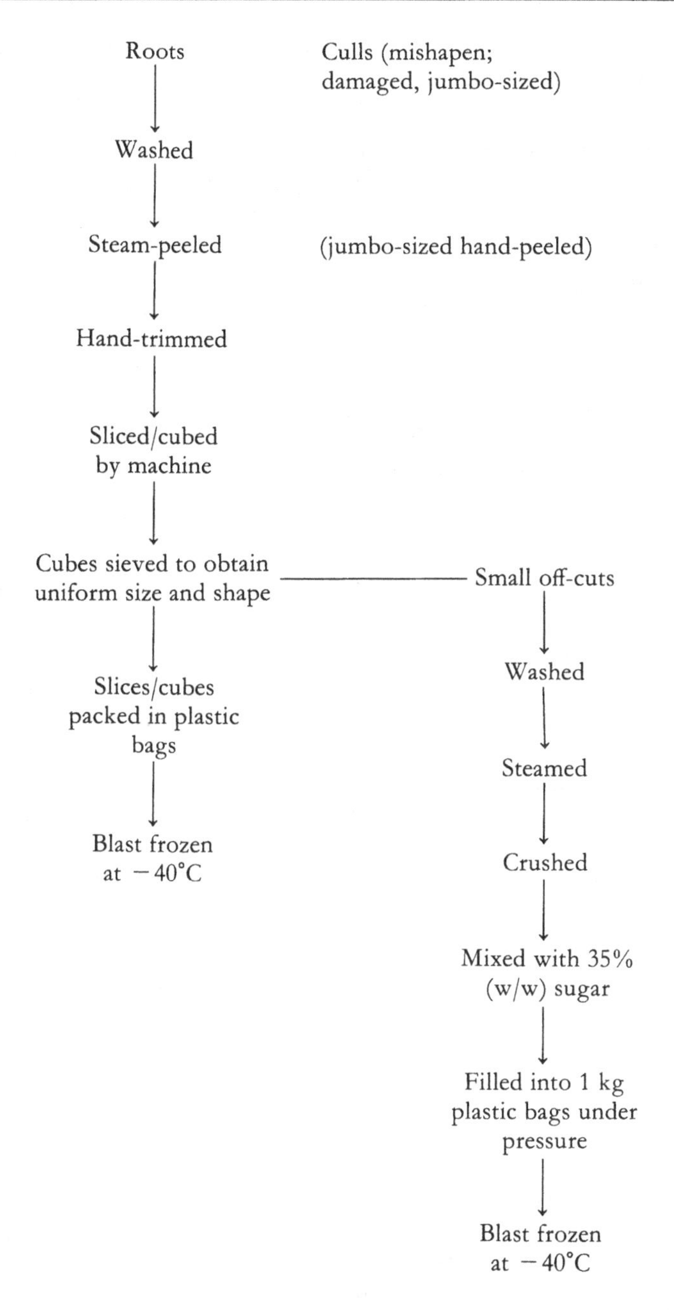

Figure 6.14. Freezing proc.sses (as carried out by Sunmoi Products, the Oyamatoshima Branch of Kumamoto Agricultural Cooperative, Japan).

is done commercially in Japan (Figure 6.14). The roots required curing and could be stored for 4 months before the production of a good quality mash.

As long ago as 1958, a french-fried sweet potato product was prepared by oil-frying slices, cooling and freezing the fried slices (Kelley, Baum and Woodward, 1958). The frozen fried slices were then oven-cooked prior to eating, and reported to be of good quality when the sweet potatoes had been cured and stored for 1 month. The quality decreased if roots were stored for 6 months. The preparation and properties of a frozen french-fry product have recently been explored again on a research basis in the United States as a possible consumer alternative to canned sweet potatoes (Walter and Hoover, 1986; Schwartz et al., 1987). Roots were washed, lye-peeled and sliced into strips 1.9 cm × 6.4 cm thick and indeterminate length. Slices were blanched in boiling water containing 1% (w/v) sodium acid pyrophosphate (to inhibit polyphenolic discoloration), partially dried in a forced air drier at 121°C, frozen and held at −34°C until being fried immediately before eating. Chemical analyses and sensory evaluation showed that the best quality product was obtained from cured roots stored not more than 24 weeks before processing. Few quality changes were observed when the french fries were stored frozen for 1 year, except for an appreciable loss in ascorbic acid and a small but significant decrease in colour score (Schwartz et al., 1987).

French fries made with sweet potatoes are used in place of potato french fries in some Taiwanese fast food establishments (Tsou, S., personal communication). There is interest in Japan in the preparation of french fries from low- or non-sweet sweet potato cultivars which would closely resemble those made from potatoes (Umemura, Y., personal communication).

Chips (crisps)

Production

In the last 20 years the production of sweet potatoes has fallen in many countries. In developed countries a similar decline in the production of potatoes was avoided by the increasing manufacture of convenience products such as dehydrated flakes, frozen french fries and chips (or crisps). The great potential for dehydrated sweet potato products has already been discussed. Fried chips could be marketed as a nutritious snack food in developing countries, especially if cultivars rich in provitamin A and ascorbic acid are used. Such chips are already being

Figure 6.15. Sweet potato chips marketed by three different companies in Peru (International Potato Center).

marketed in supermarkets in some countries, for example Peru and Japan (Figure 6.15). Chips are produced commercially in China though on a limited scale due to insufficient market demand (Wiersema, Hesen and Song, 1989). Sugar-coated sweet potato chips are a popular snack in the shape of slices, ribbons or sticks (*imokarinto*) in Japan where a variety of appearances is achieved by the use of, for example, dark brown sugar, sesame seed coatings etc. Sweet potato chips made by a rural cooperative in Tuvalu, South Pacific, are being successfully marketed in the capital city (Foss, P., personal communication). In what is a tiny group of atolls, they provide a product which is more easily shipped from the outer islands to the capital by the extremely infrequent transport system than would be the fresh roots. Salted chips have been tested in Papua New Guinea (Siki, 1979) and spicey hot chips in Bangladesh (Molla, 1973) with varying degrees of success. Chips have been made with a red-fleshed cultivar in Taiwan. A patent for the production of sweet potato chips in the United States was first granted as long ago as 1936 (Brunstetter, 1936) and the production technique has since undergone various refinements (Burton, Jones and Miller, 1958; Boggess and Woodroof, 1964; Hoover and Miller, 1973). The fact that even now, only limited local production of sweet potato chips exists (Picha, 1986) can be attributed to the necessity of overcoming problems related to quality of the finished product as well as to assurance of an all-year-round supply of suitable cultivars.

The process for production of sweet potato chips as developed in the United States is shown in Figure 6.16 and that for reconstituted chips developed in China in Figure 6.17. Reconstituted sweet potato chips were reported to contain (per 100 g) 5.7 g of protein, 33 g of fat, 1.45 mg of Fe, 1.2 mg of thiamin, 0.16 mg of riboflavin and 3.9 mg of ascorbic acid (Anon., 1989). A similar product, known as crispy sliced sweet potato and flavoured with garlic and pepper has been developed in Thailand (Prabhuddham et al., 1987).

Extruded snack products with alternative shapes to those of conventional chips are produced in Japan (Baba, T., personal communication), including one, produced from the cultivar 'Satsumahikari' which has very low sweetness, with characteristics similar to those of extruded potato snacks. This product is made by reconstituting sweet potato flakes, extruding, cutting the extruded paste into slices and frying.

Quality changes

Darkening of uncooked slices and excessive browning during frying gives rise to a problem of discoloration in the chips made from some cultivars. Phenolic oxidation results in darkening, the degree of which is correlated with phenolic substrate concentration rather than phenolase activity (Anon., 1985a). Browning is due to the Maillard reaction between reducing sugars and amino acids. An association was found between the levels of reducing sugars glucose and fructose in different cultivars or as a result of storage, and degree of chip browning, but not between sucrose levels and browning (Picha, 1986). The increase in maltose from trace levels in the raw material to more than 2% in the fried chips was also considered to be a factor in discoloration (Baba, Kouno and Yamamura, 1981). Blanching, either in boiling water or steam or in a solution of sodium acid pyrophosphate (Baba et al., 1985; Hoover and Miller, 1973), or dipping in sulphite solution (Olorunda and Kitson, 1977) helps to prevent chip discoloration.

A diffuser-extractor for removing 90–100% of the sugars in sweet potato slices before frying has been described (Hannigan, 1979). Alternatively the manipulation of pre-processing conditions has been suggested as a way of controlling reducing sugar concentrations. A period of reconditioning for 48 hours at 20°C after storage and before processing, similar to the longer period practised with potatoes to reduce sugar levels, was claimed to improve chip colour (Olorunda and Kitson, 1977). However, another author found that manipulation of postharvest storage temperatures did not prevent sugar accumulation or

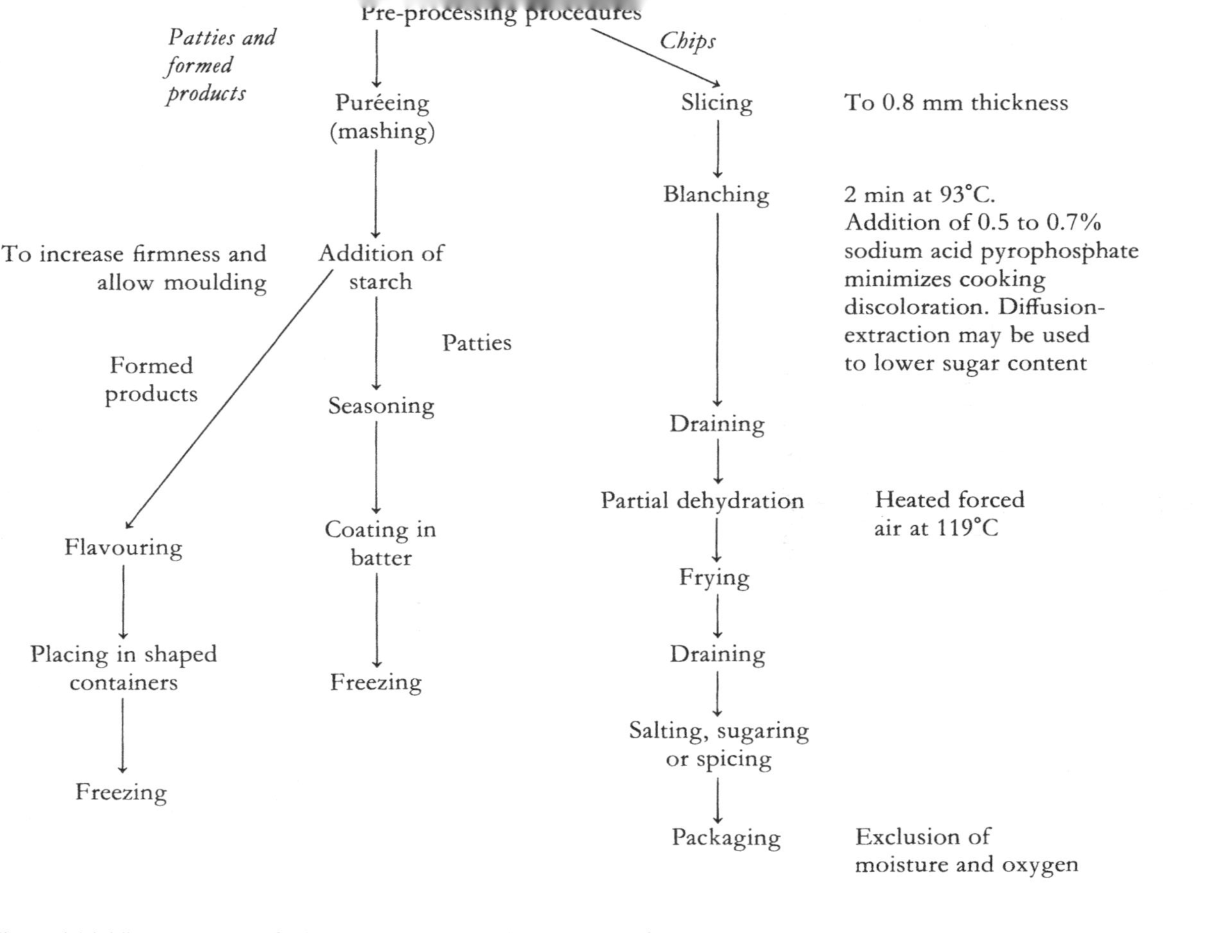

Figure 6.16. The processing of other sweet potato products (as carried out in the United States).

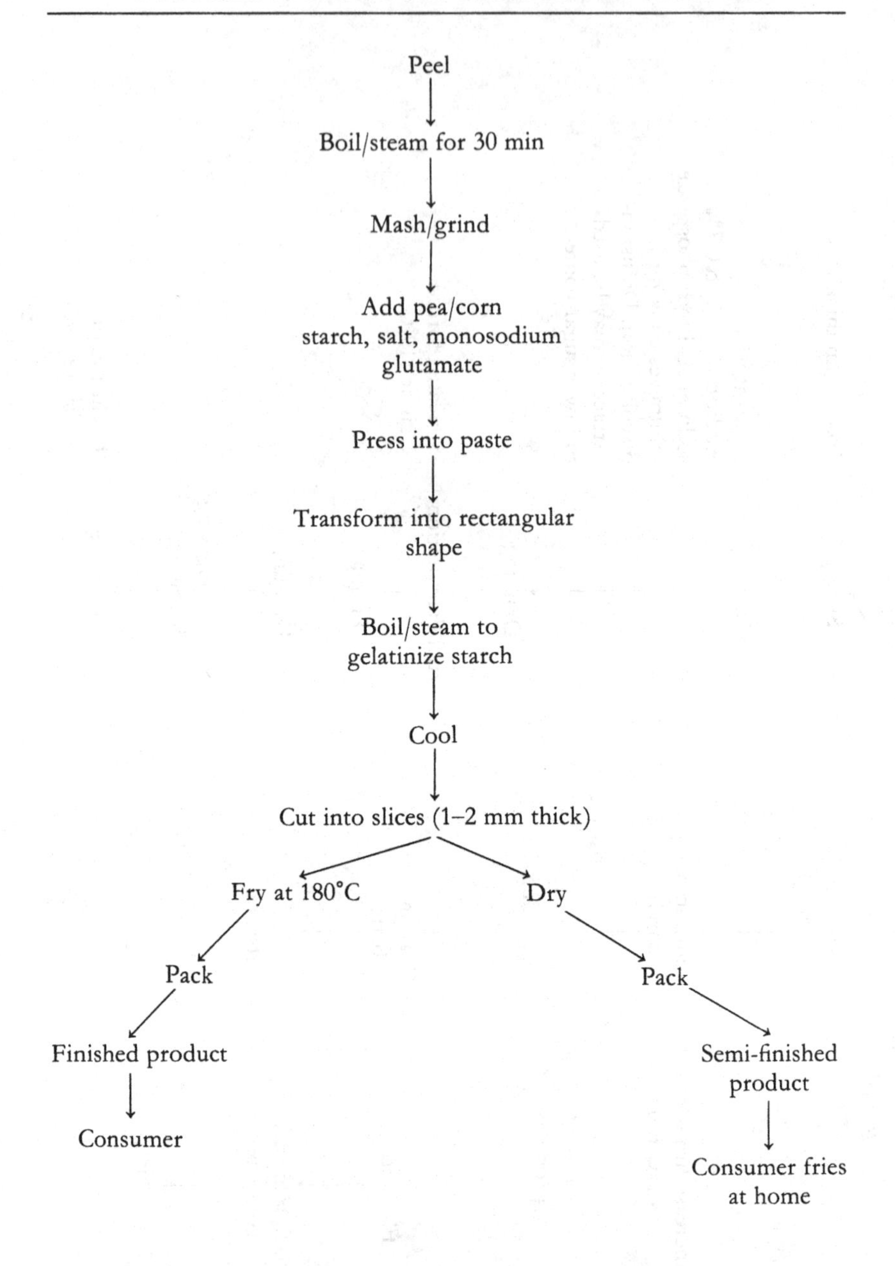

Figure 6.17. Production of reconstituted chips (as carried out at the Sichuan Academy of Agricultural Sciences and Yanting County processing plant, China) (from Wiersema et al., 1989).

chip darkening and that the lightest coloured chips were produced immediately after harvest (Picha, 1986).

Temperature and time of frying can also influence colour development (Anon., 1985a; Martin, 1987). Too low a temperature resulted in inadequate colour and excessive frying time, while too high a temperature caused over-browning to occur (Martin, 1987). Optimum frying temperatures between 143°C and 177°C have been variously recommended by authors; differences could have been related to the type of oil used.

Oiliness has been found to be one of the chief problems affecting acceptability of sweet potato chips (Anon., 1985a; Baba et al., 1981; Siki, 1979). The greater the moisture content of the uncooked slices, the greater the quantity of fat retained after frying. Blanching, which increases moisture content, can increase undesirable oiliness of chips (Anon., 1985a). This problem is exacerbated by the use of low dry matter cultivars (Anon., 1985a) and very thin slices. It was not possible to produce chips with less than 35% oil when root dry matter was lower than 30%. Oiliness increased progressively with decrease in slice thickness from 1.4 to 0.8 mm and with an increased blanch time from 0 to 60 s. Partial drying, as shown in Figure 6.16 (Hoover and Miller, 1973), after blanching can help to improve the appearance and texture of the finished chip.

Another problem which may arise is that of chip hardness and lack of crispness. This has been attributed to a small void volume or blister development (Baba et al., 1981). Blanching and/or freezing and thawing before frying greatly decreased hardness of chips (Baba and Yamamura, 1981) or sweetened deep-fat-fried sticks (*imokarinto*) (Baba et al., 1985).

Sweet potato chips have been described as of two main types (Martin and Rhodes, 1984), namely a 'cookie-like' chip with a taste influenced by sweetness and a 'potato-like' chip similar to that of the white potato. Selections of staple type sweet potatoes low in sugar both before and after cooking have been found to produce acceptable non-sweet chips (Martin, 1987, and see above). If salted chips are required, slices may be immersed in a 1% (w/v) solution of sodium chloride and drained prior to frying (Martin, 1987).

A post-process decrease in chip quality can take place if chips are allowed to take up moisture or excessive oxygen, which can result in leatheriness and rancid off-flavours, respectively. It is therefore important that chips are packaged in suitable moisture-proof and airtight material immediately after processing. Slices may be treated with an antioxidant by brief dipping in 1% (w/v) ascorbic acid solution before frying (Martin, 1987). This treatment was found to increase crispness of chips.

Candies, jam and sweets

The natural sweetness of many sweet potato cultivars lends itself to the preservation of roots by the addition of further sugar to give a variety of candied products, candies, jam and 'sweets'. The latter, known as *doces* (Portuguese) or *dulces* (Spanish) are particularly popular in Latin America. In fact *dulce de batata* is the national dessert of Argentina, consumption being in the order of 65,000 tonnes per year in that country alone (López Hernández et al., 1983). However, a recent report (Horton, 1988) states that both the consumption and quality of Argentinian *dulce de batata* are declining as its price increases relative to other desserts, and artificial vanilla flavour and maize-derived fructose replace natural vanilla and sucrose in its formulation. *Doce* or *dulce* of sweet potato is prepared in a similar way throughout Latin America: sweet potato purée is boiled with an equal weight of sugar, a little vanilla being added for flavour, and either the natural pectin content of the sweet potato or added agar is used to produce a pasty or gelled mass which is sold in shops and supermarkets. The resulting sweet, which is similar to a jam, can be used as a dessert or snack. In an attempt to improve the nutritional value of Argentinian *dulce de batata*, the protein content was increased from a negligible value to 5.2% by the addition of a purée of soy predigested with papain (López Hernández et al., 1983). The proportions of sweet potato and soy purées were adjusted so that the former furnished 62% and the latter 38% of the solids content of the mixture. Rats fed sweet potato *dulce* grew poorly whereas the growth of those fed the sweet potato-soy *dulce* more nearly approximated that of those fed with casein. The soy-enriched *dulce* was suggested to be suitable for use in institutions such as schools, hospitals, military camps etc.

Sweet potato jam has been prepared in the Philippines (Figure 6.18; Truong, 1987). Sweet potato is appropriate for jam making as it contains a suitable content of water-soluble pectin (Winaro, 1982) with gelling properties similar to that of apple pectin. The Philippine process consisted of cooking a mixture of 20.7% sweet potato, 45% sugar, 34% water and 0.3% citric acid until a solids content of 68°Brix was obtained. The jam scored highly for colour, flavour and overall acceptability on sensory testing, but only fair to good for gel consistency (Figure 6.19). The difference in consistency between sweet potato and fruit jams was thought to be due to the high starch content of sweet potato. Jams with various natural colours, yellow, orange or pinkish, could be prepared using cultivars with different flesh colours. Sweet potato jams are prepared for sale on a small scale in some areas of China (Sheng and Wang, 1987). It is interesting to note that the European Community has altered its directive on jams to include sweet potato (as a 'fruit'), largely

Figure 6.18. A range of fruit-flavoured sweet potato jams produced at ViSCA, Philippines (ViSCA).

at the instigation of Portugal, where sweet potato jam is a traditional product.

Small individually wrapped candies (known as sweets or toffees in some countries) are produced in Japan by small traditional businesses. One such establishment in southern Japan, visited by the author, prepares candies from cream-fleshed cultivars which give a light brown candy, and also from a purple-fleshed cultivar which gives the finished candies a pinkish tinge. Unpeeled steamed sweet potatoes are crushed and 2.5% (by weight) dry barley malt added. The mixture is maintained at 55°C for 1.5 hours to hydrolyse starch to maltose and dextrins. The saccharified mix is then pressed to extract the juice which is boiled for 7 hours to produce a very thick stiff syrup. The syrup is boiled at 105°C in a metal bowl over a fire and then poured into water-cooled metal trays to cool. The candy is then pulled (which used to be done by hand but is now done by machine), to introduce air and harden it. The resulting block of candy is mechanically pulled out into a long thin strip, cut into small pieces and the pieces separated on a series of conveyor belts. The candy pieces are either rolled in barley flour or individually wrapped in transparent twists before packing into 125 g plastic packs.

Figure 6.19. The gel consistency of sweet potato jam may be somewhat softer than that of fruit jams due to the high starch content of sweet potato (ViSCA).

Another type of sweet potato candy has been developed for sale in China by the processing laboratory of the Sichuan Academy of Agricultural Science (Wiersema et al., 1989). In this process, peeled sweet potato is cut, washed in a basket to remove starch, boiled for 8 min in a sugar solution, flavourings added, steeped for 3 hours in a sugar solution, dried in a traditional oven for 8 hours using moderate heat, weighed and packed by hand and machine-sealed.

In Puebla, Mexico, sweet concoctions known as the Camotes de Santa Clara are sold in shops called *camoterias* (Austin, 1973). The confection is a sweetened sweet potato paste or dough with added flavours such as orange, strawberry and pineapple, moulded into a cylindrical shape.

Halwa, a traditional sweet, has been prepared on an experimental basis in south India (Seralathan and Thirumaran, 1990) by mixing sweet potato flour with 4 times its weight of water, heating until the volume is reduced to one half, adding sugar and ghee (dehydrated rancid butter oil), stirring and cooking until all the water evaporates and the ghee begins to ooze from the mixture, and then pouring the product into trays to cool followed by cutting into cubes.

There is obviously a wide range of sweet candied products which can be produced from sweet potato. Recipes can be adapted to produce traditional candies for local tastes in different areas of the world.

Further products from primary processes

Flour

Baked goods

Bread

Yeast-raised and flat breads made from wheat flour are staple foods in almost every part of the world, including tropical regions where wheat is seldom grown and therefore has to be imported. The purchase of wheat necessitates foreign exchange spending which could be reduced if total or partial substitutes for wheat could be found. There has been a considerable amount of research into the use of composite flours, that is mixtures of wheat with flours prepared from other plant foods such as legumes, oilseeds and root/tuber crops. The possibility of utilizing wheat-sweet potato composite flours in breads and other baked goods has been investigated in several countries including Egypt (Hamed, Hussein and Refai, 1973b), Ghana (Bortey, 1982; Osei-Opare, 1987), India (Nair et al., 1987; Seralathan and Thirumaran, 1990), Israel (Plaut and Zelzbuch, 1958), Korea (Kim et al., 1973), Philippines (Tapang and Rosario, 1977; Palomar, Perez and Pascual, 1981), Peru (Luna de la Fuente, 1960; Carpio Burga, 1984), Taiwan (Lee, 1985; Tsai, W., personal communication) and the West Indies (Sammy, 1970). In some cases sweet potato paste has been found to be preferable to the flour as a partial substitute for wheat (Carpio Burga, R. del, personal communication; Lee, 1985). Sweet potato bread has been commercialized on a limited scale in some areas, for example Peru and Japan.

A commercial bakery in Peru uses the very simple method of mechanically grating peeled raw sweet potato roots and adding them to the dough at 30% substitution for the wheat flour. This would be a cheap

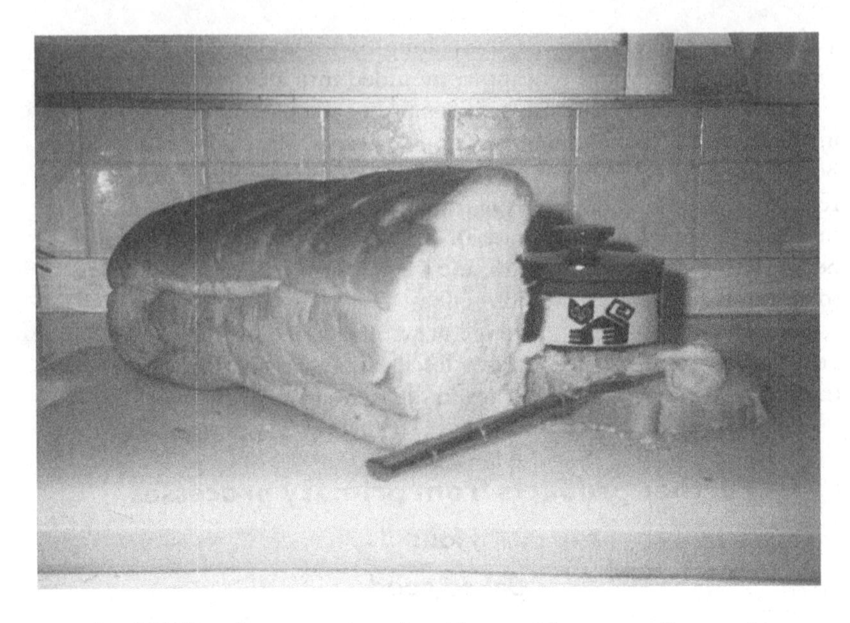

Fig 6.20. Peruvian sweet potato bread has a golden crust, a fine crumb, well-risen structure and good keeping qualities (J.A. Woolfe).

and practical solution for use in areas where fuel to dry sweet potato into flour is expensive or scarce, and where it is climatically difficult to sun-dry. The method for making this bread is shown in Appendix 1. Before applying this method, it should be ascertained that the baking process destroys trypsin inhibitors which may be present in the raw sweet potato (see Chapter 4). Otherwise, cultivars low in trypsin inhibitor concentration should be used.

Most studies have investigated the feasibility of using sweet potato flour as a partial substitute for wheat flour. Many types of bread can be prepared, not only Western-style (Figure 6.20), but also traditional breads such as Egyptian balady bread (Hamed et al., 1973b), Indian chaphatis (Seralathan and Thirumaran, 1990) or poories and Chinese steamed bread (Figure 6.21; Tsai, W., personal communication). The level of wheat flour substitution with sweet potato flour found to produce a formulation resulting in an acceptable bread with characteristics similar to those of bread made entirely with wheat differed between researchers. This could have been due to variations in recipe formulations, methods of preparation and local taste preferences. However, all reported a maximum substitution level of about 20% above which bread became unacceptable in terms of loaf volume, flavour and texture. Most researchers found 10–15% substitution the most acceptable in these

Figure 6.21. Speciality breads can also be made with a proportion of sweet potato in place of wheat, e.g. Chinese steam bread from Taiwan (W. Tsai).

terms. The characteristics of loaves prepared with varying proportions of sweet potato and wheat flour as determined by one researcher are shown in Table 6.6. The development of a wheaten loaf with an acceptable volume and texture depends on the content and quality of its gluten (protein). It is not possible to use a composite flour containing a high percentage of sweet potato, due to excessive dilution of wheat protein by sweet potato flour with a low protein content lacking the characteristics of gluten.

The use of unpeeled roots for preparation of sweet potato flour, thus eliminating waste, was found to be equal to that of peeled roots for composite flour bread making (Hamed et al., 1973a; Plaut and Zelzbuch, 1958; Sammy, 1970). At 6% substitution, unpeeled root flour increased the volume of bread and improved crumb structure (Plaut and Zelzbuch, 1958). The treatment of sweet potato with sulphite before drying has been found in some cases to improve the appearance of the flour and subsequently the bread prepared from it (Hamed et al., 1973b; Sammy, 1970), but a residual sulphite concentration of > 100 p.p.m. in the flour had an adverse effect on its baking properties (Sammy, 1970).

Not all cultivars of sweet potato are equally appropriate for bread making (Sammy, 1970, 1984). The quality of bread prepared with composite flour (containing 15% sweet potato flour) from 17 different cultivars was found to relate positively with the total protein content of

Table 6.6. *Effect of different proportions of sweet potato and wheat flour on bread loaf characteristics*

	Sweet potato flour (%)					
Character	0	5	10	15	20	25
Loaf volume						
(ml/g)	4.6	4.2	4.3	4.3	3.7	2.8
ml/kg flour	7429	7214	7286	7357	6333	5000
Colour						
crust	Pale brown	Pale brown	Brown	Brown	Brown	Brown
crumb	White	Light grey	Light grey	Light grey	Light grey	Light grey
Texture						
crust	Smooth	Smooth	Smooth	Smooth	Few cracks	Many cracks
crumb	Cells even (good)	Cells even (good)	Cells even (good)	Cells even (good)	Cells uneven (fair)	Cells uneven (fair)
Shape	Well risen	Well risen	Well risen	Well risen	Slightly fallen	Fallen
Flavour	Good	Good	Good	Good	Fair	Poor

Note:
From: Sammy, 1970.

the sweet potato flour (Sammy, 1984). Some cultivars were found to be inappropriate for composite bread making in terms of resulting loaf shape, texture and flavour.

The addition of dough improvers – glyceryl monostearate or glyceryl monopalmitate – brought about little increase in loaf volume in sweet potato–wheat bread (20%:80%). The addition (5% by weight of dry ingredients) of high protein concentrates in the form of fish protein concentrate or cottonseed flour made little difference to its baking properties (Sammy, 1970).

The use of 6% sweet potato flour substitution for wheat has been reported to improve loaf volume by 15% (Plaut and Zelzbuch, 1958), and at 6% or 10% to increase yield of loaves (Hamed et al., 1973b; Plaut and Zelzbuch, 1958) chiefly because the addition of sweet potato greatly increases the water absorption of flour. This also leads to a dough perceived on handling as wetter or stickier. However, loaf volume has generally been found to diminish, texture to coarsen, colour to darken and off-flavour to increase with increasing levels of sweet potato flour (Sammy, 1970; Tapang and Rosario, 1977; Hamed et al., 1973b; Luna de la Fuente, 1960). The development of an off-flavour noticeable in bread made from sweet potato flour can be entirely eliminated by using sweet potato paste instead of flour (Carpio Burga, R. del, personal communication; Lee, 1985), whereas the addition of flavourings such as vanillin, chocolate or peanut did not improve the flavour of sweet potato flour composite bread at all, and butter or milk flavouring or egg improved it only slightly (Lee, 1985). The composite bread made with sweet potato paste was claimed to be of a quality superior to that of 100% wheat bread (Lee, 1985). The undesirable flavour of sweet potato flour composite bread was not due to thermal changes in beta-carotene during baking as the addition of beta-carotene to 100% wheat flour did not produce an off-flavour (Lee, 1985).

The potential economic advantage of partial substitution of wheat flour with sweet potato flesh, flour or paste in baked goods has already been mentioned above. An added advantage is an improvement in the nutritional value which can be accomplished. Utilization of sweet potato flours or pastes rich in beta-carotene (absent in wheat) gives a bread which can make a significant contribution to provitamin A intakes (Carpio Burga, 1984). Such bread, *camote-pan*, has been commercialized in two coastal towns of Peru – Cañete and Chincha – and at the Agricultural University, La Molina. In Cañete, the roots are peeled, grated and added raw to the wheat flour as described above. In Chincha, where a grater is not available at present, the roots are steamed and mashed before being added to the wheat flour. Sweet potato constitutes 30% and wheat flour 70% of the mix. The bread is said to keep well for 4 or 5 days without staling and can therefore be bought occasionally by people who live isolated from a bread-making centre. It is said to be stable in price, relatively cheap and accessible to purchase by low-income workers (Carpio Burga, R. del, personal communication). The use of bread made with high beta-carotene sweet potato paste for feeding school children in Japan has already been mentioned (and see Chapter 3). In addition the use of sweet potato flours with relatively high protein concentrations, which is becoming increasingly possible with the development of new cultivars (Carpio Burga, 1984), may improve both the protein quantity and quality of composite flour. The high level of lysine in sweet potato protein relative to that of cereals has already been discussed (Chapter 3). In a rat feeding experiment a composite sweet potato-defatted soy flour was superior in terms of net protein utilization and biological value*, and equal in digestibility, to wheat flour or other composite flours composed of barley, corn, or potato and defatted soy flour (Kim et al., 1973).

It has been claimed that sweet potato bread retains its freshness for a longer period than 100% wheat bread (Carpio Burga, 1969 and personal communication), but other authors found no difference in retention of freshness between sweet potato and regular bread (Plaut and Zelzbuch, 1958).

White bread containing 6% sweet potato flour when sold along with regular white bread by stores in Israel elicited no comments from either the store or public; in other words the public did not notice any differences between the breads (Plaut and Zelzbuch, 1958). In the Philippines, two popular types of bread *pan de sal* and hot rolls containing 20% sweet potato flour were found to be acceptable when consumer-tested with 100 adults and 45 children (Palomar et al., 1981). A high percentage preferred them to samples prepared with 20% cassava flour.

In India, it has been found possible to prepare chaphatis with 50% sweet potato flour (Seralathan and Thirumaran, 1990). In Maharashtra and Bihar, sweet potato flour is sold for home preparation of chaphatis and poories (Ghosh, S.P., personal communication).

Experimental addition to white or lower extraction (standard) wheat flour of 3–10% dried flour made from by-products of sweet potato starch manufacture did not adversely affect white bread, but produced a standard bread which became increasingly grey in appearance of crumb

* Net protein utilization (NPU): the proportion of nitrogen intake that is retained for maintenance and/or growth, i.e. the product of biological value and digestibility.

Biological value (BV): the proportion of absorbed nitrogen that is retained for maintenance and/or growth.

Figure 6.22. Biscuits and cake made with 50% sweet potato flour and 50% wheat flour in India (S. Seralathan).

and crust with increasing levels of sweet potato by-product flour (Plaut and Zelzbuch, 1958).

It has recently been shown that a type of wheatless bread can be prepared from root crop flours alone (Satin, 1988, and personal communication). A recipe (for use with cassava flour or other root crop flours) has been devised (see Appendix, p. 603) and it is suggested that sweet potato flour might be equally appropriate, although this has yet to be ascertained. The quality of the resulting loaves is claimed to be good and, apart from their economic advantage in tropical countries which have to import all wheat from abroad, they are said to be particularly appropriate, being gluten-free, for people with coeliac disease.

Cakes, pastries etc.

In a large variety of other baked goods such as cakes, cookies, biscuits, doughnuts etc. researchers have found it possible to utilize composite flours with higher proportions of sweet potato flour than is possible with bread (Figures 6.22 and 6.23). For example, in India cakes and biscuits (Seralathan and Thirumaran, 1990) and in the Philippines butter cake, cookies, brownies (Montemayor and Notario, 1982) and muffins (Anon., 1985b) of acceptable quality have been prepared using 50% sweet potato–wheat flour composites. Furthermore chiffon cake has been successfully prepared from 100% sweet potato flour in the Philip-

Figure 6.23. Sweet potato cookies commercialized in Taiwan (Chia-yi Agric. Expt. Stn).

pines (Lauzon, R.D., personal communication). There were no significant differences found between the attributes of the product made with sweet potato flour and chiffon cake made with 100% wheat flour, and it was highly acceptable to consumers. Sweet potato products such as cookies, muffins and butter cake were not as acceptable to consumers as the wheat flour equivalents by reason of their pronounced, though not objectionable, sweet potato flavour (Lauzon, R.D., personal communication). In Israel, a lower proportion (10%) of sweet potato flour produced biscuits, yeast cake and cookies which were indistinguishable in general quality and nutritional value from those made with 100% wheat (Plaut and Zelzbuch, 1958). When marketed on a large scale, no adverse comments were received from customers. In Egypt, cake mixes prepared with up to 30% sweet potato as a substitute for wheat flour were acceptable to consumers (El-Samahy et al., 1980) and could be stored without a decrease in acceptability for up to 165 days in either double polyethylene–aluminium foil or aluminium foil alone (Morad et al., 1980).

As with bread, sweet potato cultivars were found to differ in their degree of acceptability when incorporated into a range of baked goods including sponge cake, raisin bread, pancakes and doughnuts at 20% substitution for wheat flour or at 30% in cookies (Sammy, 1970). One of the two cultivar flours tested was consistently rated good and the other poor in all products.

The quality of yeast-raised doughnuts, a popular food item in the

United States, was little affected by incorporation of up to 21% sweet potato as either flour or purée and some attributes, for example tenderness, were improved (Collins and Aziz, 1982). Water additions to the dry ingredients had to be reduced as the level of sweet potato increased in order to maintain a stable dough consistency. The quality of doughnuts made with purée was higher than those made with flour, but flour had the advantage of ease of handling and storability. Taste panel testers apparently consumed all the sweet potato doughnuts with relish! It is apparent from the literature therefore that a range of acceptable baked goods can be prepared. In Japan, sweet potato flakes are already produced on a large scale for addition to cakes, pies, tarts, cookies and pastries of various types. There is also much interest in the use of high beta-carotene and high anthocyanin cultivars for such purposes.

Breakfast foods

There are two reports of experimental production of breakfast foods using sweet potato. In the West Indies, preliminary work, on both a crisp flake-type product (resembling cornflakes) and an instant porridge, was carried out with a view to replacing imported breakfast cereals (Sammy, 1973, 1984). Various formulations for the flakes using levels of sweet potato flour as high as 94% were tried (Sammy, 1973). The dough was cold-rolled into flakes which were then toasted at 210°C. A quick-cooking product was also prepared from sweet potato flour made into a dough and cooked under pressure for 45 min; the cooked dough was extruded, cut into pellets and the pellets milled to the desired particle size (Sammy, 1984). If necessary, dried skim milk, vitamins, minerals and a thickener could be added to the pellets. An instant product with high acceptability could be made by adding banana pulp to mashed sweet potato before drum-drying (Sammy, G.M., personal communication). These products have not been commercialized.

In India, a breakfast dish, *iddiappam*, utilizing 100% sweet potato flour made into a dough, formed into strings, steamed and seasoned with condiments including mustard, blackgram dhal, onions and green chilis was pronounced highly acceptable in sensory evaluation (Seralathan and Thirumaran, 1990).

Noodles

Noodles are a very popular staple food in many countries especially those of the Far East. They are basically of two types: made either from wheat flour (or a mixture of wheat and buckwheat) – spaghetti-type – or from a starch source. Sweet potato starch noodles are discussed in Section C.

Noodle making consists essentially of the preparation of a dough by adding water to the chosen flour, rolling the dough into a sheet, forming noodles by cutting, forcing through a container with holes or passing through a noodle-making machine, and then cooking immediately or drying. As long ago as 1789, a recipe for preparing noodles from equal parts of sweet potato and wheat flour was published in Japan in a book by Chinkoro entitled *Imo hyakuchin* (100 unusual ways of eating sweet potato) (Duell, B., personal communication).

There has been experimental preparation of noodles, in which part of the wheat flour is replaced by sweet potato flour, in several countries including Japan, Taiwan, Korea and the Philippines. In Japan, both fresh and dried noodles containing varying levels of sweet potato flour, up to about 50%, are sold in shops and supermarkets in several regions and are also served in some restaurants specializing in sweet potato dishes. The dried noodles are attractively packaged and several packs may be sold in a wooden box as a luxury item to be given as a gift to friends or relatives.

In Taiwan, noodles with a pleasant orange colour have been prepared using high beta-carotene sweet potatoes, and green noodles can be made by adding sweet potato leaves to the dough (Liao, C.H., personal communication). In Korea, sweet potato flour could replace 70% of wheat flour in experimental noodles with the addition of glyceryl monostearate improver without affecting noodle sheet and dried noodle characteristics, 40% without affecting cooking quality, but only 20% without significantly decreasing organoleptic quality (Chang and Lee, 1974).

Sauces

Soy sauce

Soy sauce, a popular condiment used every day with Asian dishes, is traditionally prepared from a mixture of soybeans and wheat fermented by moulds, especially *Aspergillus oryzae* or *A. sojae*, to give a dark brown salty liquid used as a flavouring agent. The potential replacement of wheat flour with suitable substitutes in soy sauce manufacture, as in the case of bread could mean a considerable saving in foreign exchange at present used for wheat importations into tropical countries. With this in mind, successful experimental production of soy sauce using sweet potato flour has been carried out in the Philippines (Data, Diamante and Forio, 1986). The process is outlined in Figure 6.24. The starter culture was prepared by soaking 200 g of rice in water for 5 hours, draining and pressure cooking at 1 bar (10^5 Pa) for 15 min, followed by cooling to

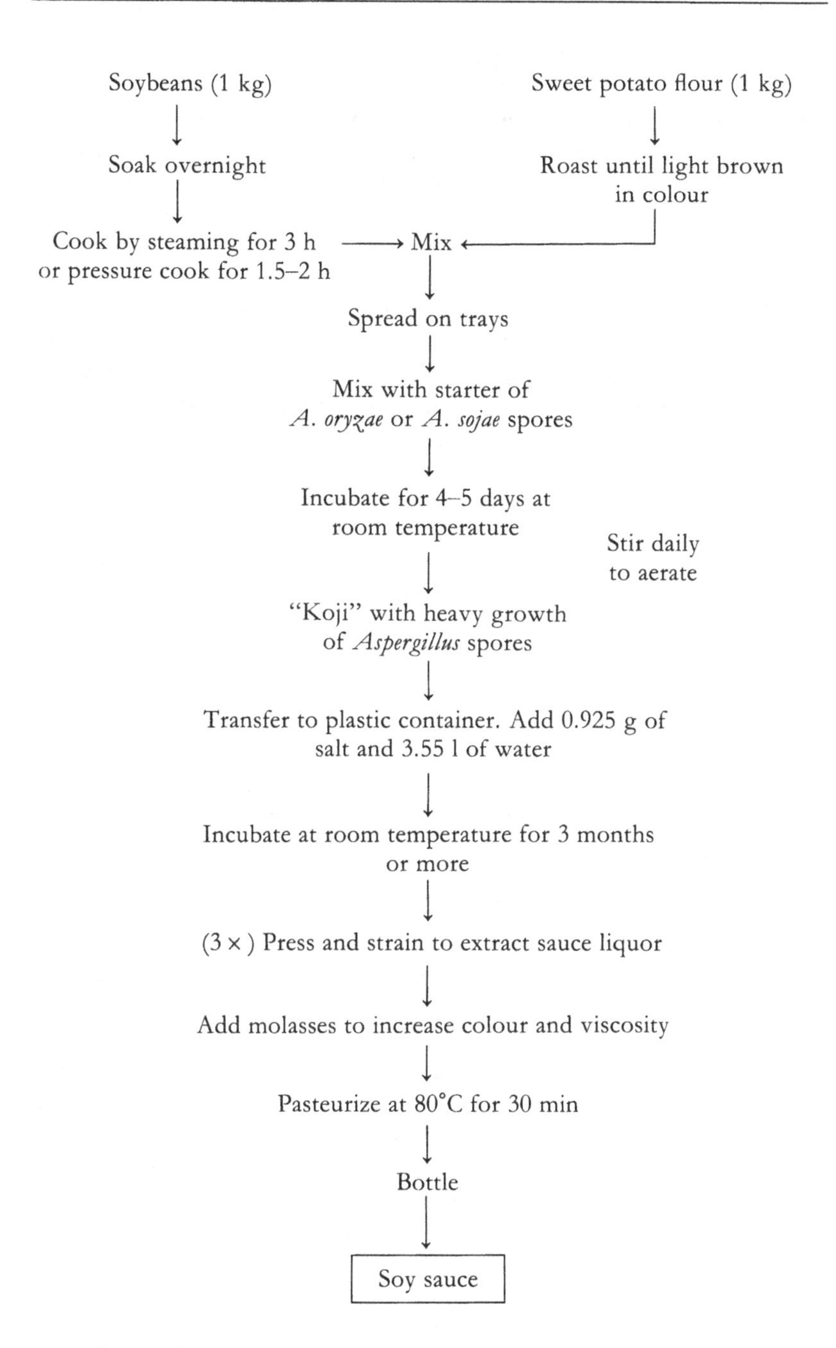

Figure 6.24. Experimental production of soy sauce using sweet potato flour (from Data et al., 1986).

room temperature, inoculating with spores of *A. oryzae* or *A. sojae* and incubating at room temperature for 4–6 days or until greenish spores were formed. During the subsequent *koji* preparation (see Figure 6.24) the healthiest growth of mould spores and eventually the highest yield of sauce were given by sweet potato flour made from cooked rather than raw roots. Cooking gelatinized the sweet potato starch thus rendering it more available for breakdown by microorganisms. Chemical properties of titratable acidity, pH, and salt, amino nitrogen and protein contents were not significantly different in cooked sweet potato and wheat flour sauces. Sensory evaluation revealed no significant difference between the soy sauce made with sweet potato flour and two commercial brands of soy sauce in terms of colour, aroma, consistency, flavour and overall acceptability. It was shown therefore that sweet potato flour could successfully replace wheat flour in the preparation of soy sauce. It also appeared to have certain advantages over cassava flour for this purpose: more prolific mould spore growth during *koji* production, a greater yield of sauce when hand pressing is used, and higher percentages of protein and amino nitrogen in the finished sauce.

Catsup (ketchup)

Another sauce which is popular throughout the world as a condiment is catsup (or ketchup), usually prepared with tomatoes. However, tomatoes are an expensive raw material in some regions. Hence cheaper substitutes, including sweet potatoes, have been sought (Figure 6.25). Sweet potato-based catsup has been commercialized in Indonesia (Watson, G., personal communication) and Malaysia (Tan, S.L., personal communication). A process for making catsup from beta-carotene-rich cultivars in the Philippines has been described (Truong, 1987). Roots are washed, trimmed, cut into chunks and steamed. The cooked chunks are blended with water, vinegar, spices and food colour and boiled to the correct consistency before bottling. The sweet potato catsup had a total soluble solids content of 25–27°Brix and a pH of 2.7–2.8. Its eating quality was found to be comparable to a popular local commercial banana-based catsup. Sensory evaluation indicates that some cultivars yield more acceptable catsup than others (Truong et al., 1986). Hence an appropriate cultivar for this purpose should be chosen.

Brewing adjunct

The experimental replacement of part of the malt (germinated barley grains) by sweet potato flour in beer brewing has been carried out in India (Dhamija and Singh, 1979). The production of quality barley malt

Figure 6.25. Sweet potato catsup (ketchup) made at ViSCA in the Philippines (ViSCA).

is a difficult, time consuming and expensive process in some countries. In the modified Indian process, up to 50% of the malt was replaced by sweet potato flour. The resulting adjunct beers were comparable both analytically and organoleptically to a control made only with barley malt, but had a somewhat darker colour than the control and equivalent American beers.

Flakes

Sweet potato flakes are already produced on a large scale in Japan and are sold to processors for incorporation into other products or are retailed as such in shops and supermarkets. Flakes can be reconstituted and used in a variety of ways to produce sweetened mashed balls, pre-formed patties, croquettes, pie fillings etc. Purple or orange flakes as mentioned above can be used to add natural colour to a variety of products.

Chips

In Japan, dried sweet potato chips are sometimes sold in speciality sweet potato shops in Kawagoe. These are purchased for grinding into flour to make dumplings. Sweet potato 'tea' is also commercialized. This consists of tiny slivers of dried sweet potato alone or as a mixture of traditional green tea and dried sweet potato pieces which are said to impart additional flavour.

Paste, mash or purée

Sweet potato mash or paste can be used as an alternative to flour or flakes in many of the processes described above. Processors may buy raw sweet potato roots and prepare their own paste for incorporation into their products, or they may purchase heat-processed or frozen paste from other sources.

A traditional product prepared in Cameroon is made from paste produced by peeling, washing, grating and pressing the roots (Numfor and Lyonga, 1987). The remaining pulp is then ground on a stone to a fine paste which is mixed with such ingredients as salt, crayfish, pepper and oil, wrapped in wilted banana leaves and steamed for about 1 hour.

Japan produces an extremely wide variety of sweet potato products, both traditional and recently devised, made from paste. Cream- and orange-coloured pastes both frozen and thermally preserved are sold to a variety of processing institutions. Cream-, orange- and purple-fleshed cultivars are all used by processors both large and small, particularly in the south of Japan, for making paste to add to cakes, pastries, pie fillings, pre-formed patties, icecream, sorbets, noodles etc. Paste can be utilized both as an outer casing for products stuffed with other savoury or sweet fillings, or as a stuffing ingredient itself. In both Japan and Taiwan sweet potato paste is frequently used to replace part of a sweet bean paste filling in desserts, and in Japan it is also used as a substitute (*imo kinton*) for the more expensive sweet chestnut (*kurikinton*) paste. A southern Japanese cake factory visited by the author produces 15 different types of cakes, sweets and desserts incorporating from 10% to 80% sweet potato, many of them employing both cream and orange paste or mash (Figure 6.26). New products are constantly being devised in the experimental kitchen and business is rapidly expanding.

Sweet potato-flavoured icecream is a popular product in some parts of Japan. A thermally processed high beta-carotene sweet potato paste produced in Kagoshima Prefecture in southern Japan is sold for incorporation at 16% (Umemura, Y., personal communication) into icecream which is being retailed nationally (Anon., 1988). In addition sweet potato-flavoured icecream is sold in small tubs in retail establishments and served, delicately coloured by the use of orange- or purple-fleshed cultivars, in restaurants in Kawagoe and elsewhere (Figure 6.27). Also in Japan, small plastic tubs with accompanying spoon containing

Figure 6.26. A range of dessert products containing 10–80% sweet potato made by a cake factory in southern Japan (J.A. Woolfe).

Figure 6.27. Enjoying sweet potato icecream in Japan. The natural pigments of sweet potato produce attractive coloured desserts and, in the case of carotene, raise nutritional value (J.A. Woolfe).

cream, orange or purple sweet potato paste to be consumed directly as a snack have been commercialized.

Swiss-style yoghurt has been experimentally prepared with baked orange-fleshed sweet potato (Collins, Ebah and Mount, 1988). Levels of up to 18% sweet potato produced yoghurts which were organoleptically satisfactory and had reduced fat and energy levels, and increased ascorbic acid and provitamin A contents as sweet potato concentrations increased. Although calcium and zinc levels decreased slightly as sweet potato concentration in the yoghurt increased riboflavin and niacin levels remained unchanged. Sweet potato also provided additional flavour, colour, dietary fibre, and stabilizer in the form of starch.

A deep-fat-fried snack/dessert item in the form of a sweet potato turnover has been developed experimentally in the United States (Jaynes and Corley, 1981). Very sweet, moist-fleshed sweet potato purée was mixed with raisins, raisins and orange juice, or crushed pineapple, plus sugar, salt and allspice for the filling to a pastry crust which was deep-fried in vegetable oil. Sensory scores were high and the packaged turnovers complete with nutritional labelling were highly acceptable to consumers, 58% of whom indicated that they would purchase the turnovers monthly. To eliminate the necessity for refrigeration, sorbate or propionate could be added to the filling to prevent mould growth.

A method for sweet potato patties, which could be produced from mash using jumbo (extra large) or misshapen roots has been described in the United States (Hoover, Walter and Giesbrecht, 1983). Cooked strips of peeled sweet potato are mixed with 8% cornstarch, 10% sucrose, 1% mono- and diglycerides, 0.3% sodium chloride, 0.05% sodium acid pyrophosphate and 0.015% yellow colouring and the mass ground finely. The mass is then cooked, cooled and filled warm and under pressure into moulds to give patties, which are finished by frying in peanut oil at 170°C for 2.5 min. This method produced high quality patties (as judged by a sensory panel) from both fresh and cured, stored roots. Sweet potato pies have been made by adding sucrose (25%), margarine (10%), condensed milk (15%) and vanilla (0.1%) to sweet potato purée and filling into a conventional pie crust (Silva et al., 1989). The quality of patties and pies made from mashes prepared in different ways (canned mash, canned whole roots subsequently mashed and frozen mash) have been compared (Silva et al., 1989) by North American sensory panellists.

Frozen products

The uses of frozen paste have already been mentioned in the previous section. In Japan, a range of high quality sweet potato desserts is produced and frozen for despatch to a Tokyo chain of coffee shops where they are served with tea or coffee or sold to be consumed at home as a tea-time snack.

A frozen product has been produced in the United States by baking roots followed by mixing with powdered sugar and cinnamon and one of each of the following ingredients: chopped pineapple and a little corn starch, raisins or imitation butter (Collins et al., 1974). The mixture was filled into small aluminium 'boats' and blast frozen. The frozen product was heated in an oven before being served warm. The products made with pineapple or raisins scored highest in sensory tests and storage of the frozen product for 2 months improved acceptability.

Other products

Sweet potato non-alcoholic beverages

Non-alcoholic beverages, from high beta-carotene sweet potato cultivars, possessing a nutritional value comparable with or superior to fruit drinks have been formulated in the Philippines (Truong and Fementira, 1990). The taste and appearance of the sweet potato beverages were claimed to be similar to those of commercial fruit drinks. The colour of the beverages varied according to the flesh colour of the roots, ranging from yellow to orange or pinkish-purple (Truong, 1987).

The process involves washing, peeling, trimming, steaming, juice extraction, addition of 0.2% (w/v) citric acid, 232 mg ascorbic acid/l and 12% (w/v) sugar. Citric acid can be replaced by lemon juice which improves flavour and enhances acceptability. Furthermore, addition of juice or pulp of other tropical fruits such as pineapple or guava prior to the extraction stage of the process also masks the sweet potato flavour and increases acceptability scores on sensory testing. The highest sensory scores were given by beverages formulated from a moist orange-fleshed cultivar. In general, panellists preferred an orange to a yellow coloured beverage. The formulated beverages were bottled in 150-ml glass containers and pasteurized at 90–95°C. The sweet potato beverage had a pH of 3.2, total soluble solids of 13°Brix and insoluble solids of 9.4 mg/100 ml, which were in the ranges of values for commercial fruit drinks. A residual starch content of 0.8–1.4 g/100 g in the juice enhanced its resistance to separation and was not found to produce flatulence in consumers drinking more than 200 ml/day for a week. Sweet potato beverages formulated with citric acid and no further flavouring (SP) or with additional 5% guava (SPG) were found to have sensory attributes comparable with or superior to commercial fruit drinks either in aseptic packs or in cans. The natural colour of sweet potato beverage compared to that enhanced artificially in most commercial fruit juices was thought to present a promotional advantage for the sweet potato drink. Consumer testing was carried out on the SP and SPG beverages with about 150 respondents in each of the following groups: elementary school, high school, college students and adults. Between 80% and 92% of respondents liked the SP beverage to some degree. These figures rose from 87% to nearly 99% for the SPG beverage. Similarly, over 50% and up to 93% of respondents rated the SP and SPG beverages, respectively, as the same or better than commercial fruit juices.

In terms of provitamin A content sweet potato beverage was nutritionally superior to all the commercial fruit juices analysed. Juice prepared from high beta-carotene sweet potato cultivars contained approximately 1 mg beta-carotene/100 g beverage, whereas the content in commercial drinks ranged from 0.8 mg/100 g in canned mango juice to only 0.02 mg/100 g in aseptically packed orange juice. One hundred grams of sweet potato beverage furnished >40% of the adult Filipino recommended daily allowance (RDA) of vitamin A, whereas the same quantity of commercial fruit juices provided between only 1% and 30% of this RDA. Sweet potato beverages also contained significantly greater quantities of calcium and phosphorus than did other juices. Marketing possibilites for sweet potato beverage are to be explored in the whole of the Philippines.

A traditional non-alcoholic sweet potato beverage *dolo* is popular in parts of some provinces of Cameroon (Numfor and Lyonga, 1987). It is made by peeling and slicing roots, drying the slices for 1–2 days, and boiling the resulting chips in water for 1 or 2 hours, keeping the volume of water constant by addition. The liquor is filtered, transferred to calabashes or bottles and cooled. The product keeps for 1–2 weeks if periodically reboiled to prevent fermentation.

Vinegar

Vinegar is a condiment made from sugar- or starch-containing materials by an alcoholic fermentation followed by the microbial oxidation of alcohol to acetic acid. If a starch-based material such as sweet potato is used as the raw material, the starch requires initial hydrolysis to sugars before alcoholic fermentation by yeasts (*Saccharomyces cerevisiae*) can take place. The next step, oxidation of the alcohol to acetic acid is carried out by acetic acid bacteria (*Acetobacter* spp. or *Bacterium* spp.). The completed vinegar must contain a minimum of 4 g acetic acid/100 ml. A vinegar manufactured from a high beta-carotene cultivar has been commercialized in Japan. The appearance and taste is similar to that of white wine vinegar. In addition, a vinegar similar in appearance to red wine vinegar has been produced from a cultivar with a high flesh concentration of anthocyanins, but this is not yet commercialized. The vinegars contain 4.3 g acetic acid/100 ml.

Pickles

Experimental work on pickling of sweet potato chips has been carried out in India (Nair et al., 1987). Various pickling media of 1%, 2%, or 5% (w/v) brine or 25% and 50% (w/v) sucrose syrups were used in trials. Only the sweet potato chips preserved in 50% sucrose could be stored for any length of time (8 weeks) at room temperature.

Pickled sweet potato petioles have been commercialized in Japan. The petioles are preserved in soy sauce with the addition of a little sugar, sesame seeds and chilis, and vacuum sealed in plastic.

B. Nutritional value of processed products

There has been little investigative work carried out on nutritional changes occurring in sweet potatoes as a result of processing. References to this work in the literature are mainly confined to investigations of changes in carotenoids and protein in products such as canned and flaked sweet potato roots. However, some information is available on the proximate composition of canned sweet potato tips, and the vitamin and amino acid changes in dried sweet potato leaves. The nutritional composition of various products is briefly discussed. Where known, details of changes in individual nutrients are then reviewed.

Roots

Nutritional composition of processed products

The proximate composition of various sweet potato root products has been given in Table 3.3 (p. 127). The vitamin and mineral composition of products manufactured in the United States, where most of the research has been carried out, can be compared by use of Table 6.7.

As might be expected, freezing appears to exert the least, and canning the greatest, effect on composition. The nutrients most adversely affected by the various processes are beta-carotene, thiamin, riboflavin and ascorbic acid. Variability in beta-carotene between batches of roots with different flesh colours is much greater than variability in other nutrients. Therefore it cannot be assumed that the differences in beta-carotene between products are mainly due to the processing technique. However, changes taking place in carotenoids during processing are fully discussed in the following section.

Table 6.7. *Vitamin and mineral composition of some processed products (per 100 g)*

Item	Vitamin A (μg) RE[a]	Thiamin (mg)	Riboflavin (mg)	Niacin (mg)	Pantothenic acid (mg)	Pyridoxine (mg)	Folic acid (μg)	Ascorbic acid (mg)	Ca (mg)	Fe (mg)	Mg (mg)	P (mg)	Zn (mg)	Cu (mg)	Mn (mg)
Raw	2006	0.07	0.15	0.67	0.59	0.26	13.8	22.7	22	0.59	10	28	0.28	0.17	0.36
Canned, vacuum pack	798	0.04	0.06	0.74	0.52	0.19	16.6	26.4	22	0.89	22	49	0.18	0.14	0.46
Canned, mashed	1513	0.03	0.09	0.96	ND	ND	ND	5.2	30	1.33	24	52	0.21	0.28	1.0
Canned, syrup pack (solids and liquid)	572	0.02	0.05	0.46	0.33	0.05	ND	10.5	15	0.80	13	27	0.19	0.12	0.51
Canned, syrup pack (drained solids)	716	0.03	0.04	0.34	0.40	0.06	ND	10.8	17	0.95	12	25	0.16	0.17	0.62
Frozen, unprepared	1864	0.07	0.05	0.60	0.52	0.18	21.3	13.3	37	0.53	22	45	0.31	0.18	0.67
Frozen, baked	1641	0.07	0.06	0.56	0.56	0.19	22.3	9.1	35	0.54	21	44	0.30	0.18	0.67
Candied	419	0.02	0.04	0.39	ND	0.04	11.4	6.7	26	1.13	11	26	0.15	0.10	ND

Notes:

ND, not determined.

From: Haytowitz and Matthews, 1984.

[a] Retinol equivalents $= \frac{\beta\text{-carotene}}{6} + \frac{\text{Other carotenoids}}{12}$.

As would be expected with a vitamin unstable to both heat and oxygen, ascorbic acid is reduced in processed products as compared to the fresh roots. However, all the products shown in Table 6.7 contain some ascorbic acid. Riboflavin, which is susceptible to oxidation, is greatly reduced by all the processes given in the table. The concentration of pyridoxine is reduced by canning and candying, but not by freezing.

There can, however, be considerable variation in the composition of a processed product such as canned sweet potatoes made either from the same cultivar, but under different growing and processing conditions, or from different cultivars grown and processed under similar conditions as shown for North American roots (Collins, 1981). Especially pronounced variations were found for calcium, iron, provitamin A and ascorbic acid. In general, however, canned roots were found to provide significant percentages of the adult United States recommended dietary allowances for iron, vitamin A and ascorbic acid. It should be remembered, however, that the yellow- or orange-fleshed cultivars are those most commonly canned in the United States.

The comparative composition of sweet potato flours, produced (with the exception of flakes from the United States) by traditional means and recorded in Food Compositon Tables is shown in Table 6.8. The flours if used for adding to other foods on a dry basis are good sources of energy, of minerals especially iron, and of beta-carotene if made from an orange-fleshed cultivar. However, autoxidation of carotenoids may take place when flours are stored for any length of time (see pp. 357–8). Flours also provide some protein. Changes in amino acids which take place during drying are discussed below.

It has been shown for potatoes that sulphiting to prevent enzymic darkening of the flesh during dehydration can reduce thiamin to very low levels (Augustin et al., 1979). The low thiamin content of sweet potato flakes, even on a dry basis, may reflect treatment of the roots with sulphite during processing. The other flours had probably not undergone sulphiting. However, many processes described for the production of flours utilize some form of sulphite for prevention of enzymic discoloration during drying (Hamed et al., 1973a; Moy and Chi, 1982; Sammy, 1970; Seralathan and Thirumaran, 1990; Thomas, 1982). Such flours are likely to have greatly reduced thiamin contents. From Table 6.8, the flaking process appears to favour retention of more ascorbic acid than drying to flours, which may have been dried in the sun, exposing ascorbic acid to heat degradation and oxidation. One sample retained no ascorbic acid after processing. The significant contributions which some processed forms of sweet potato can make to recommended daily intakes of vitamin A (if beta-carotene-rich roots are used), ascorbic acid and iron can be appreciated from Table 6.9.

Table 6.8. *Nutrient composition of some dehydrated products (per 100 g)*

Product	Moisture (%)	Energy (kcal)	Protein (g)	Lipid (g)	Total carb. (g)	Ash (mg)	Ca (mg)	P (mg)	Fe (mg)	β-Carotene equiv. (mg)	Thiamin (mg)	Riboflavin (mg)	Niacin (mg)	Ascorbic acid (mg)
Imoko (flour)[a]	13.2	337	3.3	0.6	78.3	2.7	85	120	1.4	0.420	0.23	0.16	1.4	4
Flour[b]	13.2	339	2.2	0.9	80.8	2.9	50	95	2.0	—	0.24	0.09	1.5	—
Flour[c]	11.7	334	12.5	0.5	73.1	2.2	ND	ND	ND	ND	ND	ND	ND	ND
Flakes,[d] dry	2.8	379	4.2	0.6	90.0	2.4	60	80	2.2	28.2	0.06	0.13	1.3	45
Flakes, prepared[d]	75.7	95	1.0	0.1	22.6	0.6	15	20	0.6	7.2	0.02	0.03	0.3	11
Flour[e]	11.0	335	1.6	0.8	84.4	2.2	106	99	5.3	0.54	0.12	0.15	1.1	6
Mushikiriboshi[a] (sliced and dried after steaming)	17.6	320	3.3	0.6	74.1	2.5	55	100	2.2	—	0.20	0.08	1.7	10

Notes:

A dash indicates zero content. 1 kcal = 4.184 kJ. carb., carbohydrate; equiv., equivalent; ND, not determined.

[a] Japan: Ashida, 1982.

[b] East Asia: Leung et al., 1972.

[c] Africa: Leung et al., 1968.

[d] United States: Watt & Merrill, 1975.

[e] Latin America: Leung & Flores, 1961.

Table 6.9. *Percentage contribution of 100 g of processed sweet potato to the adult (male) RDA of various nutrients*

	Protein	Energy	Vit. A[b]	Thiamin	Riboflavin	Niacin	Vit. C[b]	Fe[b]
RDA[a]	37 g	12.6 MJ	600 μg RE	1.2 mg	1.8 mg	19.8 mg	30 mg	8–23 mg
Canned (drained solids)[c]	4	4	0–100	3	2	2	40	4–12
Frozen, baked[c]	5	3	0–100	6	3	3	30	2–7
Flakes, prepared[d]	3	3	0–100	2	2	2	40	3–8
Candied[c]	2	5	0–100	2	2	2	20	5–14

Notes:

Except where indicated figures calculated using RDAs from:

[a] Passmore et al., 1974.

[b] FAO, 1988; contribution to Fe supplies depends on bioavailability of Fe in the diet.

Data for processed sweet potatoes from:

[c] Haytowitz and Matthews, 1984;

[d] Watt and Merrill, 1975.

Changes in individual nutrients

Carotenoids

Cultivars of sweet potato possessing high levels of beta-carotene have been used as examples for the study of carotene degradation in dehydrated foods both during the drying process and subsequently during storage of the dried product. Degradation of beta-carotene at three different temperatures commonly employed during air-drying has been studied. Beta-carotene concentration was reduced less at 60°C than at 70°C or 80°C at any particular drying time (Stefanovich and Karel, 1982). Kinetically, the degradation was shown to follow a first-order reaction. Model systems which could be used for detailed studies of beta-carotene degradation during drying in foods such as sweet potato have been evaluated, but did not provide totally adequate simulation. When small strips of sweet potato were dried in a cabinet beginning at 70°C and reducing to 50°C over a period of 6 hours, loss of carotenoids (80% beta-carotene) was about 20% for each of two Indian cultivars (Ranganath and Dubash, 1981). Between 18% and 20% of the remaining carotenoids were lost during storage of the dried strips in polyethylene at 24°C for 120 days.

During storage of dehydrated foods, their water activity has a significant effect on the stability of nutrients present. Beta-carotene is lipid-soluble and exhibits its highest stability at intermediate water activities (Haralampu and Karel, 1983). A model describing the effect of water activity on beta-carotene loss which could be used for prediction of shelf life was developed using sweet potato. The beta-carotene content of freeze-dried sweet potato flour maintained at different water activities by storing open over a series of saturated salt solutions at a controlled temperature of 40°C for 15 days was measured (Haralampu and Karel, 1983). The loss of beta-carotene exhibited first-order kinetic behaviour. The reaction was found to be inversely proportional to the water activity. The actual rates of decomposition varied between batches of sweet potato flour, probably through differences in state of lipid oxidation, presence of metal catalysts and physical differences in the flour such as porosity.

The oxidative degradation of sweet potato flakes produced by drum-drying (as described on pp. 312–13) has already been discussed in relation to the development of off-flavours. The highly unsaturated nature of beta-carotene and other carotenoids leads to their rapid oxidation and consequent partial loss in flakes unless these are stored in a reduced oxygen environment. When flakes are stored in air, between 20% and 40% of the carotene is destroyed within 30 days, after which the rate of destruction decreases considerably (Walter et al., 1972). Most of the carotene degraded is present as surface lipid which is released from chromoplasts as a result of processing and distributed on the surface of the flake particles, thus being readily attacked by oxygen. Bound carotene held within the flake matrix is much less susceptible to destruction.

It is not known whether a similar phenomenon occurs during simple air-drying of sweet potato to flour. However, a sample of Brazilian sweet potato flour containing 50 mg/100 g total carotenoids (dwb) lost 50% of its carotenoids in 5 months of storage (Cascon et al., 1988). The authors recommended suitable packaging in order to improve carotenoid retention. Taiwanese researchers (Chen and Chiang, 1985b) discovered that the stability of beta-carotene during storage of flour was strongly affected by storage temperature and light. Beta-carotene was retained most effectively by storage at a low temperature (4°C) in the dark. Regardless of packaging type or storage temperature beta-carotene was lost more rapidly from a sample of flour with a low initial beta-carotene content (S1) than from one with a high initial content (S2). At 4°C in polypropylene packaging (the most effective conditions for retention of beta-carotene in S2), beta-carotene content was reduced in S1 from 1.7 mg/100 g (dwb) to traces only after 4 months of storage, whereas S2 initially containing 29.0 mg/100 g (dwb) retained 67% of its beta-carotene even after 6 months of storage.

Loss of carotenoid bioactivity by oxidation and isomerization has been discussed in relation to cooking. The oxidative degradation, and subsequent loss, of total carotenoids from sweet potato has been observed to result from several processing techniques apart from dehydration. In addition, reduction in the biological activity of beta-carotene through conversion of part of the all-*trans* isomer to *cis* forms takes place during some processing procedures involving heat application. The extent of these changes depends on the severity and length of heat treatment (Chandler and Schwartz, 1988; Lee and Ammerman, 1974; Panalaks and Murray, 1970).

Table 6.10 shows the changes in total beta-carotene and the extent of all-*trans* to *cis* isomerization as a result of various processing techniques (Chandler and Schwartz, 1988). No changes in total or isomeric carotene content were found as a result of freezing or of frozen storage. Significant increases in total beta-carotene compared to the raw roots were noted for blanched, puréed and gelatinized samples (not shown in the table). These could have resulted from greater ease of carotenoid extractability due to heat-mediated changes in tissue morphology. Canning or drum-drying of the purée and microwave cooking of whole roots all caused total beta-carotene losses of about 20%. In addition, all

Table 6.10. *Isomerization of β-carotene in processed sweet potatoes*

Product	Total β-carotene (g/g DM)	All-*trans* (g/g DM)	13-*cis*	9-*cis*	% *cis*-isomer
Raw	440.0	417.5	22.5	—	5.1
Canned purée	390.3	323.0	56.7	10.6	17.2
Drum-dried	349.8	248.8	101.0	Tr.	28.9
Microwaved	340.2	283.9	56.3	Tr.	16.5

Notes:
DM, dry matter; Tr., trace only.
From: Chandler and Schwartz, 1988.

processes involving appreciable heat treatment caused the formation of the 13-*cis* isomer, and, in the case of canning, the 9-*cis* isomer also. The greatest extent of isomerization was caused by drum drying. Such changes decreased the biological activity of the total beta-carotene and it can be calculated that actual losses of effective beta-carotene were about 25%, 30% and 27% in the canned, drum-dried and microwave-cooked products respectively. The authors recommend maximizing nutrient bioavailability by using processes which minimize isomerization.

In the case of canning sweet potatoes, it has been shown that varying temperatures and procedures of retorting can significantly affect the resulting degree of beta-carotene isomerization (Lee and Ammerman, 1974). A study in which canned sweet potatoes received heat processes of comparable lethality, showed that either a temperature–time combination of 116°C for 34 min in a still retort, or higher temperatures of 127°C or 132°C for about 13 min in an agitating retort, were more favourable to retention of all-*trans* beta-carotene than still or agitated retorting at 121°C for 23 or 19 min, respectively.

Ascorbic acid

Ascorbic acid is significantly reduced by all types of processing (see Tables 6.7 and 6.8), the largest change taking place in sweet potatoes traditionally dried to flour, from which it can apparently be eliminated entirely (Table 6.8). Among the methods of drying, flaking appeared to cause the least reduction in ascorbic acid (Table 6.8) probably because of short exposure time to high temperatures. From results obtained with other crops, it seems unlikely that ascorbic acid is significantly affected by freezing, any reductions taking place during pre-processing preparations.

Cabinet drying of strips between temperatures of 70°C and 50°C for 6 hours reduced ascorbic acid by 33–39% in two Indian cultivars (Ranganath and Dubash, 1981). Storage losses in the polyethylene packed dried roots ranged from about 10% after 30 days to 50% after 120 days at 24°C. Hence the roots lost a total of about 70% of the initial ascorbic acid content during drying and storage for 120 days. Ascorbic acid is at its most stable at low water activities (Haralampu and Karel, 1983). Thus sweet potato should be dried to a low moisture content and packed in a suitable form to avoid moisture uptake during storage.

Canning has been shown to reduce ascorbic acid to a significantly greater extent than cooking by baking, boiling, steaming or microwave cooking (Junek and Sistrunk, 1978). However, the extent of reduction can vary considerably among processors (Collins, 1981). For example, values for canned samples of one cultivar provided by six different experiment stations ranged from 2 to 12 mg/100 g. Crushed (mashed) Korean sweet potatoes stored in polyethylene film for 1 month lost 15–25% of its ascorbic acid (Kim and Kim, 1984).

There has been little work on the determination of ascorbic acid in processed sweet potatoes and it would seem valuable for scientists to pursue a simple method for preparing dried sweet potato which retains the maximum amount of this important vitamin. Alternatively ascorbic acid could be added back to the dried product by manufacturers, as is sometimes carried out for dried potato products.

Other vitamins

The literature is almost devoid of studies to determine changes in the B group vitamins during processing of sweet potatoes. As in the case of other foods, however, niacin is probably the most heat-stable vitamin, being lost mainly by peeling and leaching into processing water. Thiamin is susceptible to heat and sulphite destruction and leaching, while riboflavin is readily oxidized. One study (Junek and Sistrunk, 1978) found that canned roots contained significantly less niacin, pantothenic acid and riboflavin than baked roots and less pantothenic acid than boiled roots. The contributions of 100 g portions of various sweet potato products to RDAs of B vitamins are apparently very small (Table 6.9).

Minerals

Changes in minerals and trace elements have been studied in detail only in canned sweet potatoes. However, elements are not subject to degradation by heat or oxygen and are lost mainly by leaching into processing liquids. During canning of peeled roots in syrup, losses of minerals and trace elements from the roots ranged from 8% to 47% (Lopez, Williams and Cooler, 1980). When the content of minerals and trace elements in the syrup surrounding the roots was also determined, overall losses were negligible. The amounts of Cu, Mg, P, K, Si and Zn in the canned, drained roots were significantly lower than in the fresh roots. There were no significant differences in Ca, Fe, Mn and Na between canned and fresh roots. The content of all minerals and trace elements analysed increased in the syrup during processing. The authors concluded that canned sweet potato roots are relatively good sources of K, and fair sources of the other elements studied with the exception of Ca, Na and Zn.

Nitrogenous constituents

Detailed studies of the changes in total protein and individual amino acids have been carried out in canned, dehydrated and puréed sweet potatoes. In the case of dehydrated products, both drum-dried flakes and oven-dried flour have been analysed and compared.

As might be expected, processes utilizing relatively severe heat treatments caused greater changes than cooking (baking) (Purcell and Walter, 1982; and see Table 6.11). The reduction in total protein content (26% less than baked roots) in drained canned roots compared with flaked or baked roots was assumed to be due to leaching of soluble nitrogen into the can syrup. A previous study (Purcell, Walter and Giesbrecht, 1978) showed that the supernatant fraction centrifugally separated from coagulated protein in fibre-extracted sweet potato contained a large proportion of aspartic and glutamic acids and lower amounts of the other amino acids. The leaching of such soluble compounds from roots into the can liquid could account for the apparent increase in some amino acids in the roots after canning. The most notable change in an individual amino acid took place in lysine, which significantly decreased in both canned and flaked roots. This destruction was believed to be due to carbohydrate–amino acid interactions (Maillard reaction) at the high temperatures involved. Both canned and flaked roots contained approximately 26% less total lysine than did baked roots. In addition histidine and methionine of the essential, and alanine of the non-essential, amino acids were significantly reduced in flakes, and methionine was reduced in the canned sample compared to baked roots. Within a single cultivar, drum-drying was found to reduce total lysine content more than forced-draught oven-drying at 60°C (Walter et al., 1983). Moreover, not only had drum-drying destroyed some of the lysine completely, but part of that which remained was indicated by rat

Table 6.11. *Protein and amino acids in baked and processed sweet potatoes*

	Baked	Canned	Flakes
Protein[a]	7.52	5.55	7.06
Thr[b]	4.50	4.90	4.82
Val	6.83	6.81	6.07
Met[c]	2.69[B]	2.38[AB]	2.06[A]
½Cystine	0.56	0.64	0.86
Total S	3.25	3.02	2.92
Ile	4.57	4.53	4.31
Leu	7.47[A]	9.01[B]	7.57[A]
Tyr	5.81	5.74	5.04
Phe	7.32	7.82	6.65
Total aromatic	13.13	13.56	11.69
Lys	6.60[B]	4.84[A]	4.74[A]
Trp	0.44	0.46	0.75
His	2.75[B]	2.90[B]	2.36[A]
Total essential	46.79	47.13	42.87

Notes:
From: Purcell and Walter, 1982; canned results refer to drained roots.
[a] % dwb.
[b] g/16 g N.
[c] Values in rows with different superscripts are significantly different at $p<0.05$. Otherwise differences not significant.

feeding experiments to be nutritionally unavailable. During drum-drying, amino acids and carbohydrates are brought together at a high concentration at high temperatures. If a flour or other dehydrated form is to be used in a product where protein nutritional value is of prime importance, such as a weaning food, it should be prepared with a relatively mild heat treatment. However, that having been done, the protein quality of the flour is likely to reflect that of the original sweet potato from which it was prepared. A Brazilian sample of sweet potato flour (Carvalho et al., 1980) dried in a circulating air oven at 50°C had much lower contents (g/16 g N) of all essential amino acids than the oven-dried North American samples (Walter et al., 1983) mentioned above.

Apart from the changes in total nitrogen and amino acids taking place during processing, those occurring during subsequent storage of the processed product have been investigated. For example, amino acids in the protein fraction of sweet potato purée remained stable over a 6 month storage period, whilst those in the non-protein nitrogen fraction remained stable or increased slightly (Creamer, Young and Hamann, 1983). In other words nitrogenous compounds were not lost from stored purée.

The autoxidation of unsaturated lipids taking place when dehydrated foods are stored in air can lead to the reaction of lipid oxidation products and protein, resulting in protein denaturation and loss of biological value (Labuza, 1972). Dehydrated sweet potato flakes either alone or fortified with protein concentrates or isolates were found to retain essential amino acids when stored under nitrogen at 23°C or 40°C (Walter et al., 1978a). When such flakes alone or fortified with casein or soy-methionine were stored in air, autoxidation of lipids took place and glutamic acid decreased in all three formulations, isoleucine only in the unfortified sample, lysine in the fortified samples and histidine in the soy-methionine formulation (Walter et al., 1978a). Changes were least in the sweet potato flakes alone. Storage of the flakes at 23°C did not lead to any change in their water-binding capacity. Determinations of moisture, acid and nitrogen over a 7 month period in Brazilian sweet potato flour packed in three different materials (Carvalho et al., 1980) revealed that glass or sealed plastic bags could be used for storage but that flour packed in 'craft' paper absorbed moisture to an inappropriate degree, encouraging the growth of moulds. Available lysine decreased in Taiwanese samples of oven-dried flour during a 6 month storage period regardless of storage temperature or type of packaging (Chen and Chiang, 1985b). Available lysine was reduced to about 30% of its initial value after 2 months and about 10% or less of its initial value after 4 months. There was little change between 4 and 6 months of storage time. There was an indication that available lysine decreased less at 4°C than at 25°C.

Tops

It is not surprising that there are hardly any reports in the literature on the nutritional value of processed sweet potato greens, firstly because the interest in the nutritional composition even of fresh leaves or tips is of relatively recent occurrence and secondly because there has been very little development of leaf processing. Traditionally green leaves including sweet potato leaves may be sun-dried, notably in parts of Africa, but this is not a widespread practice. The canning of sweet potato greens has been investigated in the United States.

Only the proximate composition of canned sweet potato greens has so

Table 6.12. *Carotene and ascorbic acid in fresh and dried sweet potato leaves*

Treatment	Carotene (mg/100 g (dwb))	Retention (%)	Ascorbic acid (mg/100 g (dwb))	Retention (%)
Fresh leaves	49.6	100	1374	100
Dried leaves[a]:				
Enclosed solar drier (with shade)	16.9	34	22.5	1.6
Enclosed solar drier (without shade)	5.2	10.5	19.0	1.4
Open sunlight	2.1	4.2	30.7	2.2

Notes:
From: Maeda and Salunkhe, 1981.
[a] Blanched in boiling water 50 s before drying.

far been determined (Pace et al., 1985). This differed little from fresh samples and was similar to that of canned spinach or turnip greens, having (dwb) 23.9% total protein, 6.7% lipid, 10.7% crude fibre, 21.6% ash and 34.0% nitrogen-free extract.

The cheapest way of ensuring a supply of vegetables during the dry season in the tropics is to sun-dry the surplus produced during the growing season. Traditionally this is by exposing the vegetables including green leaves to direct sunlight, which is highly destructive to ascorbic acid, provitamin A carotenoids and some B vitamins (Labuza, 1972; Okoh, 1984). When retentions of ascorbic acid and carotenoids were investigated in African sweet potato leaves (with stalks removed), either dried traditionally in open sunlight or in an enclosed solar drier with or without shade, they were found to be very low using the traditional method (Maeda and Salunkhe, 1981). Using a solar drier did not confer any advantage in the case of ascorbic acid, but the retention of carotenoids was increased significantly when leaves were dried by a solar drier with shade (see Table 6.12). The change in concentration of carotenoids when mechanically dried sweet potato foliage is stored has been determined in relation to its use as an animal feed (Garlich et al., 1974). Beta-carotene and total xanthophyll carotenoids decreased from 21.8 mg/100 g and 41.7 mg/100 g, respectively, in freshly dried vines to 14.7 mg/100 g and 30.6 mg/100 g in dried vines stored for 3 months (losses of about 30%). The addition of an antioxidant prior to storage reduced losses to only 11% and 7%, respectively, with final contents of 19.3 mg/100 g and 38.9 mg/100 g. Although the retention of riboflavin does not appear to have been determined directly in sun-dried sweet potato leaves, it is interesting to note that riboflavin was little affected by

Table 6.13. *Essential amino acid content (g/16 g N) in solar dried sweet potato leaves*

Amino acid	In dried leaves[a,b]	Requirement pattern of pre-school child[c]
Thr	4.7	3.4
Val	4.9	3.5
Met + Cystine	3.4	2.5
Ile	4.1	2.8
Leu	8.4	6.6
Phe + Tyr	9.5	6.3
Available Lys	3.6	5.8
Trp	ND	1.1

Notes:
ND, not determined.
[a] Maeda, 1985.
[b] Chemical score = 62; lysine limiting.
[c] FAO/WHO/UNU, 1985.

sun-drying in a number of Nigerian vegetables, including a sample of leaves (Okoh, 1984).

The protein quality of solar-dried African leaves has also been investigated (Maeda, 1985). Most amino acids were unaffected by drying, with the exception of total lysine, which was significantly reduced from 11 to 4.6 g/16 g N and histidine which decreased from 3.7 to 1.2 g/16 g N. Furthermore, the availability of lysine was reduced to 78% (3.6 g/16 g N) of the total value. The reduction in bioavailability of sweet potato leaf lysine was, however, much less than that in cassava, cowpea or amaranth leaves dried similarly. The lysine bioavailability in these other leaves was reduced to between 34% and 47% of its apparent total value. These differences could have reflected a difference in leaf sugar concentration as the adverse changes were assumed to be a result of Maillard carbonyl–amino acid interactions. The amino acid pattern of the dried leaves is compared in Table 6.13 with that required for the pre-school child. It can be seen that lysine is apparently the limiting amino acid (although tryptophan was not determined). The chemical score of 62 is reasonably good. However, it should be pointed out that this sample of leaves on a fresh basis had a very high lysine content compared with samples analysed in other parts of the world (see Table 3.15, p. 171). Protein quality of dried leaves should be checked before they are recommended as a protein source in a child's diet.

C. Alternative or derived sweet potato products for human food

The uses of sweet potato for human food in the form of products manufactured directly from the roots or tops of the plant are considered first. There is a multiplicity of both present and potential food products from sweet potato in what can be termed an indirect or alternative form. The two most important of these in terms of total production at the present time are starch and alcohol. Additional processing of starch or its by-products yields further products of added value. In southern Japan, for example, 95% of all starch is processed further into sugar syrups. This section discusses the alternative products potentially available from sweet potato with the view, as in Section A, of making greater use of the sweet potato as an economic crop in developing areas. One obvious indirect use of sweet potato as a human food is that of an animal feedstuff. This subject is discussed separately in Chapter 7.

Starch manufacture

Starch production from sweet potato is one of the major uses to which the crop is put in many producer countries. In the largest producer, China, about 15–20% of total sweet potato from some provinces is converted into starch (Gitomer, 1987), with a yearly production of more than 300,000 tonnes (Wang, 1984). In Japan, where some 100,000 tonnes of sweet potato starch are made each year, starch making accounted, in 1986, for 35% of all uses for sweet potato (Duell, 1990). With the burgeoning demand for industrial starch in China, not only for direct uses but for further processing into a variety of products, and the demand for starch in countries which wish to conserve foreign exchange by using local raw materials, sweet potato could play an increasingly important role. Unfortunately, the manufacture of sweet potato starch is declining in some countries (for example Taiwan and Japan) which freely import cheaper corn or cassava from abroad.

The content and properties of sweet potato starch has been fully discussed in Chapter 2. Sweet potato has the advantage of being a high starch producing crop with 30%, 30% and 49% greater starch yield than rice, corn or wheat, respectively, under the same conditions (Wang, 1984). Cultivars with relatively high starch contents have been developed especially for processing. In addition sweet potato can be grown with low inputs on marginal soils and can withstand adverse natural conditions.

In countries producing sweet potato starch, processing takes place both on a small (cottage industry) and large scale. In Sichuan Province of China, for example, processing capacity ranges from 300–500 to 10,000–15,000 tonnes of fresh roots per year among 17 plants, and only 100–300 tonnes per year in more than 200 village enterprises producing both starch and noodles (Wiersema et al., 1989; and see Chapter 8, China). Starch production is on a seasonal basis (usually 6 months or less) in some countries. Larger plants, or cooperatives (for example Bangbu starch plant, Anhui Province in China) can function all year round, however, by storing dried chips in sufficient quantity (Wiersema et al., 1989). There is a need to improve the extraction rate and quality of starch from sweet potatoes in some countries. The extraction rate from traditional and larger scale processing in China is usually not more than 12% and 15%, respectively (Wiersema et al., 1989), whereas a starch plant in Japan visited by the author claimed an extraction rate of 28%. Sweet potato starch is often less pure and of a darker colour than corn starch. This may be due to contamination with 'jalapin', the resin produced by the sweet potato latificers, and with polyphenolic compounds formed during processing.

In China, starch is made from two types of raw material. In one system, fresh roots are processed fairly rapidly after harvest due to poor storability. In the other, roots are first transformed by the farmer into dried chips or slices sun-dried in the field. The dried chips can be stored for long periods if necessary and are used by large-scale processing plants. In areas where natural climatic conditions do not permit field-drying, processing units are much smaller to reduce transport costs of fresh roots (Wiersema et al., 1989). Processes for starch manufacture in different countries vary in some details. However, most employ the same basic steps which are outlined in Figure 6.28.

In a farmer-level process near Jinan, China, roots are cut up by hand, crushed by machine and sieved to remove waste such as peel and fibre. The starch is left to settle in a tank and then removed. Water is added to the starch which is once more sieved and settled. Wet starch after removal and transfer to jute bags is allowed to 'drip-dry' for 1 day, and air-dried outside at ambient temperatures. One author describes a traditional method used in rural areas of China for enhancement of starch extraction from the slurry after sieving. An 'acid paste' naturally formed during starch production by the action of lactic acid bacteria is added to the starch slurry (Zhao and Jia, 1985). This enhances coagulation and separation of starch, fibre and protein, the starch sedimenting most rapidly and forming a bottom layer with the other constituents suspended above it. The top layer on removal can be allowed to ferment again to form an 'acid paste' for the next day's processing. The acidity not only helps to speed starch separation, but inhibits the growth of

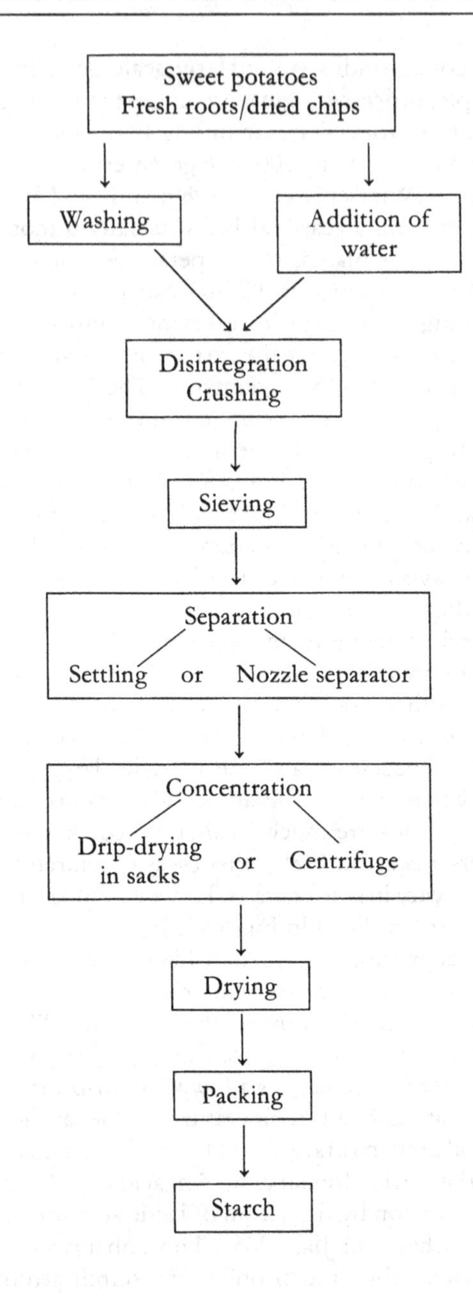

Figure 6.28. Basic methods of starch production.

Figure 6.29. Storing wet sweet potato starch in outside ponds at a starch factory in southern Japan (J.A. Woolfe).

undesirable microorganisms. In Vietnam, village-scale processing stores wet starch, after extraction, under anaerobic conditions in holes in the ground (Wiersema, 1989), covered with a layer of salt to prevent microbial growth. After closing the hole with plastic, a cover of bricks and soil is applied. This method permits storage of wet starch for 6–9 months. Sweet potato during the winter season gives a starch recovery of 12–14%, and during the spring season 18–20%.

In Japan the starch extraction process is kept alkaline throughout by the addition of lime (calcium hydroxide) (Anon., 1968). The benefits of adding lime, to increase the pH to 8, in terms of enhanced starch yield and improved whiteness have been shown experimentally in Korea (Suh, 1966). Centrifugal separators are often used in Japan now for starch separation in place of settling tanks. It is also frequent practice in Japan for roots to be processed up to the slurry stage in Autumn, for the concentrated starch slurry to be stored in outside concrete ponds (Figure 6.29) and the process of purification, concentration and drying to take place during the winter. The raw starch pulp which is retained by the sieves after the starch suspension has passed through is dried and despatched for livestock feeding or citric acid manufacture (Figure 6.30). Where processing starts with dried chips as in some large plants in China, the chips are first sieved to remove soil, crushed and soaked in water

Figure 6.30. Dried starch waste to be sent for further processing to citric acid, Japan (J.A. Woolfe).

before being processed as for fresh roots. In the larger processing plants in China and Japan starch slurry is concentrated by centrifuge and the wet starch dried by hot air before being packed or stored for secondary processing.

Starch is used in the food industry as a thickening, pasting or gelling agent in a variety of formulations. In Japan sweet potato starch is sometimes used as a 'setting' agent for a fish jelly paste called *kamaboko*. A starchy sheet jelly, eaten with vegetables, is made in China by spreading a thin layer of starch paste in the sun/oven to dry (Zhang, D.P., personal communication). Sweet potato starch has also been suggested as a replacement for other flours or starches in various traditional Korean products such as *mook* (a traditional starch gel made from acorn or buckwheat flour) or *kochujang* (a red pepper paste seasoning prepared with rice flour) and has been studied to this end (Bae, Sohn and Moon, 1984; Lee et al., 1980; Lee, Park and Lee, 1981; Yeo and Kim, 1978). Sweet potato 'sago', which can be made into a porridge by boiling in water, has been prepared in India. Sweet potato starch is moistened, and shaken in a pan over heat to form granules which are then dried by dry heating in a pan over a fire.

Secondary products from starch

Noodles

Starch noodles are a popular food in many countries of the far east including China, Korea, Japan and Taiwan. A varying proportion of sweet potato starch production is used for their manufacture. In China (Wiersema et al., 1989) and Korea (Hong, 1982) a high percentage of sweet potato starch production is utilized for noodle making, whereas in Japan, which makes *harusame* noodles from a mixture of sweet potato and potato starches, noodles are a minor product compared with sugar syrups.

At the present time sweet potato starch noodles are made mainly by a very large number of small family- or village-scale units in China. In Xuling County, Sichuan Province, for example over 70% of farmers produce noodles. Fresh roots are first processed into starch in the months immediately following harvest and the starch is stored in open sheds for subsequent noodle making. The method for farmer-level noodle making in Jinan, Shandong Province, has been described (Wiersema et al., 1989) and is outlined in Figure 6.31. There is a large demand for high quality noodles in China. Improvement of the present rather low quality starch and noodles made by traditional methods is an on-going research concern of the Sichuan Academy of Agricultural Science. However, the similarity in quality of fine noodles (vermicelli) made by a People's Commune in Laiyang County with those made from traditional mungbean or broad bean starch and the accruing economic value of this product compared with fresh sweet potato have been described (Wang, 1984). Since 1980, the vermicelli has been so successfully marketed to customers in Beijing and northeast China that demand has now outstripped supply.

The replacement of part of the wheat or buckwheat flour used to prepare Korean *kuksu* or *naeng-myon* noodles with sweet potato starch has been suggested. These noodles are made by kneading the flour with salty water, followed by pressing the dough through small holes to form long spindles, which are dried in the shade. The rheological properties of wheat–sweet potato starch noodles have been studied and compared with those from wheat flour alone (Lee and Kim, 1983a–c).

In Vietnam, noodles are prepared by adding water to wet starch which is then steamed by placing the slurry on a cloth above boiling water (Wiersema, 1989). The resulting starch film is stretched on bamboo racks for drying in the open air. Semi-dried starch sheets are then cut into strips by a hand-operated machine. The strips are dried in the open on

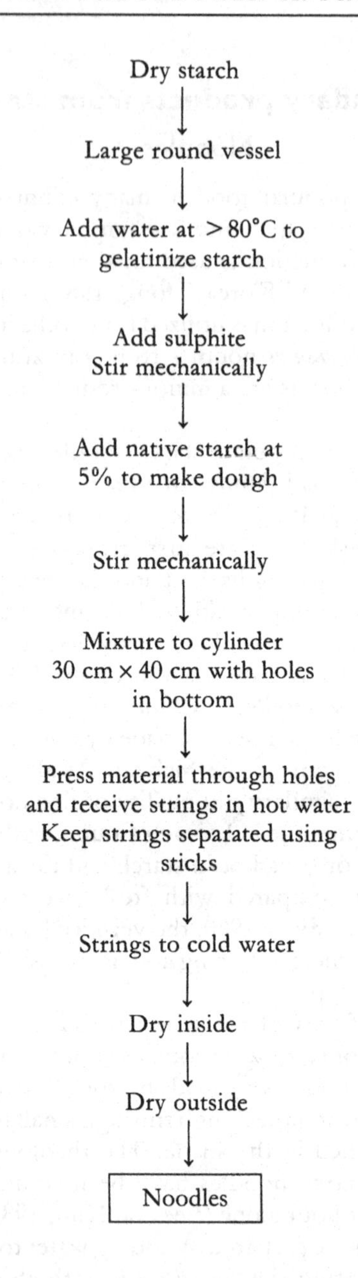

Figure 6.31. Small-scale process for producing noodles, used by farmers in China (from Wiersema et al., 1989).

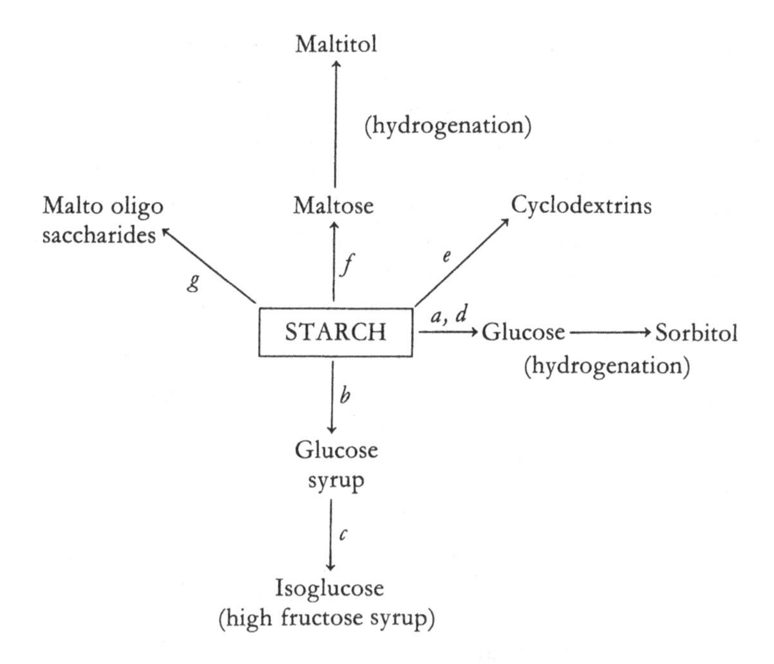

Figure 6.32. Products of enzymic starch degradation: *a*, based on Kainuma (1984); *b*, α-amylase; *c*, glucose isomerase; *d*, glucoamylase; *e*, cyclodextrin glucanotransferase; *f*, β-amylase; *g*, maltooligosaccharide-forming amylases.

bamboo mats. These sweet potato noodles are frequently dark coloured and hence of too poor quality to be marketed.

Sugar syrups

The conversion of starch into a range of sugar syrups and other derivatives (shown in Figure 6.32) is becoming increasingly important in some countries where these products can be used to replace more expensive imported sucrose extracted from beet or cane. These conversions employ a sophisticated technology based on microbial enzymes and can utilize any starch source including sweet potato, which may be particularly appropriate as it is highly susceptible to saccharification by enzymes.

Though starch can be converted into sugars by the use of acid, this method is rapidly giving way to the use of immobilized bacterial enzymes, the specific properties of which give rise to a variety of compounds useful in the dessert, bread, icecream, fermented milk products, soft drinks, brewing and other industries. Glucose syrup, for example, is produced from starches, including sweet potato starch, by bacterial amylase. However, it has only 70% of the sweetness of sucrose. The partial conversion of glucose into its isomer fructose by bacterial glucose isomerase to give high fructose syrup, or isoglucose as it is also known, gave rise to a substance with much greater sweetness than sucrose. Isoglucose (which contains at least 42% fructose) has become an important replacement for sucrose in many areas of the food industry where it can be used in a lower concentration than sucrose to provide the same sweetness, hence reducing the energy (calorie) content of the food. A further sugar, especially useful to the brewing industry, is maltose produced from sweet potato starch by the action of bacterial beta-amylase. Two novel products which have more recently been developed from starch are cyclodextrin and maltooligosaccharides (Kainuma, 1984). Cyclodextrin has a variety of uses including the stabilization to heat, light or oxygen of compounds such as fatty acids and vitamins, prevention of the volatilization of flavour compounds and spices, deodorization of fish or meat smells, inclusion to increase water solubility of substances normally hard to dissolve and many others (Kainuma, 1984). Maltooligosaccharides are used as reagents for blood testing. These products, which are expensive, could be produced from sweet potato starch thus adding to its value.

The extensive use of sweet potato starch in Japan to make glucose and fructose syrups has already been mentioned. China now has several large plants for processing of glucose and fructose syrups from sweet potato starch. The first of these, in Bangbu, Anhui Province, was established in 1975 and employs 500 labourers (Wiersema et al., 1989). Annual production of glucose and fructose from this plant is 2000 and 5000 tonnes, respectively, utilizing dried chips which are first converted into starch. The outline of the process for isoglucose as used at the plant has been described (Wiersema et al., 1989) as follows: enzymes are added to wet starch to break it down (1 hour at 92°C), further enzymes are added to form glucose (40 hours at 60°C) which is purified by filtering with active carbon and ion exchange, enzymes are added to the glucose (1 hour at 60°C to isomerize part of the glucose to fructose), the isoglucose is purified and vacuum evaporated. The finished product contains 42% fructose, but efforts to increase this to 55% and eventually to 90% are in progress. A number of plants in Sichuan, China, produce a total of 20,000 tonnes of glucose, fructose and maltose per year (Wiersema et al., 1989). The shortfall of sugar production in China and the increasing potential of sugar syrup processing from sweet potato starch to bridge this gap has been pinpointed (Wang, 1984).

A method which could be used for making maltose from sweet potato starch on a small scale in rural areas has been fully described (Zhao and Jia, 1985). It consists of mixing malted barley or *Aspergillus* spp. mould (both sources of amylase enzymes) with steamed fresh root cubes or steamed crushed dried chips in a cauldron or insulated earthernware pot. The temperature is maintained at about 60°C for several hours. The extent of saccharification is tested with tincture of iodine. The mixture is boiled to stop saccharification and prevent fermentation. Residues are removed by filtration and can be used to feed animals. To lighten the colour and remove the strong sweet potato odour, the maltose liquid may be refined using activated carbon before boiling to concentrate it.

Many households in Vietnam produce maltose from sweet potato and other root starches in a similar way. Water is added to wet starch, the resulting slurry is boiled, cooled to 60°C, amylase is added (in the form of 8 kg rice sprouts/100 kg wet starch), the mixture is incubated for 6 hours, filtered, then boiled to concentrate the maltose (Wiersema, 1989).

Citric and other organic acids

Several countries, notably China and Japan, now manufacture citric acid from sweet potato starch or from its by-products. The process necessitates the initial breakdown of starch to sugars before these sugars are fermented by moulds, for example *A. niger* to citric acid. In Sichuan Province, China, the largest sweet potato growing area of the country, citric acid is the fourth most important product from sweet potato after starch, noodles and alcohol (Wiersema et al., 1989) with a yearly output of 4500 tonnes.

In the food industry, citric acid is added as a flavour enhancer or preservative in a wide range of products particularly soft drinks. In Japan a drink consisting of a mixture of citric acid from sweet potato and ascorbic acid crystals, which is added to water to taste, has been commercialized.

Monosodium glutamate

This very important flavour enhancer of a wide range of savoury foods, originally important mainly in the countries of the east, but now employed world-wide, is made in China using sweet potato starch as one of the raw materials. In Sichuan Province it is the fifth most important product from sweet potato, almost equal in tonnage to citric acid. Bangbu sugar plant, Anhui Province, mentioned above in connection with glucose–fructose production, alone produces 3000 tonnes of monosodium glutamate per year. The starch has first to be degraded to sugars which are then converted by microorganisms such as *Brevibacterium glutamicum* to glutamic acid. This is then converted to the monosodium salt. Other organic acids such as lactic and gluconic acids could also be produced from sweet potato.

Amino acids

There is already some production of amino acids from sweet potatoes in China (Sheng and Wang, 1987). Sweet potato starch is converted to sugar which is then fermented to the requisite amino acids. The great potential for augmenting the scale of this industry in China due to the increasing demand for amino acids for food, feed, medical and chemical use has been described (Wang, 1984).

Enzymes

The production of microbial enzymes is an important and rapidly growing industry in many parts of the world. The enzymes produced find a range of applications in the food industry, including the hydrolysis of starch in mashes for alcoholic fermentation, clarification of beers and fruit juices, in the manufacture of glucose and fructose syrups etc. Hydrolysed sweet potato starch could be a potential ingredient of the culture medium or mash on which the requisite microorganism producing a particular enzyme is grown. Its use as such has been suggested for China (Wang, 1984).

Snacks

The preparation of finger-shaped snack foods from a formulation of sweet potato starch, corn starch and defatted soy flour in the proportions 2:2:1 has been described by Indian researchers (Laul, Bhalerao and Mulmuley, 1985). They illustrate the construction of a piston-type hand-operated extruder, suitable for use even at village level, employed for making the snacks. The main ingredients are mixed to a dough with water, raising agent and natural colouring agents such as turmeric or annatto, the dough is cooked, cooled, cut into pieces and fed to the extruder. Extruded tubes are sun-dried in trays to 20–25% moisture, cut into 5-cm pieces and further sun-dried to less than 6% moisture and packed in bags. The snacks are fried or toasted before eating. Sweet potato starch alone gave a snack with a hard texture when fried, but mixing it with an equal weight of corn starch overcame the texture defect. The protein content of the finished snacks, due to the addition of soy flour was 7.5–7.9%.

Another type of snack using sweet potato starch has been developed in India (Central Tuber Crops Research Institute, Trivandrum). The starch is gelatinized in hot water to form a thick liquid to which chili and salt (or salt alone) is added. The mixture is spread in spoonfuls to dry in the sun. The dried wafers are deep-fried to finish into a crisp snack.

Other food and non-food products derived from starch and alcohol

Chinese researchers have suggested that sweet potato could be used for the manufacture of products such as ascorbic acid and antibiotics (Wang Han, unpublished material; and see Table 8.21, p. 550). Sweet potato alcohol, apart from its use as a wine or spirit or direct use as industrial alcohol, could be used as a raw material for producing plastics, ether, vinyl, artificial rubber and silk, acetone, butanol and various other chemicals and pharmaceuticals (Wang Han, unpublished material). The tremendous potential of sweet potato alcohol as an alternative fuel source has been mentioned in Chapter 2.

Sweet potato starch itself can be modified chemically, in the same ways as other starches, to give a whole new range of properties useful not only in the food but also in the paper, textile and chemical industries (Wang, 1984). For example, phosphate starch, a form of modified starch, can be used as a vegetable substitute for gelatin. The paste made from it can be stored for long periods of time with great stability and can be used not only as an emulsifying agent to separate metals in mining, but also to recover materials from the waste water of food plants, hence reducing environmental pollution from the waste.

Another starch derivative (chemical modification not described by Wang (1984)) is resistant to attack by acids, alkalis or salt and has high bonding ability. It can be used as a spray on film or coating agent in the form of a thin membrane to preserve foods such as meat, fish, fruits, candies or biscuits. This excludes oxygen and prevents absorption of moisture and contamination by microorganisms. Similarly, spraying or coating metals such as iron and steel can moisture- and rust-proof them.

A further product, cationic starch, can be used as a bonding agent in paper making to improve the physical nature of the paper, increasing its strength and improving its folding characteristics. In the textile industry it can be used to starch yarn and to bond textiles made of cotton or artifical fibre. Furthermore as an ion exchange agent it separates, purifies and concentrates biological materials.

Finally attention is drawn to a starch derivative called porous cyclodextrin which can be utilized in wrapping and storing agricultural chemicals, and essences, to maintain their quality over a long period of time. The estimated profitability of starch and its derivatives in China is given in Table 8.8 (p. 498).

A non-food use for sweet potato waste products such as underground stalks and non-storage roots is conversion to charcoal for use as a fuel. This has been listed as one potential product from sweet potato in China.

By-products from starch wastes

The waste water discharged from starch factories contains compounds such as sugars, nitrogenous constituents etc. which, when the waste is discharged into the environment, cause pollution and increase the biological and chemical oxygen demands of rivers and lakes. These same compounds, however, can be a useful source of nutrients for microorganisms such as yeasts which can be cultured to provide single cell protein for food and feed purposes, at the same time reducing thc polluting potential of the waste water. This has been amply shown for sweet potato starch waste water by Japanese scientists (Sugimoto, Shiga and Goto, 1967, 1969; Sugimoto et al., 1968a,b; Takakuwa, Ikeda and Murakami, 1972). A series of experiments describes the development as far as pilot scale of the continuous production of baker's yeast, *Saccharomyces cerevisiae*, from the waste water discharged from the nozzle separator in a sweet potato starch factory (Sugimoto et al., 1967, 1968a,b, 1969). The growth of yeast which could be used for baking or in foods or feeds greatly reduced the chemical oxygen demand in the waste water. Others have also produced *Candida utilis* with a higher yield than *S. cerevisiae* on sweet potato juice and have shown that shiitake (*Lentinus edodes*), a type of mushroom popular in Japan, can be cultivated on sweet potato juice with glucose and urea added (Takakuwa and Furukawa, 1973).

In addition, the nutritional value of the protein prepared from sweet potato starch waste water by coagulating with acid and heat has been investigated (Horigome, Nakayama and Ikeda, 1972). The coagulated dried residue, contained 48.7% protein which had a biological value and digestibility of about 72% and 82%, respectively, being low in methionine and lysine compared with whole egg, but better in overall essential amino acid content than soy protein. Up to the present time the production of single cell protein or of a protein concentrate from sweet potato starch waste water has not been commercialized. Solids content of waste waters is low, making evaporation for their recovery expensive. There have been recent experiments on the concentration of sweet potato starch waste water by ultrafiltration and reverse osmosis (Chiang and Pan, 1986), techniques which can separate a large volume of a dilute solution into a small volume of concentrate and a large volume of permeate which can be reused as water or safely discharged.

The use of the waste residue, which still contains some starch, left on sieves and filters during sweet potato starch manufacture, for making citric acid has already been mentioned.

Alcoholic beverages

A mildly alcoholic beverage called masato is traditionally made from sweet potato (but more frequently from manioc) by certain Indian tribes of the Peruvian Amazon region (Austin, 1985). Even when it is made from manioc, an orange-fleshed sweet potato may occasionally be added to give the drink a more pleasant colour.

The potential importance of sweet potato as a source of biomass for large-scale ethanol production was discussed briefly in Chapter 2. The manufacture of alcohol from sweet potatoes for human consumption as well as for chemical and pharmaceutical purposes is already well established, especially in countries such as China, Japan and Korea. In the larger sweet potato cultivating provinces of China, alcohol is the most important product after starch and noodles; in Korea, however, alcohol production has been steadily rising and, in the 1980s represented more than 30% of total production, compared to only about 8% in the case of starch. Japan uses only 5–6% of its production for alcohol, most of the alcoholic beverage being made in the southern part of the country.

Chinese sweet potato alcohol plants utilize mainly dried chips as the raw material, the chips being stored in sacks in the open air with some covering as protection against rain (Wiersema et al., 1989). This use of dried chips in large quantities enables them to operate the year round. The largest alcohol plant in China, situated near Jinan, Shandong Province, uses 140,000 tonnes of dried chips (equivalent to 350,000 tonnes of fresh roots) per year to produce 50,000 tonnes of alcohol, most of which is for industrial purposes. However, 95% of the 10,000 tonnes production of a smaller, but more modern, plant is high quality alcohol for human consumption. The quality requirements for dried chips at a large plant in Anhui Province are a starch content of at least 65% and less than 13% moisture (equivalent to 30% dry matter in fresh roots). Alcohol is made by adding water to the dried chips and boiling, adding enzymes to convert starch to sugar, fermenting the sugar to alcohol and distilling (Wiersema et al., 1989). There is an increasing demand for alcoholic beverages which will have to be fulfilled by improving the quality of alcohol produced; a high proportion from some large plants is at present unfit for human consumption. A process for preparing sweet potato spirits in rural areas (Zhao and Jia, 1985) has been described. In this process, dried chips are first crushed, mixed with rice husks and distillers stillage from a previous batch. Water is added and the whole cooked by steaming and cooled to 30°C. An inoculum prepared by the growth of a mould (*Aspergillus* spp.) on a medium of wheat bran, barley meal, pea meal and a small percentage of other unspecified components is added, together with a solution of yeast, at a dosage of 10% and 10–20%, respectively, to the cooked mash. A pit cellar is dug into the ground and the mixture placed in it. The pit is made air-tight by the excavation of a furrow around the circumference of the mouth. The furrow is filled with water, which forms a seal with the rim of the cellar cover. This allows carbon dioxide to escape but prevents external air entering. The cellar is sealed with mud. Conversion of starch to sugar and fermentation of sugar to alcohol take place simultaneously and this process is permitted to continue for 7 days, with the internal temperature not allowed to exceed 40°C. The fermented material is removed from the cellar and distilled. It is claimed that, from 100 kg of dried sweet potato, 60 kg of spirits with an alcohol content of 50% can be produced by this method.

Studies on the possible use of cellulase enzyme to degrade a high percentage of the cellulose in sweet potato chips and hence significantly increase the yield of alcohol have been carried out in China (Gao and Fengyun, 1982). Forty per cent of cellulose was degraded and alcohol yield increased by 10.8% with cellulase. One hundred kilograms of sweet potato chips yielded 64 kg of Chinese spirit with an alcohol content of 65% (v/v).

The production of *shochu*, a traditional Japanese distilled liquor which may also be made from rice, barley, buckwheat or crude sugar as well as sweet potato, has a long tradition. The production of sweet potato *shochu*, which probably originated in China, has been known since the 1700s. The process is outlined in Figure 6.33. Traditional *shochu* making in which the fermentation takes place in ceramic jars set into the ground can still be seen as a tourist curiosity, but nowadays *shochu* establishments are modern, automated factories. In Kyushu, southern Japan, 36% of *shochu* production is from sweet potato, which is second only to barley as a raw material, and within Kyushu, Kagoshima Prefecture alone makes 86% of all the sweet potato *shochu* (Sugama, S., personal communication).

The preparation of *koji*, a heavy inoculum of *Aspergillus niger* or *A. kawachii* on steamed rice is to provide a source of enzymes which hydrolyse starch to sugar. *A. niger* also produces citric acid in the first *moromi* or seed mash which lowers the pH to 3.2–3.4 and thus inhibits the growth of undesirable microorganisms. Unpeeled sweet potato is trimmed of defect parts by hand before being washed, steamed and crushed and added (4:1) to the *moromi*. During fermentation of this main mash, simultaneous starch conversion to sugar by the *koji* enzymes and fermentation of sugar to alcohol by the yeast *Saccharomyces cerevisiae* takes place (Figure 6.34). The final alcohol concentration of the mash is 13–

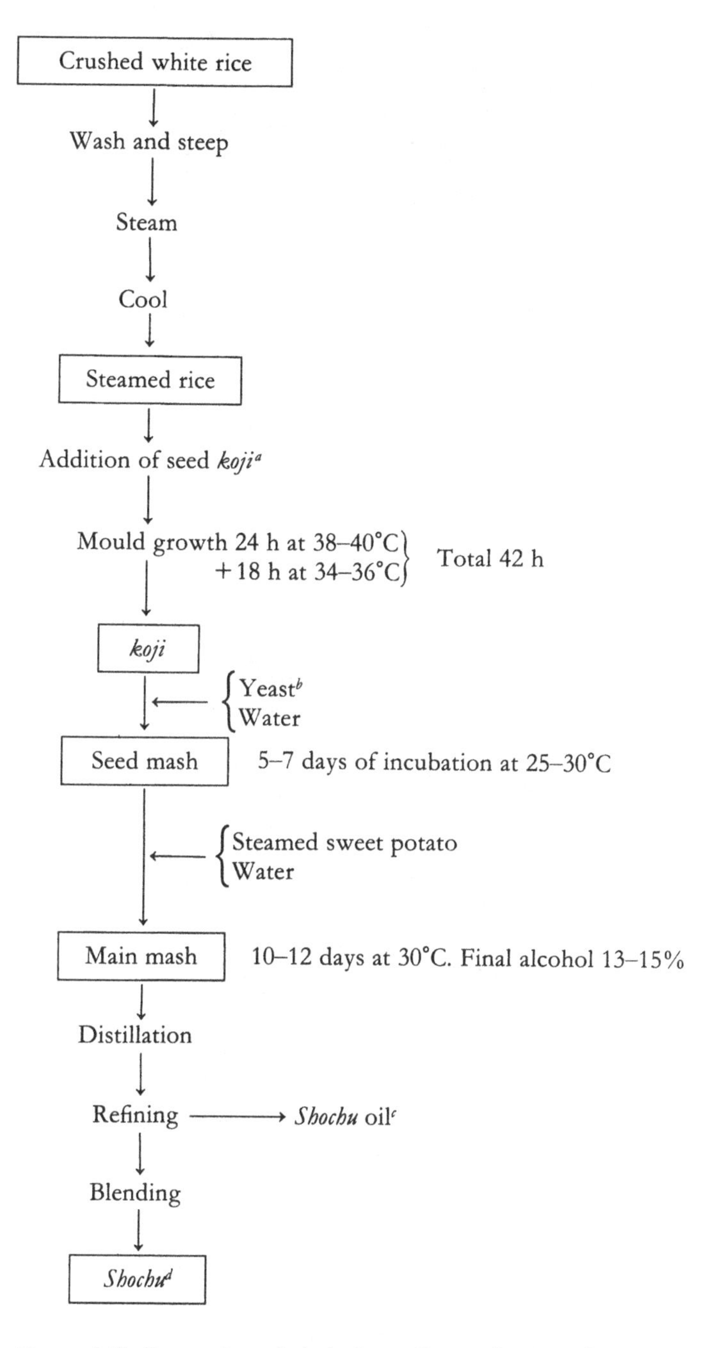

Figure 6.33. Preparation of *shochu* (according to Sugama, S., personal communication). *a*, spores of *Aspergillus niger* or mutant *A. kawachii*; *b*, *Saccharomyces cerevisiae*; *c*, this has to be removed as it can become oxidized and give rise to off-flavour in *shochu*; *d*, alcohol concentration is adjusted to 20–40% (v/v) before bottling.

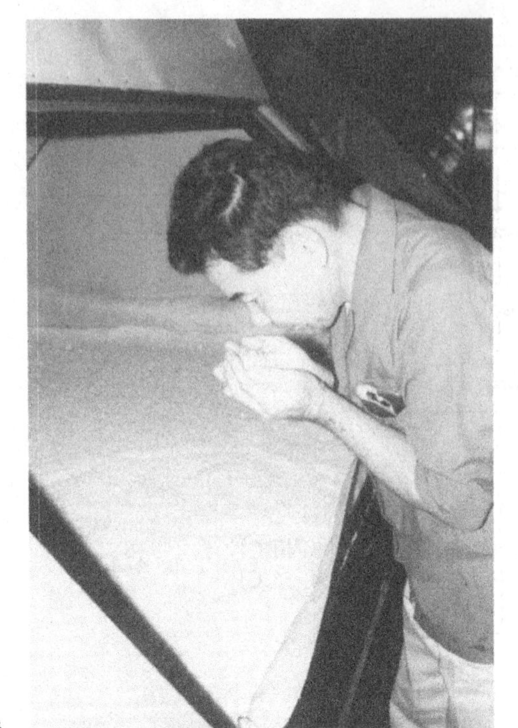

a

b

15%. The mash is then pumped to the still and alcohol distilled off. Different batches of *shochu* may be blended to give a uniform product and the alcohol content is adjusted to 20–40% (v/v) before bottling. *Shochus* are sold in a variety of bottles with very attractive shapes and labels.

There have been a number of studies designed to improve the *shochu* process by reducing the cost, improving the fermentation and reducing undesirable bitter off-flavours. Production of *koji* in traditional *shochu* making needs a large quantity of rice and entails 40% of the total cost of the process. The quantity of *koji* used could be reduced by the addition of microbial amylases, glucoamylase and cellulase to the main mash to decompose fibre and initiate saccharification before the addition of seed mash (Ogawa, Toyama and Toyama, 1982). This reduced the viscosity of the main mash and improved the quality of the final *shochu*.

Replacement of steam distillation by vacuum distillation in the final stage of the process has been suggested as a method of improving the taste of *shochu* (Kudo et al., 1979). The product from vacuum distillation had a light taste and contained less acetaldehyde, furfural and the higher fatty acid esters (all of which impart strong and characteristic flavours) than *shochu* from steam distillation. *Shochu* may also contain compounds such as ipomeamarone (see Chapter 4); these may impart a bitter flavour to the finished product if diseased sweet potatoes are included in the raw material of the main mash. It was found that 90% of these bitter compounds could be removed by treatment of the *shochu* with 0.03% activated carbon for 5 hours (Kudo and Hidaka, 1984).

There is some experimental work in progress to produce *shochu* from a very low sweet cultivar 'Satsumahikari', which is claimed to give a product with a very light taste preferred by younger drinkers (Umemura, Y., personal communication). The feasability of extracting a purple pigment, suitable for food colouring, from a *shochu* made using a cultivar with a very high level of flesh anthocyanins is being studied (Umemura, Y., personal communication).

Utilization of residues from alcohol production

The residue from alcohol production left after distillation, known as distiller's grains or distiller's stillage, is a pollutant when discharged into the environment, but is a potential source of nutrients for food or animal feed purposes, and can also be used as a raw material for citric acid production or as a culture medium for the growth of microbes

Figure 6.34. Stages in the preparation of the Japanese sweet potato spirit *shochu*. (a) determining whether the *koji* is ready for use; (b) addition of steamed, crushed sweet potato to the main mash (S. Sugama).

producing enzymes. Sweet potato *shochu* residue, after centrifugation and drying, contained 27% sugar (mainly glucose) and 26% total protein (Ogawa and Toyama, 1982). It was used experimentally to produce > 80% recovery of citric acid by fermentation with *Aspergillus niger* and for the successful cultivation of *Trichoderma viride* giving rise to cellulase. In addition it was shown that an edible mushroom, *Flammulina veltipes*, could be successfully cultivated on the residues from *shochu* and sweet potato starch manufacture.

The optimum use of fermentation residues is said to play a potentially important role in the commercial success of enterprises seeking to produce large quantities of ethanol (for example for fuel purposes) from biomass. The nutritional value of the protein from sweet potato ethanol residues has therefore been examined recently with a view to its exploitation as a food component (Wu and Bagby, 1987). Distillation residue from a sweet potato alcohol process was separated by filtration into solid filter cake and thin stillage, which in turn was separated by centrifugation into solids and stillage solubles. The addition of pectinase to the sweet potato slurry to hydrolyse pectinous materials before saccharification and subsequent fermentation and distillation was found to provide a number of advantages. It not only increased the solids content of the wet filter cake so decreasing drying costs, but also increased the total protein content of the filter cake solids. The total protein content in dry filter cake residue was between 29.6% and 32.4% in three different cultivars. Moreover, the lysine content of the protein (g/16 g N) was higher than in the original sweet potato and the overall amino acid balance was superior to that of cereals. The authors concluded that use of the fermentation residue for food might increase the economic feasibility of using sweet potato as a source of biomass for ethanol production. Concentration of stillage solubles by ultrafiltration and reverse osmosis, which are less expensive alternatives to evaporation, has been proposed (Wu, 1988) so that this fraction of the fermentation residue could also be exploited for food/feed.

The potential utilization of wastes from other sweet potato processing establishments for the production of animal feeds will be discussed in Chapter 7.

Enzymes

The production of microbial enzymes using sweet potato or its processing residues as part of the culture medium has already been mentioned (p. 376). In addition to the indirect production of enzymes, sweet potato itself is a rich source of beta-amylase, which finds many applications in the food industry. In the United States the entire commercial production of beta-amylase comes from sweet potato using the original method of isolation (Balls, Walden and Thompson, 1948). More up-to-date alternative methods of preparation and purification have been suggested (Hegde and Joshi, 1979; Roy and Hegde, 1985).

Pectin

The content and composition of pectin in sweet potato varies with cultivar (see Chapter 2), but is at least 1% of fresh weight (Reddy and Sistrunk, 1980) and may be as much as 3–5% of fresh weight (Ahmed and Scott, 1958). The gelling properties of sweet potato pectin are said to be similar to those of apple pectin (Winaro, 1982). There is a potential for extraction of pectin from the peel and trim wastes of sweet potato processing factories.

Pigments

Sweet potatoes possess two types of pigment with potential use as natural colourings in the food industry – carotenoids and anthocyanins. The chemistry of these compounds has already been described briefly in Chapter 2.

Beta-carotene, in high concentration in orange-fleshed cultivars, could be extracted from sweet potato pulp and used as a substitute for the synthetic yellow food colorants now falling into disfavour with consumers. It would be suitable for colouring margarines and various types of soft drink. However, as far as is known, extraction of beta-carotene from sweet potatoes for this purpose has not been commercialized. In leaves beta-carotene is accompanied by xanthophylls (oxygenated derivatives, including lutein). Many xanthophylls can also be used as a yellow colouring agents, particularly in hens' eggs if included in their diet. The production of a concentrate containing a high level of xanthophyll from sweet potato leaves which could be used for this purpose has been investigated (Walter et al., 1978b) and is discussed further in Chapter 7.

The direct use of sweet potato flesh, containing a high level of anthocyanins, in foods such as pie fillings, icecreams, sorbets etc., particularly in Japan, has already been mentioned above. There is increasing interest in the extraction and use of the anthocyanin pigments themselves, which, depending on pH take on a red, purple or violet hue, are stable in isolation and can be used in place of synthetic red-purple pigments in foods. A violet pigment from sweet potatoes has been patented in Japan by a large chemical company who produce it commercially as a liquid extract (San Ei Chemical Industry, 1981). The process involves immersing the sweet potato in an aqueous solution of potassium sulphite and ethanol, adding sulphuric acid to the extract, concentrating the mixture to remove suphurous acid, precipitating impurities with alcohol and concentrating the filtrate.

Methods for extracting anthocyanin pigments from purple-fleshed cultivars and concentrating the extract into a form suitable for addition to foods has been studied in Brazil (Cascon et al., 1984). Neither aqueous nor alcoholic extractions were found to be suitable. Aqueous extraction was slow, laborious and had to be repeated six to eight times due to high starch viscosity. Alcoholic extraction resulted in the deposition of fine starch granules in the concentrated extract during storage, thus lowering its quality.

The method found to be most suitable involved the breakdown (hydrolysis) of starch with acid followed by enzymes, and subsequent extraction of a clear pigment in solution which could be concentrated or dried. Cooking of roots was found to be necessary to destroy enzymes which could subsequently have decomposed the anthocyanins and to gelatinize starch thus aiding hydrolysis. Cooked, peeled roots are homogenized with an equal volume of water, acid hydrolysed with 0.5 ml of concentrated sulphuric acid/100 ml homogenate at 80°C for 2 hours and then neutralized with calcium oxide (0.5 ml/100 ml mix). The mixture is cooled to 60–65°C and hydrolysed with amyloglucosidase for 1 hour at this temperature. The mixture is then pressed to remove solids, acidified to pH 3.0 with 50% citric acid and filtered through diatomateous earth. The resulting extract could then be concentrated under vacuum to a liquid of 72°Brix, or freeze- or spray-dried to a powder. These three products contained 240 mg/100 ml, 150 mg/100 g or 120 mg/100 g anthocyanins, respectively. The powders were stable during 8 months of storage in dark glass containers at room temperature; the liquid concentrate lost 50% of its anthocyanins after 6 months at room temperature in clear glass. Acid solutions of anthocyanin extracts (at pH 1–4.5) could be kept refrigerated for 4 months with little alteration. The liquid or powders could be used in soft drinks to give clear reddish-purple solutions. In formulations made with sugar and flavour and acidified with citric acid, 0.5–1.0% pigments gave a grape-like colour at pH 4.0, and a cherry-like colour at pH 3.0. The stability of sweet potato anthocyanins has been confirmed by others (Bassa and Francis, 1987). In a model beverage stored at room temperature for 1 year in the dark, sweet potato anthocyanin pigments were more stable than those from blackberries or grapes as judged by colour measurements and pigment analysis.

Lipids

Lipids are synthesized in appreciable amounts by some microorganisms, for example certain yeasts and moulds. Mycological lipid formation using sweet potato as a growth medium for the moulds *Aspergillus oryzae* and *A. terreus* has been investigated (Naguib and Yassa, 1973, 1974). Both growth of fungus and fat content of the growth increased when sweet potato starch was previously hydrolysed to sugars and an external source of nitrogen was added to the medium. However, the potential of this process seems extremely doubtful while cheaper sources of animal and vegetable lipids are readily available.

Leaf protein concentrates

Green leaves are not often thought of as rich sources of protein, but their average protein content at about 3.5% (fwb) is actually similar to that of cow's milk, generally considered to be a good dietary protein source (Barbeau and Kinsella, 1988). Although green leaves contain the major part of the world's protein, due to their abundance, much of it is often unavailable to monogastric animals, including humans, because it is associated with large quantities of fibre. Interest in exploitation of this protein source, by its extraction from leaves and its utilization in human food, greatly increased in the 1950s, especially as much of the malnutrition in the world was felt to be due to lack of protein in the diet, rather than an energy deficiency or lack of food per se. During and since the 1950s there have been numerous studies of the content, quality and efficiency of extraction of protein from various leaf sources, among them sweet potato leaves.

A method typically used for preparing a leaf protein concentrate (LPC) from sweet potato leaves or other sources entails homogenization of the leaves in water, filtration, coagulation of protein from the filtrate with acid and/or heat, sedimentation of the coagulated protein, centrifugation and drying (Sun et al., 1979). Researchers agreed about the generally high quality of sweet potato leaf protein concentrates (see Table 3.15, p. 171). Most, however, discounted sweet potato leaves as a potential source of LPCs due to the difficulty of extraction, which is complicated by the large amounts of mucilage present in the leaves (Byers, 1961; Deshmukh et al., 1974; Liu and Yang, 1979). However, one group of workers (Walter et al., 1978b) claimed to overcome this problem by grinding the leaves alone and then adding water before pressing out the juice and coagulating the protein with heat.

In any case, despite their high nutritional value, the utilization of LPCs as human food in the form of crude powders, which have a green coloration, a grassy flavour and an unpleasant texture, was not successful. Apart from the aesthetic disadvantages, most people have no idea how to incorporate them into their traditional dishes.

More recently, it has been suggested that a protein present in all green leaves should be extracted and employed in human foods for its excellent functional properties of emulsification, gelation, foam formation and stabilization (Barbeau and Kinsella, 1988). The protein in question is the key photosynthetic enzyme ribulose biphosphate carboxylase/oxygenase (Rubisco). Rubisco has been found in all photosynthetic, chlorophyll-containing organisms examined to date, including higher plants. If a way of extracting it on a large scale could be found, it has been estimated that yields of crystalline Rubisco/ha per year from multiple harvests of leaves could exceed protein yields per hectare of corn or wheat. The nutritional value of Rubisco is very high, with only a slight deficiency of methionine and a lysine content similar to that of whole egg. Purified Rubisco is a tasteless, odourless, white powder. It is soluble in the pH range (6–8) of most food systems and forms stable foams and emulsions at low temperatures. Under appropriate conditions it also produces firm gels when heated. Research is only beginning into the potential exploitation of Rubisco in human foods. However, the inclusion of sweet potato leaves in this research as a potential source of Rubisco is a possible future development.

A leaf protein concentrate containing about 50% total protein and approximately 0.1% xanthophyll has been produced experimentally in the United States with a view to its incorporation in poultry rations, where it would act as a nutritional supplement and an egg yolk pigmentation agent (Walter et al., 1978b; and see Chapter 7). It is suggested that the concentrate, produced from locally grown sweet potato leaves could replace imported dried alfalfa in poultry feed. The concentrate is prepared by grinding sweet potato vines (or preferably leaves and petioles, without the more fibrous stems), adding water, and pressing. The press cake is extracted with a second lot of water. The two lots of juice from pressing are combined, heated to coagulate protein and centrifuged to obtain a solid pellet containing protein and xanthophyll. The optimum date for harvesting leaves, in terms of maximum protein and xanthophyll levels and minimum fibre content, was about 80 days after planting, but at the usual harvest date of 130–140 days after planting the concentrate contains about 36% good quality protein (as judged by amino acid analysis), 9% fibre and 0.1% xanthophyll. Storage of the extracted juice, which might be required in practical production of concentrate, for several hours at up to 50°C had little effect on xanthophyll content and reduced protein by about 15%.

Single cell protein

The use of sweet potato starch and alcohol wastes as media for the culture of single cell or mycoprotein has already been discussed. However, there have also been investigations into the production of this protein using whole sweet potato-based media (El-Ashwah et al., 1980). Attempts to grow six types of fungus in shaken culture on a medium containing only sweet potato blended with water failed. However, the addition of a source of nitrogen in the form of ammonium phosphate instigated a high level of mycelial growth and total protein content with *Cladosporium cladosporioides*; the addition of yeast extract gave similar results with *Dactylium dendroides* and *Linderina pennispora*. These media–organism combinations were considered to be potentially useful for production of mycoprotein.

D. Overview

If efforts in the coming years to increase production of sweet potato, and hence farm incomes, are to succeed, those which focus on the agronomic aspects must be complemented by a determination to stimulate demand by seeking new uses for the crop. The immense potential for manufacture of a wide range of food items from sweet potato is obvious from the previous sections. However, at present, production on a significant scale is confined to a few countries and to a very narrow spectrum of products. In two countries where sweet potato processing is relatively well developed, a considerable fraction of total production is devoted to starch and alcohol manufacture (Figure 6.35). Moreover, in Japan 75% of the starch production is used for further manufacture of sugar syrups. Food processing, apart from starch for food use and alcoholic beverages, occupies a greater percentage of the total uses in China than in Japan, although it is worth noting the great diversity of products which exist in Japan. China, the greatest producer of sweet potato in the world, uses about 14% of the total roots available for processing into food products. It is significant that in countries such as China and Japan which record high average yields and the development of cultivars appropriate for industrial purposes through a vigorous programme of plant breeding, sweet potato processing plays such an important role in utilization. This suggests that a significant way to diversify utilization through increased processing is to improve yields and develop cultivars tailored to processors' needs. However, in practice higher yields and increased processing have to be developed simultaneously. There is little possibility of success for one without the other. This may be the way in which yields and processing have developed together in China over a long

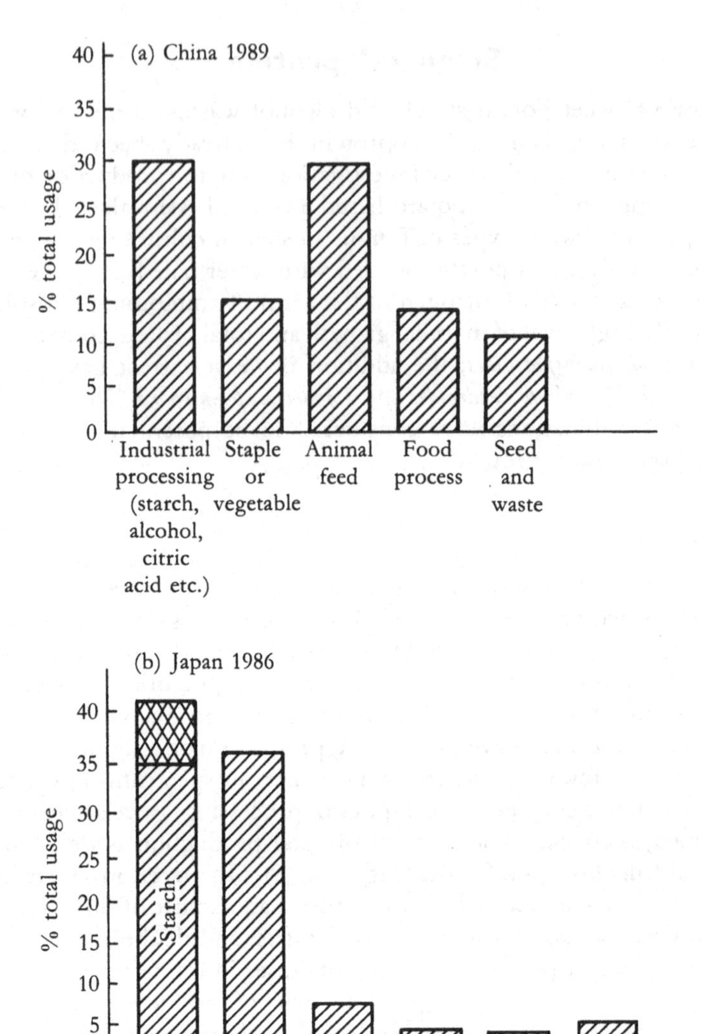

Figure 6.35. Distribution of sweet potato use in two processing countries: (a) from estimates given by Wiersema et al. (1989); (b) from Duell (1989) data originally from Ministry of Agriculture, Forestry and Fisheries Annual Report (1987).

period of time. In Japan, clearly defined breeding goals targeted to high starch yields may have resulted from an existing demand for starch, and have continued with efforts to make sweet potato competitive as a starch source with alternative raw materials.

The sweet potato products at present commercialized in varying scales of operation and the countries in which they are manufactured are shown in Table 6.14. Starch, starch-derived products, and alcohol are the main commercial products in the Far East, notably China. Other products are of far less quantitative importance.

It will be noted that a great number of potential products mentioned in the text do not appear at all in Table 6.14, because at present they are confined to the experimental or pilot scale only.

It is therefore worth considering the major reasons for constraints to sweet potato processing at the present time which may inhibit greater product development.

Highly significant

Low yields, low product recovery rates and high root production costs make sweet potato unable to compete with other raw materials yielding similar end products

The problems involved are mainly low root yields, low yields of dry matter (and hence starch, a major product or starting point for derived products) and poor conversion rates from raw material to the end product. Interlinked with these may be a further constraint of relatively high cost of harvested roots which results in uncompetitiveness of sweet potato with other raw materials. Although the relatively high cost of raw material in many areas is a direct result of low yields, factors common to other root and tuber crops also play a part. Handling costs may be high due to collection and transport problems and payment of middlemen; roots are bulky, perishable and cost more per unit weight of dry matter to transport than grains; poor quality roots result in a high degree of wastage and low product recovery rates. Farmers often obtain a better price by selling fresh roots.

A prominent example of a product with the production constraints mentioned above is starch. Low yields combined with relatively low root starch content and an inefficient recovery process result in poor starch yields per unit growing area. Hence, estimated or actual prices for sweet potato starch produced inside a country are often much higher than that of a starch made from an imported raw material. In practice, although all three constraints may operate, one alone may be of importance. In China, average root yields are relatively high and continuous efforts are in force to breed cultivars with increased starch

Table 6.14. *Commercialized food products from sweet potato and countries of production*

Product	Scale of operation	Country
Dried chips[a]	Very large, but carried out by farmers	China
Starch	Industrial	China
	Village	
	Home	
	Industrial	Japan
		Taiwan
Alcohol	Industrial	China
		Japan
Noodle[b]	Home/village	China
	Small business	Japan
		Taiwan
Citric acid[b]	Industrial	China
Monosodium glutamate[b]	Industrial	China
Glucose and fructose syrups[b]	Industrial	China
	Industrial	Japan
Amino acids	Industrial	China
Canned roots	Industrial	USA
Jam/sweet	Industrial	Argentina
		Brazil
	Home/village	China
Candy	Village	
	Industrial?	China
	Small business	Japan
	Small business	Taiwan
Snack-foods		
Fried chips	Industrial	China
	Industrial	Japan
	Small industrial	Peru
		Indonesia
Other (diverse)	Home/village	Philippines
	Home/village	Indonesia
	Village	China
	Small business	Taiwan
Purée, paste or mash		
Baby food	Industrial	USA
Frozen	Industrial	Japan
Thermally treated	Co-operative	Japan
Other	Small business	Taiwan
Flakes	Industrial	Japan
	Industrial	USA

Product	Scale of operation	Country
Granules	Industrial?	Japan
Flour	Home	Japan
	Small business	Japan
	Home/village	India
Bread[c]	Small bakeries	Peru
	Small bakeries	Japan
	Small bakeries	Taiwan
Baked goods		
Pies, pastries, cakes, etc.[c]	Industrial	Japan
	Small business	Japan
	Restaurant	Japan
	Small business	Taiwan
Frozen cubes	Industrial	Japan
Dried cubes	Small industry?	Philippines
Dried 'fruit' roll/'leather'	Small industry?	Philippines
Catsup[c]	Small industry?	Indonesia
	Entrepreneur?	Malaysia
Enzyme (β-amylase)	Industrial	USA

Notes:
Country: list not exhaustive, but includes those countries with information known to author.
[a] Dried chips are an intermediate product from which starch and alcohol are produced.
[b] Further products derived from starch.
[c] Further products containing sweet potato paste, flakes or flour.

content. However, improvements in starch extraction methods are required on a scale varying from village to large factory level (see Chapter 8, China). Japan also has relatively high average yields. Cultivars with high starch content are available and efficient starch recovery rates are claimed. However, sweet potato starch in Japan has found it hard to compete with cheaper corn starch, especially in the manufacture of sugar syrups, since restrictions on the import of sugar were lifted and imports of corn were increased. That sweet potato still exists at all as a raw material for starch manufacture in the face of competition from other sources owes everything to the vigorous Japanese programme of breeding for high starch yields per unit area. The decline of starch manufacture has stimulated efforts to increase demand for the crop as a food and as a raw material for alcohol production, and has increased the search for possibilites of product diversification. Technical problems related to production efficiency, composition and recovery can be overcome. The challenge to do so should be stimulated in countries desirous of producing industrial products from local, rather than imported, raw materials thus saving foreign exchange.

Significant

Lack of suitable cultivars for processing purposes

Cultivars with appropriate characteristics for a particular product or process have been developed, for example cultivars bred for high starch content in China and Japan. Moreover, in China and Japan strenuous efforts are continually being made to breed new clones with characteristics suitable for industrial use. Apart from starch content per se, characteristics such as low levels of polyphenolics and latex may be important in decreasing discoloration and improving quality of starch and starch derived products. However, such cultivars may not have the range of additional characteristics which allow adaptation to particular ecological or socio-economic environments. More evaluation of existing material in different geographical regions is needed as well as cooperation with producers to identify the desired complex of characteristics, including those useful for processing. Closer contacts between plant breeders and processors, even on an informal basis, could help development and promotion of cultivars tailored to specific processing needs.

Difficulties in transfer of technology to the commercial sector

This is a general problem with most root and tuber crops. In the case of sweet potato many national research systems have given low priority to the exploration of sweet potato product development. In spite of this, the range of possible products from sweet potato is very wide and many have been produced on a laboratory or even pilot scale, but have proceeded no further towards commercialization. This reflects a lack of interaction between scientists and the private sector. Collection and dissemination of information to the private sector about the relative advantages of sweet potato as a raw material and about technical aspects of using it could encourage research and development within the sector itself.

Less significant

Although the following factors are shown here as less significant than those above, this is chiefly due to their insignificance in China, the largest sweet potato producer. In areas which are ecologically, organizationally and culturally quite different from China, some of these constraints may be important.

Insufficient or unstable supply of raw material

This is a problem in several countries including Thailand, Taiwan, India, and Papua New Guinea. Some countries experience long seasonal shortages of roots interspersed with gluts. There is frequently a lack of storage or intermediate processing facilities to extend the season of supply or overcome gluts. Although this problem has been surmounted in parts of China by field drying of large quantities of root chips which are subsequently stored at factory sites until needed for processing, such a solution may be impractical in some humid tropical areas. Again, in parts of China where field-drying is not practical, sweet potato is processed immediately into dry starch which is stored for later noodle processing. There is also a storage technology for wet starch in some countries, notably Japan and Vietnam.

Poor consumer acceptability

This is all too frequently the reason for slow development or exploitation of a crop for food processing as the products which are developed cannot compete in quality with those from other crops. This may be for one of two reasons. The product developed may be trying to simulate a well-established and successful product made from another raw material and is not able to reproduce the sensory attributes expected by the consumer (for example sweet potato chips and white potato chips). One very important technical problem which will have to be overcome if sweet potato flour is to be adopted to substitute wheat flour in bread and other baked goods at any significant level is that of flavour. Detection of a pronounced sweet potato flavour may cause consumer rejection in favour of traditionally more bland products. Even an original product may suffer from poor appearance, flavour, unattractive packaging or short shelf life. A product may not be acceptable to consumers, however high its quality in the processor's eyes, if it does not fit into a local accepted dietary pattern, or if marketed at a price unaffordable by the target consumer group. Moreover, it may fulfil all other requirements and still fail through lack of appropriate promotion or marketing by advertising or extension work.

A case where these factors seem to have been of low importance or have been overcome is China, where sweet potato products fit well into rural dietary patterns and noodles, which may be considered of too low quality to sell, are made for home consumption by a great number of rural families. Furthermore low income rural consumers may not be able to afford higher quality products.

Lack of simple technology suitable for home/village scale

This is not true for China where simple village-level processing already exists on a wide scale. The technology could be improved to increase processing efficiency and up-grade product quality, but whether this is economically feasible remains to be seen depending upon consumer demand for higher quality products which are more expensive than those already existing. Meanwhile, the origins, expansion, and reasons for success of small-scale processing in China, as well as the technology, should be studied in depth to determine whether other developing countries could benefit from the Chinese experience. In Africa or India, simple home- or village-level sweet potato processing hardly exists at present and methods appropriate to local conditions could help to alleviate seasonal food shortages which occur each year as a result of poor food storage facilities. In fact in many hot, humid tropical areas, storage of highly perishable fresh sweet potato roots may be a very difficult goal to achieve using simple technology. Processing of roots into more storable forms may be a more practical alternative.

As previously indicated, the degree of importance attached to the constraints to successful processing, as described above, may vary from one country or region to another. Some of the constraints are interlinked and perhaps should not be considered in isolation. Moreover, some of the limitations mentioned are associated with production problems which vary greatly between countries and which can be resolved in the long term where a real potential for sweet potato as a raw material for processing is shown to exist.

The future, particularly in tropical developing areas, for greater use of local or regionally grown, rather than imported, crops is now under serious discussion by numerous regional and international institutions and by governments. This discussion includes the enhanced possibilites of food security which would arise from improved utilization by processing of crops such as sweet potato. In addition one of the most serious problems facing many developing countries is the external debt exacerbated by imports of food and industrial raw materials. This chapter has indicated the numerous possibilities of utilizing sweet potato as a substitute for a cereal such as wheat which has to be imported into many tropical developing countries to satisfy an increasing demand for bread, a variety of baked goods and products such as noodles. If governments desire to focus increasingly on local substitutes for wheat as a way of conserving foreign exchange they must be prepared to invest resources in research in the following areas: improvements in production; the technical aspects of utilizing sweet potato as a raw material so that equipment and methods are appropriate for making products with desired characteristics under local conditions; consumer studies and market research to ensure the success of a potential product. This can only result from close cooperation within the spectrum of disciplines engaged in farm production, postharvest technology and socio-economic research.

References

Abdulla, F.K. 1970. The influence of canning and precanning treatments on firmness of canned sweet potatoes. Ph.D. thesis, Mississippi State University.

Ahmed, E.M. and Scott, L.E. 1958. Pectic constituents of the fresh roots of the sweet potato. *Proc. Am. Soc. Hort. Sci.* **71**: 376–87.

Alcober, D.I. and Parrilla, L.S. 1987. Gender roles in sweet potato production, processing and utilization in Eastern Visayas, Philippines. Paper presented at an International Sweet Potato Symposium, 20–26 May, ViSCA, Baybay, Leyte.

Alexandridis, N. and Lopez, A. 1979. Lipid changes during processing and storage of sweet potato flakes. *J. Food Sci.* **44** (4): 1186–90.

Anon. 1968. The production of sweet potato starch and other sweet potato products. *Indian Food Packer*, Sept.–Oct.: 47–50.

1985a. Preliminary testing of sweet potato chips. *AVRDC Progress Report, 1983*, AVRDC, Shanhua, T'ainan, pp. 304–6.

1985b. Finger foods from root crops. *ViSCA Vista* **8** (2): 5, 25.

1988. World News. *Frozen Chilled Foods* **42** (10): 8.

1989. Introduction of the research and application on potato and sweet potato processing in Sichuan Academy of Agricultural Sciences (SAAS), Chendu, Sichuan. [Mimeo]

Ashida, J. (Resources Council, Science and Technology Agency) (ed.). 1982. *Standard tables of food composition in Japan*, 4th revised edn. Printing Bureau, Ministry of Finance, Japan.

Augustin, J., Swanson, B.G., Pometto, S.F., Teitzel, C., Artz, W.E. and Huang, C-P. 1979. Changes in nutrient composition of dehydrated potato products during commercial processing. *J. Food Sci.* **44**: 216–19.

Austin, D.F. 1973. The camotes de Santa Clara. *Econ. Bot.* **27** (3): 343–7.

1985. Spirits in the hills of Peru. *Fairchild Trop. Gdn Bull.* **40** (2): 6–13.

Baba, T., Kouno, T. and Yamamura, E. 1981. [Development of snack foods produced from sweet potatoes. I. Factors affecting hardness and discoloration of sweet potato chips] Japanese. *J. Jap. Soc. Food Sci. Technol.* **28** (6): 318–32.

Baba, T., Kouno, T., Tanoue, H., Maeya, Y., Tamaru, Y. and Yamamura, E. 1985. [Effects of blanching-freezing treatment on the texture of imo-karinto (product of fried sweet potato)] Japanese. *J. Jap. Soc. Food Sci. Technol.* **32** (2): 133–7.

Baba, T. and Yamamura, E. 1981. [Development of snack foods produced from sweet potatoes. II. Optimization of processing conditions in blanching and freezing for sweet potato chips] Japanese. *J. Jap. Soc. Food Sci. Technol.* **28** (7): 355–9.

Bae, K.S., Sohn, K.H. and Moon, S.J. 1984. [Structure and textural property of mook] Korean. *Kor. J. Food Sci. Technol.* **16** (2): 185–91.

Balls, A.K., Walden, M.K. and Thompson, R.R. 1948. Crystalline beta-amylase from sweet potatoes. *Biol. Chem.* **173** (9): 9–19.

Barbeau, W.E. and Kinsella, J.E. 1988. Ribulose biphosphate carboxylase/oxygenase (Rubisco) from green leaves – potential as a food protein. *Food Rev. Int.* **4** (1): 93–127.

Bassa, I.A. and Francis, F.J. 1987. Stability of anthocyanins from sweet potatoes in a model beverage. *J. Food Sci.* **52** (6): 1753–4.

Baumgardner, R.A. and Scott, L.E. 1962. Firmness of processed sweet potatoes (*Ipomoea batatas*) as affected by temperature and duration of the post-harvest holding period. *Proc. Am. Soc. Hort. Sci.* **80**: 507–14.

1963. The relation of pectic substances to firmness of processed sweet potatoes. *Proc. Am. Soc. Hort. Sci.* **83**: 629–40.

Blackwell, C. 1955. The effects of temperature and duration of storage upon the keeping quality, carbohydrate transformations and canning quality of Maryland Golden sweet potatoes. M.S. thesis, University of Maryland, College Park.

Boggess, T.S. and Woodroof, J.G. 1964. *Sweet potato chips*. Georgia Agric. Expt. Stn Leaflet No. 6.

Bortey, F. 1982. Yeast bread with three different preparations of sweet potatoes – cooked mashed, drum-dried and sun-dried. Dissertation. Home Science Department, University of Ghana, Legon.

Bouwkamp, J.C. 1985. Processing of sweet potatoes – canning, freezing, dehydrating. In: Bouwkamp, J.C. (ed.), *Sweet potatoes: a natural resource for the tropics*. CRC Press, Inc., Boca Raton, FL, pp. 185–203.

Brunstetter, B.C. 1936. How to make sweet potato chips. US Public Service Patent No. 2056884.

Burkhardt, G.J., Merkel, J.A. and Scott, L.E. 1973. Time and pressure regulated steam for peeling fruits and vegetables. *HortScience* **8** (6, section 1): 485–7.

Burton, D.L., Jones, L.G. and Miller, J.C. 1958. Sweet potato chips: a commercial potential. *La. Agric.* **1** (3): 14.

Byers, M. 1961. Extraction of protein from the leaves of some plants growing in Ghana. *J. Sci. Food Agric.* **12**: 20–30.

Carpio Burga, R. del. 1969. [*Camote pan* revolutionary national food] Spanish. *La Vida Agricola* **46** (545): 169–70.

1984. [*New advances in increasing the nutritional value of sweet potato in Chincha*] Spanish. FONAGRO, Chincha, Peru.

Carvalho, M.P.M., Jablonka, F.H., Cavalcanti, G.R.P. and Siqueira, R.S. 1980. [Age of sweet potato roots and quality of the flour] Portuguese. *Pesq. Agropec. Bras.* **15** (3): 267–74.

Carvalho, M.P.M., Moura, L.L. and Pape, G. 1981. [Process for obtaining sweet potato flour] Portuguese. *Pesq. Agropec. Bras.* **16** (4): 551–6.

Cascon, S.C., Alves, I.T.G., Bittencourt, A.M. and Leal, N.R. 1988. [Carotenoids of Jerimu sweet potato; stability on domestic cooking and industrial processing] Portuguese. *Congr. Bras. Cien. Tecnol. Aliment. Resumos* **11**: 126.

Cascon, S.C., Carvalho, M.P.M., Moura, L.L., Guimarães, I.S.S. and Philip, T. 1984. [Pigments of purple sweet potato for use in foods] Portuguese. *Bol. Pesquisa* No.9, EMBRAPA, Rio de Janeiro.

Cereda, M.P., Conceição, F.A.D., Cagliari, A.M., Heezen, A.M. and Fioretto, R.A. 1982. [Comparative study of sweet potato (*Ipomoea batatas*) varieties with a view to their use in food industries] Portuguese. *Turrialba* **32** (4): 365–70.

Chandler, L.A. and Schwartz, S.J. 1988. Isomerization and losses of trans-β-carotene in sweet potatoes as affected by processing treatments. *J. Agric. Food Chem.* **36** (1): 129–33.

Chang, K.J. and Lee, S.R. 1974. [Development of composite flours and their products utilizing domestic raw materials. IV. Textural characteristics of noodles made of composite flours based on barley and sweet potato] Korean. *Korean J. Food Sci. Technol.* **6** (2): 65–9.

Chen, K-L. and Chiang, W-C. 1985a. [Study on processing suitability of sweet potato] Chinese. *Food Sci.* **12** (3–4): 163–72.

1985b. [Preparation and storage of sweet potato flour and its application for manufacture of composite bread] Chinese. *Food Sci.* **12** (1–2): 21–8.

Chew, K.M. 1972. *Study on canning of sweet potato.* Report No. 69-T-5. Food Industry Research and Development Institute, Taiwan.

Chiang, B.H. and Pan, W.D. 1986. Ultrafiltration and reverse osmosis of the waste water from sweet potato starch process. *J. Food Sci.* **51** (4): 971–4.

Collins, J.L. 1981. Nutrient composition of sweet potatoes. *Tenn. Farm Home Sci.* No. 117: 25–6.

Collins, J.L. and Aziz, A. 1982. Sweet potato as an ingredient in yeast-raised doughnuts. *J. Food Sci.* **47** (4): 1133–9.

Collins, J.L., Ebah, C.B. and Mount, J.R. 1988. Development of yogurt with sweet potato as an ingredient. Presentation No. 650, *Annual Meeting of the Institute of Food Technology*, June, New Orleans, LA.

Collins, J.L., Sanders, G.G., Hill, A.R. and Swingle, H.D. 1974. Sweet potato products made from baked, cured roots. *Tenn. Farm Home Sci. Progr. Rep.* **91**, pp. 30–1.

Collins, J.L. and Washam-Hutsell, L. 1987. Physical, chemical, sensory and microbiological attributes of sweet potato leather. *J. Food Sci.* **52** (3): 646–8.

Constantin, R.J., Jones, L.G. and Hernandez, T.P. 1977. Effects of potassium and phosphorus fertilization on quality of sweet potatoes. *J. Am. Soc. Hort. Sci.* **102** (6): 779–81.

Constantin, R.J. and McDonald, R.E. 1968. Effects of curing, storage and processing temperatures on firmness of canned sweet potatoes. *Proc. Assoc. South Agric. Workers* p. 166 (Abstract).

Creamer, G., Young, C.T. and Hamann, D.D. 1983. Changes in amino acid content of acidified sweet potato purée. *J. Food Sci.* **48** (2): 382–8.

Data, E.S., Diamante, J.C. and Forio, E.E. 1986. Soy sauce production utilizing root crop flour as substitute for wheat flour (100% substitution). *Ann. Trop. Res.* **8** (1): 42–50.

Deobald, H.J. and McLemore, T.A. 1964. The effect of temperature, antioxidant, and oxygen on the stability of precooked dehydrated sweetpotato flakes. *Food Technol.* **18** (5): 739–42.

Deshmukh, M.G., Gore, S.B., Munigikar, A.M. and Joshi, R.N. 1974. The yields of protein from various short-duration crops. *J. Sci. Food Agric.* **25** (6): 717–24.

Dhamija, S.S. and Singh, D.P. 1979. Sweet potato as an adjunct in brewing. *Indian Food Packer* **33** (5): 3–8.

Duell, B. 1989. Variations in sweet potato consumption in Japan. *J. Tokyo Int.*

Univ. **39**: 55–67.
1990. Ways of eating sweet potatoes in Japan. In: *Tropical root and tuber crops changing role in a modern world*, Proceedings of the Eighth Symposium of the International Society for Tropical Root Crops, 30 October–5 November, 1988, Bangkok (in press).
El-Ashwah, E.T., Musmar, I.T., Ismail, F.A. and Alian, A. 1980. Fungi imperfecti as a source of food protein. II. Effect of sweet potato based media on yield and protein concentration. *Egypt. J. Microbiol.* **15** (1/2): 63–71.
El-Samahy, S.K., Morad, M.M., Seleha, H. and Abdel-Baki, M.M. 1980. Cake-mix supplementation with soybean, sweet potato or peanut flours. II. Effect on cake quality. *Baker's Dig.* **54** (5): 32–3, 36.
FAO 1988. *Requirements of vitamin A, iron, folate and vitamin* B_{12}, Report of a Joint FAO/WHO Expert Consultation, FAO Food and Nutrition Ser. No. 23. FAO, Rome.
FAO/WHO/UNU 1985. *Energy and protein requirements*, Report of a joint FAO/WHO/UNU Expert Consultation, WHO Tech. Rep. Ser. No. 724. WHO, Geneva.
Gao, P. and Fengyun, Z. 1982. [Studies on increasing the alcoholic yield of sweet potato chips by cellulase] Chinese. *Food Ferm. Ind.* **4**: 26–30.
Garlich, J.D., Bryant, D.M., Covington, H.M., Chamble, D.S. and Purcell, A.E. 1974. Egg yolk and broiler skin pigmentation with sweet potato vine meal. *Poultry Sci.* **53** (2): 692–9.
Gitomer, C.S. 1987. *Sweet potato and white potato development in China. A compendium of basic data.* International Food Policy Research Institute, Washington, DC.
Guedes, A.L. 1986. [Potential of sweet potato for industry] Portuguese. Universidade Estadual Paulista, Fac. Cien. Agron., Botucatu. [Mimeo]
Hamed, M.G.E., Hussein, M.F. and Refai, F.Y. 1973a. Preparation and chemical composition of sweet potato flour. *Cereal Chem.* **50** (2): 133–9.
1973b. Effect of adding sweet potato flour to wheat flour on physical dough properties and baking. *Cereal Chem.* **50** (2): 140–6.
Hannigan, K. 1979. Sweetpotato chips. *Food Eng. Int.* **4** (5): 28–9.
Haralampu, S.G. and Karel, M. 1983. Kinetic models for moisture dependence of ascorbic acid and β-carotene degradation in dehydrated sweet potato. *J. Food Sci.* **48** (6): 1872–3.
Haytowitz, D.B. and Matthews, R.H. 1984. *Composition of foods: vegetables and vegetable products.* Human Nutrition Information Series, USDA Agric. Handbook No. 8–11, Washington DC, pp. 428–39.
Hegde, M.V. and Joshi, P.N. 1979. A new method for the preparation of beta-amylase from sweet potato by thymol amylose adsorption. *Prep. Biochem.* **9** (1): 71–84.
Hong, E.H. 1982. The storage, marketing and utilization of sweet potatoes in Korea. In: Villareal, R.L. and Griggs, T.D. (eds.), *Sweet potato*, Proceedings of the First International Symposium. AVRDC, Shanhua, T'ainan, pp. 405–11.
Hong, S-C., Su, J-C. and Sung, H-Y. 1977. [A study on the better utilization of sweet potato] Chinese. *J. Chin. Agric. Chem. Soc.* **15** (1–2): 39–48.

Hoover, M.W. and Miller, N.C. 1973. Process for producing sweet potato chips. *Food Technol.* **27** (5): 74, 76, 80.
Hoover, M.W., Walter, W.M. and Giesbrecht, F.G. 1983. Method of preparation and sensory evaluation of sweet potato patties. *J. Food Sci.* **48** (5): 1568–9.
Horigome, T., Nakayama, N. and Ikeda, M. 1972. [Nutritive value of sweet potato protein produced from the residual products of sweet potato starch industry] Japanese. *Jap. J. Zootech. Sci.* **43** (8): 432–7.
Horton, D. 1988. Trip Report: Argentina. International Potato Center, Lima. [Mimeo]
Jayaraman, K.S., Gopinathan, V.K., Pitchamuthu, P. and Vijayaraghavan, P.K. 1982. The preparation of quick-cooking dehydrated vegetables by high temperature short time pneumatic drying. *J. Food Technol.* **17** (6): 669–78.
Jaynes, H.O. and Corley, D.T. 1981. Sweet potato turnovers. *Tenn. Farm Home Sci.* No. 120: 11–13.
Jenkins, W.F., Anderson, W.S. and Watson, W.W. 1956. Geographical location and storage affecting carbohydrates and canning quality in sweetpotatoes. *Proc. Am. Soc. Hort. Sci.* **68**: 406–11.
Junek, J. and Sistrunk, W.A. 1978. Sweet potatoes high in vitamin content but content is affected by variety and cooking method. *Ark. Farm Res.* **27** (5): 7–8.
Kainuma, K. 1984. Uses of sweet potato starch. *Farming Japan* **18** (5): 36–40.
Kattan, A.A. and Littrell, D.L. 1963. Pre- and post-harvest factors affecting firmness of canned sweet potatoes. *Proc. Am. Soc. Hort. Sci.* **83**: 641–50.
Kays, S.J. 1985. Formulated sweet potato products. In: Bouwkamp, J.C. (ed.), *Sweet potato products: a natural resource for the tropics*. CRC Press, Inc., Boca Raton, FL, pp. 205–18.
Kelley, E.G., Baum, R.R. and Woodward, C.F. 1958. Preparation of new and improved products from eastern (dry type) sweet potatoes: chips, dice, julienne strips and frozen french fries. *Food Technol.* **12** (10): 510–3.
Kim, S.K. and Kim, S.Y. 1984. [Chemical changes during the storage of crushed sweet potatoes in polyethylene film] Korean. *Res. Rep. Agric. Sci. Technol.* **11** (1): 45–52.
Kim, H.S., Lee, K.Y., Kim, S.K. and Lee, S.R. 1973. [Development of composite flours and their products utilizing domestic raw materials. I. Physical and chemical properties and nutritional test of composite flour materials] Korean. *Korean J. Food Sci. Technol.* **5** (1): 6–15.
Kudo, T., Hamakawa, S., Nakayama, K. and Hidaka, T. 1979. [Production of shochu by vacuum distillation] Japanese. *J. Soc. Brewing Japan* **74** (7): 484–6.
Kudo, T. and Hidaka, T. 1984. [Removal of bitter components from shochu produced from diseased sweet potatoes by carbon treatment] Japanese. *J. Soc. Brewing Japan* **79** (12): 900–1.
Labuza, T.P. 1972. Nutrient losses during drying and storage of dehydrated foods. *CRC Crit. Rev. Food Technol.* **3** (2): 217–40.
Laul, M.S., Bhalerao, S.D. and Mulmuley, G.V. 1985. Studies on the preparation of extruded finger shaped snack foods with protein enrichment. *Indian Food Packer* **39** (1): 68–77.

Lee, M-H. 1985. [Aroma improvement of the sweet potato flour composite bread and manufacturing of the sweet potato paste composite bread] Chinese. *J. Chin. Agric. Chem. Soc.* **23** (1–2): 133–9.

Lee, W.G. and Ammerman, G.R. 1974. Carotene stereoisomerization in sweet potatoes as affected by rotating and still retort canning processes. *J. Food Sci.* **39** (6): 1188–90.

Lee, T.S., Cho, H.O., Kim, C.S. and Kim, J.G. 1980. [The brewing of kochuzang (red pepper paste) from different starch sources. I. Proximate composition and enzyme activity during koji preparation] Korean. *J. Kor. Agric. Chem. Soc.* **23** (3): 157–65.

Lee, C.H. and Kim, C.W. 1983a. [Studies on the rheological property of Korean noodles. I. Viscoelastic behaviour of wheat flour noodle and wheat-sweet potato starch noodle] Korean. *Kor. J. Food Sci. Technol.* **15** (2): 183–8.

1983b. [Studies on the rheological property of Korean noodles. II. Mechanical model parameters of cooked and stored noodles] Korean. *Kor. J. Food Sci. Technol.* **15** (3): 295–301.

1983c. [Studies on the rheological property of Korean noodles. III. Correlation between mechanical model parameters and sensory quality of noodles] Korean. *Kor. J. Food Sci. Technol.* **15** (3): 302–6.

Lee, C.H. and Lee, S.W. 1984. [Peeling operations of root vegetables: potato, sweet potato and carrot] Korean. *Kor. J. Food Sci. Technol.* **16** (3): 329–35.

Lee, T.S., Park, S.O. and Lee, M.W. 1981. [Determination of organic acids of kochuzang prepared from various starch sources] Korean. *J. Kor. Agric. Chem. Soc.* **24** (2): 120–5.

Leung, W-T.W., Busson, F. and Jardin, C. 1968. *Food composition table for use in Africa.* US Department of Health, Education and Welfare Public Health Service, Bethesda, MD; FAO, Rome.

Leung, W-T.W., Butrum, R.R. and Chang, H.F. 1972. *Food composition table for use in East Asia.* Part I. *Proximate composition, mineral and vitamin contents of East Asian foods.* US Department of Health, Education and Welfare, Bethesda, MD; FAO, Rome.

Leung, W-T.W. and Flores, M. 1961. *Food composition table for use in Latin America.* INCAP, Guatemala; Interdepartmental Committee on Nutrition for National Defense, Bethesda, MD.

Liu, T.Y. and Yang, J.S. 1979. [Development of leaf protein. I. Study on the biochemical nutritive properties of leaf protein concentrate] Chinese. *Food Ind. Res. Dev. Inst. Tech Rep.* No. 138, Hsinchu, Taiwan.

Lopez, A., Williams, H.L. and Cooler, F.W. 1980. Essential elements in fresh and in canned sweet potatoes. *J. Food Technol.* **45** (3): 675–8, 681.

López Hernández, J., Adris, J., Fernández de Ranek, E. and Monserrat, S. 1983. [Sweet potato purée enriched in protein with predigested soy] Spanish. *Revista de la AAS*, July: 34, 36.

Luna de la Fuente, R. 1960. [Experiment in breadmaking with mixtures of wheat flour and three varieties of sweet potato] Spanish. *Infme. Mens. Estac. Exp. Agric. La Molina* **34** (393): 1–5.

Lutz, J.M. 1949. Firming the canned product made from cured and stored Porto Rico sweet potatoes. *Canner* **109** (10): 15.

Maeda, E.E. 1985. Effect of solar dehydration on amino acid pattern and

available lysine content in four tropical leafy vegetables. *Ecol. Food Nutr.* **16** (3): 273–9.
Maeda, E.E. and Salunkhe, D.K. 1981. Retention of ascorbic acid and total carotene in solar dried vegtables. *J. Food Sci.* **46** (4): 1288–90.
Magoon, C.A. and Culpepper, C.W. 1922. *A study of sweet potato varieties with special reference to their canning quality.* USDA Bull. No. 1040.
Manlan, M., Matthews, R.F., Bates, R.P. and O'Hair, S.K. 1985. Drum drying of tropical sweet potatoes. *J. Food Sci.* **50** (3): 764–8.
Martin, F.W. 1984a. Techniques and problems in small scale production of flour from sweet potato. *J. Agric. Univ. Puerto Rico* **68** (4): 423–32.
1984b. Diffusion-processed sweet potato pulp, a new product with broad appeal. *Qual. Plant. Plant Foods Hum. Nutr.* **34**: 211–20.
1985. Home-processing of sweet potato into long-lasting versatile flour. [Mimeo]
1986. Comparison of flours from thirteen low-sweet or non-sweet sweet potatoes. [Mimeo]
1987. Fried chips from staple-type sweet potatoes. *J. Agric. Univ. Puerto Rico* **71** (4): 373–8.
Martin, F.W. and Rhodes, A.M. 1984. Sweet potato variability in boiled slices and fried chips. *J. Agric. Univ. Puerto Rico* **68** (3): 223–33.
Mason, R.L. 1982. Sweet potato canning investigations. *Food Technol. Austr.* **34** (12): 574–6.
McMillen, E.C. 1981. The effects of curing, storage time and culls and jumbos versus field run for two cultivars of sweet potatoes as used in a frozen convenience food. *Diss. Abstr. Int.* **42** (11): 4349.
Molla, M.R.I. 1973. Sweet potato chips, a possible product for urban consumers in Bangladesh. *Bangladesh Hortic.* **1** (2): 77–9.
Montemayor, N.R. and Notario, J.N. 1982. Development of food products using soybean and root crop flours. *CMU J. Agric. Food Nutr.* **4** (3): 274–88.
Morad, M.M., Abdel-Baki, M.M., Seleha, H. and El-Samahy, S.K. 1980. Cake-mix supplementation with soybean, sweet potato or peanut flours. III. Effect on storability and the role of packaging materials. *Baker's Dig.* **54** (5): 34–6.
Moy, J.H. and Chi, S.P.S. 1982. Solar dehydration of sweet potato. In: Villareal, R.L. and Griggs, T.D. (eds.), *Sweet potato*, Proceedings of the First International Symposium. AVRDC, Shanhua, T'ainan, pp. 429–37.
Naguib, K. and Yassa, E.S. 1973. Growth and fat formation of *Aspergillus oryzae* and *Aspergillus terreus* on enzyme-hydrolyzed sweet potatoes. *Mycopathol. Mycol. Appl.* **51** (2/3): 163–70.
Naguib, K. and Yassa, E.S. 1974. The effect of magnesium and phosphate on mycological fat formation from sweet potatoes. *Mycopathol. Mycol. Appl.* **52** (3): 177–85.
Nair, G.M., Ravindran, C.S., Moorthy, S.N. and Ghosh, S.P. 1987. Indigenous technologies and recent advances in sweet potato production, processing and utilization in India. Paper presented at an International Sweet Potato Symposium, 20–26 May, ViSCA, Baybay, Leyte.
Nakajima, A. 1970. [Studies on the manufacturing of cooked and dried sweet potatoes. Part 1. Ingredient and exuding condition regard to the white powder of dried sweet potatoes] Japanese. *J. Food Sci. Technol.* **17** (10): 431–6.

Nanz, R.A. 1953. Sweet potato canning in Florida. *Proc. Fla State Hort. Soc.* **66**: 276–7.

Numfor, F.A. and Lyonga, S.N. 1987. Traditional postharvest technologies of root and tuber crops in Cameroon: status and prospects for improvement. In: Terry, E.R., Akoroda, M.O. and Arene, O.B. (eds.), *Tropical root crops: root crops and the African food crisis*, Proceedings of the Third Triennial Symposium of the International Society for Tropical Root Crops – Africa Branch. IDRC, Ottawa, pp. 135–9.

Ogawa, K. and Toyama, N. 1982. [Microbiological utilization of shochu fermentation residue] Japanese. *Bull. Fac. Agric. Univ. Miyazaki* **29** (2): 239–47.

Ogawa, K., Toyama, H. and Toyama, N. 1982. [Utilization of enzyme preparations in shochu making] Japanese. *Bull. Fac. Agric. Univ. Miyazaki* **29** (1): 203–11.

Okoh, P.N. 1984. An assessment of the protein, mineral and vitamin losses in sun-dried Nigerian vegetables. *Nutr. Rep. Int.* **29** (2): 359–64.

Olorunda, A.O. and Kitson, J.A. 1977. Controlling storage and processing conditions helps produce light colored chips from sweet potatoes. *Food Product Dev.* **11** (4): 44–5.

Osei-Opare, 1987. Acceptability, utilization, and processing of sweet potatoes in home and small-scale industries in Ghana. In: Terry, E.R., Akoroda, M.O. and Arene, O.B. (eds.), *Tropical root crops: root crops and the African food crisis*, Proceedings of the Third Triennial Symposium of the International Society for Tropical Root Crops – Africa Branch. IDRC, Ottawa, pp. 143–5.

Pace, R.D., Dull, G.G. and Phills, B.R. 1985. Proximate composition of sweet potato greens in relation to cultivar, harvest date, crop year and processing. *J. Food Sci.* **50** (2): 537–8.

Palomar, L.S., Perez, J.A. and Pascual, G.L. 1981. Wheat flour substitution using sweet potato or cassava in some bread and snack items. *Ann. Trop. Agric.* **3** (1): 8–17.

Panalaks, T. and Murray, T.K. 1970. The effect of processing on the content of carotene isomers in vegetables and peaches. *Can. Inst. Food Technol. J.* **3** (4): 145–51.

Passmore, R., Nicol, B.M., Rao, N.M., Beaton, G.H. and DeMaeyer, E.M. 1974. *Handbook on human nutritional requirements*, WHO Monograph Ser. No. 61. FAO/WHO, Geneva.

Picha, D.F. 1986. Influence of storage duration and temperature on sweet potato sugar content and chip color. *J. Food Sci.* **51** (1): 239–40.

Plaut, M. and Zelzbuch, B. 1958. Use of sweet potatoes for flour and bread. *Klavim* **8**: 77–92.

Prabhuddham, S., Tantidham, K., Poonperm, N., Lertbawornwongsa, C. and Tongglad, C. 1987. A study of sweet potato quality and processing methods. Paper presented during Training Course on Technology of Sweet Potato Production, July 1987, Pichit Horticultural Research Center, Pichit, Thailand. [Mimeo]

Purcell, A.E. and Walter, W.M. 1982. Stability of amino acids during cooking and processing of sweet potatoes. *J. Agric. Food Chem.* **30** (3): 443–4.

Purcell, A.E., Walter, W.M. and Giesbrecht, F.G. 1978. Protein and amino acids of sweet potato (*Ipomoea batatas* (L.) Lam.) fractions. *J. Agric. Food Chem.* **26** (3): 699–701.

Ranganath, D.R. and Dubash, P.J. 1981. Loss of colour and vitamins on dehydration of vegetables. *Indian Food Packer* **35** (4): 4–10.

Rao, C.S. and Ammerman, G.R. 1974. Canning studies on sweet potatoes. *J. Food Sci. Tehnol. (India)* **11** (3): 105–9.

Reddy, N.N. and Sistrunk, W.A. 1980. Effect of cultivar, size, storage and cooking method on carbohydrates and some nutrients of sweet potatoes. *J. Food Sci.* **45** (3): 682–4.

Rizvi, S.S.H. and Acton, J.C. 1982. Nutrient enhancement of thermostabilized foods in retort pouches. *Food Technol.* **36** (4): 105–9.

Roy, F. and Hegde, M.V. 1985. Rapid procedure for purification of beta-amylase from *Ipomoea batatas. J. Chromatogr.* **324** (2): 489–94.

S-101 Technical Committee 1980. *Sweet potato quality.* S. Coop. Ser. Bull. 249, Athens, GA.

Sammy, G.M. 1970. Studies in composite flours. I. The use of sweet potato flour in bread and pastry making. *Trop. Agric. (Trin.)* **47** (2): 115–25.

1973. The status of tropical root crop processing research at the University of the West Indies, Trinidad. In: *Proceedings of the Third Symposium of the International Society for Tropical Root Crops*, IITA, Nigeria, pp. 486–92.

1984. The processing potential of tropical root-crops. In: Dolly, D. (ed.), *Root crops in the Caribbean*, Proceedings of the Caribbean Regional Workshop on Tropical Root Crops, Jamaica, 1983. Faculty of Agriculture, University of the West Indies, St Augustine, pp. 199–206.

San Ei Chemical Industry 1981. [Violet pigment] Japanese. Japanese Examined Patent 5 617 061.

Satin, M. 1988. Bread without wheat. *New Scientist* 28 April, pp. 56–9.

Schwartz, S.J., Walter, W.M., Carroll, D.E. and Giesbrecht, F.G. 1987. Chemical, physical and sensory properties of a sweet potato french-fry type product during frozen storage. *J. Food Sci.* **52** (3): 617–9, 633.

Scott, L.E. and Bouwkamp, J.C. 1975. Effect of chronological age on composition and firmness of raw and processed sweet potatoes. *HortScience* **10** (2): 165–8.

Scott, L.E. and Twigg, B.A. 1969. The effect of temperature of treatment and other factors on calcium firming of processed sweet potatoes. *Memo Md Agric. Ext. Serv. Hort.* **77**: 69.

Seralathan, M.A. and Thirumaran, A.S. 1990. Utilization of sweet potato (*Ipomoea batatas*) flour in South Indian dishes. In: *Tropical root and tuber crops changing role in a modern world*, Proceedings of the Eighth Symposium of the International Society for Tropical Root Crops, 30 October–5 November 1988, Bangkok (in press).

Sheng, J. and Wang, S. 1987. The status of sweet potato in China from 1949 to the present. Xuzhou Institute of Sweet potato, Jiangsu. [Mimeo]

Shukor N.M. and Khelikuzzaman, M.H. 1989. Sweet potato cultivation in Malaysia: a country report. In: Improvement of sweet potato (*Ipomoea batatas*) in Asia. Report of the "Workshop on Sweet Potato Improvement in Asia", held at ICAR, India, October 24–28, 1988, International Potato Center, Lima, pp. 59–70.

Siki, B.F. 1979. Processing and storage of root crops in Papua New Guinea. In: Plucknett, D. (ed.), *Small-scale processing and storage of tropical root crops.* Westview Press, Boulder, CO, pp. 64–82.

Silva, J.L., Yazid, Md., Ali, Md. and Ammerman, G.R. 1989. Effect of processing method on products made from sweet potato mash. *J. Food Qual.* **11** (5): 387–96.

Sistrunk, W.A. 1948. Effect of calcium chloride and citric acid on the color and firmness of canned sweet potatoes. M.S. thesis, Oregon State University, Corvallis.

1971. Carbohydrate transformations, color and firmness of canned sweet potatoes as influenced by variety, storage, pH and treatment. *J. Food Sci.* **36**: 39–42.

Smith, D.A., McCaskey, T.A., Harris, H. and Rymal, K.S. 1982. Improved aseptically filled sweet potato purées. *J. Food Sci.* **47** (4): 1130–2.

Smith, D.A., Harris, H. and Rymal, K.S. 1983. New Auburn processing method could boost sweet potato market. *Highlights Agric. Res. Ala. Agric. Exp. Stn* **30** (2): 11.

Stefanovich, A.F. and Karel, M. 1982. Kinetics of beta-carotene degradation at temperatures typical of air drying of foods. *J. Food Proc. Preserv.* **6** (4): 227–42.

Sugimoto, K., Shiga, I. and Goto, F. 1967. [Treatment of waste water from the sweet potato starch factory. III. Effects of the addition of nutrients] Japanese. *J. Jap. Soc. Starch Sci.* **14** (4): 120–7.

1969. [Treatment of waste water from sweet potato starch factory. 9. Continuous treatment of waste water from nozzle separator by bakers' yeast] Japanese. *J. Jap. Soc. Starch Sci.* **17** (2): 10–16.

Sugimoto, K., Shiga, I., Suzuki, T. and Goto, F. 1968a. [Treatment of waste water from the sweet potato starch factory. IV. Comparison of yeast strains on sweet potato juice medium by shaking culture] Japanese. *J. Jap. Soc. Starch Sci.* **16** (1): 9–15.

1968b. [Treatment of waste water from the sweet potato starch factory. V. Continuous treatment of sweet potato juice by bakers' yeasts. *J. Jap. Soc. Starch Sci.* **16** (3): 100–6.

Suh, K.B. 1966. [The effect of lime treatment on the pulping operation of the current sweet potato starch manufacturing process]. Korean. *Res. Rep. RDA* **9** (1): 221–30.

Sun, C.T., Tseng, S.Y., Lu, J.J. and Chang, W.H. 1979. Leaf protein concentrates from some sources available in Taiwan. *J. Chin. Agric. Chem. Soc.* **17** (1–2): 78–92.

Szyperski, R.J., Hamann, D.D., and Walter, W.M. 1986. Controlled alpha amylase process for improved sweet potato purée. *J. Food Sci.* **51** (2): 360–3.

Taharazako, S. 1984. [Calculations of the drying efficiency in continuous drying of specimens of sweet potato] Japanese. *Bull. Fac. Agric. Kagoshima Univ.* **34**: 179–87.

Takakuwa, M. and Furukawa, E. 1973. [Utilization of by-products in manufacture of sweet potato starch. Part V. Culture of *Candida utilis* and *Lentinus edodes* using sweet potato juice] Japanese. *J. Jap. Soc. Starch Sci.* **20** (9): 405–10.

Takakuwa, M., Ikeda, K. and Murakami, T. 1972. [Utilization of by-products in manufacture of sweet potato starch. Part 4. Utilization of *Saccharomyces cerevisiae* cultivated in sweet potato juice as food yeast or materials of nucleic acids] Japanese. *J. Jap. Soc. Starch Sci.* **19** (3): 117–21.

Tapang, N.P. and Rosario, R.R. del. 1977. Composite flours. 1. The use of sweet potato, Irish potato and wheat flour mixtures in breadmaking. *Phil. Agric.* **61** (Aug-Sept): 124–33.

Taylor, J.M. 1982. Commercial production of sweet potatoes for flour and feeds. In: Villareal, R.L. and Griggs, T.D. (eds.), *Sweet potato*, Proceedings of the First International Symposium. AVDRC, Shanhua, T'ainan, pp. 393–404.

Thomas, G.S. 1982. Review of the prospects for food processing in Papua New Guinea. In: Bourke, R.M. and Kesavan, V. (eds.), *Proceedings of the Second Papua New Guinea Food Crops Conference*. Department of Primary Industry, Port Moresby, pp. 408–20.

Truong, V.D. 1984. Drying cassava and sweet potato for food. *PCARRD Monitor* **12** (3): 8–9.

1987. New developments in processing sweet potato for food. Paper presented at an International Sweet Potato Symposium, 20–26 May, ViSCA, Baybay, Leyte.

1990. Development of small scale technology for dehydrated sweet potato cubes for traditional food preparations. In: *Tropical root and tuber crops changing role in a modern world*, Proceedings of the Eighth Symposium of the International Society for Tropical Root Crops, 30 October–5 November 1988, Bangkok (in press).

Truong, V.D. and Fementira, G.B. 1990. Formulation, consumer acceptability and nutrient content of non-alcoholic beverages from sweet potato. In: *Tropical root and tuber crops changing role in a modern world*, Proceedings of the Eighth Symposium of the International Society for Tropical Root Crops, 30 October–5 November 1988, Bangkok (in press).

Truong, V.D. and Guarte, R.C. 1985. Root crop processing tools and equipment. In: Root Crops, Annual Report of the Philippine Root Crop Research and Training Center, ViSCA, Baybay, Leyte, pp. 76–9.

Truong, V.D., de la Rosa, L.S., Fementira, G.B. and Baugbog, L.A. 1986. Pilot production of Delicious S-P and sweet potato catsup. Terminal Report of the Philippine Root Crop Research and Training Center Project, ViSCA, Baybay, Leyte.

USDA 1964. *Recent research on the marketing of sweetpotato flakes. The marketing and transport situation*, Marketing and Economics Division, Economics Research Service – 194, USDA, pp. 28–32.

Walter, W.M., Catignani, G.L., Yow, L.L. and Porter, D.H. 1983. Protein nutritional value of sweet potato flour. *J. Agric. Food Chem.* **31** (5): 947–9.

Walter, W.M. and Giesbrecht, F.G. 1982. Effect of lye peeling conditions on phenolic destruction, starch hydrolysis and carotene loss in sweet potatoes. *J. Food Sci.* **47** (3): 810–12.

Walter, W.M. and Hoover, M.W. 1986. Preparation, evaluation and analysis of a french-fry-type product from sweet potatoes. *J. Food Sci.* **51** (4): 967–70.

Walter, W.M. and Purcell, A.E. 1974. Lipid autoxidation in precooked dehydrated sweet potato flakes stored in air. *J. Agric. Food Chem.* **22** (2):

298–302.
1978. Preparation and storage of sweet potato flakes fortified with plant protein concentrates and isolates. *J. Food Sci.* **43** (2): 407–10.
Walter, W.M., Purcell, A.E. and Hansen, A.P. 1972. Autoxidation of dehydrated sweet potato flakes: the effect of solvent extraction on flake stability. *J. Agric. Food Chem.* **20** (5): 1060–3.
Walter, W.M., Purcell, A.E., Hoover, M.W. and White, A.G. 1978a. Lipid autoxidation and amino acid changes in protein-enriched sweet potato flakes. *J. Food Sci.* **43** (4): 1242–4.
Walter, W.M., Purcell, A.E. and McCollum, G.K. 1978b. Laboratory preparation of a protein-xanthophyll concentrate from sweet potato leaves. *J. Agric. Food Chem.* **26** (5): 1222–6.
Walter, W.M. and Schadel, W.E. 1982. Effect of lye peeling conditions on sweet potato tissue. *J. Food Sci.* **47** (3): 813–17.
Wang, J. 1984. [The development and utilization of starch resources from sweet potato] Chinese. *Hunan Agric. Sci.* **5**: 44–6.
Watt, B.K. and Merrill, A.L. 1975. *Handbook of the nutritional content of food.* Dover Publications, Inc., New York.
Wiersema, S.G. 1989. International trip report (to Vietnam), 1–7 April, 1989, International Potato Center, Lima. [Mimeo]
Wiersema, S.G., Hesen, J.C. and Song, B.F. 1989. Report on a sweet potato postharvest advisory visit to the People's Republic of China, 12–27 January, 1989, International Potato Center, Lima. [Mimeo]
Williams, J.L. and Ammerman, G.R. 1968. Canning of stored sweet potatoes. *Miss. Farm Res.* **31** (11): 6–7.
Winaro, F.G. 1982. Sweet potato processing and by-product utilization in the tropics. In: Villareal, R.L. and Griggs, T.D. (eds.), *Sweet potato*, Proceedings of the First International Symposium. AVRDC, Shanhua, T'ainan, pp. 373–84.
Woodroof, J.G., Dupree, W.E. and Cecil, S.R. 1955. *Canning sweet potatoes.* Georgia Agric. Exp. Stn Bull. No. 12.
Wu, Y.V. 1988. Characterization of sweet potato stillage and recovery of stillage solubles by ultrafiltration and reverse osmosis. *J. Agric. Food Chem.* **36** (2): 252–6.
Wu, Y.V. and Bagby, M.O. 1987. Recovery of protein-rich byproducts from sweet potato stillage following alcohol distillation. *J. Agric. Food Chem.* **35** (3): 321–5.
Yaacob, C.M. and Raya, S. 1983. [A preliminary study on processing of sweet potato leather] Malay. *Pertanika* **6** (1): 17–21.
Yang, K. 1979. The effects of curing and storage on the acceptability of canned sweet potatoes. M.S. thesis, Mississippi State University.
Yeo, Y.K. and Kim, Z.U. 1978. [Studies on the standardization of the processing condition of kochuzang (red pepper paste)] Korean. *J. Kor. Agric. Chem. Soc.* **21** (1): 16–21.
Zhao, Z. and Jia, F. 1985. [*Storage and simple processing of sweet potato*] Chinese. The Publishing House of Agriculture, People's Republic of China (English translation available from the International Potato Center, Lima).

CHAPTER 7

Livestock feeding with sweet potato roots and vines

The feeding of animals with sweet potato, either in its immediate form or as a component of a manufactured feed, can be regarded as one alternative or indirect use of sweet potato as a human food. Both roots and vines may be used in either a fresh or dried form and are frequently fermented into silage. Roots basically represent a source of energy, and leaves a source of protein in animal diets. Cull roots (small, misshapen, or damaged roots) and vines are both by-products of sweet potato harvesting which are often discarded or ploughed back into the land as green manure, but which may be alternatively utilized as animal feed. Moreover, in countries where large quantities of sweet potato are processed into starch and alcohol, for example China, Japan and Taiwan, processing waste or by-products may also be used for feed. Sweet potatoes may be fed to all domestic animal species, both ruminants and non-ruminants, including poultry. It has been suggested that relatively recent human ecological changes in the central Highlands of Papua New Guinea came about largely through the expansion of domestic pig herds which could be foddered after the adoption of sweet potato as a crop (Watson, 1977).

In many tropical areas sweet potato tops and those roots unsuitable for marketing are used by sweet potato farmers as a supplementary feed for the animals which they also husband. However, sweet potato has certainly not attained its potential as an animal feed. Table 7.1 shows the negligible or low proportion of production utilized as animal feed in several countries of Asia and the Pacific region. In other countries, its use as an animal feed has declined severely in recent years as it has been unable to compete economically with cheaper imported feed components, especially corn and cassava. For example, in Japan the percentage of total production used as animal feed declined from about 30% in 1970

Table 7.1. *Percentage of sweet potato production used for animal feed and other purposes in some Asian and Pacific countries*

Country/Territory	Food	Feed	Industrial[a]
Bangladesh	100		
India	90	10	
Indonesia	90	10	
Malaysia	70	30	
Philippines	80	10	10
Sri Lanka	100		
Thailand	80	15	5
Papua New Guinea	85	15	
Pacific Islands[b]	91	9	

Notes:
From: Lin et al., 1983; applies to roots only. In most of the countries listed, farmers may use vines as fodder.
[a] All industrial processing is for starch production.
[b] Niue, Palau, Cook Islands, Fiji, Tahiti, Vanuatu, Tonga, Guam, and Ponape.

to less than 10% in the 1980s (Duell, 1985, 1989). Similarly in Taiwan, as incomes and foreign exchange earnings have risen, not only has total sweet potato production dramatically decreased, but the proportion in use as animal feed has declined from about 63% before 1979 (Lin et al., 1983) to a very low level at present. In the past, Taiwanese farmers prepared pig feed using a small hand-operated machine which chipped sweet potato roots and chopped the tops (Villareal, 1982). These products were sun-dried and stored. Recently, industrialization has led to a rural labour shortage leading to increased purchase of commercial feeds based on imported corn and cassava. The partial substitution of imported feed by locally produced crops such as sweet potato could be advantageous to some countries wishing to minimize foreign exchange spending. There have been various studies in tropical countries to determine the economics of using sweet potato roots and/or vines as feed ingredients (Backer, 1976; Huang and Olbrich, 1979; Pascual, Abamo and Binongo, 1987; Turner, Malynicz and Nad, 1976). On the whole these have not proved to be favourable where sweet potatoes would be grown specifically as an animal feed. Cost of animal production has been shown to be very sensitive to the cost of sweet potato roots. The energy density of sweet potato (fwb) is low compared to that of cereal grains and it has been shown in Hawaii, for example, that the cost per unit of metabolizable energy for dairy or beef cattle or for pigs is at least twice that of cereal grains (Huang and Olbrich, 1979). In Hawaii the cost of producing sweet potato with 30% dry matter was approximately equal, in 1977, to the price of imported grain with a dry matter of 89%. Drying sweet potato into meal for animal feeding adds to the cost, as does increasing the levels of protein supplements which must be added to feeds containing sweet potato (Pascual et al., 1987). Thus it was found to be less profitable in the Philippines to produce sweet potato for animal feed than to sell it as fresh roots (Pascual et al., 1987). A more promising approach might be an integrated system for the small farmer utilizing residues such as cull roots and vines for animal feed after high quality roots have been sold for direct human consumption (Backer, 1976; Backer et al., 1980; Pascual et al., 1987). In contrast there might be a large potential market for dried sweet potato as an imported feed ingredient in the European Community (Palomer, Bulayog and Truong, 1989). China is currently the sole supplier of dried and sliced sweet potatoes to the compound feed industries of the EEC countries but its shipments are irregular because of high domestic demand.

In China, the world's largest producer of sweet potato, a very high proportion of the total annual production is utilized for animal feed, especially for pig rearing (Figures 7.1 and 7.2). The percentage of production used for animal feeding in some provinces can be appreciated from Table 7.2. Before 1949, sweet potato was used mainly as a staple food for direct human consumption. This pattern changed, most notably in the last 20 years, as the direct use of sweet potato for human food declined and its utilization for feed and processing rapidly increased. The expanding use of sweet potato as a feed is due to greater demand for animal products brought about by rising incomes, improved yields of sweet potato, lack of highly developed interregional feed markets and a relatively favourable price of sweet potato versus other feed grains, especially maize and soya (Gitomer, 1987). Sweet potato roots and vines are conserved by ensilage, fermentation and mixing into fodder products. A large proportion of the production of leaves, stalks and roots are used for pig feed. Roots are used at the farm level and do not enter marketing channels (Gitomer, 1987). Stalks and leaves are harvested three or four times each growing season for use as pig feed in some areas. Vines may be marketed for this purpose and are priced very competitively with other feeds. It has been suggested that use of sweet potato as animal feed will increase very rapidly over the next 15 years, if the government pursues its present policy to increase per capita consumption of meat and livestock products (Tong, Z. as reported by Gitomer, 1987). Sweet potato vines are extensively used in other countries also,

Figure 7.1. A farmer harvesting sweet potato vines from her backyard for animal feed, Sichuan Province, China. Farmers who grow sweet potato usually have pigs as part of their farm system (M. Iwanaga).

especially for dairy cattle feed, for example in parts of Peru (Achata et al., 1988) and in Indonesia.

Use of roots and vines of sweet potato for animal feed (in common with other roots and tubers) has certain disadvantages which must be overcome if they are to compete with other feed sources such as cereals. As shown in Table I.3 (p. 4), the energy yield of sweet potato per unit land area per unit time is higher than that of cereals even with the low yields pertaining in many tropical areas at the present time. However, high production costs of sweet potato often render it uncompetitive with cereals. Added to which the roots are bulky, with a high water

Figure 7.2. Sweet potato vines being dried for storage as feed, Sichuan, China (M. Iwanaga).

content, and are consequently expensive to transport or convert into dry meal and difficult to store in tropical climates. Thus they are most often used at the farm level rather than as part of a formulated feed. On a fresh basis their energy content is less than that of a cereal and their protein content is very low. Hence relatively expensive protein supplements have to be added in greater quantities than to cereal-based diets. There is some evidence that the feeding value of roots, for example for pigs, is improved by cooking to destroy trypsin inhibitors and improve starch digestibility (see below and Chapter 4), but in certain circumstances this is impractical due to shortage or high price of fuel.

These disadvantages can be somewhat remedied by various means. For example, storage life of roots can be prolonged by fermenting them into silage, or by sun- or oven-drying them into chips or meal. Drying has the added advantage of reducing the bulk of feed necessary for an animal to achieve a certain energy intake. Simple processing of sweet potato into a dried form for animal feed has been described (Teixeira de Matos and Xavier de Almeida, 1989). The roots are grated manually or mechanically into shavings or chips measuring approximately 1 cm × 1 cm × 5 cm. Cutting the chips into as regular a form as possible facilitates the circulation of air in the subsequent drying stage, which leads to a more uniformly dried product. For natural drying in direct sunlight, the chips are spread out in a layer 2–3 cm thick and turned systematically.

Table 7.2. *Sweet potato utilization in selected provinces of China, with particular reference to livestock feed*

	% of production				
		Provinces			
Use	China total[a]	Sichuan[b]	Shandong[c]	Jiangsu[d]	Guandong[e]
Livestock feed	30	42	42	20–30	90
Fresh roots	15	22	10	5	10
Industrial	30		22	50	
Food processing	14	17[f]	10		
Seed and waste	11	11	6	25	
Other		8	10		

Notes:
See also Chapter 8.
[a] Estimates from Wiersema, Hesen and Song, 1989.
[b] Estimates from Wiersema et al., 1989. Storage losses are a severe problem in Sichuan Province, with some farmers experiencing as much as 60% loss with cave or underground storage.
[c] Estimates from the Crop Research Institute, Shandong Academy of Agricultural Sciences (International Potato Center/AVRDC/IFPRI, 1987). Of the 10% of Shandong Province's production classed as 'other', 3% may include the sale of fresh sweet potato for food and feed and 7% is exported to other provinces and abroad as dried chips.
[d] Gitomer, 1987. The 25% listed under waste and seed for Jiangsu Province is actually government purchases of sweet potato which go unused. The government controls exports of sweet potato from Jiangsu, so part of this total may be for export.
[e] Gitomer, 1987. Projections of future increased production for feed use in Guandong Province are difficult to make because at present sweet potato is not competitive with cassava. Food use is declining as rice production expands, but demand for sweet potato will continue as it is preferred to white potato. There is a little local processing of sweet potato into snack foods such as boiled, sun-dried chunks.
[f] Includes starch and alcohol production.

This method needs more man power than artificial drying and depends on climatic conditions pertaining at the time of harvest. Mechanical drying in a forced draught of hot air produces chips dried more homogeneously and rapidly and results are independent of climatic conditions. The chips are dried to a final moisture content of 12–13% (or until they are hard enough to break easily under light pressure).

Sweet potato vines are usually not fed to animals until they are cut just prior to harvesting of roots. By that time they may (although not invariably – see below) be relatively hard and dry and contain high levels of indigestible fibre in the form of lignin, cellulose and hemicellulose. A high level of fibre also reduces the digestibility of other nutrients in the vines. To try to overcome this disadvantage, experiments have been carried out to determine the extent of adverse effects on root yields of harvesting more tender vines some time before roots. Mushrooms have also been cultured on the hard dry vines (with the addition of a nitrogen source and calcium carbonate) which decreases levels of cellulose and lignin and increases the concentration of total protein in the vines (Ho and Ting, 1975). This served the dual purpose of producing a direct human food and an improved feedstuff.

Feasibility studies are needed in each area where it is proposed to encourage the use of sweet potatoes for animal feed to determine the price at which sweet potatoes would become an economically attractive item to include in feeds. Routes to achieving these price levels could then be determined. Breeding high yielding, disease-resistant clones with relatively high concentrations of dry matter and protein may encourage production thus lowering price and at the same time raising feed energy and protein levels per unit weight to levels closer to those of cereals, especially on a dry basis. It has been suggested (Tsou, S., personal communication) that a target of 30 tonnes/ha yield combined with at least 30% dry matter in the tropics should be the goal of plant breeders if roots are to be used competitively for animal feeding.

This chapter reviews present knowledge of the quality of sweet potato roots, vines or a mixture of roots and vines as feed with emphasis on three main groups of animals, namely pigs, cattle and poultry. Determined feed values of various forms of sweet potato in terms of energy and total (crude) protein for these groups are summarized later in Tables 7.21 and 7.22.

Agronomic aspects

The use for animal feed of sweet potato tops (after harvesting of roots for other purposes) or of both roots and tops, either directly on the farm or as marketable commodities, suggests the need for product optimization in the case of both plant parts. As a result, there have been various studies, notably of the effects of defoliation, during the growing period in addition to that undertaken at harvest, on the yield and chemical composition of roots. Early defoliation has been studied not only with respect to possible increments in total foliage harvested, but also to improvement in the nutritional value, especially digestibility, of vines, which may be low (Iura, Sakai and Marumine, 1958; Ho and Ting, 1975) at normal harvest date. Hard waste vines had higher levels of cellulose, hemicellulose and lignin than younger vines, but a lower concentration of total and true protein (Ho and Ting, 1975). As literature reviewed (Ruiz, Pezo and Martinez, 1980) shows a negative correlation between root production and that of leaves or total aerial parts, it has been assumed that defoliation could be one way of increasing root production. However, almost without exception researchers have demonstrated a reduction in yield and a decrease in starch content of roots with either frequent or infrequent defoliation (Ruiz et al., 1980). Japanese workers (Iura et al., 1958) showed that in the case of normal vine growth, collecting vines 15 days or less before root harvest had no significant effect on root yield, but the main advantage might be in better nutritional value of vines, as vine yield was not improved over that of normal harvest. In the case of heavy vine growth, defoliation 10 days or more before harvest significantly reduced both root yield and starch content. Others (Yamada et al., 1962) found cutting tops twice (in mid-growth and at harvest) produced 20–30% higher yield of tops, but 30% decrease in root yield. Table 7.3 shows the more recent results of Costa Rican scientists (Ruiz et al., 1980). Cutting off most of the vines (leaving only one or two intact) at two or three months and again at harvest significantly increased the foliage yield (and rate of growth, which is not shown) over cutting at harvest alone, but significantly reduced root yield. This could have resulted from poorer solar energy use for root development due to reduction in leaf area, or because regeneration of the aerial part took place at the expense of root constituents (Ruiz, 1982). However, it is suggested (Ruiz, 1982) that these results offer the possibility of managing the crop to increase foliage (protein) yield at the expense of roots (energy) or vice versa, depending on the objectives set.

Table 7.3 also shows the insignificant effect of plant density on either root or foliage yield, although there was a trend towards higher yield with decreasing plant density, contrasting with results of other investigators cited by the author. The yields of foliage and roots (fwb) in this experiment were about 36 tonnes/ha and 11 tonnes/ha, respectively. Such quantities could be doubled in tropical areas where two crops per year are possible.

The same authors (Ruiz et al., 1980) demonstrated a significant difference of foliage total protein content between clones and a higher protein content in the tops pruned off at 2 or 3 months after planting than at harvest (as has already been described in Chapter 2). An *in vitro* digestibility test, simulating rumen and gastric digestion, showed not only high foliage dry matter digestibility of 70–80% but also that the only variable apparently affecting digestibility was genetic. The clone with higher foliage protein content also had a significantly higher

Table 7.3. *Variations in foliage and root yields with cultivar, planting distance and foliage pruning*

Variable	Total foliage yield (kg DM/ha per cycle)[a]	Root yield (kg DM/ha)
Clone		
C-15	5979	3153
C-1	4305	3962
Planting distance		
10 cm	4779	3342
20 cm	5090	4465
30 cm	5254	2978
40 cm	5222	2850
Defoliation (pruning)		
Once (at harvest)	4136[A][b]	5634[B]
Twice (at 2 months + harvest)	5212[B]	3706[A]
Twice (at 3 months + harvest)	4845[B]	3409[A]
Intensity of pruning		
Leaving 1 vine	5035	3416
Leaving 2 vines	5250	3699

Notes:
DM, dry matter.
From: Ruiz et al., 1980.
[a] 165 days for C-15; 131 days for C-1; growth rates not significantly different.
[b] [A,B] figures with different capital letter superscripts significantly different ($p < 0.05$).

digestibility, suggesting that clones can be selected for animal feeding from nutritional as well as agronomic data. Interestingly, these results showed no differences in percentages of cell wall constituents (fibre) or *in vitro* digestibilities between tops pruned early or those cut at normal harvest, in contrast to those of other investigators mentioned previously.

Pigs

General background

It is possible to graze pigs on sweet potato fields, although there is a great deal of waste with this method (Waugh, 1963). Alternatively pigs may be used to eat 'leftovers' after the sweet potato has been lifted as a cash crop.

In Papua New Guinea pigs are often grazed on old sweet potato gardens, thus serving the dual purpose of 'cleaning up' and fertilizing the gardens. Occasionally, when supplies are plentiful, they may even be allowed to forage in unharvested gardens (Kimber, 1972). It is also common practice to feed pigs sweet potatoes, in the form of discarded roots or peelings from the houshold, by hand. Better quality roots may be given when in abundance. Pigs which are fed to appetite only on fresh sweet potatoes, however, gain very little weight – only 109 g and 136 g/pig per day when grazed or pen-fed, respectively (Waugh, 1963). This is because roots are bulky and intakes by young pigs are insufficient to satisfy energy and protein requirements. In Papua New Guinea, where Highland village pigs gain as little as 52–80 g/day, an attempt to improve their performance entailed tethering pigs so that they could forage either on harvested sweet potato mounds or on roots and vines gathered and left within reach (Rose, 1981). They were each supplemented in addition with 20 g/day of a protein concentrate (52% total protein). Their weight gains were about 200 g/pig per day compared to 140 g for another group foraging on grassland with no access to sweet potato. The suggested system of feeding with sweet potato could be applied in parts of Highland Papua New Guinea.

The assessment of sweet potato quality for pig feeding has normally, however, been carried out under controlled indoor conditions using diets formulated to provide apparently adequate quantities of all nutrients. Various assessments have been used: the availability of nutrients, pig growth performance when sweet potato is used to replace corn in the diet, the effects of dietary inclusion of sweet potato on pig carcass quality and the necessity of processing treatments to effect improvements in feed efficiency. An FAO reference publication (Gohl, 1981) states that fresh sweet potato can replace 30–50% of the grain in pig diets.

Digestibility of nutrients

Chemical analysis represents the first step in determining the nutritive value of a feed. The chemical composition of sweet potato roots and leaves has been detailed in Chapter 2. Major nutrients in fresh and dehydrated roots and vines are given in Table 7.4, from which they can also be compared with those in ground corn, a prominent constituent of many animal feeds. The most obvious difference between corn and sweet potato roots is the lower content of protein in the latter even on a dry weight basis. Dried vines are moderately high in total protein, but also high in indigestible fibre. The much lower dry matter content of fresh sweet potato roots than of corn means that a diet containing the former

Table 7.4. *Composition of sweet potato roots and vines compared with corn in terms of major nutrients*

Component	Fresh roots (%)	Sun-dried root chips (%)	Fresh vines (%)	Dried vines (%)	Corn meal[a] (%)
Dry matter	17.8–40.7	83.3–85.0	9.1–19.5	80.2–88.2	86.6
Total (crude) protein	1.1–1.9	2.5–3.5	2.5–3.8	11.7–18.6	9.0
Lipid	0.3–0.7	2.1	0.2–1.4	2.8–5.6	3.9
Crude fibre	0.4–1.0	2.9–3.0	1.7–4.3	10.3–17.0	2.0
Ash	0.4–1.0	3.2–3.3	1.4–2.8	7.4–9.9	1.1
N-free extract	14.7–36.1	72.5–73.1	1.7–7.6	38.5–49.3	70.7
Ca	0.02–0.04[b]	0.05–0.10[b]	0.07–0.2[b]	1.0–1.78	0.04
P	0.06–0.07	0.02–0.15	0.05–0.07	0.03–0.27	0.27

Notes:
From Taiwanese feed composition tables as given by Yeh and Bouwkamp (1985).
[a] Han et al., 1976a.
[b] See Chapters 2 and 3.

as the main energy source is necessarily much bulkier than a corn-based diet.

The actual value of nutrients, however, depends on their digestibility. One means of determining the quality of sweet potato as a feed has been by ascertaining the digestibility of major feed factors such as energy and protein and by comparing the digestibility of energy with that of corn. Table 7.5 shows the apparent digestibilities (allowing for losses in faeces only) in pigs of various weights, of sweet potato root dry matter, energy, total (crude) protein and fibre. The apparent digestibilities of total dry matter and energy of sweet potato are high and similar to that of corn. However, not only is the total content of protein low in sweet potato, but the digestibility of protein also appears to be rather low in raw roots and compares unfavourably with that of corn. The presence of a trypsin inhibitor (see p. 431) may be partly responsible, but uncooked starch may also interfere with protein digestibility. It has been noted that most root/tuber starches are inferior to cereal starches in their effect on protein digestibility (Dreher, Dreher and Berry, 1984). Rat feeding experiments using corn or sweet potato starch and casein, as energy and protein sources respectively, showed that protein digestibility in an uncooked sweet potato starch-based diet was lower than that in a corn starch-based diet (Tsou and Hong, 1989). A significant improvement in protein

Table 7.5. *Apparent digestibilities for pigs of dry matter (DM), energy, total protein and crude fibre in sweet potato roots and corn (%)*

Source	DM	Energy	Total protein	Crude fibre
Fresh sweet potato[a]	95.1	93.8	42.3	71.5
Fresh sweet potato[b]	95.6	94.8	57.2	73.6
Dried sweet potato meal[c]	84.7	86.3	77.1	
Dried sweet potato meal[d]	77.66		32.98	45.18
Corn[d]	88.92		88.42	37.18
Dried sweet potato meal[e]	82.4		63.3	67.9
Dried sweet potato meal[f]	75.5		50.1	54.9
Dried sweet potato meal[g]		92.4	46.8	
Corn[g]		89.2	73.3	
Basal diet + 30% sweet potato[g]		82.5	65.9	
Basal diet + 30% corn[g]		81.6	71.3	

Notes:

[a] Rose and White, 1980: 29 kg pigs, Papua New Guinea.
[b] Rose and White, 1980: 90 kg pigs, Papua New Guinea.
[c] Wu, 1980: 6.5 kg pigs, Taiwan, sweet potato meal approx. 40% of ration.
[d] Han et al., 1976a: growing-finishing pigs, Korea.
[e] Magay, 1972: 30–50 kg pigs, Philippines, sweet potato meal 56% of ration.
[f] Magay, 1972: 50–90 kg pigs, Philippines, sweet potato meal 60% of ration.
[g] Lee and Yang, 1981: Taiwan.

digestibility could be attained by pre-cooking the sweet potato starch. Furthermore, the effect of starch on protein digestibility varied genotypically. This result implies that the feed quality of sweet potato starch could be improved through breeding.

The digestibility of raw sweet potato starch itself may vary between sweet potato genotypes and, for the same genotype, between harvesting seasons (Tsou and Hong, 1989). A rapid laboratory method of screening lines for the extent of starch digestibility has been developed (Tsou and Hong, 1989). This is based on carbon dioxide gas production from fermentation of sugars obtained by pancreatic digestion of starch. Sweet potatoes harvested during the wet season were found to be more resistant to this *in vitro* digestion than those harvested during the dry season. Preliminary findings on starch characteristics, such as that higher amylose content, lower phosphorus content and higher gelatinization temperature may be associated with greater susceptibility to *in vitro* digestion (Tsou and Hong, 1989), deserve further investigation. Diges-

Table 7.6. *Energy values for sweet potato chips (MJ/kg DM)*

Gross energy	16.84
Digestible energy[a]	14.64
Metabolizable energy	14.14
Net energy[b]	8.45

Notes:
DM, dry matter.
Wu, 1980: values for weanling pigs; calculated from the values expressed originally as kcal/g DM.
[a] Digestible energy of sweet potato about 82% that of corn.
[b] Net energy of sweet potato about 79% that of corn.

tibility of sweet potato fibre, in contrast to starch and protein, appears to be somewhat superior to that of corn (Table 7.5).

Sweet potato roots are generally employed in pig diets chiefly as an energy source. The availability of energy derived from dried sweet potato chips has been examined in somewhat greater detail for weanling pigs (Wu, 1980). The results are shown in Table 7.6. Definitions of the various forms of energy derived from the chips are given at the end of this chapter. The net energy of sweet potato was found to be about 79% of that of corn (dwb). A Japanese experiment found the digestible energy of sweet potato roots for growing piglets to be comparable to that of corn (Furuya and Nagano, 1986). Hence sweet potato is a reasonably good source of energy for pigs.

Examination of the differences in apparent digestibility of energy between sweet potato cultivars by an *in vitro* method and by direct measurement in pigs showed that the percentage of digestible energy tended to be higher in cultivars used for starch production than in those used for the table, although the differences were not significant in the pigs (Table 7.7) (Furuya and Nagano, 1986). There was also some evidence of apparent protein digestibility differences between cultivars. Type of processing also influenced the percentage of digestible energy (Furuya and Nagano, 1986), boiled and oven-dried roots being significantly better than freeze-dried roots (Table 7.8).

Fresh sweet potato vines had a lower coefficient of apparent digestibility of organic matter than sweet potato roots, cassava roots or green papaya fruits (all of which may be fed to pigs in the Philippines) (Zarate, 1956). This was due to their higher level of crude fibre. There was some

Table 7.7. *Variation of energy values and digestible total (crude) protein in pigs with sweet potato cultivar*

Cultivar	Gross energy (MJ/kg DM)	Digestibility of energy (%)	Digestibility of total (crude) protein (%)
Kyukei 20[a]	16.86	91.0	−2.2[AB][b]
Kyukei 40[a]	17.03	90.9	−18.9[B]
Kyushu 82[c]	17.36	93.2	43.2[A]
Kogane-Sengan[c]	17.28	92.5	35.4[AB]

Notes:
DM, dry matter.
From: Furuya and Nagano, 1986; energy values converted to MJ/kg from kcal/g.
[a] Food cultivars.
[b] [A,B] Means in same vertical column with different superscripts differ significantly ($p < 0.05$).
[c] Starch cultivars.

indication of variation between breeds of pigs in their ability to digest the fibre or nitrogen-free extract of sweet potato vines.

In Korea, an experimental sweet potato meal made from a mixture of roots and tops (in a ratio of 3:1), pressed and sun-dried (Han et al., 1976b; Kim et al., 1976) had significantly lower digestible energy, and apparent digestibilities of protein and fat than corn. Substitution of corn by the meal in pig diets progressively lowered the availability of various dietary nutrients as the percentage of meal in the diets was increased from 0% to 60%. Presumably the high fibre content of the sweet potato vine in the meal had influenced digestibility of other nutrients.

Determining feed performance

When determining the suitability of a feedstuff it is incorporated into a test diet which is kept iso-nitrogenous and iso-energetic with the diets to which it is to be compared. All diets are formulated to contain adequate amounts of other essential nutrients such as vitamins and minerals. An example of diets formulated to compare corn and sweet potato is given in Table 7.9. Various pig performance factors are measured: the average daily weight gain, the total weight gained by the pig over the course of the experiment (these should be similar to the expected weight gains on a properly balanced diet), the number of days taken to reach the required final weight (if this is excessive the cost of the pig may increase), and the

Table 7.8. *Variation of* in vitro *digestible energy with variations in method of sweet potato processing*

Process	Gross energy (MJ/kg DM)	Digestible energy (MJ/kg DM)	Digestibility of energy (%)
Freeze-dried[a]	17.78	13.26[A]	74.6[A]
Sun-dried	17.45	13.68[A]	78.3[AB]
Oven-dried (105°C 16 h)	17.74	14.56[B]	82.2[B]
Boiled[b]	18.37	15.27[B]	83.0[B]

Notes:
DM, dry matter.
From: Furuya and Nagano, 1986: energy values converted to MJ/kg from kcal/g. Cultivar 'Kogane-Sengan'.
[a] Means as in Table 7.7.
[b] Digestible energy 15.27 MJ/kg DM comparable to that of corn at 16.19 MJ/kg.

feed efficiency (weight of feed administered per unit gain in weight, kg/kg; the lower this measurement, the greater the feed efficiency and the lower the cost).

Replacement or partial substitution of corn with sweet potato roots and vines

There has been a considerable research effort in various countries to determine the extent to which corn can be replaced by fresh or dried sweet potato roots as the main carbohydrate (energy) source in pig rations. In some cases both dried roots and vines have been incorporated into diets. One of the major drawbacks to the use of sweet potato roots as a pig feed is their generally low protein content even on a dry weight basis. This means that relatively expensive protein sources such as soymeal and/or fishmeal have to be added to raise the level and quality of dietary protein to that required for satisfactory growth at any stage of pig development. Although these sources have to be added to corn-based diets also, they are needed in smaller quantities because the protein content of corn is about two to four times higher than that of even dried sweet potato roots. Experiments have been carried out to incorporate relatively high protein sweet potato meal into diets as a substitute for corn with varying success. Although such a meal would need lower additions of protein sources to adjust dietary protein to the desired level, it appears that it might also require adequate heat treatment to give a

Table 7.9. *Example of composition of experimental diets for a pig feeding trial with corn or dried sweet potato*

Feed ingredient	100% corn		50:50 corn:sweet potato		100% sweet potato	
	Grower	Finisher	Grower	Finisher	Grower	Finisher
Yellow corn (%)	68.8	75.0	30.4	33.0	—	—
Dried sweet potato chips (%)	—	—	31.6	34.0	57.6	62.2
Soybean meal (%)	22.3	16.6	26.2	20.9	29.2	24.2
Wheat bran (%)	5.0	5.0	5.0	5.0	5.0	5.0
Tallow (%)	—	—	2.1	2.9	3.5	4.4
Dicalcium phosphate (%)	1.2	1.2	1.2	1.2	1.2	1.2
Limestone (%)	0.9	0.9	0.9	0.9	0.9	0.9
Salt (%)	0.5	0.5	0.5	0.5	0.5	0.5
Vitamin and mineral premix (%)	1.3	0.8	1.3	0.8	1.3	0.8
L-Lysine powder[a] (%)	—	—	0.8	0.8	0.8	0.8
Total	100	100	100	100	100	100
Total protein (%)	16.50	14.51	16.51	14.50	16.53	14.53
Ca (%)	0.72	0.72	0.78	0.77	0.82	0.81
P (%)	0.58	0.57	0.58	0.56	0.58	0.55
DE MJ/kg	13.64	13.93	13.64	13.98	13.68	13.98

Notes:
DE, digestible energy.
From: Chen et al., 1980; DE converted from Mcal/kg.
[a] Provided additional 0.2% lysine to the diet.

satisfactory performance (see below). Normal sweet potato meal or dried chips are usually added to the diet without any form of heat treatment.

Taiwanese workers (Koh, Yeh and Yen, 1976) showed that sweet potato either in the form of fresh roots or dried chips (and with some addition of dehydrated vines) could provide the major carbohydrate source in the diets of growing (20 kg to 50–60 kg) and finishing (60 kg to market) pigs if the average daily protein and gross energy intakes are maintained at 334 g and 35.5 MJ (8.5 Mcal), respectively. Others (Lee and Yang, 1974) also showed no significant difference in pig growth performance or feed efficiency between corn–soymeal and sweet potato chips–soymeal diets when these were iso-nitrogenous. However, the percentage of soymeal in the sweet potato diet had to be increased to meet this requirement.

It has been concluded (Yeh, 1982) from a survey of experiments carried out in Taiwan that in general the performance of growing-finishing pigs fed on diets containing dried sweet potato roots with or without dried sweet potato vines was not comparable to those containing corn. This is illustrated by the results given in the bottom section of Table 7.10 (Lee and Lee, 1979) and in Table 7.11 (Lee and Yang, 1979a, 1980). A lower weight gain and feed efficiency is particularly noticeable for the diet containing dried sweet potato vines. However, Table 7.10 indicates that others have found the complete substitution of corn by dried sweet potato roots (in the form of meal or chips) to have little effect on growth or feed efficiency of growing-finishing pigs provided the concentration and quality of protein in the diet is adjusted to equal that in the corn diet. Some (Tai and Lei, 1970; Han, Yoon and Kim, 1976a) even found that substitution of part of the corn in the diet with dried sweet potato roots (so that sweet potato contributed about 20% by weight of the diet) slightly improved weight gains and feed efficiency over those with corn alone. An experiment in the Philippines (Saure, 1972) showed that overall performance of pigs from starting (5 kg live weight) to slaughter (85 kg live weight) fed on rations (with appropriately adjusted protein contents for the age groups) in which all the corn (50% of the diet) was replaced by dried sweet potato meal did not differ significantly from the corn diet. The results indicated that sweet potato meal can completely replace corn in pig rations, being used at up to 50% of starter, 52–56% of growing, and 60% of finishing rations. However, it has also been shown by other workers (Chen et al., 1980) that whereas partial or complete substitution of corn by dried sweet potato (with added lysine – see Table 7.9) could provide a satisfactory overall performance (not significantly different from corn alone) in growing (25–60 kg) pigs, finishing (60–90 kg) pigs on sweet potato (partially or wholely substituting for corn) gained significantly less weight and showed lower feed

Table 7.10. *The effect of replacing corn by dried sweet potato on pig performance*

Corn in diet (%)	Sweet potato in diet (%)	Corn:SP ratio	Average total wt gain (kg)	Days of feeding	Feed/gain (kg/kg)
Taiwan[a]					
63–81	0	100:0	64.14	99	3.38
45–58	15–20	75:25	65.14	99	3.37
29–37	29–37	50:50	64.02	103	3.54
14–18	42–54	25:75	64.67	111	3.74
0	54–68	0:100	64.51	115	3.81
Korea[b]					
50	0	100:0	49.5		4.38
40	10	80:20	60.3		4.47
30	20	60:40	68.5		4.25
10	40	20:80	62.8		4.72
Philippines[c]					
50–60	0	100:0	64.7	106	2.98
25–30	25–30	50:50	63.8	112	3.00
0	50–60	0:100	64.1	105	2.95
Taiwan[d]					
72–84	0	100:0	66.60	120	3.04
35–41	35–41	50:50	56.92	120	3.58
0	69–81	0:100	52.91	120	3.86
0	52–64 + 20% dried vines		44.25	120	4.23

Notes:
SP, sweet potato.
[a] Tai and Lei, 1970.
[b] Han et al., 1976a.
[c] Saure, 1972.
[d] Lee and Lee, 1979.

efficiency than those on corn alone. Japanese workers (Kurihara et al., 1957), many years ago, concluded that fresh raw sweet potato roots could be used at 35–50% of the diet for growing pigs, at 45–60% for adults, at 45–60% for a sow with babies and at 60–70% for fattening. They also obtained satisfactory results with no diarrhoea or palatability problems when incorporating sun-dried root slices at 48% by weight of the diet mixed with 12% barley bran. An FAO reference source (Gohl,

Table 7.11. *Comparison of high protein sweet potato meal (HPSPM) with/without dried sweet potato vines and normal sweet potato meal (NSPM) with corn for pig feeding*

Ration	Treatment					
	80°C steam-heated and pelleted[a]		Heated and pelleted[b]		90°C steam-heated and pelleted[c]	
	Average daily gain (kg)	Feed efficiency (kg/kg)	Average daily gain (kg)	Feed efficiency (kg/kg)	Average daily gain (kg)	Feed efficiency (kg/kg)
100% corn-based	0.54	3.11	0.54	3.29	0.65	3.35
50% HPSPM–50% corn	0.45	3.85	0.38	3.64	0.49	4.03
100% HPSPM	0.46	3.67	0.35	3.95	0.44	4.37
100% HPSPM + vine	0.38	4.16	—	—	—	—
50% NSPM–50% corn	—	—	0.48	3.52	0.50	3.96
100% NSPM	—	—	0.50	3.23	0.47	4.26

Notes:

[a] Lee and Lee, 1979.

[b] Lee and Yang, 1979a.

[c] Lee and Yang, 1980.

1981) states that dehydrated sweet potato roots have 90% of the feed value of maize when used at up to 60% of the pig ration.

In Papua New Guinea, a different system of feeding which consisted of giving pigs a fixed amount of a protein concentrate and allowing them to eat fresh raw roots to appetite did not produce daily weight gains or feed efficiencies equal to those on a sorghum-protein concentrate diet, but was economically similar, allowing a reasonable return on the ration (Malynicz, 1971).

Variations in responses of pigs to sweet potato shown by different groups of researchers could be due to several factors including the use of different breeds of pig, variations in composition or production of dried sweet potato, formulation of experimental diets etc.

The extensive use of fresh and dehydrated vines for pig feeding in China and elsewhere has already been mentioned above. To sweet potato farmers who also keep animals they are an asset which might otherwise be discarded or ploughed back into the land. When marketed they are often one of the cheapest feeds available.

Japanese workers (Kurihara and Imamura, 1959) introduced sweet potato vines into pig feed at a maximum rate of 30% and found that the only disadvantage was a 10 day delay (compared to a control diet) in reaching the desired slaughter weight of 100 kg. In addition the saving in use of concentrate feed reduced the cost by 10%. In Taiwan, growth and feed conversion efficiency of pigs fed diets containing high protein sweet potato (HPSP) meal with dehydrated sweet potato vines added were even lower than with HPSP alone (see Table 7.11). This was partly due to the high fibre and lower digestible energy content of vines (Lee and Yang, 1979a). However, the trypsin inhibitor content was also high (see below). Korean researchers have evaluated the nutritive value of a whole sweet potato meal made by mixing roots and vines in a ratio of 3:1, pressing and sun-drying (Han et al., 1976b; Kim et al., 1976). This meal when included in diets at 0, 10%, 20%, 40% and 60% by weight to substitute for corn, generally depressed weight gains and reduced feed consumption and feed efficiency in comparison to corn alone (although the decrease was only significant with the 60% sweet potato diet). The whole sweet potato meal was more economical than corn. Pigs in Papua New Guinea allowed free access to sweet potato vines after receiving a restricted amount of a concentrate had lower weight gains and feed conversion efficiencies than a control group on restricted concentrate alone, but the differences were not significant (Malynicz and Nad, 1973). In other words there was no benefit to pig performance of giving sweet potato vines.

High protein sweet potato chips

The possibility of improving the value of sweet potato for pig feeding by breeding to increase its total protein content has already been mentioned. The results of using a relatively high total protein (7.3%) sweet potato (HPSP) meal prepared from a high protein clone as compared to normal sweet potato (NSP) meal to replace corn (at 50% or 100% substitution) are shown in Table 7.11. The NSP meal was equal, or even slightly superior, to the HPSP meal in terms of weight gains and feed efficiency. The advantage of using HPSP meal would be the need for less soymeal or other protein source in the diet compared to that used to supplement NSP meal.

A Korean experiment (Han et al., 1976a) also used a relatively HPSP meal. The total protein at 7.6% (dwb) was similar to that of the corn used at 9.0%. The lysine content of the sweet potato meal was superior, and the methionine content inferior, to that of the corn. The meal was used to replace corn and incorporated at 0, 10%, 20% and 40% by weight of the diet. Growth rates were more rapid and weight gains greater with sweet potato than with corn alone but there was no significant difference in pig performance between treatments. The corn and sweet potato proteins may have complemented each other to improve the biological value of the mixed protein.

Addition of protein source to sweet potato

There is some evidence that the type of protein source added to a sweet potato-based diet to raise the total protein content can affect pig growth and feed efficiency. Japanese workers (Hosoyamada and Kodama, 1959) compared fish and soybean meals as the additional sources of protein and found that pigs fed for 150 days (from 16 kg to about 80 kg live weight) gained least weight and consumed most feed per gain with fish meal. However, there was no difference in performance between the soymeal and a mixture of the soy and fishmeals.

Pigs in Papua New Guinea fed a fixed amount of a protein concentrate or raw peanut kernels, and sweet potato to appetite performed significantly worse on the peanut kernels; average daily weight gain and feed per gain were 0.33 kg and 4.92 kg/kg with protein concentrate and 0.09 kg and 8.53 kg/kg with raw peanut kernels (Malynicz, 1971). One of the pigs on the peanut–sweet potato ration died during the experiment and others after its completion.

When soymeal was replaced by petroleum yeast as the protein source in iso-nitrogenous diets based on sweet potato, pig weight gains and feed efficiencies were somewhat lowered, but carcass quality was unaffected

(Lee and Yang, 1974). It appears that the protein source to be added to sweet potato-based pig diets should be chosen with care, to obtain the best results while keeping costs as low as possible.

The effects of sweet potato on pig carcass quality

There appears to be little effect on pig carcass quality in terms of carcass weight and length and dressing percentage (the percentage of total carcass remaining after removal of head, feet, tail, heart, innards etc.) when sweet potato is included as the main carbohydrate source to replace part or all of the corn in a diet (Kim et al., 1976; Lee and Yang, 1979a; Saure, 1972). However, there have been suggestions that inclusion of sweet potato in diets can influence the thickness and characteristics of back fat. The back fat thickness of pigs fed dried chips was found to be less than that of those fed raw roots (Koh et al., 1976). Some research shows a trend towards decrease in back fat thickness with an increase in the percentage of dried sweet potato in the diet (Tai and Lei, 1970; Lee and Yang, 1980), but others have noted no significant difference between corn- and sweet potato-based diets in this respect (Saure, 1972; Takahashi and Furuya, 1968). A decrease in back fat thickness would be advantageous in producing a leaner carcass, preferred by consumers in many countries at the present time. The melting point and the iodine number of pig back fat were higher and lower respectively on sweet potato-based than on corn- and barley-based diets, suggesting a harder and more saturated fat with sweet potato (Takahashi and Furuya, 1968). Back fat firmness scores of pigs fed on diets in which 50% or 100% of the corn was replaced by sweet potato meal were higher than on corn alone, probably due to a higher content of saturated fatty acids, as shown by the lower iodine numbers (Saure, 1972). A significant increase in pig fat melting point accompanied by a significant decrease in the unsaturated fatty acid content of the fat took place in only two out of six breeds of pig fed on sweet potato silage (Kawaita et al., 1979). In some countries, for example Japan, such hard fat is a desirable attribute in the eyes of consumers. Experiments are being conducted in which pigs are transferred from corn to sweet potatoes for the last 1–2 months of their life in order to change the body fat characteristics (Furuya, S., personal communication). Pigs are allowed to eat in a sweet potato field during the day and are then transferred to the sty for feeding with a protein–vitamin–mineral supplement in the evening. Harvesting of sweet potato is unnecessary and pigs fertilize the field during feeding. The production of harder, more saturated fat might not, however, be regarded favourably in countries sensitive to the effects of diet on the incidence of heart disease.

Influences on the nutritive value of sweet potato in pig feed

Rural farmers who feed sweet potato roots to their pigs may boil or steam the roots before feeding as they have long known that this can achieve an improved growth performance. Selected farmers in the Philippines who fed their pigs on cooked sweet potato at a level as high as 50–60% of the ration found that growth performance and feed conversion were comparable to those of pigs fed on a commercial ration (Montanez, 1982). However, results do not invariably show an advantage of boiled over raw sweet potatoes as a feed. Some Japanese workers showed that raw, chopped sweet potato roots were as palatable as cooked roots to young pigs and that there was no significant differences in weight gain or feed efficiency between groups fed raw or cooked roots (Kurihara and Imamura, 1958). Others maintain that cooking of roots is particularly desirable for young pigs to increase the digestibility and palatability of sweet potato-based diets (Hosoyamada and Kodama, 1959; Furuya, S., personal communication). The resulting improvement in performance is no doubt due to a combination of increased starch digestibility, a decrease in the adverse effect of starch on protein digestibility and elimination of trypsin inhibitor activity (TIA). The last two factors may adversely affect the digestibility of dietary proteins other than sweet potato protein.

The presence and properties of a trypsin inhibitor in raw sweet potato roots has been described in Chapter 4. It should be recalled that there is considerable genetic variation in TIA. Hence there may be variation in the effects on pig performance manifested by the inclusion of different raw sweet potato clones in pig diets. The presence of TIA in roots causes an increase in TIA in diets with increasing levels of sweet potato and adversely affects pig performance as shown in Table 7.12. The TIA of diets with and without sweet potato roots and dried vines is given in Table 7.13. This appears to show that TIA greatly increases with the presence of sweet potato vines. Table 7.13 also illustrates the ineffectiveness of steam-heat treatment of dried sweet potato at 80°C, which did not reduce TIA, nor did it increase the feeding value of diets (Lee and Lee, 1979). The efficacy of boiling or steaming at 100°C in eliminating TIA has been described in Chapter 4. Dry heating of meal is not effective, however, and may destroy sensitive amino acids such as lysine, thus lowering the quality of sweet potato protein (see Table 7.14). Roasting sweet potato meal at 100°C for 30 min reduced its feeding value for pigs, so that whereas it was possible to replace corn completely with non-roasted meal, it was only possible to replace 50% of the corn with roasted meal (Saure, 1972). Most feeding experiments have been carried out with raw or dehydrated roots which have not been heat-treated. The

Table 7.12. *The effect of trypsin inhibitor on pig performance*

Variable	100% corn–soymeal	50% corn + 50% SP–soymeal	100% SP–soymeal	100% SP + vine–soymeal
% Inhibition of trypsin by feed	23	44	52	49
Average daily feed intake (kg)	1.85	1.83	1.78	1.74
Average daily gain (kg)	0.603	0.482	0.437	0.355
Feed per gain (kg/kg)	3.08	3.84	4.09	4.99
Days required to reach 90 kg	118	147	163	201

Notes:
SP, sweet potato.
From: Yeh, T.P., Wu, M.C. and Chen, S.Y., unpublished data, Animal Industry Research Institute, Taiwan Sugar Corporation, as given by Yeh and Bouwkamp (1985).

Table 7.13. *Trypsin inhibitor activity of pig diets with and without sweet potato roots and vines*

Diet	Trypsin inhibitor (units/g DM)	Inhibition of trypsin (%)
1. Corn–soy control		
Normal temperature pelleted	5.9	13.4
80°C steam-heated and pelleted	5.2	11.6
2. 50% corn–50% HPSPM[a]		
Normal temperature pelleted	13.6	18.4
80°C steam-heated and pelleted	8.7	16.8
3. 100% HPSPM		
Normal temperature pelleted	12.5	17.0
80°C steam-heated and pelleted	9.2	12.2
4. HPSPM + dried SP vine		
Normal temperature pelleted	27.0	79.9
80°C steam-heated and pelleted	72.2	82.4

Notes:
DM, dry matter; SP, sweet potato.
From: Lee and Lee, 1979.
[a] High protein sweet potato meal (7.3% crude protein).

Table 7.14. *Trypsin inhibitor activity (TIA), starch availability and lysine in sweet potato chips treated by dry heat*

	Sweet potato chips		
	Sun-dried	Heated dry at 390°C, 1 min	Heated dry at 430°C, 1 min
TIA units/g DM	1320	1100	630
Starch availability ml CO_2/g DM per h	10.58	11.35	15.15
% Improvement in starch availability	—	7.3	43.2
Total lysine (%)	0.20	0.17	0.17
Available lysine (%)	0.19	0.15	0.15
Available/total lysine (%)	95.0	88.2	88.2

Notes:
DM, dry matter.
From: Yeh, T.P., Chen, S.Y. and Chang, Y., unpublished data, Animal Industry Research Institute, Taiwan Sugar Corporation, as reported by Yeh and Bouwkamp (1985).

desirability or necessity of heat-treating roots is dubious in those areas where farmers suffer from a scarcity of fuel wood and purchase of fuel is expensive. It is also undesirable in feed mills as it may make sweet potato even more uncompetitive with other feed ingredients. It would seem preferable to select clones lacking or low in TIA for feeding purposes.

The performance of pigs fed raw sweet potatoes may also be affected by the ingestion of raw sweet potato starch. Attempts to increase starch availability and enhance nutritive value of dried sweet potato have been made in Taiwan (Yeh et al., 1977, 1978). Microwave cooking of dried chips at 80–90°C for up to 4.5 min effected no change in starch availability or nutrient digestibilities (Yeh et al., 1977). Heating of meal by dry air for 1 min at 390°C increased starch availability by only 7% and did not improve pig performance compared to that on diets containing unheated chips (Yeh and Bouwkamp, 1985). The most effective method of improving the nutritive value of dried chips which combines elimination of TIA with a large improvement in starch availability was by 'popping' the dried chips in the same way in which popped rice is made (Yeh et al., 1978). Popping chips with 9–10% moisture at 6–8 kg/cm^2 pressure at 164–175°C completely eliminated TIA and improved starch availability by up to 194%. Pigs fed on diets in which popped chips were substituted for corn performed significantly better than on

Table 7.15. *Improvement of nutritive value of dried sweet potato chips by 'popping'*

Performance factors	Diet			
	Corn	Untreated chips	Chips 'popped' at 6 kg/cm^2, 164°C	Chips 'popped' at 8 kg/cm^2, 175°C
Average daily weight gain (kg)	0.276^{A}[a]	0.254^{B}	0.268^{A}	0.267^{A}
Average daily feed intake (kg)	0.437	0.414	0.427	0.426
Feed per gain (kg/kg)	1.582^{A}	1.631^{B}	1.594^{A}	1.596^{A}

Notes:
From: Yeh et al., 1978.
[a] A,B figures in horizontal rows with different superscripts significantly different ($p < 0.05$).

untreated chips and equally as well as those on a corn diet (Table 7.15). Popping therefore appears to be an effective method of raising the nutritive value of dried sweet potato, but it has never been commercialized as it was not found economically feasible in Taiwan. The possibility of breeding sweet potato lines with highly digestible starch which per se would be more available and also would have little effect on protein digestibility has already been mentioned above. An attempt to commercialize pellet forms of roots and tops in Taiwan also failed because production costs proved prohibitive (Yeh and Bouwkamp, 1985).

Sweet potato silage in pig rations

The difficulties of storing fresh roots in tropical situations has already been mentioned above (and see Chapter 5). Sweet potato foliage also tends to be available in large quantities only at limited times of the year after harvest, which often does not coincide with the period of pasture scarcity. One means of preserving both roots and vines (or a mixture of the two) so that sweet potato can be used as a feed all year round is by ensiling. This is of particular benefit in the humid tropics, where drying of roots to chips or foliage to hay in the sun is impractical. Succulent preserved feeds in the form of silage also do not require so much supplemental water as dried feeds when given to animals. Silos may be constructed of simple materials and sealed with plastic bags which are both available to, and affordable by, rural populations. The construction of silos in the Philippines for preparation of sweet potato root silage has been described (Castillo et al., 1964). It was suggested that four concrete drainage pipes, of the type used by construction companies, should be cemented together, a galvanized roof placed on top, and a small pipe placed at the bottom for drainage of silage juices. Efficient production of silage requires anaerobic conditions to enable bacteria to ferment carbohydrates to organic acids, particularly lactic acid, which act as preservatives. This will be discussed further in relation to feeding cattle with sweet potato silage.

Variable results have been reported when sweet potato root or vine silage was fed to pigs as part of a mixed feed, but on the whole silage appears to be as satisfactory a feed source as untreated sweet potato. Philippine workers (Castillo et al., 1964) reported that corn can be entirely replaced by silage made from chopped sweet potato roots. Feeding 60% by weight of a basal diet plus 100% by weight of sweet potato silage (which replaced all the corn in the basal diet) produced the same weight gains and feed conversion efficiency in pigs as 100% of the basal diet alone. In Japan, pig feeding with silage prepared using 80% sweet potato roots and 20% barley bran produced a feeding value similar to that of a dried sweet potato and barley bran mix or dried sweet potato alone (Kurihara and Imamura, 1956). Others (Hosoyamada, Nakashima and Suzuki, 1962) found poorer initial rates of growth with sweet potato plus bran silage than with a standard feed, but after 220 days pigs reached expected weights and produced good-quality pork. In Korea, the use of sweet potato silage prepared from 60% roots, 30% vines and 10% barley bran was also found to be inadvisable for feeding younger pigs below 50 kg weight, but could be included in the diet of 50–70 kg pigs at 38% and in that of 70–90 kg pigs at 68% (Jung and Lee, 1968). A later experiment (Jung, Paik and Lee, 1976) showed little difference in weight gains, feed efficiencies or carcass characteristics between pigs fed on a basal diet containing corn and diets in which up to 100% of the corn was replaced by silage. However, whereas earlier workers (Lee, Jeung and Jeung, 1964) found that the cost of feed per kg weight gain rose with increasing incorporation of sweet potato silage to a level at which farmers would experience a net loss on selling pigs at market, later workers (Jung et al., 1976) found little difference in cost per gain between a corn-based diet and one containing 60% sweet potato silage.

Apparent digestibilites of dry matter, organic matter, total protein and energy in root silage were 91%, 91%, 32% and 89%, respectively (Tomita, Hayashi and Hashizume, 1985). The digestibility of total protein is somewhat low compared to most of the figures for fresh or dried sweet potato shown in Table 7.5. The palatability of the silage was found to be low unless it had been chopped finely. Only three quarters of the offered coarsely chopped silage was actually ingested.

Poultry

General background

There has been a considerable amount of research carried out into the possibility of using sweet potato, particularly the roots in the form of dried meal, as a carbohydrate source to replace partially or completely the corn in poultry diets, often with a view to utilizing local feed materials. As in the case of pigs and cattle, results of different authors with regard to the amount of sweet potato which can be introduced into the diet without affecting performance vary widely. Two groups of poultry have been considered – broilers and layers – with effects on weight gain and feed efficiency being of major importance in the former and egg-laying performance (number, weight and quality of eggs) in the latter. Most of the research has been carried out in Asian countries especially Japan, Korea and Taiwan, with the emphasis in many cases being on the feeding of young broiler chicks. Differences in findings between researchers could be due not only to the use of various breeds of chick, reared under differing laboratory conditions, but also to variations in the formulation of diets, and in the nutritive value and quality of the sweet potato brought about by variability in genetic characteristics, processing (dehydration) techniques and application of heat or otherwise to the resulting meal.

The comparatively favourable amount of metabolizable energy supplied to 2-week-old chicks by dried sweet potato compared with that supplied by other dried roots and tubers has been demonstrated (Fetuga and Oluyemi, 1976). Results are shown in Table 7.16. Only cassava was comparable to sweet potato in terms of the percentage of gross energy as metabolizable energy. Moreover, sweet potato produced chick weight gains and feed efficiency significantly superior to all the other roots and tubers.

Broilers

Researchers are generally agreed that dried sweet potato meal can replace a small percentage of corn in chick diets without affecting their growth. Early studies carried out by Japanese workers in the 1930s are reported (Yoshida and Morimoto, 1957) to have recommended the introduction of cooked sweet potato into the diet at less than 30% on a dry weight basis. This was later confirmed in Korea (Kim, 1973) where it was recommended that sweet potato should not exceed 20% of the ration,

Table 7.16. *Comparison of sweet potato with other roots and tubers in terms of metabolizable energy and resulting growth performance of leghorn chicks*

Diet	Wt gain (g)	Feed efficiency (g/g)	ME test diets (MJ/kg)	Energy of substituted materials		
				GE (MJ/kg)	ME (MJ/kg)	ME/GE (%)
Reference[a]	294.8[B][b]	1.70[A]	16.61[A]	—	—	—
Sweet potato[c]	345.4[A]	1.97[A]	16.19[AB]	18.91[A]	16.49[A]	87.2[A]
Cassava[c]	204.1[C]	2.43[B]	16.11[BC]	18.20[B]	16.23[A]	89.2[A]
Yam[c]	153.7[D]	3.03[C]	15.02[CD]	17.91[B]	13.56[B]	75.7[B]
Plantain[c]	213.4[C]	2.33[B]	14.90[CD]	18.99[A]	13.22[B]	69.6[C]
Cocoyam[c]	112.9[E]	3.57[D]	14.43[D]	17.49[C]	12.05[C]	68.9[C]

Notes:
ME, metabolizable energy; GE, gross energy.
From: Fetuga and Oluyemi 1976. Energy values converted to MJ/kg from kcal/g.
[a] Glucose 48%; basal diet of maize, fish meal, blood meal, groundnut meal etc. 51.5%, vitamin-mineral mix 0.5%.
[b] A,B figures in vertical columns with different superscripts significantly different ($p<0.05$).
[c] Dried root/tuber meal replaced 40% of glucose in the reference diet.

and in New Guinea (Springhall, 1964) where a diet containing 20% dried milled sweet potato could be fed without adversely affecting poultry performance. Korean workers have also indicated that, at higher levels of sweet potato, chick feeding becomes unprofitable (Kim, 1973; Kwack and Ahn, 1975). In 1943 (Tillman and Davis, 1943) it was shown that, in terms of growth and feed efficiency, sweet potato could replace only 10–20% of the yellow corn which originally made up 45% of chick diets. A substitution level as low as 13% of corn (7.5% of the total ration) was further reported to produce poorer growth and feed performance than corn in day old to 3-week-old chicks (Rosenberg and Seu, 1952). Others, however, have reported replacing 50–75% of corn basal rations with dried sweet potato meal without adverse effects on growth of broilers. Replacing up to 60% of maize (originally 57% of the diet) with sweet potato did not significantly affect body weight, feed intake or feed conversion in either Rhode Island Red or indigenous Nigerian chicks (Job, Oluyemi and Entonu, 1979). Fat contents of body parts at 10 weeks were significantly decreased in comparison to corn-fed chicks, but since the nature of the fat deposited was not determined, the significance of this finding to human nutrition remains to be determined. A Filipino study (Saure, 1972) concluded that raw or roasted sweet potato meal could replace 50–75% corn (to become 25–37% of the ration) in broiler diets (see Table 7.17). In fact one trial reported by the same author showed significantly higher weight gains for 42-day-old broilers when 50% of corn in the diet was replaced by sweet potato meal than for those on the basal corn diet. However, storage of the meal for 5 months caused a large decrease in carotene content from 610 mg/kg to only 100 mg/kg and the development of an undesirable smell which decreased its palatability. This disadvantage was partially overcome by roasting the meal which could then replace 25–50% of corn in broiler diets. A more recent Filipino study (Pagtan, 1984) raised the level of sweet potato to as much as 45% of the ration without significantly affecting weight gains or feed conversion efficiencies. An Egyptian experiment (Mohamed, Omar and Kamar, 1974) showed that replacing 75% of white corn in chick diets with sweet potato gave results for growth and egg production which approached those of a control composed of white corn supplemented with vitamin A, but that hatchability of eggs was poorer. Feeding broilers on sweet potato gave rise to meat with a good flavour.

In the Philippines, experimental fattening of muscovy broiler ducks with diets in which corn was substituted by increasing levels of dried sweet potato meal (20–40% of the diet) was unsuccessful due to decreasing weight gains and feed efficiencies with increasing levels of substitution (Gerona, 1983). Although at the 20% level, performance was not substantially reduced, the increased cost of feed involved rendered the operation uneconomic.

Table 7.17. *Performance of 42-day-old broilers fed with raw or roasted sweet potato meal at various levels of replacement for corn*

Treatment	Wt gain (g)	Feed efficiency (g/g)	Mortality (%)
50% YC (control)	1102[A a]	2.05[A]	0
50% SPM (replacing all corn)	907[B]	2.36[AB]	6.6
50% RSPM (replacing all corn)	966[AB]	2.40[B]	3.3
37.5% SPM (75% corn 12.5% YC replaced)	999[AB]	2.23[AB]	0
37.5% RSPM (75%) 12.5% YC	1023[AB]	2.26[AB]	6.6
25% SPM (50% corn 25% YC replaced)	1004[AB]	2.40[B]	3.3
25% RSPM (50%) 25% YC	1030[AB]	2.26[AB]	0

Notes:
YC, yellow corn; SPM, raw sweet potato meal; RSPM, roasted sweet potato meal.
From: Saure, 1972.
[a] A,B Means in same vertical column with similar superscripts not significantly different.

Supplementation of sweet potato

There are several indications in the literature that the addition of various supplements (which remedy sweet potato nutrient deficiencies) to chick diets in which raw sweet potato meal has replaced 100% of the corn can achieve parity with corn in terms of growth and feed efficiency. It has been shown, for example, that the lower digestibility of raw sweet potato starch in young chicks compared to that of corn leads to a lower available energy level and consequent overestimation of the available energy of raw sweet potato meal for young chicks (Yoshida, Hoshii and Morimoto, 1962a,b). When the available energy levels of corn and sweet potato diets are made equal (for example by the addition of a fat source to the latter) growth of chicks on the sweet potato diet is improved (Matsuda, 1969) and can equal that of corn (Yoshida et al., 1962b). The improvement in a sweet potato diet (100% replacement of corn) brought about by tallow addition could be enhanced by supplementation with methionine and glycine, raising chick growth and feed efficiency to that on a corn diet (Matsuda, 1970). However, adding rice hulls to increase fibre levels of a diet in which all the corn had been replaced by sweet potato improved the growth, feed efficiency and egg-laying performance of chickens above those on sweet potato without hulls, but did not raise them to the standard of corn (Matsuda, 1968). This result was not confirmed by other workers (Susaki and Hamakawa, 1959) who found no effect of rice hull addition, and also indicated that the small difference in growth rates of young chicks on a formula feed and on a feed containing 50% sun-dried sweet potato was expected to decrease with older chicks. Others indicated that raw sweet potato starch digestibility and availability in older chicks (6 months to adult) is superior to that in young chicks and even equal to corn (Yoshida et al., 1962b). Cocks of about 2 kg weight were found to digest almost entirely (98%) raw sweet potato starch (Szylit et al., 1978).

In diets without the addition of a vitamin supplement, vitamin A was the only vitamin found to improve the performance of chicks fed 50% white-fleshed sweet potato rations (Yoshida et al., 1960). The use of a sweet potato clone high in carotene was as effective as vitamin A in improving chick performance.

Effect of heat treatment

Cooking the sweet potato before drying can also improve growth and feed efficiency, by improving the digestibility of starch and increasing its available energy value (Yoshida et al., 1962a). However, improvements in the nutritive value of sweet potato through cooking have been variable. Dry roasted sweet potato meal when replacing 50% or 100% of corn produced weight gains similar to that produced by corn in 42-day-old broilers, whereas meal prepared from steamed roots produced the lightest birds of all the rations (Saure, 1972). Steam-pelleting of dried sweet potato, fed as the sole source of carbohydrate, improved nitrogen retention and feed efficiency in growing cocks, but the improvements were not significant (Szylit et al., 1978). A very high degree of mortality reported for chicks fed on raw sweet potato meal (which has never been found by other authors) was decreased by cooking at 100°C, but almost entirely eliminated by heating at 120°C. Moreover, although weight gains of chicks fed on meal heated to 100°C were improved compared to those on raw meal, meal heated to 120°C produced weight gains almost equal to those on the corn control diet (Yoshida and Morimoto, 1959). Though the authors ruled out the presence of a toxic substance in sweet potatoes, it seems possible that their sample of sweet potato contained a high level of trypsin inhibitor which was eliminated only by the higher heat treatment. Taiwanese workers (Lee and Yang, 1979c) reported that 39% of the trypsin inhibitor activity in a crude extract of a high protein sample of sweet potatoes was retained after heating for 30 min at 100°C, whereas only 10% was retained after treatment for the same period at 150°C. Intestinal lesions in chicks fed raw sweet potato meal have been reported (Saure, 1972; Yoshida and Morimoto, 1957). These were reduced but not eliminated by cooking or storage of the meal (Saure, 1972).

Others have also reported that cooking produces some improvement in the growth of chicks fed on sweet potato-for-corn substitutes, but that cooked sweet potato still did not achieve growth parity with a corn diet (Rosenberg and Seu, 1952) unless the energy level of the diet was raised by the addition of tallow (Matsuda, 1969). Steaming sweet potato before drying was shown to increase its available energy (determined as percentage total digestible nutrients) from 47% to about 65% and, when steamed meal completely replaced corn, to raise weight gains and feed efficiencies of young chicks to levels similar to those produced by corn (Yoshida et al., 1962a). It is possible that cooking improves digestibility of the diet not only directly by increasing starch digestibility, but also indirectly by reducing the adverse effect of starch on protein digestibility (see pp. 419–20).

Effect of pelleting

There are contrasting results about the positive effects of pelleting on a feed in which part of the basal corn has been replaced by sweet potato. In Taiwan, the replacement of 23% out of the original 60% of corn in the diet by 20% high protein sweet potato, 2% tallow, 1% soybean meal and 0.1% methionine had no effect on the growth, feed conversion efficiency or carcass quality of broilers from 2 to 8 weeks of age whether the diets were in meal or pellet form (Lee and Yang, 1979b). The high protein sweet potato contained 7.3% total protein (dwb) compared to figures of 3–4% total protein given by other authors for samples of sweet potato meal. Thus the only adjustment to the dietary protein content which had to be made to render the sweet potato diets iso-nitrogenous with the basal corn diets was as indicated above in the form of soy and methionine.

In contrast, chicks fed up to about 18 weeks of age in Korea with diets containing various levels of sweet potato substituting for corn were significantly heavier on pelleted diets than on meal diets, whilst feed conversion efficiencies were similar for pellet and meal forms (Kwack and Ahn, 1975). The differences in weight gains between the two major treatments was particularly pronounced at the earlier stages of growth. Moreover, the chicks fed on the highest level of substitution (50% of the corn substituted with sweet potato) in a pelleted form gained significantly more weight than those fed on the basal corn diet in meal form. In each group there were no significant differences in weight gain between birds fed varying levels of sweet potato. A trend towards lowered feed conversion efficiencies with increasing substitution of sweet potato for corn, leading to slightly higher costs per unit body weight gain, led the authors to conclude that locally produced sweet potato at up to 25% of a pelleted diet (about 40% substitution for corn) was more profitable for chick feeding than corn in meal form.

The literature is almost devoid of references to the use of sweet potato foliage in the diet of chickens. However, a Filipino experiment reported on the results of adding finely chopped fresh green leaves and young shoots of sweet potato, or two other types of green (*centrosema* and *ipil-ipil*), to a basal feed on the growth of chicks (Diñgayan and Fronda, 1950). The chicks given sweet potato greens were the smallest of all treatments after 12 weeks of feeding. They also had the highest mortality rate and were the least economical to produce. There was no attempt to explain this result.

Layers

Much less research is reported in the literature on the effects of sweet potato in the diets of layers than in those of broilers. The results of an experiment (Lee and Yang, 1979c) to incorporate a high protein (7.3%) sweet potato meal at two levels of substitution for corn into Leghorn chick diets, and to determine the subsequent effects on egg production and quality, are shown in Table 7.18. The highest level of substitution was such that sweet potato constituted 20% by weight of the ration. As a result of the significantly lower weight gains and total egg production at the higher level of substitution, the authors recommended the inclusion of no more than 10% high protein sweet potato meal in layer diets. They also show, however, a considerable amount of trypsin inhibitor activity (TIA) in an extract of the raw meal. This activity could be reduced to a low level only by heating at 150°C. It is possible, therefore, that a different sample of sweet potato meal prepared from a clone genetically low or lacking in TIA might have given more favourable results at the higher level of substitution. Low levels of substitution using average protein content yellow sweet potato meal have also been tested in the Philippines (Villalva, 1972). The egg production of hyline layers when either 10 or 15 parts of corn were replaced by yellow sweet potato meal was higher than that with a corn control ration, the lower substitution level giving the highest egg production. Substitution of corn to give 30% by weight of sweet potato meal in the diet of layer mallard ducks has been shown to improve egg production and decrease the amount of feed

Table 7.18. *Effect of inclusion of sweet potato meal as a substitute for corn in layer rations on egg production and quality*

Treatment	Wt gain (g)	Feed efficiency (g/g)	Age at 1st egg (days)	Wt at 1st egg (g)	Wt of 1st egg (g)	Total eggs produced	Feed/ dz eggs[a] (g)	Egg shell (%)	Yolk (%)	Thick white (%)	Thin white (%)	Hatchability (%)
Corn 60% of ration	1388.1	4.47	161.7	1744.3	43.4	3314	1647.6 (2472.0)	11.3	29.2	29.8	29.7	84.8
HPSPM 10% of ration[b]	1398.6	4.52	161.7	1722.7	42.6	3176	1668.5 (2493.6)	11.1	28.5	30.3	30.1	88.3
HPSPM 20% of ration[c]	1345.1[d]	4.63	161.5	1691.5	43.3	2931[d]	1862.8 (2824.3)	11.3	29.3	29.4	30.1	88.9

Notes:
HPSPM, high protein sweet potato meal.
From: Lee and Yang, 1979c.
[a] Numbers in parentheses are Feed/kg eggs.
[b] HPSPM replacing approximately 15% corn.
[c] HPSPM replacing approximately 30% corn.
[d] Significantly lower value than other data in same vertical column.

required per unit of eggs laid without affecting egg quality (Gerona, 1983). However, feed costs were substantially increased.

Korean workers (Kim, Oh and Lee, 1959; Lee and Moon, 1969) have reported on the effects of inclusion of sweet potato silage (prepared from roots) in layer diets on the resulting egg production. Both found that silage included at up to 40% of the diet at the expense of corn and other cereals resulted in a similar or even greater total egg production per chick over the period of the experiment than with the unsubstituted cereal-based diets. One group (Kim et al., 1959) also noted that boiled or steamed sweet potato was preferable to raw sweet potato in layer diets as it was more palatable and digestible.

Sweet potato as a source of pigment for broiler skins and egg yolks

Dried orange-fleshed 'Centennial' roots have been used successfully in poultry diets for the pigmentation of egg yolks (Weber, 1969). A more recent approach, however, is to use sweet potato vines, which might otherwise be discarded or ploughed back into the soil, for this purpose. Sweet potato greens contain carotenoid pigments including beta-carotene and xanthophylls which can be exploited as a source of natural pigments for colouring foods (Chapter 6). They have been found useful when introduced into poultry diets to enhance the yellow pigmentation of broiler skins and egg yolks (Garlich et al., 1974) and through their protein content to act as a nutritious supplement. Sweet potato vine meal for this purpose had the highest concentration of protein, beta-carotene and xanthophyll when prepared from vines collected early in the growing season (Garlich et al., 1974). A sample harvested late in the season (at the time of root harvest) was of poor quality with a high percentage of stems to leaves and, when introduced into a poultry diet, its pigments were ingested, absorbed and deposited in egg yolk only 79% as well as those of alfalfa meal. In comparison, the vine meal from vines harvested early and introduced into a broiler diet as the sole source of pigment (37.5 mg xanthophyll/kg diet) appeared to be equally as good as alfalafa meal for pigmenting the broiler skins.

To overcome the disadvantages (high fibre and low protein and carotenoid levels) of the direct use of vines harvested at the normal time, others have suggested the preparation of a leaf protein concentrate containing about 50% protein and approximately 0.1% xanthophyll for use as a pigmenting agent in poultry diets (Walter, Purcell and McCollum, 1978). It was suggested that, in the United States, such a concentrate could replace imported dried alfalfa in poultry feed. The concentrate is prepared by grinding sweet potato vines (or preferably leaves and petioles, without the more fibrous stems), adding water, and pressing. Grinding the leaves before addition of water, rather than grinding the leaves with water, was found to overcome the difficulties of concentrate preparation engendered normally by the high leaf mucilage content (see Chapter 6). The press cake is extracted with a second sample of water. The two lots of juice from pressing are combined, heated to coagulate protein and centrifuged to obtain a solid pellet containing protein and xanthophyll. The optimum date for harvesting leaves, in terms of maximum protein and xanthophyll levels and minimum fibre content, was about 80 days after planting, but at the usual harvest date of 130–140 days after planting the concentrate contains about 36% good quality protein (as judged by amino acid analysis), 9% fibre and 0.1% xanthophyll. Storage of the extracted juice, which might be required in practical production of concentrate, for several hours at up to 50°C had little effect on xanthophyll content but reduced protein by about 15%.

Cattle and other ruminants

Cattle

Cattle may be fed on both roots and tops of sweet potato, but the use of vines particularly has been shown to be successful as a cattle forage. A study of the sweet potato in the Cañete Valley of Peru, for example, showed that 70% of sweet potato foliage is commercialized as animal feed by a highly organized and efficient system and concluded that the creation, and significant growth, of dairy farming in the valley was largely due to the use of sweet potato vine forage (Achata et al., 1988; and see Chapter 8, Peru). As a result of promotion of a zero grazing (stall) system of cattle rearing in Kenya, demand for sweet potato vines by cattle farmers has increased. This has led to a vigorous programme of breeding for cultivars with high vine yields (Baulch, B., personal communication), and has provided a source of roots for human consumption.

Roots

The older literature pertaining to the use of sweet potato roots as cattle feed has been reviewed (Yeh and Bouwkamp, 1985). Most of the older work reported was carried out in the United States in the 1940s. In the case of dairy cattle, various researchers who replaced all the corn in a concentrate ration with dehydrated sweet potato root meal are reported to have found milk production 88% to 98% that of cows fed the corn concentrate. Replacement of only 50% of the corn resulted in 97% to 100% of the milk production with corn alone. Feeding of dried roots of orange-fleshed clones resulted in 22% more vitamin A and 30% more beta-carotene in the milk. Dehydrated sweet potatoes were reported to be equal to corn–soybean silage on a dry weight basis for milk production. Another researcher found that fresh chopped sweet potato roots could substitute for sorghum silage in dairy rations at the rate of 1 kg of roots for 1.8 kg of silage, resulting in less than 1% reduction in milk production and a slightly higher butter fat content. Cattle fed sweet potatoes or sorghum gained in weight (kg/day) by 0.42 or 0.13, respectively. Higher weight gains in cows fed dehydrated sweet potato than in those fed alternative rations were also reported by others.

The review also reports various results for the fattening of beef cattle. Researchers who compared the value of corn and dehydrated sweet potato roots replaced either half or all the corn in a fattening ration with dried sweet potato. They are reported to have found that animals fed the standard corn ration gained 1.07 kg/day compared to 1.17 and 0.98 kg/day when sweet potato replaced half or all of the corn, respectively. The respective feed to gain ratios were 9.51, 9.31 and 9.22. Animals fed on the standard corn or the half-replacement rations were reported to have fetched a similar market price, whereas cattle fed the whole-replacement ration commanded only 92% of that price. One group of researchers found that corn and sweet potato were approximately equivalent in terms of total digestible nutrients in a digestibility trial with beef cattle. This was confirmed in a later study (Chen and Chen, 1979). A Japanese researcher (Ando, 1962) noted that sweet potato could be included in feed at a rate of 50% for early fattening of beef cattle without affecting feed palatability or meat colour (see Figure 7.3). Cooking of roots was found to improve palatability, but not feed efficiency. Another investigation in the 1960s (Bond and Putnam, 1967) into the potential use of sweet potato trimmings, a by-product of canning operations, for fattening beef cattle found that trimmings represented only 80% of the value of corn for this purpose, resulting in somewhat lower weight gains, higher feed conversion ratios and poorer carcass quality.

More recent studies carried out in Taiwan on the feeding of beef cattle with dried sweet potato root chips are summarized in Table 7.19. The chips were found to be palatable and increased the average daily gain above that of cattle fed on a corn-based diet when replacing 50% or 100% of the corn, as long as the nitrogen content of the diets was maintained by additions of soymeal and urea.

The influence of sweet potato root starch on the rumen fermentation pattern and on urea utilization by rumen microflora has been studied by simulation of rumen fermentation *in vitro* (Szylit et al., 1978). Sweet potato starch was a good source of energy for rumen microbial growth as shown by the non-protein nitrogen (urea) utilization and volatile fatty

Figure 7.3. A farmer supplementing concentrate feed with fresh sweet potato roots for rearing cattle, Japan (J.A. Woolfe).

Table 7.19. *Experimental feeding of beef steers with corn- or sweet potato chip-based diets*

Feed					
Corn (kg/day)	2.517	1.235	—	1.060	—
Sweet potato chips (kg/day)	—	1.282	2.517	1.213	2.133
Soybean meal (kg/day)	0.984	0.984	0.984	1.253	1.406
Urea (kg/day)	—	0.038	0.066	—	—
Hay (kg/day)	2.883	2.883	2.883	2.883	2.883
Vitamin and mineral supplement (kg/day)	0.5	0.5	0.5	0.5	0.5
Cattle					
Total protein intake (g)	830.8	851.6	847.6	847.1	840.9
Total digestible nutrient intake (kg)	4.54	4.42	4.31	4.43	4.35
Average daily gain (kg)	0.802	0.891	0.851	0.736	0.698

Note:
From: Chen and Chen, 1979.

acid production which took place. In this respect sweet potato starch was similar to cassava starch, and significantly superior to yam (*D. cayenensis*) starch. However, steam-pelleting of the sweet potato roots and of the other roots and tubers improved starch availability and increased urea utilization. The digestibility of a given starch was felt to be a reflection of its crystalline structure as characterized by its X-ray diffraction pattern.

Foliage

In the 1970s, Taiwanese researchers investigated the nutrient composition and value of sweet potato vines for feeding both dairy and beef cattle (see also Figures 7.4, 7.5, 7.6). The milk yield of cows fed fresh sweet potato vines as the sole source of roughage plus a concentrate, was higher (but the milk fat percentage lower) than that of cows fed sweet potato vine/hay mixtures (Chen, Yi and Hsu, 1977). The fresh vines were palatable and produced no adverse effects on the health of the cows nor on milk acidity or specific gravity. If sweet potato vines alone are fed to beef cattle, without supplementation, they do not support growth (Chen et al., 1977). However, fresh vines can be used as the sole source of roughage, normal growth being obtained by the addition of concentrate feed. This concentrate included soybeans. Other workers also showed improvement in cattle growth rates where sweet potato vines were

Figure 7.4. Sweet potato vines for sale as fodder for dairy cattle in China where dairy farmers do not grow sweet potato and have to buy fodder (M. Iwanaga).

Figure 7.5. Vines being collected for dairy cattle feeding in West Java, Indonesia (G.A. Watson).

supplemented by soybean meal (Ffoulkes and Preston, 1978), not only to increase total dietary protein, but also to provide a source of insoluble or undegradable protein which escapes rumen fermentation. Mechanically dehydrated vines were found to be less palatable than fresh vines (Chen et al., 1977) and required higher levels of concentrate supplementation to achieve similar growth rates (Chen, Chen and Din, 1979). Moreover the cost of dehydrated vines was high, making them uneconomic.

The substitution of sugar cane forage by sweet potato vines for feeding beef steers has been investigated by several workers and reviewed briefly (Ruiz, 1982). In spite of its high digestibility, sugar cane is not consumed by cattle at levels which support high rates of growth. Sweet potato vines and sugar cane have similar dry matter digestibilities (about 70%) and substituting part or all of the sugar cane by more palatable sweet potato vines not only increases total dry matter intake, but also that of digestible dry matter (Ffoulkes, Hovell and Preston, 1978). Sweet potato vines also supply higher percentages of dietary true protein as their level of substitution increases, thus enabling a reduction in supplementary non-protein nitrogen in the form of urea. It has been suggested that sweet potato forage promotes higher voluntary feed intakes by stimulating rumen function and increasing the efficiency of microbial fermentation. In support of this theory, various researchers (cited by Ruiz, 1982) found that sweet potato foliage is extensively degraded in the rumen and that its half-life digestion rate was only about 13 hours as compared to 24 hours for sugar cane. This resulted in an increase in ruminal outflow from 40 to 120 litres/day when sugar cane was substituted by sweet potato foliage. The more rapid rumen turnover rate would result in a greater flow of protein to the duodenum. A number of workers have shown that voluntary intake and weight gain are closely and positively related (Ruiz, 1982).

Costa Rican workers (Backer, 1976; Backer et al., 1980) have used a different approach to the utilization of sweet potato in feeding beef cattle, by basing the whole diet on various combinations of foliage and roots, using urea to maintain the rations as iso-nitrogenous. These combinations they found not only highly beneficial for livestock gains but also very profitable provided only unsaleable (cull) roots are used as the root source. The proportion of cull roots, which might normally be considered as 'waste', can be substantial. For example, it has been estimated as a minimum of 12% in Costa Rica (Ruiz, 1982). Feed intakes of young bulls, in terms of dry matter and total protein, did not vary significantly among dietary treatments which ranged from 100% roots to 100% foliage (with various combinations in between) but metabolizable energy intake and *in vitro* dry matter digestibility increased as the percentage of roots increased. Although there was no significant difference in growth rates between groups, weight gain tended to increase with increasing percentage of roots until the diet consisted of 50:50 roots:foliage, and then decreased again. This could have been due to an increasing proportion of non-protein nitrogen as urea. Such experiments suggest the possibility of developing new animal feeding systems for the small farmer on the basis of the use of crop residues such as sweet potato cull roots and foliage (appropriately supplemented with urea, vitamins and minerals), left over from the main harvest of roots destined for their traditional purposes (Ruiz, 1982).

Silage making

One of the constraints to the use of the large quantities of sweet potato foliage suitable for forage is that its availability is concentrated in one or two seasons of the year after harvest, which in general does not coincide with pasture scarcity (Ruiz, Lozano and Ruiz, 1981). Preservation as hay (dehydration) or silage may therefore be needed. The latter method is more practical in the humid tropics where quick drying in the field is not possible. Furthermore, succulent feeds such as silages do not need so much supplemental water when given to livestock (Brown and Chavalimu, 1985). Simple small silos can be made with discarded sugar bags, other by-product plastic films or new plastic bags all of which are affordable by rural farmers in the tropics (Brown and Chavalimu, 1985).

The technical limitations to the production of good quality, palatable silages for cattle have been briefly reviewed (Ruiz et al., 1981). Production of good quality silage depends upon the rapid fermentation under anaerobic conditions of silage raw material carbohydrates to increase acidity and lower the pH to the point at which further microbial activity ceases, thus preventing putrefaction and preserving nutrients. Ideally the organic acids formed contain a high proportion of desirable lactic acid, some acetic acid, and little or no butyric acid, which imparts a foul smell to silage causing animals to reject it. In practice urea may be added to silage to raise total protein content, although it can also promote butyric acid production and increase silage losses. An additional source of rapidly fermentable carbohydrates such as molasses, cereal grains or roots may be added where appropriate.

The production of silage from sweet potato vines has been studied in Japan (Konaka, 1962; Sutoh, Uchida and Kaneda, 1973) and the Philippines (Gerpacio et al., 1967) and more recently in Costa Rica (Ruiz et al., 1981) and Kenya (Brown and Chavalimu, 1985). Some of these studies were made in relation to the feeding of small ruminants such as goats and sheep rather than cattle, but their findings may conveniently be discussed in this section.

In Japan, the addition of either inorganic acids (hydrochloric and sulphuric in the ratio 19:1), mashed sweet potato roots or molasses was found to be necessary to produce good quality vine silage with high lactic acid and no butyric acid content (Konaka, 1962). In contrast, others (Ruiz et al., 1981; Sutoh et al., 1973) showed that silage of excellent fermentative quality was obtained from sweet potato foliage when no additives such as urea or sweet potato roots were used. Increasing additions of urea caused a linear increase in pH and increasing losses of dry matter by putrefaction (Ruiz et al., 1981). The addition of roots had no noticeable effect on dry matter losses or lactic acid production, but increased acetic and butyric acid concentrations. The vine silages without additives, on the other hand, had excellent characteristics with an average pH of 3.9, a low concentration of butyric acid and a low (11%) loss of dry matter by putrefaction. It remained to be seen whether these results were valid for in-field as well as laboratory silages (Ruiz, 1982). A Filipino study of silage made from fresh or 24-hour wilted vines (Gerpacio et al., 1967) found acetic acid to be the organic acid in the highest concentration and lactic acid that in the lowest. Butyric acid was in a higher concentration than lactic acid. However, no comment was made about the resulting quality of the silage.

The degree of nutrient preservation by either dehydration to hay or by ensiling has been compared in sweet potato vines and other forages in Kenya (Brown and Chavalimu, 1985) for potential feeding to small ruminants such as goats. For production of silage, vines were chopped and manually packed into double nested plastic bags in a way similar to that described in the Costa Rican (Ruiz et al., 1981) experiment. For hay preparation, bundles of vines were hung from a shed roof in indirect sunlight. Although the total protein content of ensiled vines was not different to that of either dried or fresh vines, ensiling was associated with a greater part of the protein being converted to damaged insoluble biologically unavailable nitrogen than was the case for dehydration. It can be calculated that only about 8% of total protein was lost (by conversion to an insoluble form) from the vines by drying whereas 24% was lost by ensiling. Of the five forage species investigated this result occurred only in sweet potato vines, but no tentative explanation of the result was offered. This might be investigated further for use by regions where drying and ensiling are viable alternatives for sweet potato vine preservation. A Japanese study revealed 11–35% losses of provitamin A carotenoids due to ensiling of fresh vines of varying ages (days after planting) at the time of ensiling (Sutoh et al., 1973).

Vines wilted for 24 hours before ensiling were found to give silages with higher levels of dry matter as fed than fresh vines (Gerpacio et al.,

1967; Ruiz, 1982). The digestibility of vine silage dry matter and of total and individual nutrients has been investigated both *in vitro* and in various animal species. The *in vitro* digestibility of dry matter in Costa Rican prepared silage was 59%. This might have represented a decrease in digestibility as a result of ensiling as different batches of fresh forage of the same sweet potato clone had 62% (Backer et al., 1980) and 72% digestibility (Ruiz et al., 1981). Dry matter digestibility as determined with goats was 66–55%, with an indication that digestibility of silage dry matter decreased as the age of vines used to prepare it increased (Sutoh et al., 1973). The digestibilities of individual silage nutrients rose as the age of vines increased and then fell again as the age increased further, but the digestibilities of total and true protein remained higher in the oldest vine silage than in the youngest. In a trial with mature sheep to compare the digestibilities of silages prepared from fresh and wilted vines, wilted vine silage had higher protein and lower fibre digestibilities than the fresh equivalent, although these differences were not significant (Gerpacio et al., 1967). In contrast, the digestibility of lipid was significantly higher and that of the nitrogen free extract lower in fresh than in wilted silage. Mean percentages of total digestible nutrients and digestible energy were approximately equal at 43% and 46%, respectively, in fresh and wilted silages. Nitrogen balance was negative in the sheep fed fresh vine silage and positive in those fed wilted vine silage, but this difference was not reflected by any difference in the weight changes of the animals. There was an indication that different batches of silage having differing proximate compositions have varying coefficients of digestibility for energy and total nutrients.

Results of feeding values for sweet potato vine silage suggest that there may be decreases in vine nutrient values brought about by ensiling, but results are somewhat patchy and fragmented. A more systematic investigation of the quality of vine silages for different animal species might be appropriate for an area where the preparation of such silages could be especially useful.

Sheep and goats

Investigations into the properties and use of sweet potato silage for species such as sheep and goats have already been discussed in the previous section. Early studies in the United States into the use of dehydrated sweet potato meal for sheep noted that gestating ewes gained equal amounts of weight whether fed on a ration containing sweet potato meal or one containing corn (Massey, 1943). The ewes receiving the sweet potato ration produced more milk after parturition and so gave their lambs a faster rate of growth at the start. The digestibility of dried sweet potato was, however, found to be lower than that of corn for 28 kg lambs (Briggs et al., 1947). The digestibility of total protein was particularly low (19.8%) in comparison to corn (61.2%). A digestion trial with sheep showed that alkali treatment (in the form of NaOH or $Ca(OH)_2$) of dry sweet potato vines and other forages significantly increased the digestibility of fibre and increased the availability of starch (Abou-Raya et al., 1966).

A much more recent investigation in the Philippines involving native goats compared their fattening performance when fed a 1:1 ratio of sweet potato leaves to grass fodder with that of goats fed either all grass fodder or a 1:1 ratio of grass fodder to brassica waste (Carpio, 1984). Although no significant differences in weight gain were noted between groups, feed conversion efficiencies were highest in the group fed sweet potato leaves and grass.

Aquaculture

Sweet potato leaves are sometimes used for feeding fish of the *Tilapia* spp. in Ugandan fish ponds (Mwanga and Wanyera, 1988). Vines, and root trimmings from a sweet potato cannery have been investigated as a possible supplemental feed source for farmed crayfish (*Procambarus clarkii*). Dehydrated sweet potato vines and leaves fed daily *ad libitum* produced the highest weights and longest lengths of mature crayfish, followed by sweet potato trimmings, rice stubble and rye hay (Goyert and Avault, 1977). Soybean stubble and dried sugar cane stalks were not suitable as supplementary feeds for crayfish. Sweet potato vines and trimmings could be fed directly to crayfish after a very short compost period (partial decomposition in water to allow microbial build-up and an increase in total protein:carbon ratio). Sweet potato vines aged in water were not as effective as fresh vines, suggesting that some of the products of partial decomposition of the vines may be unpalatable to crayfish.

Fermented feeds and the use of sweet potato wastes as livestock feed

Fermented feeds

Fermented sweet potato feedstuffs are reported to be produced and utilized for livestock in China (Gitomer, 1987), although this author is unable to give details of the processing involved. In Korea fermented feeds have been produced experimentally by 'kojification' (inoculation with mould spores) of sweet potato roots. Investigations have entailed

Figure 7.6. Native cattle feed on sweet potato vines in West Java, Indonesia. Farmers often use sweet potato vines as fodder at the end of the dry season, since they are one of the few drought-resistant crops available (M. Potts).

determining the most appropriate microorganism (normally a particular strain of *Aspergillus oryzae*) to carry out the fermentation and the types and quantities of supplements to be added to the sweet potato. One group of workers demonstrated that the protein content (dwb) of sweet potato roots could be increased from about 3% in unfermented material to 14.5% by the addition of 4% ammonium sulphate and 4.2% calcium carbonate, inoculation with *A. oryzae*, and a further addition of 2% ammonium sulphate after 48 hours followed by a second period of 48 hours of fermentation (Kim, Cho and Park, 1962). Others mixed dried ground sieved sweet potato roots with rice straw, ammonium sulphate, superphosphate of lime, calcium carbonate and water and increased protein content of the mixed wastes from 3% to 8% (Chang and Oh, 1963) by fermentation. These products have not been adopted by farmers due to a significant decrease in sweet potato production in Korea since the 1960s (Park, E., personal communication).

Use of waste products

The subject of utilizing sweet potato wastes (normally by-products from processing operations to produce alcohol or a food product) has already been discussed in relation to human foods (Chapter 6). As such this subject partly overlaps with the production of animal feeds as some products from waste are potentially useful for both purposes. This section reviews work on the utilization of sweet potato wastes which is chiefly concerned with livestock feed production.

In some circumstances the vines of sweet potato, when simply discarded after cutting for harvest, can be regarded as a waste or by-product of sweet potato root production. Alternative uses of these vines directly as an animal forage or indirectly as a source of pigments have already been discussed in previous sections.

Cannery wastes

In areas where sweet potatoes are grown and canned in very large quantities (for example the southern United States) extremely large amounts of waste, especially from peeling and trimming operations, are generated. Approximately 40–50% of the raw material becomes waste (Sistrunk and Karim, 1977). When released into the environment this waste constitutes a source of pollution which is expensive to treat. For example one sweet potato canning factory situated in a town of 5000 people had a biological oxygen demand (BOD) equivalent of 70,000 people (Raines, 1973). Landfill disposal may be opposed on the grounds of high alkali concentration in the case of lye-peeling wastes. The high BOD, solids and nitrogen contents of wastes at various stages of processing found in a conventional sweet potato cannery are shown in Table 7.20.

An alternative to disposal or chemical treatment, which would result in the potential use of lye peelings as a livestock feed, has been suggested. This entails bacterial fermentation by lactobacilli or streptococci to reduce the pH (Sistrunk and Karim, 1977). In practice waste pH was reduced in 7 days from 11.8 to 4.9, protein and mineral contents were little affected, starch and cellulose decreased and pectin and hemicellulose contents increased. The resulting product was suitable for ruminant feed, but for other species necessitated the addition, during fermentation when the pH reached 7, of enzymes such as hemicellulase, pectinase and amylase to improve availability of nutrients. Waste was found to ferment most rapidly when seeded with 20% of an inoculum from a previous fermentation. Aeration of the waste is required during fermentation to prevent off-odour development, and a temperature of 25–30°C may be used for most rapid reduction of pH. Due to the short duration of the sweet potato processing season, generated waste for feed purposes would require storage. A system of batch fermentation pits with continuous inoculation of the waste was suggested, followed by storage

Table 7.20. *The biological oxygen demand (BOD) and nutrient content of sweet potato wastes from a cannery*

Process	BOD (mg/l)	Total N (mg/l)	Total solids (mg/l)	Soluble solids (mg/l)
Cleaning and washing	990	12	2,100	1,200
Preheater	3,700	45	8,400	1,600
Lye peeler	13,000	320	35,000	7,700
Snipping	5,900	140	13,000	3,800
Scrubbing (abrasive peeler)	14,000	330	23,000	4,400
Brush washer	3,500	71	4,300	1,200
Retort	76	—	300	—
Cleanup	2,200	—	2,700	870

Note:
From: Smallwood, Whitaker and Colston, 1974.

of the waste, with a preservative to prevent mould growth and hence mycotoxin contamination, in covered pits or silos.

Other authors have suggested that the Symba process developed in Sweden for potato processing waste water utilization could be applied to sweet potato wastes (Beszedits and Netzer, 1982). In the Symba process, *Candida utilis* (torula yeast) is grown on effluents in association with another microorganism. *C. utilis* is deficient in amylase. However, a slow-growing microorganism, *Endomyopsis fibuliger*, converts starch to simple sugars while the fast-growing *C. utilis* grows on the substrate provided by *E. fibuliger*. The resultant mixed culture of yeasts ($>90\%$ by weight of *C. utilis*) is dried and used as animal feed.

Stillage waste

Stillage, the non-volatile fraction of the material used for distillation of spirits or industrial alcohol is also a by-product of sweet potato processing with potential for use as animal feed. As indicated in Chapter 2, sweet potato is increasingly regarded as a promising crop for energy production from biomass. For example sweet potato can yield 1000 gallons ethanol/acre (~11,200 l/ha) compared with only 250–300 gallons/acre (~3300 l/ha) for corn (Wu and Bagby, 1987). Only the carbohydrates (starch and sugars) are utilized in alcohol production, leaving nutrients such as nitrogenous compounds and minerals in the stillage or liquid residue. The protein content and quality of these residues have been of interest to researchers for some time. In 1964, the protein content and quality of sun-dried distiller's residue from sweet potato was investigated in Japan (Kubayashi, 1964). Total protein, at 26% by weight of residue, had a fairly good amino acid composition and a digestibility of 60%. The residue could be added to a basic feed for chickens at the rate of 5% without problems. The residue from *shochu*, a traditional Japanese liquor (see Chapter 6), has not, however, so far been exploited for feed purposes. Its only use at present is as a fertilizer. The reason for this is said to be problems of high moisture (making it expensive to dry) and excessive acidity (rendering it unsuitable for animal feed). Nor has it been utilized on a commercial basis for the manufacture of single cell protein, perhaps because of a high sorbate content (Sugama, S., personal communication).

Removing or concentrating the nutrients from sweet potato stillage by less energy-intensive and expensive methods than direct dehydration, so that they can be utilized as an animal feed, has been investigated (Sweeten et al., 1983; Wu and Bagby, 1987). Maximum recovery of stillage solids entails various practical difficulties which must be overcome. For example, whole stillage from sweet potatoes was found to contain a high percentage of very fine particles which resulted in rapid clogging of the screen when an attempt was made to separate solid and liquid fractions with a screw press (Sweeten et al., 1983). Centrifugation of whole stillage proved to be more successful, and concentrated dry matter and nitrogen in the solid residue which even so only contained 6.8% dry matter. This would appear to be very expensive to dehydrate. In contrast, others (Wu and Bagby, 1987) improved ethanol yield and recovery of fermentation residues (stillage) by reducing the viscosity of sweet potato raw material slurries with a pectinase. After distillation, solids were removed from stillage by filtration under suction which yielded a filter cake and a thin stillage which was then centrifuged to give centrifuged solids and stillage solubles. Fermentation residues from three cultivars – 'Jewel', 'Sumor' and 'HiDry' – accounted for 36%, 32% and 25% of the original sweet potatoes (dwb). Filter cake accounted for the largest weight of fermentation residue and centrifuged solids the smallest. Percentage of dry matter of the filter cakes ranged from 14.9 to 18.0 for the three cultivars. This compares with only 3.2% found in whole stillage by other authors (Sweeten et al., 1983). Total protein contents of filter cake (dwb) were greatly increased compared to the original sweet potato and lysine contents in the protein increased from about 4.5 g/16 g N to about 6.3 g/16 g N (average of the three cultivars). Thus a protein-rich by-product which needs less energy to dehydrate than whole stillage may be produced. Such a by-product from sweet potatoes also has an advantage over corn distillers' grains in having a lower fat content and may therefore be easier to store (Wu and

Bagby, 1987). The concentration (and subsequent potential food/feed use) of stillage solubles by ultrafiltration and reverse osmosis, which are less expensive alternatives to evaporation, has been proposed (Wu, 1988).

Starch waste

The use of dehydrated starch pulp (see Figure 6.10, Chapter 6) as an ingredient in animal feeds and the cultivation of single cell protein in the form of yeasts, which could be a potential feed, using starch factory waste waters have already been mentioned in Chapter 6 (Section C).

The wet residues from village starch processing in Vietnam are reported (Wiersema, 1989) to be converted into pig feed by fermentation under aerobic conditions for 4 to 5 days. After fermentation, the feed contains about 12% protein. In addition, a fermented feed has been experimentally produced in Korea by 'kojification' (see Fermented feeds, above) of starch factory waste and tested in chicken diets. The crude and true protein contents, respectively, of a sample of untreated waste were 3.17% and 2.76%, but were increased by fermentation to 9.72% and 6.67% (Shin, You and Chang, 1964).

Fermented feed from sweet potato starch waste, when introduced into the diet of growing chicks at the expense of either 10% wheat bran (10% of the diet) or 10% wheat bran and 5% rice bran (15% of the diet), increased body weight gains and improved feed efficiency compared to a control diet (Kang, 1963); the best results were gained with the 10% level of substitution. Feeding a *koji* or fermented starch waste at 10% of the diet of laying hens in place of 5% wheat bran and 5% rice bran was also shown to increase significantly egg production, individual egg weight, feed efficiency (a decrease in the feed required per kg eggs laid) and net income gained (Shin et al., 1964) compared to a control ration. In this respect the 10% level of substitution was superior to 20% or 30% levels.

Summary

The energy and protein values of various forms of sweet potato as recorded by different authors are summarized in Tables 7.21 and 7.22. It was hoped that these would provide meaningful average values which could be used in ration formulation. However, they actually serve to emphasize the high degree of variation in data obtained, and the many gaps in knowledge which still exist to be filled by further research.

That sweet potato can be used as a partial substitute for other feed ingredients, especially corn, has been demonstrated in practice for

Table 7.21. *Energy and protein values of sweet potatoes recorded by various authors for non-ruminant livestock species*

Animal	Value	Fresh roots	Dried roots/chips/meal	Fresh vines	Dried vines
Young pigs (20–40 kg live weight)	*Energy*				
	DE (MJ/kg (dwb))	16.1[aw] 15.9[ax] 16.1[ay] 15.8[az]	15.4–16.2[b] 14.6[c]		
	ME (MJ/kg (dwb))	15.3[aw] 15.1[ax] 15.2[ay] 15.2[az]	14.1[c]		
	NE (MJ/kg (dwb))		8.5[c] (79% that of corn)		
	TDN (%)		74.3[d]		
	Total protein				
	Digestibility coefficient (%)		−18.9– +43.2[b] (Range for 4 cvs)		
	Digestible protein (%)		−0.4– +2.3[b] (Range for 4 cvs)		
	Digestibility coefficient (%)		−4.5– +91.2[e] (Range for 3 breeds)		
Growing-finishing pigs (up to 90 kg live weight)	*Energy*				
	DE (MJ/kg) (as fed)	4.52[f]	11.2[g]	0.79[h]	4.76[h]
	ME (MJ/kg) (as fed)	4.31[f]		0.74[h]	4.45[h]
	TDN (%)		72.3[i]		
	Total protein				
	Digestibility coefficient (%)	42.3[j] 57.2[jj]	33.0[g] 59.7[i]	56.2[ee]	
	Digestible protein (%)		2.64[i]		

Poultry	*Energy*			
Young chicks	ME (MJ/kg)	16.5–17.2[k] (90% of GE)		
Chickens	DE (MJ/kg)		1.7[b]	10.5[b]
				9.5[bb]
	ME (MJ/kg)		1.4[b]	8.6[b]
				7.7[bb]
	Total protein			
	Digestibility coefficient (%)		53.0[b]	53.0[b]

Notes:

DE, digestible energy; ME, metabolizable energy; NE, net energy; TDN, total digestible nutrients; GE, gross energy.

[a] Oyenuga and Fetuga, 1975; [aw] raw, peeled; [ax] raw, unpeeled; [ay] cooked, peeled; [az] cooked, unpeeled.

[b] Furuya and Nagano, 1986; range for four cultivars.

[c] Wu, 1980.

[d] Lee and Yang, 1981.

[e] Zarate, 1956; Range for three breeds of pig; [ee] Mean for three breeds.

[f] Atlas of nutritional data, 1971.

[g] Han et al., 1976a.

[h] Angeles, 1977; [hh] Mature vines.

[i] Koo and Kim, 1973.

[j] Rose and White, 1980, 10 month-old pigs; [jj] 15-month-old pigs.

[k] Fetuga and Oluyemi, 1976.

Table 7.22. *Energy and protein values of sweet potatoes recorded by various authors for ruminant livestock species*

Animal	Energy/Total protein value	Fresh roots	Dried roots	Fresh vines	Dried vines	Vine silage
Cattle	*Energy*					
(unspecified)	DE (MJ/kg) (as fed)	4.7[a]		2.4[b]		
	ME (MJ/kg)	3.8[a]		2.0[b]		
Beef	ME (MJ/kg)		10.5[c]		8.6[c]	
	TDN (%)		79[d]			
(*In vitro*	DM digestibility (%)	92.0[c]		62.0[c]		
determination)	DM digestibility (%)			75.1[e]		
Dairy	TDN (%)		77–80[f]			
	Total (crude) protein					
Dairy	Digestibility		3.2[f]			
	coefficient (%)		−25.2[f]			
Beef	Digestibility coefficient (%)		50.8[f]			
Unspecified	Digestible	0.3[a]		1.1[g]	6.9[g]	
	protein (%)			1.8[a]	7.6[gg]	
Sheep	*Energy*					
	DE (MJ/kg) (as fed)	4.8[a]		2.2[a]		
	ME (MJ/kg) (as fed)	3.9[a]	11.3[b]	1.8[a]		
	TDN (%)					41.3–56.9[i] (fresh)
						42.0–54.3[i] (wilted)
	Total protein					
	Digestibility	37.5[f]	14.0[b]	80.0[b] (leaves)	8.6–11.7[i] (fresh)	
	coefficient (%)				23.0–52.7[i] (wilted)	
Lambs	Digestibility coefficient (%)	19.8[f]				

Goats	*Energy*					
	TDN (%)					56.0–65.1[j]
	Total protein					
	Digestibility coefficient (%)					49–68[j]
	Digestible protein (%)	0.7[a]		1.9[a] 1.3[g]	8.1[g]	1.2–1.8[j]
Rabbits	*Total protein*					
	Digestibility coefficient (%)					44–69[j]
	Digestible protein (%)			1.9[a]		
Unspecified	*Energy*					
	TDN (%)	27.6[k]	72.7[k] 69.0[kk]		51.7[k]	
	Digestibility coefficient (%)		14[k]			
	Digestible protein (%)	0.2[k]	0.7[k] 0.4[kk]	8.9[k]		

Notes:
DE, digestible energy; ME, metabolizable energy; TDN, total digestible nutrients; DM, dry matter.
[a] Atlas of nutritional data, 1971.
[b] Average of values from Atlas of nutritional data (1971) and Angeles, 1977.
[c] Backer et al., 1980; MEs calculated by authors.
[d] Chen and Chen, 1979.
[e] Ruiz et al., 1980.
[f] As given by Yeh and Bouwkamp (1985).
[g] Angeles, 1977; [gg] mature vines.
[h] Gohl, 1981.
[i] Gerpacio et al., 1967.
[j] Sutoh et al., 1973.
[k] Morrison, 1957; [kk] sweet potato residue from starch manufacture.

several livestock species including pigs, cattle and poultry. Both sweet potato roots and tops can be used in various forms, not only as fresh materials but in the more convenient and storable forms of dried meal and fermented silage. In addition, wastes from sweet potato processing plants can also be converted into useful feedstuffs. However, in many cases the energy value of sweet potato, even on a dry basis, was not found to equal that of the most common feed ingredient, corn. Substitution of part of the corn with sweet potato above a moderate level often led to a decrease in livestock weight gains and feed efficiencies. Heat treatment of roots may be needed to improve nutritional values, but might be avoided by selection or breeding of cultivars with better starch digestibility and low or absent trypsin inhibitor activity.

The chief disadvantages to increased use of sweet potato as a livestock feed is its cost due to its poor storability in a fresh form, its low energy density on a fresh basis and, in the case of roots, low protein content. This has been shown in several local tropical situations to limit its profitability and render it uncompetitive with cereal grains. However, such disadvantages might be overcome by encouraging greater production of sweet potatoes by farmers, particularly if greatly improved yields could be combined with significantly increased dry matter contents. Alternatively, increasing use of 'waste' products in the form of foliage and damaged unsaleable roots could be utilized by farmers in an integrated system of sweet potato and animal production.

Note

Gross energy is the total energy released as heat when the feed is completely burned in a bomb calorimeter.

Digestible energy allows for loss in the faeces (incomplete digestion).

Metabolizable energy makes an allowance for the energy of urea excreted in the urine as the physiological end-product of protein metabolism, and for digestibility.

Net energy makes an allowance for the energy lost by *specific dynamic action* (thermogenesis) stimulation of heat output by food.

References

Abou-Raya, A.K., Abou-Hussein, E.R.M., Ghoneim, A., Raafat, M.A. and Mohamed, A.A. 1966. Effect of $Ca(OH)_2$ and NaOH treatments on the nutritive value of maize stalks, sorghum stalks and dry sweet potato vines. *United Arab Republic J. Anim. Prod.* **4** (1): 55–65.

Achata, A., Fano, H., Goyas, H., Chiang, O. and Andrade, M. 1988. [*The sweet potato (Ipomoea batatas* L. (Lam.)) *in the Cañete valley of Peru. Analysis of the present situation and of the research requirements of producers and consumers*] Spanish. International Potato Center/INIAA, Lima.

Ando, H. 1962. [Use of sweet potato for fattening beef cattle] Japanese. *J. Japan. Grassland Sci.* **8** (1): 67–71.

Angeles, H.J.B. 1977. [*Dietary evaluation of the principal flora and of some species of the fauna of Peru. (Table of feed composition)*] Spanish. Min. Aliment. Bol. Tec. No. 83, Lima.

Atlas of nutritional data on United States and Canadian feeds. 1971. National Academy of Sciences, Washington, DC.

Backer, J. 1976. [Integral utilization of sweet potato (*Ipomoea batatas* (L) Lam.) in the production of meat] Spanish. M.Sc. thesis, University of Costa Rica, Turrialba.

Backer, J., Ruiz, M.E., Muñoz, H. and Pinchinat, A.M. 1980. The use of sweet potato (*Ipomoea batatas* (L) Lam.) in animal feeding. II. Beef production. *Trop. Anim. Prod.* **5**: 152–60.

Beszedits, S. and Netzer, A. 1982. *Protein recovery from food processing wastewaters.* B & L Information Services, Toronto, Ontario.

Bond, J. and Putnam, P.A. 1967. Nutritive value of dehydrated sweet potato trimmings fed to beef steers. *J. Agric. Food Chem.* **15** (4): 726–8.

Briggs, H.M., Gallup, W.D., Helier, V.G., Darlow, A.E. and Cross, F.B. 1947. The digestibility of dried sweet potatoes by steer and lambs. *Okla. Agric. Expt. Stn Tech. Bull.* T-28.

Brown, D.L. and Chavalimu, E. 1985. Effects of ensiling or drying on five forage species in western Kenya: *Zea mays* (maize stover), *Pennisetum purpureum* (Pakistan napier grass), *Pennisetum* sp. (bana grass), *Ipomoea batata* (sweet potato vines) and *Cajanus cajan* (pigeon pea leaves). *Anim. Feed Sci. Technol.* **13** (1/2): 1–6.

Carpio, W.N. 1984. A comparative study of feedlot fattening performance of native goats fed with sweet potato leaves and combination with kikuyu stargrass forage. BS thesis, Mountain State Agricultural College, La Trinidad, Beneguet.

Castillo, L.S., Aglibut, F.B., Javier, T.A., Gerpacio, A.L., Garcia, G.V., Puyaoan, R.B. and Ramin, B.B. 1964. Camote and cassava tuber silage in swine rations. *Agric. Los Baños*, April–June, pp. 11–3.

Chang, Y.H. and Oh, J.S. 1963. [Studies on fermented feedstuffs. I. Determination of optimum levels for various supplements in koji fermentation using sweet potato and rice straw] Korean. *Res. Rep. RDA* **6** (3): 61–5.

Chen, M.C. and Chen, C.P. 1979. [Studies on the utilization of sweet potato chips and cassava pomace for cattle] Chinese. *J. Agric. Assoc. China* **107**: 45–54.

Chen, M.C., Chen, C.P. and Din, S.L. 1979. [The nutritive value of sweet potato vines for cattle. 5. Fresh and dehydrated sweet potato vines] Chinese. *J. Agric. Assoc. China, New Ser.* **107**: 55–60.

Chen, Y., Yeh, T.P., Yang, Y.S., Chang, T.C., Wu, M.C., Siao, C.M. and Wang, W.L. 1980. [Studies on the formula feeds with sweet potatoes as diets for growing-finishing pigs] Chinese. *Anim. Ind. Res. Inst., Taiwan Sugar Corp.,*

Ann. Rep. pp. 95–101.
Chen, M.C., Yi, J.J. and Hsu, T.C. 1977. [The nutritive value of sweet potato vines produced in Taiwan for cattle: (3) feeding value for milk production, (4) feeding value for growth at different forms] Chinese. *J. Agric. Assoc. China, New Ser.* **99**: 39–45.
Diñgayan, A.B. and Fronda, F.M. 1950. A comparative study of the influence of the leaves and young shoots of centrosema, ipil-ipil, and sweet potato as green feed on the growth of chicks. *Philipp. Agric.* **34**: 110–5.
Dreher, M.L., Dreher, C.J. and Berry, J.W. 1984. Starch digestibility of foods: a nutritional perspective. *CRC Crit. Rev. Food Sci. Nutr.* **20**: 47–51.
Duell, B. 1985. Post-World War II change in sweet potato production and use in Japan. *J. Int. Coll. Commerce Econ.* **32**: 43–51.
1989. Variations in sweet potato consumption in Japan. *J. Tokyo Int. Univ.* **39**: 55–67.
Fetuga, B.L. and Oluyemi, J.A. 1976. The metabolizable energy of some tropical tuber meals for chicks. *Poult. Sci.* **55** (3): 868–73.
Ffoulkes, D., Hovell, F. de B. and Preston, T.R. 1978. Sweet potato forage as cattle feed: voluntary intake and digestibility of mixtures of sweet potato forage and sugar cane. *Trop. Anim. Prod.* **3** (2): 140–4.
Ffoulkes, D. and Preston, T.R. 1978. [Cassava or sweet potato forage as a combined source of protein and forage in molasses diets: effect of supplementation with soy meal] Spanish. *Trop. Anim. Prod.* **3**: 188–94.
Furuya, S. and Nagano, R. 1986. [Differences of nutritive values among the variety of sweet potato and among the processing methods including drying and boiling of sweet potato, potato and cassava] Japanese. *Jap. J. Swine Sci.* **23** (2): 62–7.
Garlich, J.D., Bryant, D.M., Covington, H.M., Chamble, D.S. and Purcell, A.E. 1974. Egg yolk and broiler skin pigmentation with sweet potato vine meal. *Poult. Sci.* **53** (2): 692–9.
Gerona, G.R. 1983. *Studies on the utilization of rootcrops as energy sources in duck rations.* (Available from ViSCA, Baybay, Leyte).
Gerpacio, A.L., Aglibut, F.B., Javier, T.R., Gloria, L.A. and Castillo, L.S. 1967. Digestibility and nitrogen balance studies on rice straw and camote vine leaf silage of sheep. *Philipp. Agric.* **51** (3): 185–95.
Gitomer, C.S. 1987. *Sweet potato and white potato development in China. A compendium of basic data.* International Food Policy Research Institute, Washington, DC.
Gohl, B. 1981. *Tropical feeds. Feed information summaries and nutritive values.* FAO Animal Production and Health Ser. No. 12, Rome.
Goyert, J.C. and Avault, J.W. 1977. Agricultural by-products as supplemental feed for crayfish, *Procambarus clarkii. Trans. Am. Fish. Soc.* **106** (6): 629–33.
Han, I.K., Ha, J.K., Yoon, D.J. and Kim, C.S. 1976b. [Studies on the nutritive values of dried sweet potato meal. 2. Replacement of corn by dried sweet potato meal in the growing swine ration] Korean. *Seoul Natl. Univ. Coll. Agric. Bull.* **1** (1): 75–80.
Han, I.K., Yoon, D.J. and Kim, C.S. 1976a. [Studies on the nutritive values of dried sweet potato meal. 1. Partial substitution of dried sweet potato meal for

corn in the ration of growing swine] Korean. *Seoul Natl. Univ. Coll. Agric. Bull.* **1** (1): 67–74.

Ho, H-K. and Ting, I-N. 1975. [Properties and utilization for feeding of hard waste vines of sweet potato] Chinese. *J. Chin. Agric. Chem. Soc.* **13** (1–2): 15–30.

Hosoyamada, H. and Kodama, S. 1959. [On the feeding of sweet potatoes as a hog feed (3,4,5)] Japanese. *Kyushu Agric. Res.* **21**: 220–2.

Hosoyamada, H., Nakashima, T. and Suzuki, Y. 1962. [On the feeding value of sweet potatoes and starch feed as a hog feed] Japanese. *Kyushu Agric. Res.* **24**: 182–3.

Huang, W-Y. and Olbrich, S. 1979. *Feed potential of sweet potatoes in Hawaii.* Hawaii Agricultural Experimental Station Departmental Paper 57, University of Hawaii.

International Potato Center/AVRDC/IFPRI 1987. *Sweet potato research in the People's Republic of China.* International Potato Center, Lima.

Iura, M., Sakai, K. and Marumine, S. 1958. [Effect of early cutting of sweet potato tops used as feed on yield and quality of tubers] Japanese. *Kyushu Agric. Res.* **20**: 43–5.

Job, T.A., Oluyemi, J.A. and Entonu, S. 1979. Replacing maize with sweet potato in diets for chicks. *Br. Poult. Sci.* **20** (6): 515–19.

Jung, C.H. and Lee, J.H. 1968. [A study on substitution of sweet potato ensilage for concentrates in pork production] Korean. *Res. Rep. RDA* **11** (4): 35–43.

Jung, C.Y., Paik, B.H. and Lee, K.H. 1976. [Studies on feeding value of sweet potato silage for growing finishing pigs] Korean. *Res. Rep. RDA* **19**: 87–91.

Kang, H.S. 1963. [A study on the feed value of fermented starch waste for growing chicken] Korean. *Jinju Agric. Coll. Bull.* **2**: 63–6.

Kawaita, H., Fukumoto, M., Kusumoto, S., Miyauchi, Y., Tomita, Y. and Kojima, M. 1979. [Effects of sweet potato feeding on porcine fat melting point and fatty acid composition] Japanese. *Jap. J. Swine Sci.* **16** (2): 184.

Kim, C.S., Lee, N.H., Han, I.K., Ha, J.K. and Yoon, D.J. 1976. [Studies on the nutritive value of the sweet potato for feed. III. Feeding performance in pig grower] Korean. *Korean J. Anim. Sci.* **18** (3): 220–4.

Kim, H.S., Cho, D.H. and Park, T.J. 1962. [Selection of superior strains for the kojification of sweet potato starch] Korean. *Res. Rep. ORD* **5**: 188–93.

Kim, H.U. 1973. [Effect of dried sweet potato supplemented with tallow on the growth of broiler chicken] Korean. *Korean J. Anim. Sci.* **15** (4): 283–8.

Kim, S.C., Oh, S.J. and Lee, K.S. 1959. [A study on the value of sweet potato ensilage fed to laying poultry] Korean. *Res. Rep. RDA* **2**: 101–8.

Kimber, A.J. 1972. The sweet potato in subsistence agriculture. *Papua New Guinea Agric. J.* **23** (3/4): 80–95.

Koh, F-K., Yeh, T-P. and Yen, H-T. 1976. [Effects of feeding sweet potatoes, dried sweet potato chips and sweet potato vine on the growth performance of growing-finishing pigs] Chinese. *J. Chin. Soc. Anim. Sci.* **5** (3–4): 55–67.

Konaka, N. 1962. [Use of sweet potato stolon for silage] Japanese. *J. Japan. Grassland Sci.* **8** (1): 64–7.

Koo, N.W. and Kim, D.K. 1973. [The feeding value of dried sweet potato pulp for growing-finishing swine] Korean. *Kor. J. Animal Sci.* **15** (4): 313–17.

Kubayashi, K. 1964. [Nutritional studies on the utilization of distillers' stillage. II. Insolubles of sweet potato-alcohol distillers' stillage] Japanese. *J. Agric. Chem. Soc. Japan* **38** (7): 317–22.

Kurihara, T. and Imamura, T. 1956. [Experiments on feeding of sweet potatoes-barley bran silage, sweet potatoes-barley bran meal and dry sweet potatoes for growing pigs] Japanese. *Kyushu Agric. Res.* **17**: 140–1.

1958. [The effect of boiling on sweet potatoes and starch feeds as a hog feed. II.] Japanese. *Kyushu Agric. Res.* **20**: 168–70.

1959. [On utilization of grasses in feeding of hog. Part 1. The effect of sweet potato vines on growing swine] Japanese. *Kyushu Agric. Res.* **21**: 216–18.

Kurihara, T., Imamura, T., Sakai, Y., Nagano, R. and Nagao, T. 1957. [On utilization of sweet potatoes in feeding of hog] Japanese. *Kyushu Agric. Res.* **19**: 5–8.

Kwack, C.H. and Ahn, B.H. 1975. [Studies on the utilization of sweet potato as a poultry feed. I. Substitution level of pelleted sweet potato for corn in growing chick rations] Korean. *Korean J. Anim. Sci.* **18** (3): 261–70.

Lee, J.M., Jeung, S.Y. and Jeung, C.H. 1964. [Studies on the marketing age of hogs fed by sweet potato ensilage] Korean. *Res. Rep. ORD* **17** (3): 15–21.

Lee, P.K. and Lee, M.S. 1979. [Study on hog feed formula of using high protein sweet potato chips and dehydrated sweet potato vines as the main ingredient] Chinese. *Taiwan Livest. Res.* **12** (1): 49–71.

Lee, B.O. and Moon, S.S. 1969. [Effects of sweet potato silage on the productivity of layers] Korean. *Korean J. Anim. Sci.* **11** (3): 310–13.

Lee, P.K. and Yang, Y.F. 1974. [Comparative study on petroleum yeast, soybean meal as protein supplement, and dried sweet potatoes, corn meal as basal feed in the nutrition of growing fattening pigs] Chinese. *T'ai wan Nung Yeh Chi K'an* **10** (1): 165–85.

1979a. [Comparative study of high protein sweet potato chips and common sweet potato chips as substitute for corn grain in diet on growth, feed efficiency and carcass quality of the growing fattening pigs] Chinese. *Taiwan Livestock Res.* **12** (1): 31–48.

1979b. [Nutritive value of high protein sweet potato chips as feed ingredients for broilers] Chinese. *J. Agric. Assoc. China* **106**: 71–8.

1979c. [Nutritive value of high protein sweet potato meal as feed ingredient for Leghorn chicks] Chinese. *J. Agric. Assoc. China* **108**: 56–65.

1980. [Comparison of high protein sweet potatoes and common sweet potatoes on nutritive value of the growing fattening pigs] Chinese. *Taiwan Livestock Res.* **13** (1): 97–112.

1981. [Study on digestibility of crude protein and energy with the pigs fed on diets containing locally produced corn meal, sorghum grains, sweet potato chips or cassava meal] Chinese. *Taiwan Livestock Res.* **14** (1): 65–74.

Lin, S.S.M., Peet, C.C., Chen, D-M. and Lo, H-F. 1983. Breeding goals for sweet potato in Asia and the Pacific: a survey on sweet potato production and utilization. *Proc. Am. Soc. Hort. Sci. Trop. Reg.* **27B**: 42–53.

Magay, E.J. 1972. Utilization of carotene from yellow sweet potato by broilers and digestibility of sweet potato rations by swine. M.Sc. thesis. University of the Philippines.

Malynicz, G.L. 1971. Use of raw sweet potato, raw peanuts and protein concentrate in rations for growing pigs. *Papua New Guinea Agric. J.* **22** (3): 165–6.

Malynicz, G.L. and Nad, H. 1973. The effect of level of feeding and supplementation with sweet potato foliage on the growth performance of pigs. *Papua New Guinea Agric. J.* **24** (4): 139–41.

Massey, Z.A. 1943. Sweet potato meal as livestock feed. *Ga. Agric. Expt. Stn Press Bull.* No. 522.

Matsuda, Y. 1968. [Studies on poultry feeding with sweet potato. I. The effect of casein and rice hull] Japanese. *Bull. Fac. Agric. Miyazaki Univ.* **15** (2): 176–85.

1969. [Studies on poultry feeding with sweet potato. III. The effect of tallow] Japanese. *Bull Fac. Agric. Miyazaki Univ.* **16** (2): 126–36.

1970. [Studies on poultry feeding with sweet potato. V. The effect of methionine and glycine] Japanese. *Bull Fac. Agric. Miyazaki Univ.* **17** (2): 241–7.

Mohamed, O.E., Omar, E.M. and Kamar, G.A. 1974. The use of sweetpotato in chicken rations for growth and egg production. *Egypt. J. Anim. Prod.* **14** (2): 229–35.

Montanez, E.A. 1982. The growth performance of growing-finishing pigs fed with cooked sweet potato as a source of energy by selected farmers of Baybay, Leyte. BSA thesis, ViSCA, Baybay, Leyte.

Morrison, F.B. 1957. *Feeds and feeding*, 22nd edn. The Morrison Publ. Co., Ithaca, NY.

Mwanga, R.O.M. and Wanyera, N.W. 1988. Sweet potato growing and research in Uganda. In: *Improvement of sweet potato (Ipomoea batatas) in East Africa, with some references of other tuber and root crops.* Report of the 'Workshop on sweet potato improvement in East Africa', held at ILRAD, Nairobi, Kenya, 28 September–2 October 1987. (UNDP Project CIAT-CIP-IITA). International Potato Center, Lima, pp. 187–96.

Oyenuga, V.A. and Fetuga, B.L. 1975. Chemical composition, digestibility and energy values of some varieties of yam, cassava, sweet potatoes and cocoyams for pigs. *Nigerian J. Sci.* **9** (1): 63–110.

Pagtan, J.Y. 1984. Broiler response to different levels of sweet potato. BS thesis, Mountain State Agricultural College, La Trinidad, Benguet.

Palomar, M.K., Bulayog, E.F. and Truong, V.D. 1989. Sweet potato research and development in the Philippines. In: *Improvement of sweet potato (Ipomoea batatas) in Asia.* Report of the 'Workshop on sweet potato improvement in Asia', held at ICAR, India, 24–28 October 1988. International Potato Center, Lima, pp. 79–85.

Pascual, N.P., Abamo, A.P. and Binongo, M.S.G. 1987. Economic tests for profitability, marketability, and alternative uses of sweet potato in the Philippines. Paper presented at an International Symposium on Sweet Potato, 20–26 May, ViSCA, Baybay, Leyte.

Raines, B.W. 1973. Characterization of sweet potato processing wastes. Thesis. North Carolina State University.

Rose, C.J. 1981. Preliminary observations on the performance of village pigs (*Sus scrofa papuensis*) under intensive outdoor management. Part I. Dietary intake and liveweight gain. *Sci. New Guinea* **8** (2): 132–40.

Rose, C.J. and White, G.A. 1980. Apparent digestibilities of dry matter, organic matter, crude protein, energy and acid detergent fibre of chopped, raw sweet potato (*Ipomoea batatas* (L.)) by village pigs (*Sus scrofa papuensis*) in Papua New Guinea. *Papua New Guinea Agric. J.* **31** (1/4): 69–72.

Rosenberg, M.M. and Seu, J. 1952. Sweet potato root meal vs. yellow corn meal in chicks' diet. *World Poult. Sci.* **8** (2): 93–8.

Ruiz, M.E. 1982. Sweet potatoes (*Ipomoea batatas* (L.) Lam.) for beef production: agronomic and conservation aspects and animal responses. In: Villareal, R.L. and Griggs, T.D. (eds.), *Sweet potato*, Proceedings of the First International Symposium. AVRDC, Shanhua, T'ainan, pp. 439–51.

Ruiz, M.E., Lozano, E. and Ruiz, A. 1981. Utilization of sweet potatoes (*Ipomoea batatas* (L.) Lam) in animal feeding. III. Addition of various levels of roots and urea to sweet potato forage silages. *Trop. Anim. Prod.* **6**: 234–44.

Ruiz, M.E., Pezo, D. and Martinez, L. 1980. The use of sweet potato (*Ipomoea batatas* (L.) Lam.) in animal feeding. I. Agronomic aspects. *Trop. Animal Prod.* **5**: 144–51.

Saure, R.V. 1972. Sweet potato meal as a replacement for corn in iso-nitrogenous and iso-caloric broiler and swine rations. M.Sc. thesis, University of the Philippines.

Shin, O.Y., You, H.U. and Chang, Y.H. 1964. [Effect of sweet potato starch waste supplemented with ammonium sulphate on egg production] Korean. *Res. Rep. ORD* **7** (3): 7–13.

Sistrunk, W.A. and Karim, M.I. 1977. Disposal of lye-peeling wastes from sweet potatoes by fermentation for livestock feed. *Ark. Farm Res.* **26** (1): 8.

Smallwood, C., Whitaker, R.S. and Colston, N.V. 1974. *Waste control and abatement in the processing of sweet potatoes.* Environmental Pollution Agency – 660/2-73-021, Corvallis, OR.

Springhall, J.A. 1964. Locally available ingredients for poultry rations in New Guinea. *Proc. Austr. Poult. Conv.*, pp. 123–6.

Susaki, S. and Hamakawa, H. 1959. [Feeding experiment of sweet potato on male chicks] Japanese. *Bull. Fac. Agric. Miyazaki Univ.* **4** (2): 236–43.

Sutoh, H., Uchida, S. and Kaneda, K. 1973. [Studies on silage making. XXII. The nutrient content of sweet potato (*Ipomoea batatas* L. var. *edulis*) at the different stages and the quality of sweet potato vine silage] Japanese. *Sci. Rep. Fac. Agric. Okayama Univ.* **41**: 61–8.

Sweeten, J.M., Etheredge, R.S., Lawhon, J.T., Egg, R.P., Coble, C.G., Schelling, G.T. and McBee, G.G. 1983. Stillage processing for nutrient recovery. Paper No. 83–3059 presented at the 1983 summer meeting of the American Society of Agricultural Engineers, Montana State University.

Szylit, O., Durand, M., Borgida, L.P., Atinkpahoun, H., Prieto, F. and Delort-Laval, J. 1978. Raw and steam-pelleted cassava, sweet potato and yam (*cayenensis*) as starch sources for ruminant and chicken diets. *Anim. Feed Sci. Technol.* **3**: 73–87.

Tai, N-L. and Lei, T-S. 1970. [Determination of proper amount of yellow corn and dried sweet potatoes in swine feed] Chinese. *J. Agric. Assoc. China* **70**: 71–6.

Takahashi, S. and Furuya, S. 1968. [Influence of carbohydrate sources in swine rations on the quality of carcass fat] Japanese. *Bull. Natl. Inst. Anim. Ind. (Chiba)* **16**: 45–9.

Teixeira de Matos, A. and Xavier de Almeida, E. 1989. [Processing and use of grated manioc and sweet potato] Portuguese. *Agrop. Catarinense* **2** (1): 5–6.
Tillman, A.D. and Davis, H.J. 1943. Studies on the use of dehydrated sweet potato meal in chick rations. *Louisiana Expt. Stn Bull.* **358**.
Tomita, Y., Hayashi, K. and Hashizume, T. 1985. [Palatability to pigs of sweet potato-silage and digestion trial by them] Japanese. *Bull. Fac. Agric. Kagoshima Univ.* No. 35, pp. 75–80.
Tsou, C.S. and Hong, T-L. 1989. Digestibility of sweet potato starch. In: *Improvement of sweet potato (Ipomoea batatas) in Asia.* Report of the 'Workshop on sweet potato improvement in Asia', held at ICAR, India, 24–28 October, 1988. International Potato Center, Lima, pp. 127–36.
Turner, W.J., Malynicz, G.L. and Nad, H. 1976. Effect of feeding rations based on cooked sweet potato and a protein supplement to broiler and crossbred poultry. *Papua New Guinea Agric. J.* **27** (3): 69–72.
Villalva, L.D. 1972. The effect of yellow sweet potato meal as a supplement to corn meal on the egg production of Hyline layers. BSA thesis, Palawan National Agricultural College, Aborlan, Palawan.
Villareal, R.L. 1982. Sweet potato in the tropics – progress and problems. In: Villareal, R.L. and Griggs, T.D. (eds.), *Sweet potato*, Proceedings of the First International Symposium. AVRDC, Shanhua, T'ainan, pp. 3–15.
Walter, W.M., Purcell, A.E. and McCollum, G.K. 1978. Laboratory preparation of a protein-xanthophyll concentrate from sweet potato leaves. *J. Agric. Food Chem.* **26** (5): 1222–6.
Watson, J.B. 1977. Pigs, fodder, and the Jones effect in postipomoean New Guinea. *Ethnology* **16** (1): 57–70.
Waugh, W.F. 1963. Sweet potatoes as pig feed. *Fmg S. Afr.* **38** (10): 12–13.
Weber, C.W. 1969. Trends and forecasts. In: *Proceedings of the 13th Annual Poultry Industry Day*, Department of Poultry Science, University of Arizona, Tucson, AZ, pp. 55, 58.
Wiersema, S.G. 1989. International trip report (to Vietnam), 1–7 April. International Potato Center, Lima. [Mimeo]
Wiersema, S., Hesen, J.C. and Song, B.F. 1989. Report on a sweet potato postharvest advisory visit to the People's Republic of China, 12–27 January. International Potato Center, Lima. [Mimeo]
Wu, J.F. 1980. Energy value of sweet potato chips for young swine. *J. Anim. Sci.* **51** (6): 1261–5.
Wu, Y.V. 1988. Characterization of sweet potato stillage and recovery of stillage solubles by ultrafiltration and reverse osmosis. *J. Agric. Food Chem.* **36** (2): 252–6.
Wu, Y.V. and Bagby, M.O. 1987. Recovery of protein-rich byproducts from sweet potato stillage following alcohol distillation. *J. Agric. Food Chem.* **35** (3): 321–5.
Yamada, T., Moriya, N., Yoshihara, K. and Hoshino, M. 1962. [Studies on growing sweet potato crop as a forage in view point of utilization of both top and root] Japanese. *J. Jap. Grassl. Sci.* **8** (1): 45–54.
Yeh, J.F. 1982. Utilization of sweet potatoes for animal feed and industrial uses: potential and problems. In: Villareal, R.L. and Griggs, T.D. (eds.), *Sweet*

potato, Proceedings of the First International Symposium. AVRDC, Shanhua, T'ainan, pp. 385–92.

Yeh, T.P. and Bouwkamp, J.C. 1985. Roots and vines as animal feed. In: Bouwkamp, J.C. (ed.), *Sweet potato products: a natural resource for the tropics.* CRC Press Inc., Boca Raton, FL, pp. 235–53.

Yeh, T.P., Wung, S.C., Koh, F.K., Lee, S.Y. and Wu, J.F. 1977. [Improvements in the nutritive values of sweet potato chips by different methods of processing. *Ann. Res. Rep. Anim. Ind. Res. Inst.*, Taiwan Sugar Corp., pp. 89–95.

Yeh, T.P., Wung, S.C., Lin, H.K. and Kuo, C.C. 1978. [Studies on different methods of processing some local feed materials to enhance their nutritive values for swine. I. Popping sweet potato chips] Chinese. *Ann. Res. Rep. Anim. Ind. Res Inst.*, Taiwan Sugar Corp., pp. 25–31.

Yoshida, M., Hoshii, H. and Morimoto, H. 1960. [Nutritive value of sweet potato as a carbohydrate source for poultry feed. III. Effect of vitamin A supplementation on chick growth] Japanese (with extensive English summary). *Bull. Natl. Inst. Agric. Sci.* **19**: 133–42.

1962a. Nutritive value of sweet potato as carbohydrate source in poultry feed. IV. Biological estimation of available energy of sweet potato by starting chicks. *Agric. Biol. Chem.* **26** (10): 679–82.

1962b. Nutritive value of sweet potato as carbohydrate source in poultry feed. V. Reliability of available energy value of sweet potato estimated by feeding experiment. *Agric. Biol. Chem.* **26** (10): 683–8.

Yoshida, M. and Morimoto, H. 1957. [Nutritive value of sweet potato as a carbohydrate source for poultry feed (I).] Japanese (with extensive summary in English). *Bull. Natl. Inst. Agric. Sci.* **13**: 123–32.

1959. Nutritive value of sweet potato as carbohydrate source in poultry feed. II. Effect of sweet potato feeding on day-old chicks. *Bull. Natl. Inst. Agric. Sci.* **18**: 7–14.

Zarate, J.J. 1956. The digestibility by swine of sweet potato vines and tubers, cassava roots, and green papaya fruits. *Philipp. Agric.* **40**: 78–83.

CHAPTER 8

Consumption and utilization patterns and trends

Previous chapters of this book detail the chemical and nutritional aspects of sweet potato as a human food, based on scientific studies carried out over many years. They also deal with the technological and scientific aspects of processing and feeding to livestock. The data thus far presented indicate that sweet potato can make a significant nutritional contribution to human health and that a wide range of food and non-food products is potentially available. The present chapter discusses further dimensions: patterns and trends in human consumption and utilization and the factors which influence them. These factors, which include socio-economic as well as chemical and technical aspects, have so far been little researched and are poorly documented.

In spite of the sweet potato's high productivity and its potential contribution to food supplies especially in poor areas, it is presently underexploited as a direct human food, and as a feedstuff for animals. In some traditional sweet potato growing areas production is decreasing as food consumption patterns change to imported cereal-based foods. However, sweet potato production is increasing elsewhere. If the sweet potato is to be fully exploited as a food resource the factors influencing consumption and use need to be analysed and understood. The potential use of sweet potato as an industrial raw material, whether on a family or factory scale, also needs to be explored particularly in tropical areas.

Although historically sweet potato has received little attention in crop improvement programmes, this situation is now being remedied by its inclusion in the programmes of international research organizations such as the International Potato Center. Efforts to raise the quantity of sweet potato available for consumption or other uses may proceed through expansion of planting area, improvements in yield, development of storage facilities and reduction in wastage. However, put simply, these efforts could be fruitless if consumption and utilization do not proceed apace. Expansion of production will not be sustained unless demand for sweet potatoes is stimulated by diversification of uses. In some areas, expanded sweet potato production could help to reverse trends towards increased consumption of imported food crops, especially cereals. This would result in foreign exchange savings and greater food security. In countries where rice or other cereals are the preferred staple, sweet potato could be used to fill gaps in food consumption caused by seasonal scarcities or high price of the major staple. Effective sweet potato improvement programmes must be based on a knowledge of where growth in production will be used: for consumption, livestock feed or processing on a home, village or industrial scale. Markets for sources of storable, nutritious and affordable sweet potato products may already exist, or may need to be created. Different uses may be most appropriate for alleviating food shortages and improving nutritional status in different countries.

Appropriate policies to enhance production and utilization can be formulated by targeting priority areas or groups for sweet potato promotion. The major factors influencing sweet potato consumption and utilization must be known. Consumer demand may respond to changes in the availability, price, quality or sensory characteristics of sweet potato. Thus consumer preferences and the constraints to increased consumption have to be considered in the design of sweet potato promotion or production programmes. By means of a review of the sparse and fragmented existing information I sought tentative answers to the following questions:

What levels and patterns of sweet potato consumption and utilization are found in different countries, and among which groups of people?
Which major factors influence sweet potato consumption and utilization?
What is the potential for expanding sweet potato consumption and utilization?
How can agricultural research and development best contribute to greater utilization of the sweet potato?

Several approaches to answering these questions are used. First, sweet potato availability and methods of utilization are compared for major regions and countries. Secondly, consumption and utilization in selected countries are examined in case studies. For the third type of approach, results of surveys conducted among sweet potato researchers are described. On the basis of information obtained from the case studies and from the few descriptions in the literature, the main factors affecting sweet potato consumption and utilization are then reviewed. Finally, future prospects are discussed.

The information presently available is incomplete. More detailed studies are fortunately underway or are planned by researchers. It is hoped that the information given here will stimulate further investigation into all aspects of sweet potato consumption and utilization.

A typology of sweet potato consumption

In considering sweet potato consumption patterns in different countries, particular attention was paid to how sweet potato is used in the diet. Poats (see Woolfe, 1987) developed a useful typology of potato consumption that refers to the quantity consumed at a 'typical' meal, the frequency of consumption, and the relationship of potato to other items in the meal. A typology of sweet potato consumption has been developed based on Poats' typology for potato and including additional factors.

Poats defined four roles for the potato: staple or co-staple, complementary vegetable, luxury or special food and non-food. A typology of sweet potato consumption was defined as shown in Table 8.1. The various roles of sweet potato as defined here will be referred to further in the country case studies. As a *main staple* a food occupies the entire meal plate, is regarded as a source of energy, having high satiety value and is an item without which a meal is incomplete. The sweet potato plays this role in the Highlands of Papua New Guinea. A *co-staple* may be a staple which occupies a place on the main meal plate equal to that of one or more other staples. This can occur with sweet potato (for example in some Asian countries where sweet potato and rice are cooked together). Other categories of co-staple in relation to sweet potato contain a temporal dimension. Sweet potato may occupy the position of staple alternating with other staples at different times of the week. It may also attain importance as a staple only on a seasonal basis when, for example, other foods are considered to be unsuitable in consistency for the time of year, are in short supply, or are too expensive. For sweet potato a preferable name for this role might be *seasonal or supplementary staple*.

A *complementary vegetable* serves as a major or minor side dish or addition to the principal staple(s). It may occur in the meal in relatively smaller quantities than the staple, but with a fairly high frequency. For sweet potato the distinction between co-staple and complementary vegetable can be somewhat blurred. Sweet potato may be present in a main dish as only one of several components, even though it is at least partly fulfilling the role of energy source. In Argentina and Uruguay it is typically included in a dish called *puchero* along with potatoes, carrots,

Table 8.1. *A proposed typology of sweet potato consumption*

Role	Frequency	Rate	Beliefs
Major staple	Daily or almost daily	> 100 kg per year	Sweet potato = food/energy Meal incomplete without sweet potato
Co-staple	Daily or almost daily	?	As above, but shares this position in the meal with other staples
Co-staple	Once or twice a week	?	Eaten as the staple, alternating with other staples on the plate
Supplementary or seasonal staple	Daily, but only for a limited period	?	Sweet potato = food/energy or is a major staple only at times of scarcity or high prices of the other preferred staples. Sweet potato = low status food
Complementary vegetable	1–7 meals per week	3–10 kg per year?	Sweet potato ≠ food/energy. Sweet potato = vegetable. Sweet potato = source of vitamins, minerals or dietary fibre
Special food	1–7 meals per week	?	Sweet potato considered to be particularly suited to 1 meal of the day, for example breakfast
Luxury/festival food	1 or more meals per year	0.1–2 kg per year?	Sweet potato ≠ food/energy. Sweet potato = food for a particular occasion Sweet potato = gift or souvenir
Non-food	Never	—	Sweet potato = food for others or unknown

squash, cabbage and meat all boiled together. Whether sweet potato in this role would be considered a co-staple or complementary vegetable is difficult to say. Farmers consume *puchero* in the southern cone of South America as a solid nourishing food to fend off the winter cold, whereas in the hot summer it is considered to be rather 'heavy'. This illustrates that the role of sweet potato either as a co-staple or complementary vegetable may be seasonal in nature.

The role of *special food* can take on more than one meaning for sweet potato. Where it is eaten very frequently but at one particular meal only, for example breakfast, sweet potato may be fulfilling the role of specialized food, but can hardly be said to be a luxury or celebratory item. It can also be a special or festival food for example for Thanksgiving dinner in the United States (and see Figure 8.1) or as a special food for carnival eaten with *chicharon* of pork in Peru. Sweet potato would normally be regarded as anything but a luxury food. However, it has been processed and packaged into forms in Japan which would certainly put it into this category. Such items are purchased mainly as gifts for friends or as tourist souvenirs.

The final role is as a food which is never consumed either because it is a food for other people or because it is unknown. Sweet potato is little known at present among much of the population of northern Europe or is still largely considered as a food for immigrants.

Information on consumption and utilization

Before describing consumption and utilization patterns, it is necessary to point out the difficulties and inaccuracies involved in collecting and analysing sweet potato statistics. Consumption estimates are usually derived from food balance sheets published by the FAO. These equate domestic production plus imports minus exports and net changes in year-end stocks to total domestic availability. The last is the sum of the quantities available for human consumption, industrial products, animal feed, seed and waste. Human consumption is often estimated as total production minus the other four quantities. The balance is then divided by the estimated total national population to obtain per capita consumption. Although this figure is often presented as per capita *consumption*, it should actually be referred to as per capita *availability for consumption*. The disadvantages of this approach to estimating root and tuber crop consumption levels have been discussed by Horton (1988) and Poats (see Woolfe, 1987).

Total production estimates are often faulty and are frequently revised by the FAO. They are often based on official estimates generated by government agencies which tend to underestimate production and consumption because sweet potatoes are often grown in isolated areas and/or on small irregular plots (Figure 8.2), for example as intercrops, relay crops, secondary crops or backyard/garden crops. As a result, they can be overlooked in national estimates. Official yield estimates may underrate the actual yields achieved by many small farmers.

Utilization data are also suspect. The quality of food balance sheet data is poor. Waste factors of 10% or 20% may be applied indiscriminately without regard to indigenous practices in peeling, cooking or other utilization. Mention is made, for example, in the literature on Papua New

Figure 8.1. Sweet potatoes being sold as an exotic vegetable in an English market. Increasing interest in sweet potatoes in non-producer countries could present export opportunities for producers (J.A. Woolfe).

Figure 8.2. Sweet potato growing along the roadside, planted at the edge of a Kenyan farmer's corn field. Reliable production statistics are difficult to gather in countries where sweet potato is grown in small or scattered plots (P. Ewell).

Guinea that certain groups of people roast unpeeled roots in the ashes of a fire, dust off the ash when roots are cooked and consume them whole (Sinnett, 1975). Waste determination is further complicated by use of peels or small or damaged roots for feeding to animals which are subsequently consumed. Farmers' practice of cutting off weevil-damaged parts of the roots and consuming the remainder has been noted in Chapter 4.

The quantity which is perhaps the most difficult to estimate is that utilized for processing. This may not be reported or is underreported. For example, balance sheets (Horton, 1988) show no sweet potato processed products in Argentina and the United States, but in practice *dulce de batata* and canned roots, respectively, represent a small but significant proportion of total production in the two countries. Roots which are reported as sold for fresh consumption may later be channelled into processing. Where the percentage of total production used for processing is small or mainly in the form of home-produced items, it may be overlooked and included under 'food'. In this respect it may be significant that, in some developing countries, many hand-crafted products are produced for sale from home or in the immediate neighbourhood by women, with consequent official disregard of their economic significance. In the particular case of China, with a very large number of small rural processing units, quantities utilized may be very difficult to assess with any precision.

National population estimates may also be inaccurate. If, in spite of all the drawbacks to its collection, the input data used is accurate, there are still further disadvantages to the use of balance sheet information. The term 'per capita availability' implies equal distribution of food among all groups of people within a country regardless of differences in race, religion, socio-economic circumstances, age and sex. In practice this does not occur. Moreover, it discounts seasonal or regional differences in consumption among these groups.

Despite all their limitations, food balance sheets provide a general indication of consumption levels and serve as a useful starting point for estimating sweet potato utilization. For many countries, the FAO food balance sheets are the only estimates of sweet potato use.

Differences among countries

In *Underground crops*, Horton (1988) gives tables presenting the FAO's most up-to-date country level estimates of root crop (including sweet potato) production and use. World trends in production have already been discussed in Chapter 1. Table 8.2 (from Horton, 1988) shows the average per capita quantity of root crops available for consumption by major region in 1984. Of the average world per capita root and tuber crop consumption availability of 70 kg, 19 kg were accounted for by sweet potato. From this table it appears that per capita consumption of sweet potato is low in all regions, with the exception of Oceania and Asia. However, a more detailed appraisal of the per capita availability of individual countries reveals that, in the two regions with average or above-average consumption, the spread of consumption levels between countries is not at all even. In Oceania, Papua New Guinea and the Solomon Islands each have well over 100 kg of sweet potato per capita per year available for consumption. Tonga, which uses about half its production for animal feeding has 65 kg available, but the other 16 countries have low or negligible (<0.5 kg) amounts available. Within Africa, for example, the group of neighbouring countries which includes Burundi, Rwanda and Uganda all have more than 100 kg per capita available for consumption. Among other African countries quantities range from negligible to 35 kg per capita per year. In Asia the largest producer of sweet potato, China, also has the highest per capita availability of sweet potato for food with 71 kg per year (77% of per capita production). Other Asian countries come far below, with negligible to 23 kg available. However, in Table 7.2 (p. 414) it can be seen that

Table 8.2. *Average quantity of root crops available for consumption by region, 1984 (kg per capita per year)*

	Sweet potato	Cassava	Aroids	Potato	Yam	Total
World	19	16	1	32	3	71
Africa	9	78	5	8	28	132
North and Central America	3	2	0	40	1	46
South America	4	41	0	32	1	78
Asia	29	8	1	19	0	57
Europe	0	0	0	81	0	81
Oceania	19	6	10	46	8	91
USSR	0	0	0	110	0	110
Developing nations	25	22	1	17	4	70
Developed nations	1	0	0	75	0	76

Notes:
From: Horton, 1988; derived from FAO data.

other estimates show much lower percentages of roots produced available for food in China as a whole. Moreover, provinces differ in the percentages of sweet potato utilized for different purposes. In Guangdong, for example, where 90% of production is used for animal feed it would therefore appear that the per capita consumption of fresh roots would be low.

As already noted, estimates for percentages of sweet potato utilized for different purposes are only approximate. For the world as a whole, FAO data put the principal use of sweet potato as direct human consumption; seed and industrial uses are insignificant and less than 15% is utilized as livestock feed (see Table 8.3). The situation in Asia, where more than 90% of the world's sweet potato is grown, is shown to mirror that in the world as a whole (Table 8.3). However, the distribution of utilization in China and Japan (see Figure 6.14, p. 392) belies these statistics. In both these countries, more than 50% of total production is utilized for purposes other than direct human consumption. A further example is the Republic of South Korea which, in 1984, utilized 17% and 39% of production for feed and processing, respectively (Horton, 1988). There is thus an urgent need to revise the FAO estimates of utilization on the basis of both the country case studies presented here and further studies from elsewhere. With revisions of these estimates in mind, the figures provided above for quantities of sweet potato available for consumption should be regarded with reservation.

Table 8.3. *Sweet potato utilization in major areas of the world, 1984*

	Utilization (% availability)				
	Food	Feed	Processed	Seed	Waste
World	77	13	3	0	6
Africa	83	3	0	1	13
North and Central America	83	4	0	3	9
South America	67	21	0	0	12
Asia	77	14	4	0	5
Europe	53	37	5	0	5
Oceania	84	2	0	0	14
Developing nations	77	13	3	0	6
Developed nations	55	13	25	4	3

Notes:
From Horton, 1988; derived from FAO data.

Some suggestions have been made (see Chapter 7 and under China below) about the reasons for the high level of livestock feed usage, especially in the form of foliage, in China. However, the origins and development of processing in China, Japan and Korea have apparently not been investigated. In Japan and some parts of China, starch making at the home or village level, which preceded today's larger-scale manufacture, was in operation many years ago. It is possible that the popularity of noodles in the diet of Asian peoples created this demand for starch. However, research into the circumstances surrounding the development of these small-scale enterprises might provide useful insights into the possibilities of establishing locally adapted processing units in other parts of the developing world.

The extensive use of sweet potato foliage for livestock feeding in some countries has already been mentioned (see Chapter 7), while it is very difficult to obtain statistics on the use of sweet potato foliage as a green vegetable. It is known, however, that in some countries sweet potato leaves or tips are as important a human food as are the roots. In other countries, where sweet potatoes are grown and the roots consumed, eating sweet potato leaves is unknown or considered to be unthinkable. The reasons for these variations should be studied for pointers to improving promotion of leaf consumption. Quantities consumed are also mostly educated guesses at present and should be investigated by field studies using direct weighing. By this means the nutritional significance of leaves in diets could be more readily assessed.

Differences within countries

Sweet potato consumption often varies considerably among areas or racial communities within a country. In the Cook Islands, sweet potato (*kumara*) is considered to be a minor crop and a filler for times when there are insufficient supplies of major crops such as taro, bananas etc. However, in the southern group of the Cook Islands, *kumara* is one of the most important root crops (Tavioni, 1982). The Solomon Island diet has traditionally been based on staples of taro, yam and sweet potato. In eastern Malaita (one of the Solomon Islands' group), the sweet potato has become established as the primary food staple, assuming the former role of yam and more recently of taro (Strahan et al., 1989). The predominance of sweet potato in the Highlands of Papua New Guinea in contrast to the greater importance of sago or taro in other areas has been documented. A study of an indigenous Fijian village and an Indian village in the Sigatoka valley of Fiji found that Fijian males were deriving 1553 kJ (371 kcal) per day of energy from sweet potato, out of a total of 6032 kJ (1441 kcal) from all roots and tubers, which represented 51% of the total daily energy intake (Chandra and de Boer, 1975). In contrast, Indian males derived only 63 kJ (15 kcal) per day from sweet potato out of a total of 5969 kJ (1426 kcal) (55% of total energy intake) from rice, pulses and roots and tubers. The Fijians' preference for sweet potato and other root and tuber crops and the Indians' for rice and pulses was derived from both dietary preferences and economic factors such as differences in land tenure.

Another starting point for indicating variations in sweet potato consumption within a country can be a nutrition survey. This uses the total consumption of a large sample of people for a set period of time (1 day, several days or a week) to calculate daily average intakes and annual consumption rates for the area in question. Surveys conducted with the same sample population at different times of the year can be used to indicate seasonal differences in consumption. They may also give information about variations in consumption among different socio-economic groups. However, nutrition surveys have certain disadvantages. They often group several foods such as root and tuber crops together, making it impossible to distinguish sweet potato consumption levels. Informants estimating their consumption of a food by recall may not remember a food which played only a nominal or minor role in the overall meal. Moreover, a nutrition survey may miss foods consumed only on holy days, holidays or festivals when the survey was suspended. Although omitting consumption taking place in the last two categories may not seem to introduce a large error or be of nutritional significance, estimating apparently small or periodic demand, so as to plan for production, storage and marketing, is important.

Table 8.4 presents survey results on sweet potato consumption in relation to household expenditure levels in Indonesia. (The more detailed tables from which these data are taken show that compared to Java itself monthly expenditure on food outside Java represented a higher proportion of total expenditure.) Sweet potato consumption outside Java is higher than in Java and thus also higher than the national average. Outside Java, where sweet potato consumption levels more nearly approximated those of rice and other cereals, there was little variation in sweet potato consumption with expenditure. In Java, where cereals and other roots and tubers are more important than sweet potato, the highest consumption levels were reported for the middle expenditure groups, the lowest and highest groups having lower consumption levels. Determining the reasons for such variations could help to indicate ways of increasing consumption among the various groups. For example, the lowest expenditure group might respond to price decreases for fresh roots and the high expenditure group to improved processed products. Further examples of within-country differences in sweet potato consumption are presented in the following case studies.

Case studies

In examining the factors affecting sweet potato consumption and use and in answering the questions posed above, a few countries were chosen for detailed scrutiny, partly because meaningful data were available for immediate review, but other factors were also considered. Countries were chosen from different major regions of the world where sweet potato is grown – Asia, Oceania, Africa and Latin America. They were also selected to present contrasting situations demonstrating a variety of consumption and utilization patterns. As such they range from a situation in **Papua New Guinea** where sweet potato is a dominant staple and still to some extent part of a subsistence economy, to that of **Japan** where it is one small item, available in many interesting processed forms, among a wide variety of food choices. **Rwanda** and **Uganda** have been combined in a single case study owing both to their similarities of usage and the scarcity of information for each individually. They represent a case where there is little discrimination against sweet potato in its role as major staple food.

No study would be complete without the inclusion of **China**, the world's largest sweet potato producer. It is a developing country in which sweet potato processing is well established and ranges in scale from home to factory level. However, in spite of many efforts to obtain

Table 8.4. *Estimated consumption of sweet potato in Indonesia (urban and rural areas) by groups with different levels of expenditure (100 g per capita per month)*

Area	Average	Expenditure levels (in 10^3 Rp)							
		Up to 1999	2000–2999	3000–3999	4000–4999	5000–5999	6000–7999	8000–9999	15,000+
Nationwide	1.1	0.6	0.9	0.9	1.4	1.3	1.2	1.4	1.0
Java	0.7	0.4	0.6	0.7	1.0	0.8	1.1	1.0	0.5
Outside Java	1.7	1.8	2.2	1.3	1.9	1.7	1.2	1.8	1.9

Notes:
Biro Pusat Statistik. National Socio Economic Survey. Series B. Jakarta, April 1981. No data given for 10,000–14,999.

information, it was not found possible to present as detailed a picture of China as was desired. The **Philippines** might be described as a country in transition, with sweet potato moving from the role of a subsistence staple food to a more market-oriented and complementary vegetable item. It also affords an example of a country in which vigorous research is pursuing possibilities of simple improved postharvest technology. **Peru** forms part of the area in which sweet potato originated and is a country with a long history of appreciation for and use of the crop. It is an example of an area in which the sweet potato has developed from a little used subsistence item to a highly commercialized crop. In the process sweet potato became increasingly important as a food for low income urban groups and as fodder for a successful dairy cattle industry. The information presented here is intended to provide thought-provoking insights into the present use of sweet potato which could be used as the basis for research in other countries.

China

In China the sweet potato has made the transition from a staple food to a major industrial crop. Very large processing factories have been established, but numerous traditional small processors, in the rural areas, also continue. The role of small-scale processors and of means to up-grade their methods and efficiency need to be studied. The study results would be of importance and interest to many developing countries who seek solutions to root crop preservation and utilization problems through small-scale rural-based industries.

Like Japan, China is an excellent example of a country which has diversified into many products, thus demonstrating the versatility of sweet potato as a raw material for industry and food processing. More information is needed of past and present uses including medicinal applications (see below). The quality of sweet potato products varies greatly, matching the market needs of many groups. Processing, even by simply converting fresh roots into dried chips, significantly increases the value of the crop. However, there is some risk involved with field drying when climatic conditions are variable.

There is immense scope for research into factors affecting sweet potato consumption patterns and levels. China has the greatest number of sweet potato consumers in the world, but little is known about their attitudes and preferences. Some of the difficulties involved in the sweet potato marketing system which affect production levels and potential utilization are described, but much more information is needed.

Historical aspects

A brief historical sketch of the places and dates of sweet potato introduction to China (Gitomer, 1987) suggests that the crop may have been cultivated for more than 400 years. Although its earliest introduction may have been into Yunnan Province from India or Burma before 1563, the spread and popularization of sweet potato cultivation is thought more likely to have taken place from Fujian after 1594. According to one source, sweet potatoes were introduced there from the Philippines as a famine relief crop after a typhoon. Stories tell that, because of a prohibition by the Spanish on exportation of the crop from the Philippines, Chinese sailors smuggled vines out by hiding them in shipboard water barrels. So well received were they that the people called them *jinshu* (golden potatoes) to honour Inspector General Jin, who requested the presentation of suitable famine relief crops. However, *jinshu* may also indicate that sweet potatoes have so many uses that growing them is like growing gold.

From Fujian sweet potato spread to other provinces until, in 1786, during the reign of Emperor Qian Long, an order was given that sweet potatoes should be popularized and cultivated throughout China. They probably continued to relieve hunger during periodic food shortages and wars. Yen (1974) mentions that this function was well remembered even among Cantonese expatriates.

Sweet potato, therefore, had several historical uses: as a food security crop, as a staple crop in poor mountain areas and as a livestock feed. Even though the crop remains of major importance in China today, several recent sources state that it is still associated in the Chinese mind with poverty and low status (Sheng and Wang, 1988; Wang Han, Beijing Agricultural College, unpublished material). Yen (1974) cites a Chinese researcher who, in 1945, could divide the population of a Shangtung (Shandong) village into economic levels on the basis of food consumption. The lowest strata had sweet potato as the main component of the diet while for the wealthiest, who ate mainly wheat, sweet potato was undoubtedly converted into meat by feeding to animals. In the Tonkin delta, the root was an 'evil' but necessary rice substitute for the poor. The southern Chinese of Kwangtung (Guandong) were described as having a low regard for the crop, but as savouring certain desirable cultivars which were enjoyed even by the wealthy (Yen, 1974).

Sweet potato has also been regarded as having medicinal properties (Duke and Ayensu, 1985, p. 261). Sliced and dried roots are made into a tea to allay thirst. Sweet potatoes are believed to benefit the kidney, spleen and stomach. A recent source (Wang Han, unpublished material) also claims that clinical experiments show sweet potato to be effective in

Table 8.5. *Total area devoted to sweet potato growing in China, 1946–88*

Year	Growing area (in 10^6 ha)
1946	3.3
1950	5.8
1963	9.6
1978	9.3
1982	6.9
1983	6.8
1985	6.4
1986	6.2
1987	6.4
1988	6.5

Source: National Year Book as given by Wiersema et al., (1989).

preventing constipation, stopping bleeding, increasing blood platelet levels, lowering blood sugar and stimulating the immune system.

Sweet potato in China today

There is no doubt that sweet potato was an important staple food of the mass of Chinese people up to 1949. Although it still played a major role in subsistence after the change of government, its function was to alter radically in the 1970s as a result of a government initiative to increase wheat and rice production. The area devoted by farmers to sweet potato consequently declined (see Table 8.5). At the same time, however, an impressive increase in yields had been taking place (from an average of 8 tonnes/ha in 1961 to 18 tonnes/ha in 1985; Horton, 1988). These yield increases are the result of introducing high yielding cultivars from abroad and also of a vigorous breeding programme for new cultivars (Gitomer, 1987). At present China plants on average about 6.5 million ha of land per year and produces about 100 million tonnes of sweet potato. This represents about 80% of the world's total production (see Chapter 1). The five agro-ecological zones where sweet potatoes are grown and the five main growing provinces are shown in Figure 8.3.

The changes in production characteristics described above were accompanied by an alteration in the major function of sweet potato from a staple food crop to an industrial material. Before the 1960s, 50% of the

Figure 8.3. Sweet potato agroecological zones of China and the five major growing provinces. 1 (northern spring region): sweet potato grown mainly as a dry-land crop in plain, hill and mountain regions. 2 (Huang-Huai basin spring and summer region): sweet potato distributed on plains and grown in rotation with other dry-land crops; this is the main sweet potato growing region. 3 (Changjiang basin summer region): sweet potato grown mainly on the hills, as rice is grown on the plains. 4 (southern summer and autumn region): a rotation of rice and sweet potato with alternating wet and dry fields; the proportion of sweet potato area increases from north to south. 5 (southern autumn and winter region): as region 4. (From Gitomer, 1987).

crop was used as a subsistence staple food, its second most important role being as an animal feed (30%), with only about 10% used for starch and alcoholic drink processing (Qiu Rui Lian, Lin Chang Ping and Dai Ti Wei, Jiangsu Academy of Agricultural Sciences; and Hu Jian Xun and Liu Xiao Ping, Anhui Academy of Agricultural Sciences, unpublished material). After the late 1970s sweet potato's importance as a food declined, with only 15% now being used as fresh roots for this purpose, but with 44% and 30% of production being used for industrial processing and livestock feed, respectively (Wiersema, Hesen and Song, 1989). Of the 44% of production utilized for processing, 14% is devoted to food products and 30% to industrial products including starch, alcohol and others. Other authors have given similar utilization figures of 12% for use as food, 45% for processing and 30% for feed (Qiu Rui Lian et al., unpublished material).

Breeding programmes at research institutes, recognizing the importance of cultivars adapted for industrial, animal feed or food uses, have goals which vary according to these uses. These include: improving starch yields per unit area for industrial use by increasing yield and starch content; producing lines with high vine and root yields for animal feeding; breeding for high nutritional value (for example at least 5 mg beta-carotene and 10 mg ascorbic acid/100 g fresh roots), taste (3% soluble sugar/100 g fresh roots), and good shape, for direct human consumption. The specific breeding targets of starch content, dry matter, yield, disease resistance and a number of important agronomic characters for each of the utilization types have been established by the national sweet potato programme and are therefore common to all sweet potato research institutions throughout the country. New lines for all three types of use should have good adaptation for environments and growing conditions of the area in question (International Potato Center, 1987).

Consumption levels and patterns

FAO statistics show that per capita production of sweet potato in 1984 was 92 kg (Horton, 1988). The FAO balance sheet for 1984 (Horton, 1988) indicated that 77% of total production was used for food, 15% for feed, only 3% for processing and that 5% was waste. From this distribution therefore, some 71 kg of sweet potato was available for per capita consumption. However, the percentage distribution of utilization given by FAO would seem to be somewhat out of date as judged by the figures given in the previous section, and by the fact that sweet potato is utilized more as an industrial crop than a staple food crop nowadays. The consumption figure of 71 kg per capita per year (although including both fresh roots and those processed as food) therefore seems high. However, it has been pointed out that little is known about private plot cultivation statistics as they are not reflected in national statistics (Gitomer, 1987). Even though appreciable quantities are grown privately they fall outside the national grain procurement and reporting system. There may be higher than average consumption of fresh roots among certain groups, therefore, but levels have not been measured.

The generally perceived shift away from fresh root consumption, however, has been attributed to a bias against root and tuber crops versus other staples. Rising incomes with accompanied changes in consumption patterns favour preferences for higher-priced fine grain crops (Gitomer, 1987). I was unfortunately unable to obtain any data which would provide a picture of consumption level variations among different regions or groups of people. Therefore it was not even possible to ascertain whether the description of sweet potato as a rather despised food of poor people was accurate. The opinion of Chinese researchers (Sheng and Wang, 1988) that sweet potato has also declined in popularity due to flatulence problems, difficulty of storage, and undervaluation of its nutritional value, has yet to be confirmed.

Sweet potato's role in the Chinese diet is changing from that of a staple to a supplementary food and snack in areas with rising incomes, but retains its function as a staple in poor and mountainous districts. In both cases it remains a food security crop (Gitomer, 1987). Where farmers still use part of their crop as a subsistence staple food it is often cooked by adding fresh roots or dried root chips to rice and boiling them together (Sheng and Wang, 1988). However, most fresh sweet potato roots are now marketed and commonly boiled, baked or fried in the countryside and towns as a supplementary food, while in a baked form are a favourite between-meal snack on sale in the city streets (see also Chapter 5, p. 263). Apart from its uses as a partial substitute for wheat flour in steamed bread (see Chapter 6, p. 334), sweet potato flour can be used alone for making pancakes (Sheng and Wang, 1988).

Farmers may use part of their sweet potato harvest as food, part as feed and a further part for processing into food products. The proportion used for each of these purposes depends upon their production scale. In the major producing provinces most roots are used for trade, largely in the form of dried chips. Production of these chips is described in Chapter 6. Fresh roots and chips are used *in situ* by home- or village-scale processors for starch (and noodle) production or are sold to large factories for a similar purpose (see below). Although generally grouped with industrial products for statistical purposes, the noodles manufactured from sweet potato starch are an important staple food. The decline of sweet potato use as fresh roots has been accompanied by an increased demand for non-staple processed food products. Even so, the overall trend is a decline in sweet potato for direct food use.

Sweet potato tops are eaten in the countryside in the form of tender tips or petioles as a vegetable green, but are not very popular for this purpose due to toughness, and browning on cooking, as described in Chapters 2 and 5. They are utilized, fresh or dried, as a feed for livestock, chiefly pigs (see Chapter 7).

Marketing

Great changes have occurred in marketing of sweet potato since 1978 (Gitomer, 1987; International Potato Center, 1987). Between the late

1950s and 1970s staple food markets were severely restricted in most places, with punishment of private traders in some areas. Consequently, integration and development of markets on a private basis in China is at a low level. The government acted in many areas as the sole procurement and marketing agent for sweet potatoes. Since economic reform in 1978, the government has reduced or eliminated purchases in some provinces while in others it may never have purchased the crop.

Marketing structure has changed as a result of government withdrawal. Prices have also increased. Individual households and traders have replaced the government purchasing organization. Those farmers owning a bicycle or tricycle market sweet potato themselves. Otherwise they sell their crop to traders having transport. The competition introduced by this plethora of very small 'firms', and sales of sweet potatoes at low fixed prices in government stores reduces price volatility.

While the area planted to sweet potato is declining in some parts of China it is rising in others. Changes in marketing arrangements may be important in explaining differences in sweet potato area trends among provinces. For example, in Shandong Province production is described as insufficient and profitability as favourable. In spite of this, the sown area has been declining steadily. This may be because government purchasing organizations had virtually ceased buying sweet potatoes by 1980. The sudden elimination of government marketing channels in an environment with depressed marketing infrastructure could partly explain the short supply of sweet potatoes leading to rising prices in some parts of the province and a decline elsewhere due to lack of marketing opportunities (International Potato Center, 1987). In Sichuan Province, government agencies may never have purchased sweet potatoes to any extent. However, they acted as intermediaries between counties with surpluses or acute shortages. This role has remained relatively unchanged with respect to sweet potatoes over the last ten years. Hence other factors may be playing a part in the decline of sweet potato production area. The province's especially underdeveloped infrastructure, coupled with topographic barriers may limit private food market development which was legalized, as in other provinces, only recently.

The prices which farmers charge for fresh roots are usually about 20% less than those charged by traders. The wholesale cost to traders is also about 20% below the retail price charged by farmers (Gitomer, 1987). Even modest processing will substantially increase the value of the crop, for example by as much as eight times for simple food processing in Sichuan (Gitomer, 1987; and see Table 8.8, below). Sales of sun-dried chips to industrial users will normally also bring a higher price than selling fresh roots. However, there is some risk involved in the drying process, where loss of the entire crop may result from consistently rainy weather during harvest. Traders may also risk losses if chips bought from farmers are rejected or reduced in price at the factory due to poor quality.

Prices for fresh roots, however, have also increased over time. Sweet potato prices in selected areas of Henan Province rose by 79% between 1986 and 1987, a much greater rise than for other major grains such as maize, wheat and soy beans (Gitomer, 1987). Farmers provided contradictory evidence when questioned about seasonal price changes of fresh roots (Gitomer, 1987). Some claimed price changes were large and others that they were fairly stable. In general prices were highest just before harvest, dropping sometimes by as much as half with the appearance on the market of the new season's crop. Prices may also rise when sweet potatoes are in great demand at the time of certain holidays or festivals. In Chengdu, Sichuan Province, for example, the price just before harvest could be 3.5 times higher than that just after harvest.

Processing

The degree to which sweet potato is used for processing in China has increased tremendously in recent years, not only in scale, but also in diversity of products. This has been of great economic and social benefit to a wide range of trades and professions (Qiu Rui Lian et al., unpublished material). Perhaps the most interesting facet of sweet potato processing is the importance still maintained by a very large number of rural small-scale (home or village) processors especially for the manufacture of starch and noodles. Some of them receive additional income by using traditional low cost methods, with mainly manual operations. Other village units employ more technologically advanced equipment. These different levels of technology result in variable product quality. There is a market for products of all qualities (Wiersema, S., personal communication). However, small industries near cities produce items of sufficiently high quality to satisfy urban consumer requirements.

Present utilization trends suggest that processing will increase in importance. This acknowledges that a further decrease in growing area is not expected as sweet potatoes are grown mainly on marginal soils, and that production may increase as a result of improved cultivation practices and the adoption of improved clones (Wiersema et al., 1989).

The necessity for increased utilization of sweet potato as a starch resource in major growing areas such as Hunan Province where as yet industry is poorly developed, has been expounded (Wang, 1984). In Hunan, according to Wang, large quantities of surplus dried roots have been unused leading to waste and a disincentive to the farmers to plant sweet potato. Meanwhile sweet potato farmers' economic development in production areas has been very slow. This situation could be remedied by increased industrial usage. There is evidence of underutilization of sweet potato in Jiangsu Province also. For six counties, previous government purchases of 25% of the crop remained unused, with consequent losses. Government purchase was therefore greatly reduced, which was at least partly responsible for a significant reduction in sweet potato growing area (International Potato Center, 1987).

Processing of sweet potato in China can basically be divided into three areas: industrial processing, food processing and feed production.

Industrial processing

A wide range of light industrial and chemical products is produced and many more are envisaged. Already in manufacture are starch, starch derivatives such as amylophosphate and amyloacetate, noodles, alcoholic drinks, industrial alcohol, maltose, glucose, high fructose syrup, citric acid, lactic acid, monosodium glutamate, amino acids and enzymes (see Table 8.9, below).

Starch is a major product made on a scale varying from home processors to large factories. The primary on-farm processing of sweet potato roots into dried chips, and their sale to processing factories by farmers directly or through middlemen (Gitomer, 1987), has already been described in Chapter 6. Chips are subsequently processed into starch (see Figure 8.4) and noodles. The importance of small processors in the overall scale of starch and noodle production in one major sweet potato area, Sichuan Province, can be appreciated from Table 8.6. The main reason for small-scale production in Sichuan is that climatic conditions are generally unsuitable for field drying of fresh roots into chips. Fresh roots are therefore used as raw material on the spot at farm level or in smaller processing units, to reduce transport costs and problems of bulky, perishable roots. Some noodle processing also takes place using stored starch during the off-season when fresh roots are not available. Farmers in Jiangsu Province have been making starch in the household for more than 100 years (International Potato Center, 1987). Now, although village level facilities are rapidly taking over from family processing, large factories still remain of less importance. In 1985, county machinery factories began producing noodle-making equipment.

Starch extraction rates are often low and starch quality poor from processors using traditional methods with low cost equipment (Wiersema et al., 1989). Such shortcomings are most likely remedied by the

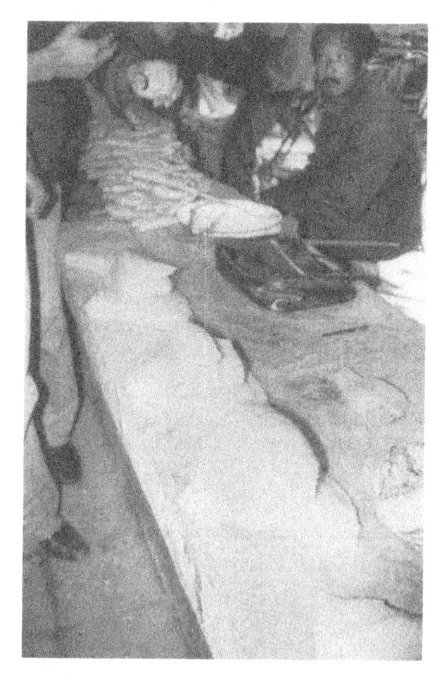

a

b

Figure 8.4. (a) A sweet potato starch gel ('tofu-like') product and (b) a dried sheet starch product at a market in Jinan, China (M.Iwanaga).

Table 8.6. *Capacity of sweet potato processing units in Sichuan Province, China, 1989*

Number of processing units	Processing capacity (Fresh roots/year per plant)[a]	Final product	Total output/ year[a]
3	10–15	Starch	
1	5	Starch	92.7
9	1–2.2	Starch	
4	0.3–0.5	Starch	
>200	0.1–0.3 (village scale)	Noodle & starch	67
>2000	Less than 0.1	Mainly noodle	>100
2	10	Chemical products for industrial use	0.9

Notes:
From: Wiersema et al., 1989.
[a] In thousands of tonnes.

application of better processing equipment, which could mean studying the feasibility of starch and noodle processing in larger cooperative units using more advanced equipment (Wiersema et al., 1989). However, the socio-economic implications of such a development would need careful study. More efficient processing equipment is needed in some larger-scale plants where low extraction rates and unsatisfactory product quality were also mentioned as the main problems. Hence plant modernization efforts should be stimulated (Wiersema et al., 1989). Although starch residues are already utilized, for example for citric acid and fodder manufacture, more could be done in this direction including the treatment of organic waste to avoid environmental pollution problems. Starch is sold not only for food purposes, but also for use in the textile industry.

Alcohol has been produced from sweet potatoes since 1919. The first and largest alcohol plant is near Jinan in Shandong Province. The plant produces 50,000 tonnes of alcohol per year mainly from sweet potato. This represents a use of about 140,000 tonnes of dried chips per year, equivalent to 350,000 tonnes of fresh roots (Wiersema et al., 1989). There are other large and more modern alcohol plants in Shandong and other provinces. Plants produce all the year round as sufficient supplies of dried chips, delivered by farmers or middlemen, are stored at the plant. A large proportion of the alcohol produced at the oldest plant is

Table 8.7. *Capacity of sweet potato processing plants in Sichuan Province, China, which derive other products from starch*

Number of plants	Raw material	Final products	Total output/ year[a]
5	Sweet potato starch	Fructose Glucose Maltose	20
3	Sweet potato starch	Monosodium glutamate	4
3	Sweet potato starch	Citric acid	4.5

Notes:
From: Wiersema et al., 1989.
[a] In thousands of tonnes.

used for industrial purposes such as the pharmaceutical industry. About 60% of the alcohol made at the third largest plant in China, in Anhui Province, is sold for human consumption and about 40% for industrial use. Quality requirements for dried chips are a starch content of at least 65% and a moisture content of less than 13% (Wiersema et al., 1989). The demand for alcohol, which appears to be increasing in China, could be partially fulfilled by sweet potato.

Starch and alcohol are also the starting materials for a plethora of derivative products (see also Chapter 6). The scale of production of major starch derivative products in Sichuan is shown in Table 8.7. Anhui Province produced glucose from sweet potato starch by the early 1960s (Sheng and Wang, 1988). The total output from four glucose plants is now 3200 tonnes per year. The construction of ten high fructose syrup plants producing 300,000 tonnes has been projected by 1990. Production of pyroacemic acid from root residues began in the early 1970s (Sheng and Wang, 1988). Citric acid production from sweet potato has been given as 18,000 tonnes per year (Qiu Rui Lian et al., unpublished material).

An International Potato Center team (Wiersema et al., 1989) noted that the large-scale processing industry uses a considerable amount of sweet potato as a raw material and therefore appears to have a significant positive effect on farm gate prices for fresh roots and dried chips. It is in the interests of the many small producers of the crop that these processing plants operate as efficiently as possible and make high quality products to ensure that sweet potato remains a competitive raw material. Chinese scientists appear to be hopeful for the prospects of sweet potato starch or alcohol and their derivatives given the increased demand for

Table 8.8. *Products, their value and profit estimated to be derivable from sweet potato in China*

Product	Weight (tonnes)	Price[a]	Estimated profit[a]
Dried sweet potato	1000	158	
Crude starch	500	260	33
Fine starch	400	360	54
Vermicelli (noodles)	410	328	64
Alcohol	370	388.5	163.7
Common spirit	500	850	140.4
Monosodium glutamate	100	900	450
Citric acid	295	1475	457.5
Lactic acid	500	2000	750
Glucose, sorbitol, ascorbic acid	?	4000	1870
Starch residue:			
Calcium citrate	100	280	56

Notes:
Science and Technology Commission, Bengbu City, Anhui Province, as given by Wang (1984).
[a] In thousands of yuan.

starch and alcohol which is unlikely to be met only from cereal sources (Wang, 1984; Sheng and Wang, 1988; Qiu Rui Lian et al., unpublished material). They also point out the relatively high starch yields (kg/ha) and low production costs of sweet potato starch compared to cereal starches (Wang Han, unpublished material). The added value to be gained by processing 1000 tonnes of dried chips into various products has been estimated by the Science and Technology Commission of Bengbu City, Anhui Province (Wang, 1984). This is illustrated in Table 8.8.

The main factors affecting industrial use are the growth in demand for industrial starch and alcohol and the price competitiveness of sweet potato against the other major sources, maize and cassava (Gitomer, 1987). Demand is increasing rapidly, from light chemical industries in the case of starch and its derivatives, and from the pharmaceutical, beverage and chemical industries in the case of alcohol. As domestic and international markets develop, demand for these products should increase further. This should promote production of sweet potato in most parts of the country, with the exception of those where it is subsidiary to cassava.

Food processing

Sweet potato food products are many and varied as in the case of Japan. They include sweet potato preserves, jam or jelly, frozen fried sweet potato chips, canned and candied sweet potato, paste, refined dried root products called 'Liangcheng dried red flesh sweet potato' or 'Red Heart brand' dried sweet potato pieces and strips, vermicelli (fine noodles), noodles, sheet starch jelly, soy sauce and vinegar (Qiu Rui Lian et al., unpublished material; Sheng and Wang, 1988; and see Table 8.9). As was indicated in the previous section (and see Table 8.6) there is a large production of vermicelli and noodles at the home and village level. Products such as jams, preserves, candied sweet potato, dried sweet potato bits and other handicrafted items are made in the countryside at the farm and village level.

In the processing laboratory of the Crops Research Institute, Sichuan Academy of Agricultural Science, a programme of food product development started in 1985 (and see Figure 8.5). Products include starch and noodles, reconstituted chips and candy (the methods for which were described in Chapter 6), jam, canned roots and sweet potato flour (Wiersema et al., 1989). In addition, processing research is carried out on a pilot scale in order to test processing techniques and product quality. Good quality dried and candied sweet potato chunks have been produced. Little equipment is needed and the items are produced and packaged by farm families (International Potato Center, 1987). By making these items it has been estimated that farmers can increase their net income by 50% over selling sweet potatoes direct. Extension is carried out by the Sichuan Academy of Agricultural Science through demonstrations and visits, but extension workers are few. Improved processing technology and new products have been successfully transferred to six cooperative plants in the province.

One cooperative processing plant in Yanting County, Sichuan, has been described (Wiersema et al., 1989). It was built with assistance from the Processing Laboratory and completed in 1988. The plant has 123 farmers participating on a cooperative basis. It requires 300 tonnes of fresh roots per 6 months of processing season (November to April). Each farmer contributes a sum of money as initial investment and has the right to send one labourer to work in the plant and to supply 1.5 tonnes of fresh sweet potato roots per year.

The main products are starch, candy and reconstituted chips. Starch is processed from small and reject roots. The processing methods for candy and semi-finished and finished reconstituted chips as used at the plant were described in Chapter 6. For 1989 a total of 195 tonnes of product was sold by contract. This includes 100 tonnes of starch, 70

Table 8.9. *Actual and potential products from sweet potato in China*

A. Leaves and stalks

1. Leaves
   a. Fodder products: silage, fermented fodder, mixed fodder
   b. Vegetables: tender tips and stalks used as greens
2. Underground stalks and roots *used for charcoal*
   a. Fodder products
   b. Fermentation products: alcohol, acetic acid, fodder
   c. Sugar products: maltose, sugared grain for fodder

B. Roots

1. Food products
   a. Unprocessed fresh, dried and boiled roots as food
   b. Canned and candied roots, cakes, frozen fried roots, icecream, sherbet, pancakes, paste
   c. Refined dried roots including Red Heart Brand dried roots and strips
2. Refined starch products
   a. Starch
      (i). Vermicelli, sheet jelly, noodles
      (ii). Refined starch
      (iii). Starch derivatives: soluble starch, amylophosphate, positively ionized starch, amyloacetate
   b. Starch processing residue
      (i). Fodder
      (ii). Saccharification products: maltose and sugar residues
      (iii). Brewing products: alcohol, acetic acid, calcium citrate, ethanol for fodder
3. Fermentation products (using fresh or dried roots)
   a. Alcohol, acetic acid, ethanol
   b. *Lactic acid*
   c. *Butanol, butyric acid, and acetone as chemical product precursors*
   d. Citric acid, amino acids (lysine, glutamic acid etc.)
   e. Monosodium glutamate
   f. *Antibiotics*
   g. Enzymes
4. Saccharification products
   a. Fructose, glucose, maltose, amylose, amylose paste
   b. Sugar residues

Notes:
From: Wang Shudian et al., 1984, cited by Gitomer (1987).
Those products given in italics are not manufactured in China at present as far as can be ascertained.

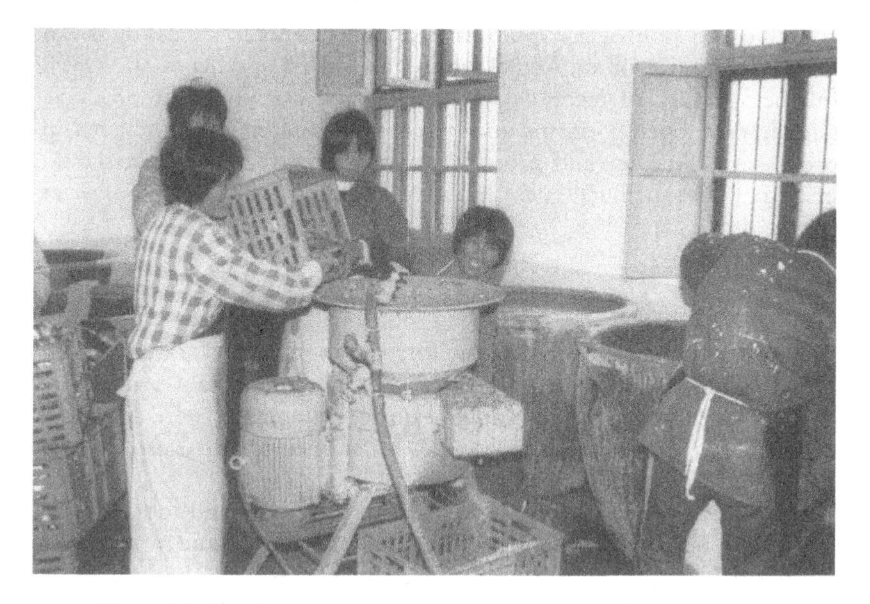

Figure 8.5. Experimenting to formulate many types of processed product in the Xuzhou Institute of Sweet Potato, China (M. Iwanaga).

tonnes of candy, 15 tonnes of semi-finished chips and 10 tonnes of reconstituted (finished) chips. By present returns on invested capital it would appear that the plant will have no difficulty in repaying a bank loan (of about 80% of total initial investment) within a stipulated 3 year period. Products are marketed through exhibition in commercial 'fairs', visited by salesmen and representatives of the food industry. Products are sold by contract. The optimum production capacity of the plant was estimated at 1000 tonnes of fresh roots per year. A larger production capacity would result in high transport costs as a result of poor infrastructure. A number of factors were identified for study in order to increase present production capacity and enhance product quality (Wiersema et al., 1989). These included identifying the most suitable cultivars for individual products, installing storage facilities at the plant for raw material and improving production line activities. In the case, for example, of candy processing, hand peeling and cutting was too slow and the resulting pieces too irregular in shape. Slicing equipment resulted in irregular thickness. Frying equipment did not maintain a constant temperature and mixing and shaping by hand was too slow. To optimize quality and cost, it was suggested that the optimum sugar concentration of candy be studied with respect to cost, shelf life and market requirements.

Sichuan is one of the poorest provinces, where processing is an important source of additional family income (Wiersema et al., 1989). Further transfer of technology, by extension activities, for products already developed to farms and villages, as well as improvements in processing efficiency and product quality for established plants could help to boost product sales, improve incomes and encourage sweet potato production.

Livestock feed

Utilization of fresh material in the form of roots, stems and leaves for livestock feeding, and the proportion of total sweet potato production devoted to this purpose in various major growing areas were described in Chapter 7 (p. 411, and see Table 7.2). Most sweet potato is used for pig feed.

In addition to their feed use in the fresh form, vines and roots may be converted into silage, fermented feed or dried tops and root chips. Although a high proportion of sweet potato is still no doubt directly used on farms, especially in the case of roots, an increasing amount is being used for industrial processing of feed (Qiu Rui Lian et al., unpublished material). Apart from the direct use of roots and tops for incorporation into commercially produced feeds, there is some utilization of by-products from starch and alcohol processing. Unfortunately no figures for the proportion of sweet potatoes used in commercial feed have been obtained. The dry root chips or stems and leaves are said to be incorporated into mixed feeds at the rate of 10–25% (Qiu Rui Lian et al., unpublished material).

The main factors affecting use of sweet potatoes for livestock feed are rapidly increasing demand for livestock products brought about by rising incomes, high yields of sweet potato in China, a lack of highly developed interregional feed markets and favourable relative pricing of sweet potato versus other feed grains (Gitomer, 1987). The latent demand for feed in the country is very high and a rapidly increasing use of sweet potato for feed has been predicted over the next 15 years (Tong, Z., as cited by Gitomer, 1987).

Sweet potato breeding objectives at various research institutes include those geared to animal feed purposes, taking into account not only root yields and root chemical composition (International Potato Center, 1987) but also the importance of high vine yields. During a growing period, farmers usually harvest vines three times or more. Thus cultivars should have a vigorous vine regrowth after cutting. Goals of high total dry matter content in tops and roots are set and, in one area, a top:root ratio of 1:1 is desired for new cultivars.

Conclusions

In the last ten years the role of sweet potato in China has moved from that of a staple food crop to an industrial raw material, although it continues as a staple food in poor and mountainous districts. In higher-income areas it has become a supplementary food and snack. As fresh root use has declined, consumption of processed food products has increased.

The marketing system in China deserves a detailed study. Difficulties associated with recent historical changes, alterations in purchasing channels, lack of infrastructure and price instability play a part in creating uncertainties regarding marketing opportunities in various parts of the country. This at least partly illustrates the importance of an efficient marketing system in stimulating demand for fresh roots.

The transformation of sweet potato into an industrial crop for the production mainly of starch and alcohol and their derivatives has been accompanied by the construction of large processing plants and factories, but the important role of farm and village starch processing units in rural areas remains. This is particularly true in areas where climatic conditions make use of fresh roots as a raw material obligatory. China's experience in the field of small-scale processing could have relevance to other developing country situations and should be studied further.

Expansion of fresh root use for industrial purposes and food processing could increase if storage problems were overcome and storage facilities provided and improved (see Chapter 5). In other areas the success of large enterprises no doubt owes much to the conversion of bulky, perishable roots into root chips by sun-drying at the farm level. These chips can be stored for extended periods at both farm and factory. Tropical countries with climatic conditions suitable for on-farm drying might adopt this system of root preservation for later processing. There are some encouraging examples of the transfer of technology for new products from the laboratory to cooperative enterprises through extension work. At the same time it appears that more emphasis could be placed on the importance of extension by employing more personnel.

In many areas, a steady decline in sown area, even though accompanied by impressive yield increases, has meant a net result of no growth trend in production. However, an expansion of sweet potato use for both industrial purposes and animal feed has been predicted due to constantly increasing demand for starch and alcohol both in China and elsewhere, and for livestock products in China itself. Production may be stimulated to meet this demand by the introduction of new cultivars tailored to specific needs, by improvements in process efficiency and quality, and by the establishment of sweet potato processing facilities in growing areas where farmers could benefit from value-added products.

Japan

Japan is a highly developed country and one of the most economically successful nations in the world. As such it may seem to have little relevance to the situation of many tropical developing nations. However, it is a case where the results of research into yield improvements and the development of specialized cultivars has enabled the sweet potato to demonstrate its flexibility and adaptation to changing economic circumstances. At a time of decreased direct use as a human food it evolved from a subsistence staple into a major industrial raw material and a component of livestock feeds. Moreover, the challenge which sweet potato is facing from economic competition with other crops in these two spheres is being met with imagination by an attempt to diversify into a wide range of food products. Perhaps here more than in any other country, except China, the potential of sweet potato as a component of highly presentable, tasty and interesting processed items attractive even to consumers with a wide range of food choices can be seen clearly. The type, level of technology and expensive packaging of some products may be inappropriate for direct adoption by many developing countries. Nevertheless, their existence demonstrates what can be achieved and might offer a fund of ideas which could be tapped by food technologists for adaptation elsewhere. The introduction of a well-organized grading and marketing system with emphasis on quality as a vegetable item has also served to stabilize average consumption (even at a fairly low level) of fresh roots. Moreover, the nutritive value of sweet potato roots is well recognized and is being exploited in some areas. Japan is also one of the few countries at present for which an attempt has been made to analyse variations in sweet potato consumption with socio-economic and seasonal factors. Though these results are deficient in detail they serve to demonstrate the type of information which could lead to targeting groups potentially responsive to promotion of sweet potato.

Historical background and changes in utilization

As a case study in human consumption and utilization patterns of sweet potato, Japan presents a complete contrast to the situations described for Papua New Guinea or Rwanda and Uganda, but is similar in some ways to that pertaining in China. The rapid economic development which has taken place in Japan since World War II has led to an urbanization of life in the broadest sense and has had a profound effect on consumption patterns of food in general.

Sweet potatoes are the third most important staple crop after rice and potatoes (but only the ninth most important when other vegetables and fruit are included) and provides 100% self-sufficiency (Duell, 1985). However, the per capita quantity of sweet potato available for consumption is now much lower than it was in the years following the Second World War and a high percentage of sweet potato production is used for industrial purposes, particularly starch and alcohol manufacture (see Figure 6.14, p. 323). Only a small percentage of total production is utilized for food processing (apart from starch and spirits), but within that sphere, a great diversity of products has arisen catering to increasing sophistication of consumer tastes. Many of these products were described in Chapter 6. In addition about 12% of the total production is used for animal feed purposes. Some of the waste from starch factories is also processed into feed.

The sweet potato was introduced into Japan in the early 1600s. From then on its most important role was repeatedly as a famine relief crop due to its resistance to drought and typhoons. Thus it provided a source of food when rice was in short supply. However, its introduction and spread into all parts of Japan was not always easy to accomplish. Parallels have been drawn between the promotion of sweet potato in Japan and that of the potato (*Solanum tuberosum* L.) in Europe in terms of beliefs about its poisonous nature, resistance to its cultivation and efforts to overcome these obstacles (Duell, 1983). The claim made by at least one major Japanese historical figure to have eliminated starvation by encouraging sweet potato cultivation is still remembered today in the form of a yearly sweet potato festival at a Tokyo temple (Duell, 1983). The popularity of sweet potato apart from its role as a famine food eventually became established. In Edo (Tokyo), roasted sweet potatoes became a popular winter snack, especially among merchants and the working class (Duell, 1984). Labourers and students liked them because they were cheap and filling. The varied ways of eating sweet potato in present day Japan have been ably described (Duell, 1990).

The most recent period when sweet potato was called upon to alleviate a major food shortage was during World War II. However, after the war, the popularity of sweet potato declined due in part to its association with war-time hardship and following a general improvement in food supplies. Consequently its direct production for food decreased and an increasing proportion of the supply became used for starch, a high percentage of which was in turn converted into sugar syrups. Before the Second World War, in the 1930s, over 70% of the crop was used as a subsistence food on the farms where it was produced. Thirty years later, the sweet potato had become a major cash crop and only about 10% of the crop was retained for human consumption on the farm, the rest being used for animal feed or sold either as fresh food, or

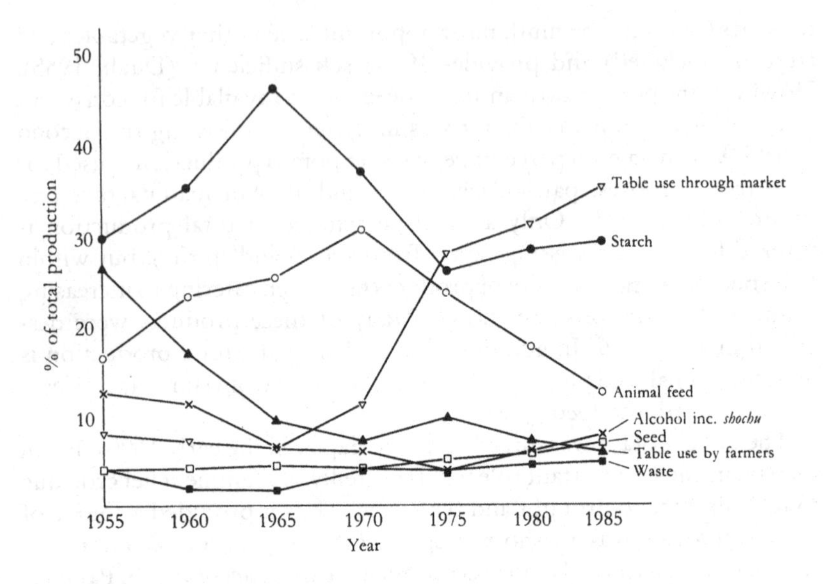

Figure 8.6. Trends in sweet potato utilization in Japan, 1955–84. (Source: Ministry of Agriculture and Forestry, 1955–77, Production and distribution of sweet potato; Ministry of Agriculture, Forestry and Fisheries, 1978–86, Production and distribution of sweet potato.)

for production of starch, alcohol, livestock feed or as processed food. Figure 8.6 illustrates the changes in sweet potato utilization which have taken place during the 30 post-war years, 1955–1984. As can be seen, until the 1970s, as the percentage of the crop used by farmers for their own consumption fell, that sold as fresh roots through the markets rose; these changes stabilized in the mid-1970s and have remained fairly stable since. Starch production, which greatly expanded after the war, rose until the mid-1960s and then dropped sharply as a result of competition from corn starch and imported sugar. Thus the sweet potato farmers' income became uncertain and acreage devoted to sweet potato decreased significantly. While the planting area of the crop declined, its yield increased as the result of the introduction of high yielding cultivars. Yield increased from 10 tonnes/ha in 1900 to 15 tonnes/ha in about 1950 and has reached more than 20 tonnes/ha at present (Sakamoto, 1984). Today, efforts are increasing to safeguard sweet potato farmers' livelihoods in Kagoshima Prefecture, the southern traditional sweet potato growing area of Kyushu, where 70% of production is used for starch, by expanding the range and type of sweet potato products available to the consumer.

Changes in levels of consumption

During 1955 to 1984, as mentioned above, the percentage of total production consumed on producer farms declined from about 27% to only 6%, while the percentage sold through the market and destined both for vegetable use and for food processing rose from about 8% to 35%. In spite of this percentage rise, the actual weight of sweet potato sold through the market remained remarkably stable. By 1986, the percentage of production sold by the farmer for fresh vegetable use was over 35% and only a further 5% was utilized for food processing (Ministry of Agriculture, Forestry and Fisheries, 1987, as shown by Duell (1989)). However, a portion of the sweet potato sold for vegetable use actually goes on to food processing or other uses (Duell, 1989). Hence the percentage of total production ending up as food products may be somewhat higher than is shown in official statistics.

The Japanese Food Balance sheets prepared by the Ministry of Agriculture, Forestry and Fisheries for 1955 to 1984 show that the quantity of sweet potato per capita per year available for consumption in the country as a whole fell from 27 kg in 1955 to only 4.3 kg in 1984. This represents the portion of the crop reaching consumers as fresh sweet potatoes and processed food. The major decline occurred between 1955 and 1967, but consumption has remained quite stable ever since. This decrease represents a fall in daily per capita consumption levels from 75.4 g in 1955, supplying 379 kJ (90 kcal) of energy and 1.0 g of protein, to only 11.8 g in 1984, supplying 61 kJ (14.5 kcal) of energy and 0.1 g of protein. It also indicates a basic change in consumption pattern from a co-staple food to a complementary vegetable and snack food (Ono, M. and Sakamoto, S., personal communications).

Among farmers sweet potato consumption continued to decline until the early 1970s, before finally stabilizing several years after consumption had stabilized in the general population (Duell, 1989). The gap between the greater amount farmers grew and the lesser amount they purchased, for self use, also narrowed significantly between 1967 and 1986. Unfortunately, figures available do not separate sweet potato farmers from farmers not growing sweet potatoes. The Ministry of Health and Welfare National Nutrition Survey, since 1972, has annually measured directly the amount of foods consumed by people throughout Japan, from all strata of society over 3 days each November. Statistics from these surveys have shown (Duell, 1989) that per capita daily sweet potato consumption levels were much higher among full-time farmers than the general population, at least for November. The amount salaried workers apparently ate was slightly less than the national average. These levels are shown in Figure 8.7.

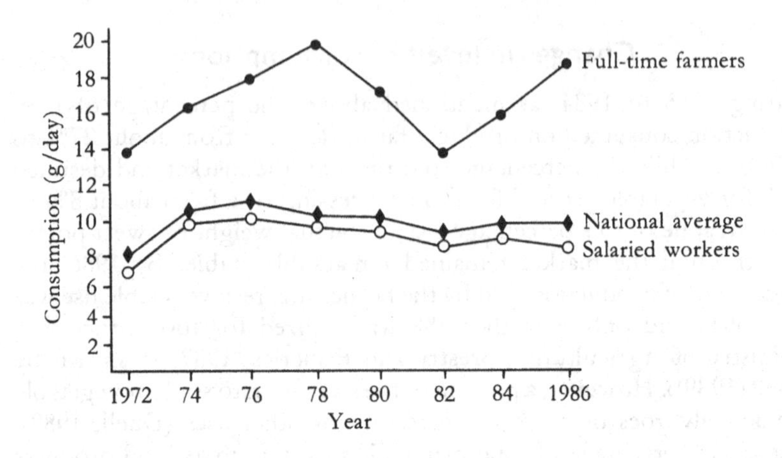

Figure 8.7. Sweet potato consumption in November per person, over 15 years in Japan (based on Duell, 1989; data originally from the Ministry of Health and Welfare Annual Reports 1974–88).

Variables associated with differences in consumption levels

Duell (1989) has used various statistical assessments relating to Japanese domestic sweet potato consumption to show how consumption fluctuates. Variables included region, size of town or city, season of the year, occupation, age of family household head, income level and level of household expenditure. Unfortunately the reasons for the variations demonstrated have not yet been researched. The statistics utilized were from several government sources which carry out surveys varying in their purpose and the way in which the statistics are measured. The Ministry of Agriculture, Forestry and Fisheries estimates per person consumption from total crop production and subsequent crop utilization. These are the official figures submitted to the FAO. The Management and Coordination Agency's annual survey of income and expenditure shows not only the amount of money spent on sweet potatoes, but also the quantity of sweet potatoes purchased per family. The Ministry of Health and Welfare carries out an annual nutrition survey as described in the previous section. These different surveys do not invariably show the same basic trends and may differ with regard to relative quantities of sweet potatoes consumed. A detailed analysis of these sets of statistics was carried out by Duell for the year 1986; some of the findings are presented here. There is considerable variation in consumption with the factors considered, in spite of the overall country consumption stability already mentioned for the last 20 years.

The trends in consumption by city size as shown in Figure 8.8 were

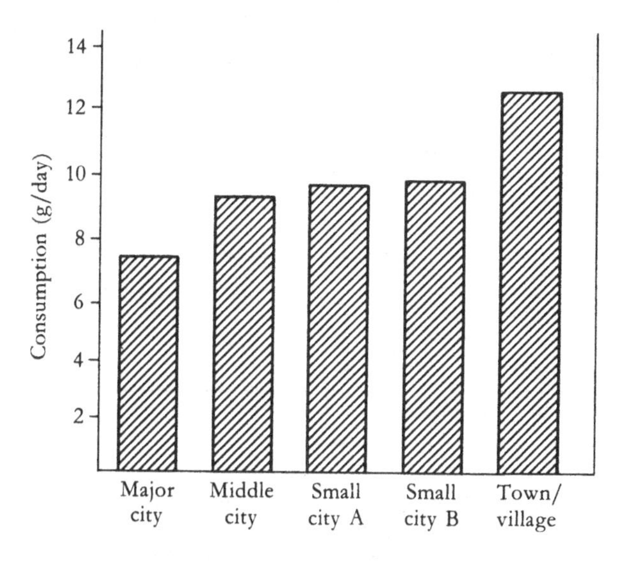

Figure 8.8. Per capita sweet potato consumption, by size of city in Japan, 1986 (based on Duell, 1989; data originally from the Ministry of Health and Welfare Annual Report 1988).

obtained by direct measurement of food consumption. They indicate that the more urbanized an area is, the lower its sweet potato consumption. Another set of data (not shown) based on purchases of sweet potatoes portrays the reverse picture. There was a consistently lower quantity of sweet potato purchased over a number of years by small town and village dwellers compared to city inhabitants (not illustrated). Duell concludes that more residents in small towns and villages may be growing their own sweet potatoes than in other areas. Thus they might actually be consuming more sweet potato than city dwellers, as shown in Figure 8.8.

Individual consumption is demonstrated to increase with increasing age of the household head in Figure 8.9 and to decrease with the level of monthly expenditure (Figure 8.10). This last finding is in agreement with Figure 8.11, which shows also that sweet potato consumption is highest in the lowest income bracket. However, the two figures disagree with regard to the amounts of sweet potato consumed by the groups with the highest expenditure and income levels.

The next set of figures deals with statistics relating to sales of sweet potatoes. Figure 8.12 shows the monthly variations in sweet potato sales at the Tokyo and Osaka wholesale markets, which together account for more than 10% of the fresh sweet potato market. A seasonal variation in sales (and by inference consumption) can clearly be seen. Further data

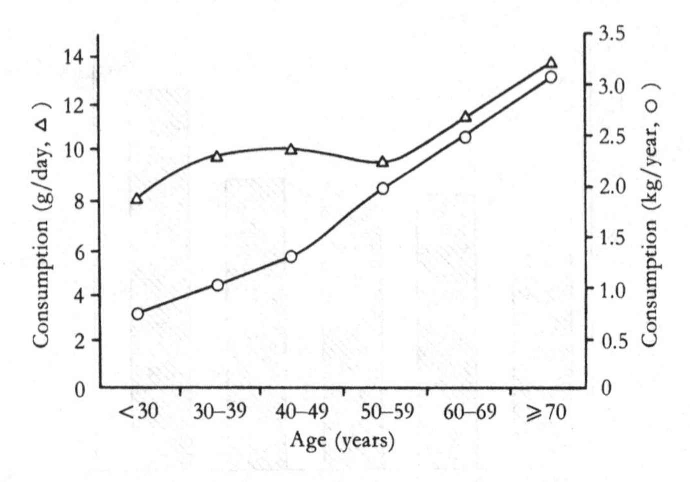

Figure 8.9. Per capita sweet potato consumption, by age of the head of household in Japan, 1986 (based on Duell, 1989: (△) Ministry of Health and Welfare Annual Report 1988; (○) Management and Coordination Agency Annual Report 1987, data is for heads of households aged <30, 30–39, 40–49, 50–59, 60–64 and >65 and does not include farm families.

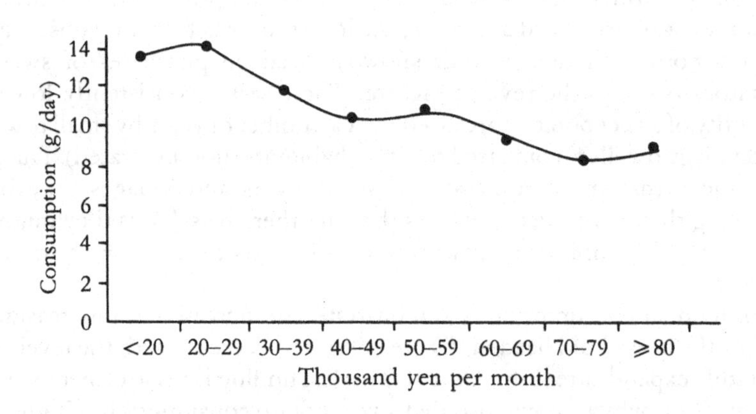

Figure 8.10. Per capita sweet potato consumption in relation to per capita household expenditure in Japan, November 1986 (based on Duell, 1989; data originally from the Ministry of Health and Welfare Annual Report 1988).

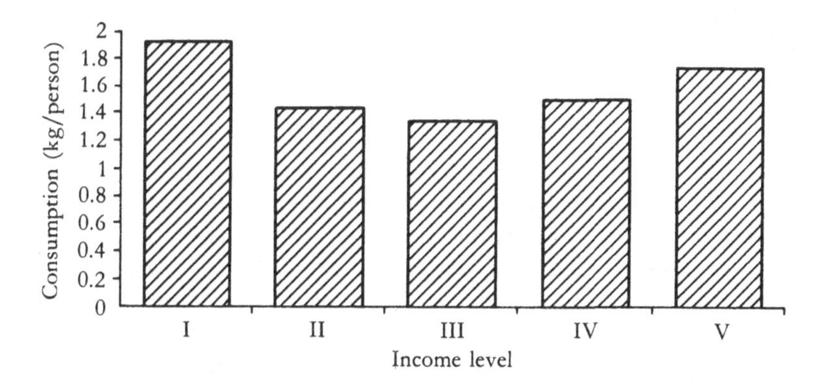

Figure 8.11. Per capita sweet potato consumption, by income in Japan, 1986; I to V denote groups of income levels (low to high). (Based on Duell, 1989; data originally from Management and Coordination Agency Annual Report 1987.) The statistics do not include farm families.

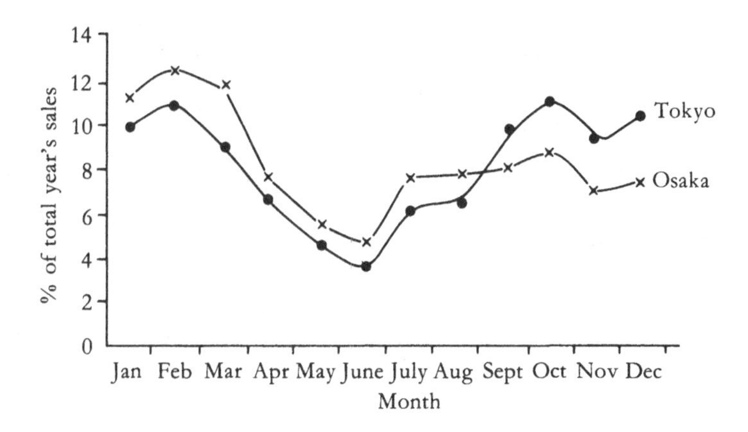

Figure 8.12. Monthly variation in sweet potato sales in two major markets in Japan, 1986 (based on Duell, 1989; data originally from Tokyo Central Wholesale Market Annual Report 1987 and Osaka Central Wholesale Market Annual Report 1987).

(not shown) for variations in purchases in different regions with the retail price of sweet potato exhibit much variation but generally show lower purchases in areas with higher prices.

Results of a questionnaire

I sought further information on attitudes to sweet potato consumption in Japan in two contrasting communities by means of a questionnaire. This was answered by 53 families in each of two areas: Ibusuki City which is a town of 50,000 people in a rural, traditional sweet potato growing area; and Nagoya City, population 2,150,000, which is a large industrial urban area where sweet potato is consumed but not cultivated. Sweet potato was cultivated as a staple food around Ibusuki from about 1750 onwards as the volcanic soils were unsuitable for rice cultivation. Now, however, rice is transported in from other areas and has replaced sweet potato as the staple food. Of the household heads questioned in Ibusuki, 47% were farmers, 42% salaried men and 11% business or professional men. In contrast, the corresponding figures for Nagoya were 4%, 70% and 26%. For Ibusuki families most sweet potato was grown in their own fields or received as a gift, whereas for Nagoya families most sweet potato was purchased.

All families cooked and ate sweet potato roots. Very few families also used the tops, with twice as many using tops in Ibusuki than in Nagoya. However, only the petioles are used as the leaves are considered too 'harsh' tasting.

Though most respondents in both areas preferred a red or purplish-skinned root, with a dry or intermediate textured flesh, it was interesting to note (Table 8.10) the differences in preferences for flesh colour. This at least partly reflects the greater exposure and access to a wider range of cultivars in Ibusuki, the growing area, than in Nagoya. For example, the purple-fleshed cultivars grown around Ibusuki are not yet available in Nagoya.

Almost 100% of respondents in both areas liked sweet potatoes, the major reasons for high acceptability being their flavour and sweetness, their nutritional value (specifically content of dietary fibre, vitamins and minerals) and their ready availability, in that order. Factors such as energy and satiety value and cheapness were also important in Ibusuki, but not in Nagoya. No one from Nagoya rejected sweet potato for any reason and only 3.8% and 1.9% of respondents from Ibusuki rejected them because of heartburn or flatulence problems, respectively.

The consumption patterns of roots, petioles and leaves are shown in Table 8.11. There is an interesting contrast between the two areas in that although a significant percentage of respondents from Ibusuki still use

Table 8.10. *Preferences for sweet potato according to skin and flesh colour and consistency in Ibusuki and Nagoya, 1988*

	Ibusuki (n=53)		Nagoya (n=53)	
Character	No.	%	No.	%
Colour of skin				
Reddish purple	27	51	40	76
Light yellow	25	47	4	8
Light red	15	28	10	19
Light brown	4	8	2	4
Deep purple	3	6	4	8
Colour of flesh				
Cream	31	59	33	62
Yellow	17	32	20	38
Light purple	7	13	0	0
White	6	11	6	11
Deep orange	6	11	1	2
Light orange	4	8	5	9
Deep purple	3	6	0	0
Texture (consistency)				
Moist	6	11	7	13
Dry	36	68	41	77
Intermediate	16	30	12	23

Notes:
Number (No.) refers to families in each area which answered to one or more characteristics in each category. n is number of families questioned.

sweet potato as a co-staple, almost no-one from Nagoya does so. Twice the percentage of people in Nagoya as in Ibusuki use sweet potato as a complementary vegetable and a high percentage in both areas use it as a snack. It is obviously not considered to be a food for special occasions, at least at home, but there was no question concerning the consumption of sweet potato processed products, which are likely to be eaten on a special occasion or received as a gift. Though petioles were used by hardly anyone, two respondents from Ibusuki considered them to be suitable for special occasions, whereas no-one in Nagoya did.

The most popular methods of cooking in both areas were deep-frying, baking and steaming, followed by boiling. This reflects the popularity of a dish called *tempura* (see Appendix 1) where sweet potato slices are dipped in batter, deep-fried and served as a complementary vegetable with a main meal. Petioles are usually boiled in soy sauce or deep-fried.

Of respondents in Ibusuki, 94% considered sweet potatoes to be a suitable food for children, whereas in Nagoya 85% considered them suitable and 15% unsuitable. Twice as many respondents in Ibusuki gave sweet potato to a child at under 1 year of age than in Nagoya. In both areas a high percentage gave sweet potato as a snack only or as a main food and a snack, but hardly any gave it only as a main food.

More than 90% of all respondents considered the sweet potato to be beneficial to health mentioning as reasons its fibre content being good for maintaining bowel function and its high nutritive value. This surely reflects the increased awareness through education in Japan of the nutritional value of sweet potatoes and their image as a 'healthy' food.

The traditional role of sweet potato in Ibusuki was reflected in the answer to the question of how sweet potato rated compared to other major foods.

	Ibusuki	Nagoya
Superior to other major foods such as rice and bread?	43%	6%
Inferior to other major foods?	6%	34%
Equal to other major foods?	43%	60%

The reasons for sweet potato's superiority were again given as nutritive value, taste and fibre content as well as ease of cooking. The reasons produced for its inferiority were that it was found to be cloying, unsatisfying and that not enough recipes were available for it to be used as a staple food. The last point was thought to be associated with its sweet taste in that sweet foods are often considered to be unsuitable as staples to be combined with other savouries.

Questions were also included to compare sweet potato with two other root/tuber crops, namely the potato (*S. tuberosum*) and taro (*Colocasia esculenta*). The potato was consumed with about the same frequency all the year round while the sweet potato was consumed more frequently in the autumn and winter than in the spring and summer. Taro was consumed less frequently than either sweet potato or potato. A number of characteristics of the three root/tuber crops were compared by means of a points system (see Table 8.12). It is interesting to note that in all characteristics sweet potato was placed equal to, or scored more highly than, either potato or taro by the Ibusuki sample. However, the Nagoya families rated sweet potato below potato in terms of varieties of ways available for preparation, ease of preparation and economy, but they still rated it above taro in almost all respects.

Table 8.11. *Consumption patterns of roots, petioles and leaves in Ibusuki and Nagoya, 1988*

	Ibusuki ($n=53$)		Nagoya ($n=53$)	
Pattern	No.	%	No.	%
Roots				
Staple	1	2	1	2
Co-staple	6	11	1	2
Complementary vegetable	20	38	38	72
Snack	47	89	40	76
Special occasion	1	2	1	2
Not eaten	0	0	0	0
Petioles				
Complementary vegetable	3	6	2	4
Snack	0	0	0	0
Special occasion	2	4	0	0
Not eaten	49	93	51	96
Leaves				
Complementary vegetable	1	2	0	0
Not eaten	52	98	53	100

Notes:
See note to Table 8.10 for abbreviations.

Table 8.12. *Comparison of sweet potato with potato and taro by families in Ibusuki (I) and Nagoya (N), 1988*

Root/tuber		Taste	Nutritive value	Ways of cooking	Ease of cooking	Economy	General acceptance
Potato		3.0	3.0	3.0	3.0	3.0	3.0
Sweet potato	I	3.9	4.1	3.1	2.9	4.4	3.7
	N	3.5	3.7	2.5	2.1	2.6	2.9
Taro	I	2.7	2.4	2.3	2.3	2.7	2.5
	N	2.2	2.7	2.1	2.2	2.3	2.3

Notes:
Assessment system: 5, very favourable compared with potato; 4, favourable compared with potato; 3, equal to potato; 2, unfavourable compared with potato; 1, very unfavourable compared with potato.

Only 13% of respondents replied affirmatively to the question about feeding animals with sweet potato. However, the difference between this percentage and the only 1.9% who answered affirmatively in Nagoya is no doubt due to the greater presence of livestock and domestic pets in the rural area of Ibusuki.

The answers to this questionnaire are interesting in that they highlight a number of major differences in sweet potato consumption patterns, attitudes and preferences which exist between two contrasting areas, one largely rural with ready access to home grown sweet potatoes of varying genetic characteristics and the other urban with access mainly to purchased sweet potatoes. Although in general sweet potato is perceived to have changed roles from a staple or co-staple to a complementary vegetable, it appears to maintain its original role among some sweet potato farm families. A significant percentage of rural respondents still preferred sweet potato over other staples. The importance of high nutritive value in the expressed appreciation of consumers for sweet potato is notable. The comments made by the city dwellers on lack of recipes for sweet potato as a main food raises the question of whether consumption would be increased if such recipes were forthcoming via the media.

Utilization

Variety in Japanese sweet potato usage is not a new development. A classic cookery book devoted exclusively to sweet potatoes and printed by Chinkoro in 1789 includes the use of starch and pickled sweet potatoes (Duell, 1990). Yanaka's later (1905) cookery book also describes how to prepare sweet potato powder and use it in a number of recipes (Duell, 1990). The existence of these books (containing 142 recipes) indicates a traditional fund of diverse uses many of which have been forgotten. However, some are being revived by noodle makers and restaurateurs.

Cultivars

Production of several older cultivars for human consumption declined after the Second World War whereas there was a rapid increase in the cultivar 'Norin 2' grown primarily for industrial use, chiefly starch and ethanol manufacture (Duell, 1985). This cultivar also declined in production as the use of sweet potato for starch was challenged by alternative sources. At the same time two newer cultivars and an older one rose to prominence. The most popular cultivar for fresh vegetable use is 'Kokei 14', a red-skinned, creamy-fleshed cultivar with a long thin shape and a smooth skin. It owes its popularity to a particular suitability for preparing the popular dish *tempura* (see above). New cultivars, with high dry matter contents and high concentrations of starch in the dry matter, have been developed for starch and alcohol production. There is also a much smaller production of orange- or purple-fleshed cultivars, the latter being confined to limited areas in the south. However, the imaginative use of such cultivars in processing establishments to make cakes, pastries, desserts, icecreams and sorbets for sale in shops, coffee houses and restaurants which is beginning in some parts of the country may lead to their increased popularity. One farmer's wife visited includes a recipe leaflet with serving suggestions in every box of a deep purple-fleshed cultivar she despatches to market. Moreover, a cultivar with very low sweetness, 'Satsumahikari' (see Chapter 6), now commercially available, may stimulate both table and processing use. However, it appears to be very difficult to obtain sought-after speciality cultivars outside the production zones. Supplies of purple-fleshed cultivars, for example, are limited by low yields.

Fresh roots

The decrease in sweet potato consumption levels which has taken place since World War II has been accompanied by an increase in the market price. There has also been an improvement in the quality of fresh roots, and the appearance in shops and supermarkets in some areas of a wide variety of processed products. Japan has placed special emphasis on growing cultivars with characteristics suitable for different purposes. In the case of roots for fresh vegetable use, these are good appearance and fine eating quality. Whether despatched from the farm or a large cooperative packing station, roots are selected for freedom from disease and blemishes, washed and dried, graded by size (often by automatic weighing machines) and packed into uniform cardboard boxes holding a stipulated weight of roots, before transport to market. They therefore arrive at the urban markets in prime condition and command a high price. Seasonal supplies of sweet potato are extended for home use, marketing and processing by both on-farm traditional, and large-scale temperature-controlled storage facilities.

Processing

Starch production

This process has already been described briefly in Chapter 6. Storing wet starch in outside reservoirs enables extension of the processing season beyond the period immediately after harvest. However, starch plants still do not operate all year round. The difficulties which sweet potato faces as raw material for starch production are described in Chapter 2. In 1963, sweet potato starch accounted for 67% of Japan's total starch production. By 1982, this percentage had fallen to 5% as sweet potato had been overtaken by corn as the major raw material (Kainuma, 1984). A high proportion (more than 70%) of sweet potato starch is processed further into sugar syrups for the food industry. Other uses include starch noodles, fish paste binder, adhesives, beer and modified starches. Japanese scientists have been in the forefront of research for alternative starch-derived products. Although these are diverse, sweet potato starch is too expensive for them to become a commercial possibility at present. Research into increased yields and higher starch concentrations are continually pursued to increase relative competitiveness of sweet potato. By-products of starch manufacture are exploited for animal feed and citric acid production.

Food products

Although a relatively large variety of sweet potato processed food products are available in Japan, the production of many is on a fairly limited scale. Furthermore, they represent a very small fraction of total food products available to the consuming public. Processing establishments of all capacities often exploit the sweet taste of sweet potato roots to produce dessert items. One major food firm which handles 10,000 tonnes of roots per year claims to produce 80% of all the sweet potato processed products in Japan. They manufacture a range of deep-fried, sweetened sweet potato snacks, the most popular of which – *imo-karinto* – consists of crunchy french-fry-shaped sticks and also sweet potato paste. Sweet potato processing establishments range from very small 'back-room' businesses making handicrafted items and such things as sweet potato snacks, noodles or candies to large factories and agricultural cooperatives producing flakes or frozen paste and cubes.

The processing laboratory of the Kagoshima Agricultural Experiment Station carries out product development and processing technology testing. New products include sweet potato granules and an extruded snack made from a low-sweet cultivar (see Chapter 6) and a purple food colouring agent. There appears to be a close relationship between plant breeders and food technologists, which promotes development and utilization of new cultivars, and between scientists and industry, which eases the transfer of technology from the laboratory or pilot plant to the commercial sector. A study of the means of transferring sweet potato processing technology to the private sector, which appears to be more successful in Japan than in any other country, could be of benefit to research institutions in other lands.

Promotion

Public interest in sweet potato nutritional value, cookery and products is stimulated from time to time by articles in national newspapers and magazines, cookery booklets and leaflets distributed by newspapers, talks on television and radio, and promotion of sweet potato products at food fairs and in supermarkets and departmental stores. Judging by the comments of some of the sample families from Nagoya mentioned above, more could be done in this direction to promote sweet potato. There are a few restaurants specializing in sweet potato dishes, where sweet potato forms at least a part of every course served, and others in whose menu it figures prominently. In addition, as described in Chapter 6, a high beta-carotene cultivar is being used by a cooperative in one area of Japan for transformation into a paste utilized in bread making for a schools' lunch programme. Possibilities of introducing baked sweet potato or croquettes into school lunches have also been explored. High beta-carotene cultivars for this purpose could be an attractive alternative to carrots, which are unpopular with many Japanese children.

The need to diversify into products which are alternatives to starch has led in the southern prefecture of Kagoshima to the opening of a Sweet Potato Trading (Merchandizing) Information Centre in the Union of Agricultural Cooperatives of Kagoshima. The centre aims to provide information on various aspects of sweet potato to producer and processor alike. Seven cultivars are featured: two popular cultivars; a starch/processing type; a high carotene orange-fleshed type; a high anthocyanin purple-fleshed type; the new non-sweet cultivar; and a new red-skinned, yellow-fleshed cultivar with a moderate content of beta-carotene. On display is a range of processed products. The centre has disseminated much information about sweet potato through the medium of television and newspapers and is receiving interest from merchants and processors. However, its success has still to be determined.

The town of Kawagoe, Saitama Prefecture, although no longer the centre of a major sweet potato growing area, maintains its traditional association with sweet potato. There is a small, but highly prized, production of fresh roots and processing of both older-style and newly developed products being made in a series of small shops and businesses. The products which have nostalgia value are sold to tourists and as gifts for friends. A few years ago only three of these businesses existed, whereas now they are numerous. The supply of local roots being limited, roots are now brought from other parts of the country for processing purposes. Some small shops sell as many as 12 different sweet potato products. In addition there are several restaurants featuring sweet potato dishes. A few years ago, a citizen's activity group, known as the Friends of Kawagoe Sweet Potato, was formed to improve the image and encourage consumption of sweet potato. Its members – sweet potato farmers, processors, restaurateurs, teachers, dieticians and other members of the public – have many activities including growing 15–20 cultivars of sweet potato, sweet potato history studies, seminars, cookery sessions, printing of sweet potato publicity materials and so on. They have been featured in a national magazine and have encouraged the formation of similar groups elsewhere in Japan. One of the group members has opened a sweet potato museum which will display information and objects associated with sweet potato from Japan and all over the world. All this may seem very far away from the problem of feeding the world's hungry, but it is one way in which concerned and influential people in any country could go about publicizing the nutritional advantages of sweet potato and improving its image in the eyes of consumers. There has been some interest in forming a similar group in the Philippines.

Conclusions

Although it appears that in Japan many of the constraints implicit in the factors affecting consumption and utilization of sweet potato have been tackled effectively, consumption of fresh roots is at a low level and utilization as starch and animal feed is waging an economic battle for survival. However, the development of high yielding cultivars adapted to a variety of purposes, and emphasis on quality, has no doubt been responsible for preventing a greater decline than might otherwise have taken place, faced as the sweet potato was with a situation of competition from a great variety of other foods. As it is, the consumption of sweet potato has now been relatively stable for many years. It has even been suggested that sweet potato is riding on a wave of gourmet food preferences as well as benefiting from health concerns connected with dietary fibre intakes.

There would appear to be some possibility of expanding the market for processed food products especially by exploiting their nutritional benefits for institutions such as schools. There is also a possibility of increased utilization of speciality cultivars, such as purple or orange-fleshed or low-sweet clones, if the public are educated in their use and processors are able to obtain them outside the production areas in greater quantity.

Although production of sweet potato starch does not suffer from the constraints of poor yields and quality which bedevil the industry in other countries, it still suffers from price competition because Japan chooses to import relatively cheap supplies of corn and sugar. Sweet potato starch has no unique characteristics which would set it apart from other starches for specialist uses and cannot therefore command a premium price. However, from the point of view of developing countries wishing to exploit local raw materials rather than import corn or other starch sources, sweet potato has the advantage that its non-specialized characteristics in fact enable it readily to replace other starches. The high starch yields which Japan achieves from sweet potato should encourage others to develop high starch cultivars adapted to local conditions.

The Philippines

Although in recent years in the Philippines the role of sweet potato has largely changed from that of a subsistence staple to that of a complementary vegetable and snack food, with a consequent decline in consumption levels, it presents various signs of encouragement that this decline could be halted or even reversed. There is a tremendous potential for increasing production by increasing yields which have remained static for many years, but this potential will be realized only by stimulating demand. Studies have made some progress towards identifying variables associated with consumption levels. Country-wide investigations, which have helped to pin-point problems involved in marketing and suggest solutions which might positively influence consumption could be reproduced elsewhere. Above all, vigorous research, oriented towards simple processing technology and the development of products which not only fulfil local dietary needs and habits but are highly nutritious could be studied and emulated by other tropical developing countries.

General background

The sweet potato is well established in the Philippines, having most probably been introduced there in the sixteenth century by the Spanish, trading between Mexico and Manila (O'Brien, 1972; Yen, 1982). It is now cultivated all over the Philippines both as a subsistence and as a cash crop. Besides its role as a farm crop, it is found growing in most communities as a backyard or garden crop, and as such helps to supplement family food intake. It is especially reliable in times of drought and typhoon. Sweet potato roots helped to stave off famine during the Second World War and, in the case of American prisoners of war fed sweet potato leaves, vitamin A deficiency. Both the roots and tops are still consumed. Indeed, sweet potato tops are one of the most important green, leafy vegetables in the Filipino diet. The roots are consumed as a staple in a few communities, especially in remote mountainous unirrigated areas of the eastern Philippines, and more generally as a vegetable and snack. Many families, however, use the roots as a substitute for cereal staples such as rice or corn when these are in short supply due to bad harvests. In this respect sweet potato is acting as a supplementary or seasonal staple (see typology of consumption). Peelings and roots unfit for human food are mixed with other waste vegetables and fed to pigs. A balance sheet for 1984 records that 90% of total utilization is as food, and only 5% as livestock feed, the remaining 5% being wasted (Horton, 1988). No industrial processing is recorded. Although most sweet potato is eaten as fresh roots, which are commonly boiled or baked, there is a low level of conversion to flour and starch and some traditional processing especially into snack foods, such as *camote cue* (a sweetened fried item) and *guinata'an* (a sort of soup made from a mixture of root crops).

Production of roots has increased in recent years, from 657,000 tonnes in 1970 to 1,005,000 tonnes in 1985 (Horton, 1988). This increase has been achieved by increasing the area under cultivation, the average yields for the whole country having remained at a low level of 5 tonnes/ha for the last 25 years. Since 1981, 11 new cultivars of sweet potato have been developed at various institutions and released to farmers by the Philippine Seed Board (Villamayor, 1989). Some of these cultivars are capable of yielding more than 20 or 30 tonnes/ha under experimental conditions, and have a shorter growing season than traditional cultivars. The extent of research into agronomic aspects of the crop, complemented by investigations into postharvest handling and storage, processing, livestock feeding and marketing, is reflected in a state-of-the-art bibliography which presents abstracts of Philippine sweet potato research from 1921 to 1985 (Philippine Root Crops Information Service and Philippine Council for Agriculture and Resources Research and Development, 1986).

Variations in consumption levels

According to Philippine Food Balance Sheets there was a steady decline in the annual per capita quantity of sweet potato available for consumption from 31 kg in 1957 to 21.5 kg in 1965. After 1965, however, consumption appeared to stabilize and was still recorded as 20 kg in 1980. There may have been a further decline in consumption levels in the 1980s. The amount of sweet potato roots available for consumption per capita in 1985 was estimated as 14 kg (Horton, 1988). The annual average per capita consumption of sweet potato roots for the whole Philippines has been estimated by several different agencies over the last 30 years. There are considerable differences between these estimates. The 21.3 kg per capita available for consumption as estimated from food balance sheets (Alkuino, 1983) for 1978, for example, is a much higher figure than that of 5.1 kg determined by weighing food directly during the National Nutrition Survey of the same year. The latter represents an average consumption of 14 g per capita per day. In urban areas the average per capita consumption of roots was 9 g per day, which is within the range 5–18 g per capita per day found during a very recent survey in two cities of Visayas region (International Potato Center/ViSCA, 1988). The 1978 survey estimated the average per capita consumption of sweet potato tops to be 5 g per day or 1.8 kg per year. This was slightly less than the 2.2 kg per year of *kangkong* (*Ipomoea aquatica*) consumed. Although sweet potato has been described as the most important root crop in the Philippines, it appears to have been overtaken in recent years by cassava in terms of production and amount available for per capita consumption (Horton, 1988). Intakes determined by direct weighing are likely to be more accurate than those estimated from balance sheets, but may still not reflect seasonal or away from home consumption. Though levels appear to be low there are striking variations in consumption with a number of factors, as illustrated by surveys which have been carried out.

Regional differences

According to a series of quarterly food consumption surveys during 1974–6, the average sweet potato consumption was 11 kg per capita per year (Aviguetero et al., 1976). However, great regional differences appeared, from a high of 31.8 kg in Eastern Visayas to a low of 2.9 kg in Southern Luzon. Differences between urban and rural areas for the whole Philippines and between major regions of the country were also found during the 1978 Nutrition Survey (see Table 8.13). Consumption levels of both roots and tops can be seen to be generally higher in rural than in urban areas. The lower level of root consumption in Visayas is surprising in view of the importance of sweet potato in Eastern Visayas which has been noted by several sources (e.g. Aviguetero et al., 1976). It is also interesting to note that although Visayas apparently had the lowest per capita root consumption it did not also have the lowest per capita consumption of tops.

The proportion of families using sweet potato roots and leaves in the diet in 12 major regions of the country was studied in 1974–6 (Aviguetero et al., 1977). The mean proportion of families in the whole country

Table 8.13. *Per capita consumption of sweet potato roots and tops in various regions of the Philippines in 1978 (kg/year)*

	Philippines	Urban	Rural	Luzon	Visayas	Mindanao
Roots	5.1	3.3	6.2	5.5	3.3	6.9
Tops	1.8	1.1	2.2	1.5	2.2	2.6

Notes:
From: Food and Nutrition Research Institute, 1981; First Nationwide Nutrition Survey, 1978; 2800 households. Data collection by food weighing during the summer months.

Table 8.14. *Percentage of families in the Philippines using sweet potato roots or tops by income group in 1974–6*

	Less than P400	P400–799	P800–1499	P1500 & above	All
Roots	42	40	38	34	38
Tops	68	65	65	57	64

Notes:
P, pesos.
From: Aviguetero et al., 1977.

using roots was 38%, but the mean proportion using leaves was 64%. There was considerable variation between regions with highs of 65% and 78% of families in Eastern Visayas using roots and tops respectively to lows of 25% and 50%, respectively, in East Mindanao.

Income differences

The same study also revealed how the proportion of families using roots and tops changed with income level. The proportion of users of both roots and tops decreased as income levels rose, as may be seen in Table 8.14. In 1970–1 a country-wide survey showed that weekly consumption per 1000 people increased from 138 kg at the lowest income level to 148 kg at the second lowest income level and then decreased to 125 kg and 84 kg at the second highest and highest income levels, respectively (Urbino, Torres and Darrah, 1972). However, coefficients of income-quantity elasticity were found to be only −0.28 and −0.15 for roots and tops, respectively, indicating that there was no dramatic change in consumption with change in income. The National Nutrition Survey of 1978 showed that daily per capita consumption of roots decreased from 20 g in the lowest income group to 6 g in the highest income group. There was much less variation in the quantity of tops consumed by income group. There was no difference in consumption levels between income groups (5 g per capita per day) except in the two highest income groups, who consumed slightly less (3 or 4 g per capita per day).

Frequency of consumption of sweet potatoes and other root and tuber crops has been studied in Laguna province among 20 faculty members of the University of the Philippines and their families, and 20 farmers and fishermen and their families. The results in Table 8.15 show interesting differences between the frequency of consumption in the two groups of people. Whereas 14 of the 20 farmers and fishermen ate sweet potato roots sometimes or very often and only 2 never consumed them, 15 of the 20 professional group ate them rarely or never. Sweet potato leaves were hardly consumed by the latter group whereas 15 out of 20 of the former consumed leaves every day or very often. There was also a slightly greater frequency of consumption of processed products among the professional group than among the farm/fishing group. The mean frequency of sweet potato root consumption was a little lower in both groups than that of cassava, but equal to or greater than that of taro. However, it was much lower in both groups than that of either rice or bread. The 1978 Nutrition Survey also shows daily per capita consumption of sweet potato roots and tops by occupation of household head (Table 8.16). The highest daily consumption of both roots and tops occurred among the farm owners or farm workers, presumably reflecting the greater degree of accessibility that these people have to the crop and the fact that they may not have to purchase it. Among employed people, the professional, technical or entrepreneur group consumed the least quantity of either roots or tops, but there was little difference between them and other occupations, excluding farm people.

Factors affecting demand for sweet potato

A consumer survey in two regions of the Philippines, one with a rice-based and the other a corn-based diet, determined the major factors affecting consumer demand for sweet potato (Alkuino, 1983). The most significant variable affecting demand was income of the household head, followed by the retail price of sweet potato. Size of the household was significant in one region, but not the other. Age of the principal shopper was only significant in higher income groups in one of the regions. In the lower income group demand increased with increased income, but in the higher income group, demand decreased with increased income. This indicated that sweet potato was regarded more highly by the low income

Table 8.15. *Frequency of consumption of sweet potato in two contrasting communities of the Philippines*

	University personnel						Farmers and fishermen					
	Every day	Very often (twice a week)	Sometimes (once a week)	Rarely (once a month)	Never		Every day	Very often (twice a week)	Sometimes (once a week)	Rarely (once a month)	Never	
	No. of persons using					Mean frequency	No. of persons using					Mean frequency[a]
Roots			5	6	9	1.8		5	9	4	2	2.9
Flour		1	2	3	14	1.5			2	2	16	1.3
Starch		2	4	2	12	1.8			3		17	1.3
Leaves			1	1	18	1.2	7	8	4		1	4.0

Notes:
From: Kawabata et al., 1984.
[a] Frequency computed from 5 every day; 4 very often; 3 sometimes; 2 rarely; 1 never.

Table 8.16. *Daily per capita consumption (in grams) of sweet potato roots and tops in the Philippines by occupation of household head in 1978*

	Occupation						
	Professional, technical, entrepeneur, skilled	Farm owners/ managers	Farm workers	Fishermen	Other (mostly skilled)	Housewives, students, retired	None
Roots	12	18	17	14	12	10	11
Tops	3	7	6	5	5	1	5

Notes:
From: Food and Nutrition Research Institute, 1981.

group than by the high income group. For both low and high income groups, demand for sweet potato decreased with increased retail price. However, the demand price for sweet potato was found to be slightly inelastic for all groups. In other words, consumption was not very responsive to price change. This makes it unrealistic for farmers to increase production and lower price for the fresh root market. The alternative suggested is a programme to stimulate an increased demand for sweet potatoes. This has also been suggested by others (International Potato Center/ViSCA, 1988; and see below).

Consumer perceptions and preferences

Consumer perceptions of, and preferences for, sweet potato were recently studied in two cities of Visayas region as the initial part of a project seeking to develop acceptable processed sweet potato food products for low- and middle-income consumers (International Potato Center/ViSCA, 1988). The respondents interviewed were households and food service firms in Tacloban City and Cebu City. The food service outlets were small restaurants, fast food stalls, canteens or cafeterias and pavement eateries. All the household respondents were sweet potato eaters with a frequency of at least once a week or a few times a month. On the other hand, about 83% of the Tacloban and 73% of the Cebu food service firms did not use sweet potato in preparing dishes. Between 80% and 97% of the non-user food service firms said that the main reason for this was lack of suitable recipes, or disintegration of roots on cooking.

Householders' perceptions related to status, food values and price are shown in Table 8.17. Contrary to the often expressed view that sweet potato has a poor image in the Philippines due to its status as a 'poor man's food', 65% of respondents in Tacloban and 92% in Cebu considered sweet potato to be a food for all. A very high percentage of respondents liked sweet potato for its sweetness and described it as having a good taste, while 60% thought it was also nutritious. However, only 30% considered it to be filling. Only 1% of respondents in Tacloban complained about flatulence, whereas 20% in Cebu did. The reason for this difference is not known. It would be interesting to know whether this result was related to the use of different cultivars in the two places (see below).

The number of cultivars being marketed in Tacloban and Cebu cities was about 16 and 40, respectively. The most preferred cultivars were different in the two cities. These preferences were normally translated into higher prices for the choice cultivars. An informal survey revealed that the most preferred characteristics are sweetness, light yellow coloured flesh and a slightly dry texture. However, some consumers

Table 8.17. *Consumers perceptions of sweet potato as a food in two cities of the Philippines*

Opinion	Percentage of respondents
Food for all	65–92
Poor man's food	8–13
Rich man's food	—
High nutritive value	60
Very filling	30
Good taste	98–100
Flatulence causing	1–20
Price:	
Relatively expensive	—
Relatively cheap	52–63
Neither expensive nor cheap	25

Notes:
From: International Potato Center/ViSCA survey, 1988.

particularly liked the slightly moist orange-fleshed cultivars. For sweet potato tops as vegetables, consumers preferred leaves with a purple colour and a digitate shape.

Utilization

The International Potato Center/ViSCA survey also determined the ways in which people consume sweet potato. In the households, 31–33% of respondents boiled or steamed the roots. This was followed by fried *camote cue* 15–26% or mixed in *guinata'an* 13–18%. While 28–34% of household respondents would substitute sweet potato for potato when necessary, only 3% of food service firms would do so. The food service firms mainly utilized sweet potato in a fresh form and to a small extent as flour (4–11%), chips (2%), flakes (2%) and canned (2%). Sweet potato was presented by the food service outlets as snacks such as *guinata'an* and *camote cue* or mixed with vegetables or meat dishes and as boiled roots or fried chips. Sweet potato flour is used as a coating in several food preparations.

Preference of households and food service firms for sweet potato were in the order (1) chips as a ready-to-eat snack, (2) cubes for ready-to-mix dishes such as *guinata'an* and (3) powder for instant food and soup thickeners. Research on product development and refinement will therefore be focused on these. Utilization of sweet potato in the formulation of instant noodles, the most popular food item in the surveyed area will also be attempted. Processing treatments to minimize the problem of disintegration during cooking of sweet potato chunks or cubes should encourage food service outlets to use more sweet potato, as should the development of recipes and improved snack items.

Particularly vigorous and effective research into simple improved technology and product development (see Chapter 6) has already been pursued for some time at ViSCA. Products such as a dried fruit-like snack item (see Table 6.1, p. 302) and a non-alcoholic beverage are made from orange-fleshed cultivars and are very nutritious. The substitution of sweet potato flour for imported wheat flour in bread and other baked goods and items such as soya sauce, which have also been successfully accomplished on an experimental basis, could mean significant savings in foreign exchange.

Marketing studies

There are two major marketing studies known to the author. They were conducted in 1974–6 and 1977–9. The reports for the individual area surveys, of which they are composed, are available separately. The main results have been consolidated in two reports (Santos, 1977; Medina et al., 1979). In addition, an informal marketing survey was conducted in 1988 as part of the International Potato Center/ViSCA project mentioned above.

The two formal marketing surveys interviewed both producers (farmers) and buyers (middlemen and consumers). Farms sold a high proportion of their sweet potatoes – from 98% in Bataan to 61% in Northern Mindanao (Medina et al., 1979). Some farmers cleaned sweet potato by trimming or washing with water, but this was infrequent. Roots were sorted on the farm by removal of diseased and badly bruised specimens and graded by size. Farmers used mostly sacks or baskets in packaging roots and tied them with ropes. Roots either were delivered (about 30% of cases) to the outlet by the farmer in a variety of means of transport including carts, buses and mini-buses, jeeps and boats, or were picked up by the purchaser (70%). Major purchasers, apart from consumers, were middlemen of at least six types including wholesalers, assembler-wholesalers, retailers, wholesaler-retailers, contract buyers, and agents. Much smaller purchases were made by processors (*camote cue* makers). Middlemen performed one or more market tasks of harvesting, hauling, cleaning, grading, sorting, packaging, repackaging, picking up, delivery and selling. For the different regions prices paid to farmers, costs incurred by middlemen and dealers' markup margins differed widely.

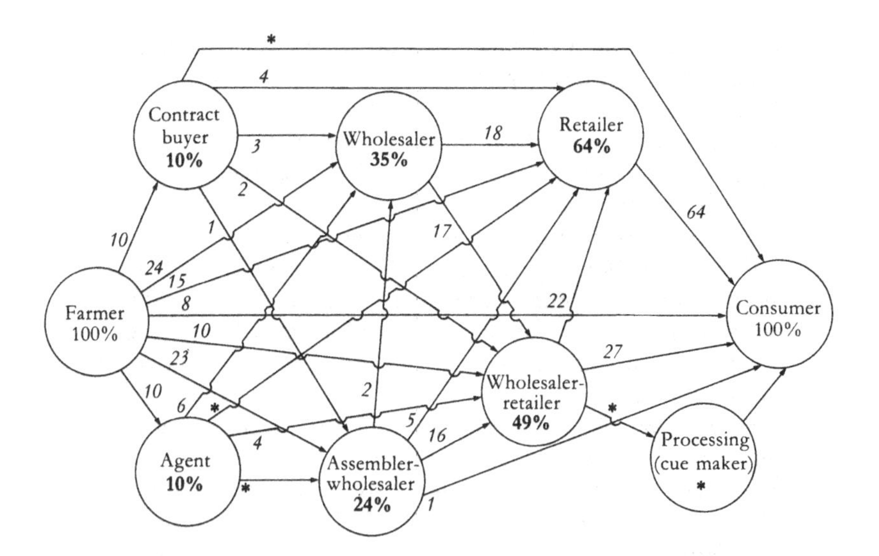

Figure 8.13. Marketing channels of sweet potatoes in the Philippines, 1977–9 (from Medina et al., 1979). An asterisk indicates <0.5%. Bold percentage figures are maximum held at sometime, by a middleman, from all sources before he sells on to another middleman. Italic numbers are percentages of farmers' sales going to each middleman.

It is beyond the scope of this book to detail all the findings of the surveys. However, it is pertinent to consider the major problems identified by the researchers.

1. *Roads and transport.* Sweet potato farms were often located in isolated areas where there was a lack of good roads or farm-to-market roads and a lack of transport facilities. This made it difficult for other than local buyers to penetrate the production areas and resulted in relatively high handling costs. The 1974–6 survey also noted that rainy days during the harvest period caused feeder roads to become so muddy that they were impassable by most means of transport. As a result, harvested products were left in the fields and deteriorated. Improved farm-to-market roads would facilitate marketing and reduce costs which are passed on to the consumer.

2.(a) *Proliferation of channels.* One of the major problems in several areas was the proliferation of agencies through which sweet potatoes passed before they finally reached consumers. This is illustrated in Figure 8.13. Since a number of the agencies involved performed similar functions, some could be eliminated without detriment to consumers. In so doing, the time and cost of moving sweet potatoes to consumers could be reduced.

2.(b) *Intra-agency movement.* Dealers were often buying sweet potatoes from other dealers. There were instances, for example, of wholesalers purchasing sweet potatoes from other wholesalers, or retailers from other retailers. This resulted in additional marketing costs and may have prolonged the time needed for sweet potatoes to reach the consumer. A system of market intelligence was required to allocate sweet potatoes more efficiently to different outlets and improve the flow to the consumer.

3. *Price information.* Due to the geographical location of farms, many farmers were not aware of prevailing prices of sweet potatoes. Thus they had no alternative but to accept the price dictated by dealers. This resulted (1977–9 survey) in relatively low farm prices, wide marketing margins and a relatively high price to the consumer. A price information system for farmers could help to allocate prices more fairly.
4. *Market intelligence.* The 1974–6 survey found that sweet potatoes generally moved from major production areas to nearby consuming areas or market outlets. However, there were instances when the roots moved to more distant trading centres and then back to areas which were closer to the original production areas than the trading centres. Some cases were reported of roots moving from an area and then back into the area. The problem reflected a lack of information regarding potential markets on the part of producers and perhaps the first buyers. It must have resulted in increased time for the roots to reach consumers and additional cost which must have been passed on to the consumer.
5. *Quality.* Some farmers with rodent or weevil problems did not cull affected roots. Poor and good quality roots were mixed together, packaged and sold. This practice down-graded quality and affected prices and consumption levels. Control information and assistance was needed. Furthermore, cleaning of sweet potatoes was seldom practised, even though clean roots commanded better prices. This was reportedly due to lack of cleaning facilities and space and additional costs of water and labour. However, cleaning facilities such as space and water pipes could be installed in city markets.
6. *Lack of standard sales units.* Sweet potatoes were normally retailed on a per *tumpok* or pile basis, with no standard

> quantity of sweet potatoes in a *tumpok*. Customers had to buy a *tumpok* consisting of mixed quality sweet potatoes, with no right to select those of better quality. Standard units of sale, for example on a per kilogram basis, would improve marketing procedures and make price information more meaningful.

During the 1988 informal market survey (International Potato Center/ViSCA; and see Figure 8.14a) sweet potatoes were still retailed by the pile in Tacloban City, with price varying according to pile size (see Figure 8.14b). During lean periods when the price of sweet potato rises, the number of roots in the pile is reduced, but pricing remains the same. In Cebu City, sweet potato was sold by the kilogram with prices in the lean months about double those in the peak period. During the dry season when there is a prevalence of weevil infestation, prices increase because of decreased supply of good quality roots.

Resolving the problems highlighted above could have effects on two major factors affecting consumption: price and quality. Resolution of inefficiencies in the market system could eliminate extra, unnecessary costs which consumers eventually have to bear. Quality control, which was described as very poor, could improve the image of sweet potato in consumer eyes. Enhancing product quality is desirable for all consumers, so that waste in terms of excess peeling or removal of bruised or diseased areas is reduced to a minimum, thus giving better value for money. However, improvement in quality should not be allowed to raise prices to levels unaffordable by the lower income consumer in whose diet sweet potato plays an important part.

Marketing strategy trials

To ascertain whether fresh roots of sweet potato could penetrate the supermarket, thus elevating its status, new marketing strategies were tried by researchers at ViSCA (Pascual et al., 1987). The first innovation, in 1984, was to pack sweet potatoes in white nylon net screen bags of three sizes (small, medium and large). The pre-packed sweet potato roots with labels were advertised and sold at two big stores, one in Ormoc City and one near ViSCA. In 1985–6, another form of packaging in the shape of a cellophane bag with 1 kg capacity, punched with holes for ventilation and labelled, was tried. The pre-packed sweet potato in cellophane bags was advertised and sold in five department stores of Cebu City.

After 6 months of selling experience with the net bag pre-packed sweet potatoes in Ormoc and ViSCA, the following observations were

a

b

Figure 8.14. (a) Researchers from the International Potato Center interviewing sweet potato sellers in a Baybay market, Philippines (Truong Van Den). (b) Local people in the Philippines use volume units, not weight, to sell sweet potatoes in the market. These volumes change in size, but not in price over the seasons (G.A. Watson).

made: of the buyers questioned 52% preferred to buy sweet potato in pre-packed form for (a) convenience, (b) time saving and (c) hygiene. A 3 kg pack was generally preferred by those buying pre-packed sweet potatoes. Of the total volume sold, however, 75% was bought unpacked.

Although sweet potato in cellophane was sold at five department stores in Cebu and was therefore exposed to more buyers, sales in a 6 month period were very much lower in Cebu than in Ormoc. The main reason for this was that buyers generally purchased the roots only in 1 kg bags. Most of the small quantity purchasers used sweet potato as a snack item and bought only once a week. The majority of small buyers were college graduates in gainful employment. They were not very particular about the characteristics of the roots purchased, in contrast to the Ormoc buyers. The latter expressed preferences for different cultivars depending on such factors as the race of the consumer and whether they wanted the roots for direct consumption or for processing. Further investigation revealed that larger-quantity Cebu buyers preferred to buy sweet potato unpacked from the public market. As the majority of buyers in both cases preferred to buy unpacked roots, it might be preferable to concentrate on improving the quality of unpacked roots as a means of enhancing the image of sweet potato.

Processing

The important role of women in the processing of sweet potato into a variety of traditional products has been observed in at least one area of the Philippines (Alcober and Parrilla, 1987). Although in the area in question (Eastern Visayas) the majority of fresh roots were not marketed, but used for home consumption, home-processed sweet potato products made by the women were sold at school canteens, cock fighting pits, town markets, local stores and from the processors home. It was suggested by the researchers, therefore, that development of new cultivars and technologies designed to help end-users should seek reliable preliminary information on preferred characteristics of sweet potato from female respondents.

During the International Potato Center/ViSCA survey in 1988, in Visayas, it was found that the bulk of sweet potato trading was in the form of fresh, unprocessed roots. Some primary processing was noted in the form of peeled roots sold in cellophane bags, catering to the needs of consumers who prefer sweet potato in a more ready-to-cook form. These pre-peeled roots were priced a little higher than unpeeled roots. The peelings are sold as animal feed. Dried sweet potato chips which are used for soup preparation by Chinese people were also available in Cebu

City. However, drying of sweet potato into chips did not appear to be attractive to farmers who could easily sell their fresh produce at a higher price. This observation had previously been made by other researchers (Pascual et al., 1987) who investigated the profitability of processing fresh roots into dried chips.

Other processed products found in the market by the International Potato Center/ViSCA survey included fried sweet potato chips (crisps), *camote cue* and sweet potato candy. Processing of sweet potato in the surveyed area was insufficient to necessitate an increase in production of the crop. However, there is a great potential for increasing production as the existing areas planted to the crop are small and could be further expanded. Moreover there is a large gap between potential and actual yields. As the demand for fresh roots has been declining, increased production would depend on the demand for the crop from processing. Although processing for dried chips was unattractive to farmers, they appeared to view other types of product with more favour, opining that there would be added value in the finished products. The potential profitability for farmers or processors of two sweet potato products was found by researchers (Pascual et al., 1987) to depend on the price of fresh sweet potato roots. Processors will only be interested in sweet potato for food or feed manufacture when prices are lower than at present. One way of changing this situation could be at the commercial level where through contract growing at guaranteed prices farmers (supply) and processors (demand) work together. In this system farmers could use improved cultivars and production methods with little risk, since the price (although lower than before) would be guaranteed. Increased yields would mean that prices could gradually decrease while farmers' incomes were maintained. The key to this approach is to identify sweet potato products which are of interest to industry.

The International Potato Center/ViSCA project aims to develop simple technology to process sweet potato into nutritious, acceptable, stable and low-priced food products to meet the needs and preferences of low- and middle-income urban consumers. If successful, this could stimulate an increased demand for sweet potato benefiting farmers and consumers alike. Much has already been accomplished by reseachers at ViSCA in designing simple processing equipment and in formulating a variety of products (see Chapter 6, pp. 298–302).

However, in spite of strenuous efforts to transfer processing technologies to the private sector this step has not been as successful as desired (Truong, V.D., personal communication). Researchers at ViSCA divide their potential clients for technology transfer into two categories: (1) farmers and small-scale processors and (2) medium and large-scale companies. For the former, initial efforts involve surveys to identify communities, farmer associations and processors who are interested and have the resources to adopt the developed/improved technologies. Subsequent steps are community organization and technology demonstrations, setting up of pilot processing facilities, training, monitoring, and marketing assistance. These projects are implemented in collaboration with local government offices and non-governmental organizations. Major problems identified with this type of project involve management and finance; poor packaging and lack of advertising or promotion lead to stiff competition with similar products being made and advertised by commercial companies.

Technology with industrial potential, such as the non-alcoholic sweet potato beverages (described in Chaper 6), was patented before it was publicized through announcements in investment fora, scientific seminars etc. Several companies showed interest in the high provitamin A sweet potato beverages. After several meetings, ViSCA and a large food company signed a Memorandum of Agreement whereby the technical know-how for the beverages was made available to the company in return for a donation of processing equipment from the company to ViSCA for research and development purposes. Working out the terms of the Agreement was the critical factor causing delay in the technology transfer process. Companies usually prefer to have exclusive rights to the developed technology. However, the request for exclusivity is not in line with the mandate of government institutions such as ViSCA. A non-exclusive agreement was finally made over the beverages through a provision that ViSCA will not make use of the technology, if any, generated during pilot-scale production or commercialization of the product, for the benefit of a third party during the inclusive period of time. Though this is only one example it illustrates the problems involved in reconciling commercial interests with the mandate of a government institution to develop local food utilization and improve nutritional status.

Animal feed

Although farmers in many parts of the Philippines feed home-produced fresh or cooked sweet potato roots and the tops to animals, especially pigs, incorporation of dried sweet potato chips or meal into processed feeds has not been found to be profitable (Pascual et al., 1987). Using sweet potato as feed was recommended only when the market price of roots was very low or there was a surplus production which would result in storage loss. It remains to be seen whether this state of affairs could be remedied by encouraging farmers to grow high yielding cultivars, high in starch and dry matter, appropriate for incorporation into animal feeds, with a guaranteed outlet and price (as suggested above for food processing). There might also be a positive discrimination on the part of the government towards locally produced feeds which would save a substantial amount of foreign exchange by reducing imports of alternative energy sources.

Conclusions

Quantities of sweet potato available for consumption on a per capita basis in the Philippines are small and declining, even though total production, through expansion of the planted area, has increased in recent years. Most sweet potato is consumed in the form of fresh roots as a vegetable or snack, maintaining its role as a regular staple food only in remote or mountainous areas. However, sweet potato still remains a backyard or garden crop for many families, acting especially as a seasonal or supplementary food in times of need. Even though the consumption of sweet potato has been shown to decline at the higher income levels, it seems doubtful that the classification of sweet potato as a 'poor man's food' of low status is entirely correct. Consumption of fresh roots might be stimulated by paying increasing attention to quality factors associated with more efficient methods of grading, presentation and marketing. In addition food service outlets and institutions might be future sources of demand if certain sensory qualities could be improved and a wider range of preparation methods made available.

A further promising approach aimed at stimulating demand would focus on the use of sweet potato as a raw material for industry in the form of food products and livestock feed. Active research into food and feed product development from sweet potato is being pursued. However, the economic feasibility of such an approach would depend on placing equal emphasis on production factors such as yield increases and breeding appropriate cultivars; there is a great potential for both of these developments. The presentation of detailed studies into problems involved with the transfer of technology from the research to the commercial sector, in an environment where simple improved processing equipment and a variety of potential products are available, could be of benefit to other countries planning similar strategies.

Sweet potato leaves are an important and dependable green vegetable eaten in small quantities by most classes of consumer. Increased consumption of leaves especially by low-income groups is probably highly desirable from a nutritional perspective. However, there do not seem as yet to have been any formal consumer or marketing studies which might indicate the most important aspects of consumer preferences, marketing practices, quality factors and the major constraints to increased consumption.

Papua New Guinea

Papua New Guinea serves to illustrate the (admittedly unusual) case where sweet potato plays a central role in society. Not only is there no discrimination against it as a major staple, due to preconceived images or dislike of its sweet taste, but it also forms a vital part of the foundation upon which the economic, political and cultural organization of society is based. Far from being despised, it is a food item of primary, year-round significance, rather than of only secondary or seasonal importance.

General background

Subsistence agriculture, which is still extremely important in Papua New Guinea, is based on the production and consumption of several staples: namely sweet potato, yams, taro, sago, cassava and bananas. Depending on the environment, and with the exception of cassava, one of these becomes the main staple, with the others assuming varying degrees of secondary importance (Kimber, 1972). However, sweet potato is by far the most important food crop (Bourke, 1985) with a value estimated at over US$200 million per year. The only other crop approaching this value is coffee, which is grown as an export crop. Although sweet potato is traditionally a subsistence crop, nowadays its role as a cash crop is increasing. It is also the most important crop grown at schools and other institutions (Bourke, 1985). Both roots and leaves are frequently fed to pigs and other livestock; this use often represents a high proportion of total production. In some communities living on the fringe of the Central Highlands, sweet potato is used exclusively as a pig fodder, with taro as the human staple food (Watson, 1977). Leaves are consumed only rarely by humans, as a green vegetable or as part of a relish.

Sweet potato was probably introduced to Papua New Guinea between 300 and 500 years ago, although the date and manner of its introduction are still highly controversial (see Chapter 1, p. 18). Although sweet potato is grown in many parts of lowland Papua and holds a position there as co-staple with other crops, it has attained a virtually monopolistic position as a staple in the Highlands, at altitudes of 1200 to 2700 metres above sea level (m.a.s.l.). However, production of sweet potato is also increasing in the Lowlands (0–600 m.a.s.l.) and intermediate altitude zone (600–1200 m.a.s.l.) at the expense particularly of the traditional staple *Colocasia* taro. The displacement of taro by sweet potato has been seen to a large degree on the islands of New Ireland and Bougaineville. This change is taking place due to loss of fertile forest land necessary for taro cultivation, the relative ease of sweet potato cultivation, pest and disease problems of taro and the loss of traditional values associated with other staples (Bourke, 1985). Whereas ritual and magical significance and ceremonies are or have been associated with staples such as taro and yams, there is no special or ceremonial significance attached to sweet potato. Sweet potato is a relative newcomer compared to yam and taro, which have been embedded in local cultural practices for millennia. However, its apparent failure to become culturally integrated is surprising given other known cases where new cultural elements have been rapidly and totally integrated into symbolic systems. One possible explanation is that sweet potato has taken on a different kind of symbolic significance in terms of political and social relations rather than in ritual or magic. The importance of sweet potato in the politics of pig herding, for example, is described briefly below. Moreover, sweet potatoes are also used as food exchanges, between women and men, which help to bind together a fragmented society (Pospisil, 1963).

The trend towards increasing dominance of sweet potato in areas traditionally devoted to alternative staples may represent a continuation of a prehistoric shift from taro to sweet potato which, in the Highlands, allowed a dispersion of cultivation on to less favoured soils, a less stringent approach to planting times and a decreased labour input compared with those of the traditional taro staple (Clarke, 1977). There has been disagreement among researchers as to the effect on the Highland population of the introduction of sweet potato, some (Watson, 1965, 1977; Golson, 1982) claiming that a subsistence revolution occurred with, among other radical changes, a population explosion, an increase in the population and importance of pigs, changes in gardening patterns and in the social structure of the people. Others suggest that changes were not of pronounced importance, except in so far as sweet potato allowed expansion of cultivation to higher altitudes and resulted in clearing of montane forest and the increased planting of casuarina trees to provide wood in deforested areas (Brookfield and White, 1968).

That reliance on sweet potato rather than another major staple, sago, can affect the pattern of people's lives has been noted among the long house communities of the Etolo people of the Southern Highlands Province (Dwyer, 1985). Those people with a high reliance on sweet potato spent more time trapping wild sources of animal protein and less time hunting, whereas the reverse was true of people reliant on sago. The sweet potato group were committed to spending many days gardening and constant harvesting while sago eaters spent a short but concentrated period processing their sago, which could then be stored for many months, thus freeing them for successive days of hunting.

The tremendous number of cultivars in Papua New Guinea has been mentioned by many authors. There may be more cultivars in Papua New Guinea than from any other area of the world (Yen, 1974), as many as

5000 probably still being grown. According to a table constructed by Bourke (1985), the recorded number of cultivars grown by any one group ranges from 6 to 71, with a mean of 33. In addition, a fairly recent nutrition survey of a Highland village in Simbu Province (Harvey and Heywood, 1983) noted 51 cultivars of sweet potatoes growing in the gardens of sample households. However, the two most popular types accounted for 74% of all the sweet potato eaten. In the Lowlands the number of cultivars held by any one group is generally lower than in the Highlands. Cultivars introduced since European contact are rapidly replacing traditional types and many of the latter are now being lost (Bourke, 1985). One author (Bailey, 1963) has suggested that an apparently monotonous diet based largely on a starchy staple is not monotonous to the Papuans due to the divergence of cultivars providing variety to their sensitive palates. He, and others (Kanua and Rangat, 1987), also suggest that the colour, taste and texture of a cultivar are more important to farmers than yield.

The almost unique position which sweet potato occupies in the diets of Papua New Guinea peoples, compared to those of other countries, has resulted in many studies concerned with the extent of its contribution to diets and nutrient intakes over the last 30 to 40 years. The percentage contribution to total weight of food intake as assessed by many of these studies has again been summarized by Bourke (1985). This ranged from only 0.3% in East Sepik Province (where sago is the traditional staple) to more than 90% in parts of the Highlands. Moreover sweet potato contributed over 50% of total food intake in all the Highland sites. However, as recently as 1986 (Thomason, Jenkins and Heywood, 1986), sweet potato was found to play an important part in the diet of the Au people of West Sepik, where sago is the main staple. It often formed part of the morning meal and was one of the first foods given to children, either mashed or premasticated by the mother. The extremely high contribution which sweet potato makes to diets in the Highlands of Papua New Guinea has given cause for concern with regard to the nutritional status, especially of children (see Chapter 3), and it has been suggested that cultivars with higher than average protein contents should be sought and promoted (Bailey, 1963; Oomen et al., 1961; Heywood and Nakikus, 1982).

Dietary surveys

The degree of importance of sweet potato in the diets of various groups of Highland people has been illustrated by surveys in villages. A dietary survey carried out at the end of the 1960s in a clan of the people of Murapin (Sinnett, 1975) who intensively cultivated sweet potato as their

Table 8.18. *Mean daily food intake (g/day) of the people of Murapin, Papua New Guinea Highlands, 1966–7*

	Age group						
Food	0–4	5–9	10–14	15–19	20–29	30–39	40+
Males (no.)	7	7	2	3	4	11	8
Sweet potato	441	843	1117	1266	1712	1566	1583
Other[a]	72	113	46	149	136	135	69
Total	513	956	1163	1415	1848	1701	1652
Females (no.)	6	6	8	5	8	9	6
Sweet potato	593	872	909	1236	1206	1202	1067
Other[a]	46	126	162	116	139	128	100
Total	639	998	1071	1352	1345	1330	1167

Notes:
From: Sinnett, 1975; estimated over 7 consecutive days for each subject.
[a] Potato, pumpkin, marrow, corn, cabbage, kumu, pandanus, sugar cane and pig meat. The contribution of sweet potato is more than 90% by weight of total food.

single subsistence staple illustrates the extremely high intakes which pertained among males and females of all age groups, as shown in Table 8.18. These average daily intakes were assessed by weighing food over a 7 day period. Sweet potato contributed about 90% of the food intake. The most common method of preparation was by roasting the whole root in hot ashes. A less frequently used mode of preparation was steaming in an earth oven known as a *mumu*. This consists of a hole dug in the ground and lined with leaves. Alternating layers of sweet potato, leaves and preheated stones are built up inside and the hole closed with a layer of leaves and earth. Water is poured in and the sweet potatoes are allowed to steam for 1–1½ hours before removal and consumption. Three meals were eaten daily, the main ones in the morning and evening with cooked sweet potato or other food being carried to the bush or garden as a lunch time snack. As still takes place in Highland communities today, sweet potato gardens were prepared by the men, but maintained and harvested by the women. Harvesting was progressive and sweet potato was not normally stored for longer than 24 hours in the house. As has been described for other communities, pig herding was the other most important basis of the clan's subsistence economy. Pigs were allowed to forage and were also fed on sweet potatoes and potatoes. Typically although extensive herds of pigs were maintained, they were killed and eaten only infrequently, usually for a special occasion or ceremony and contributed only in a minor way to average daily food intakes. This remains true for most communities today.

Although sweet potato retains its importance in Highland diets, a change in its contribution to dietary patterns has been noted by several authors. This has taken the form of an increasing dependence on imported foods as the amount of cash cropping expands, changing traditional cropping patterns and raising people's disposable income. The changes in Beha, an area of the Eastern Highlands Province, have been described (Orr-Ewing, 1983). Although the village life-style remains fairly traditional and similar in many respects to that described above, a major change from total subsistence to a situation where some crops are now grown for marketing has occurred. The major staple, sweet potato, which is eaten two or three times every day, is also grown for selling in the town markets. The sale of this and other crops now brings in an income which is used to purchase foods such as canned fish and meat, sugar and rice. The traditional ways of preparing sweet potato described above have been joined by an increase in boiling since people acquired saucepans.

Another study (Harvey and Heywood, 1983) assessed the changes in food and nutrient intake taking place over a 25 year period in one village of Simbu Province (Highlands) where there had been a rapid switch from a solely subsistence economy to one in which there is now a substantial cash income. This was done by comparing the results of three surveys in the same village, one in 1956, one in 1975 and by carrying out a further survey in 1981. Individual intakes of food were assessed in the 1981 survey by weighing food consumed in the village and by recall of foods consumed away from home. The high average daily adult consumption levels of sweet potato, which was still the most important item in the diet, are shown in Table 8.19. However, there was a marked decrease in the contribution which sweet potato made to energy and protein intakes in 1981 compared to either 1956 or 1975; the figures for 1956 and 1981 are shown in Table 8.20. This was largely due to an increase in the contribution made by imported foods especially cereals, such as rice, wheat flour and maize, and canned fish. There was a significant increase in the contribution of trade store goods to energy and protein intakes in 1981 compared with 1975. Pigs kept by most households were fed on sweet potato, a mean of 42% of production being used for this purpose. This may seem high because, as mentioned previously, pigs make little contribution to food intake except during feasts, but they do play an extremely important part in the social, cultural and ceremonial life of Papua New Guinea people (see below). Harvey and Heywood note that the 1981 survey took place during a period of food shortage. The shortfall in subsistence production (which would

Table 8.19. *Mean adult daily intake (grams) of sweet potato and other foods in Yobakogl, Simbu Province, Papua New Guinea Highlands, 1981*

Food	Adult male	Adult female
Sweet potato	1107	875
Dark green leaves	125	102
Pumpkin	18	32
Corn	42	44
Rice	181	117
Flour	42	61
Canned fish	30	27
Pandanus spp.	78	101
Sugar cane	37	18
Pig meat	12	15
Other	107	83
Total	1779	1475

Notes:
From: Harvey and Heywood, 1983. The contribution of sweet potato is about 60% by weight of total food.

have included sweet potato) resulted from a lack of gardening activity in the previous 5–6 months due to plentiful cash from sale of the coffee crop, disruption from tribal fighting and the mens' failure to clear and prepare new gardens. Thus it appeared that available cash in one season to cover gaps in the food supply had led to a lack of planning with regard to the following season's subsistence food supply.

Consumption patterns

An attempt to obtain an up-to-date picture of sweet potato consumption patterns in various parts of the country was made by means of a simple questionnaire, compiled by myself and completed by a number of Papua New Guineans. Unfortunately only 10 replies were received. However, they serve to illustrate the important role which sweet potato maintains in the Papuan diet.

The adult respondents, from both urban and rural backgrounds and of varying educational status, originated from different provinces in the highland and lowland areas of Papua New Guinea. Only one respondent, from North Solomons Province, said that leaves as well as roots

Table 8.20. *Percentage contribution of sweet potato and other foods to energy and protein intakes in Yobakogl Highland village over 25 years*

	% total energy		% total protein	
Food	1956[a]	1981[b]	1956[a]	1981[b]
Sweet potato	76.0	53.2	56.3	34.2
Other roots/tubers	7.0	1.4	4.9	0.9
Cereals and grains	4.0	22.2	5.0	25.4
Pandanus spp.	—	7.8	—	7.8
Fish and meat	4.0	6.0	11.3	18.4
Leafy vegetables	1.0	2.0	5.6	9.6
Other vegetables and fruit	8.0	2.1	16.9	1.5
Fats and oils	—	1.9	—	—
Alcohol	—	2.0	—	1.3
Other	—	1.4	—	0.9
Total	100.0	100.0	100.0	100.0
All trade-store goods	—	25.8	—	34.6

Notes:
From: Harvey and Heywood, 1983.
[a] All age groups.
[b] Adults only.

were consumed. Leaves were further said to be rated lowest of all green leafy vegetables. Preferences for a particular type of root in terms of sensory qualities did not seem to be very pronounced, although most respondents favoured white flesh, and in the case of those from the Highlands, a dryish texture.

The Highland respondents all specified sweet potato as being the major staple on an every day basis – the *numba wan kaikai* – or most preferred food, but all said that taro and yams are considered as more special, and for special occasions sweet potato would be considered number two. However, several people also said that sweet potato would be used on special occasions and all used it also as a snack. In several cases it was mentioned particularly as a snack for children. Lowland rural respondents said that sweet potato is used as a co-staple and the one urban lowland respondent who replied to the question stated that it was only used as a complementary vegetable. Rural lowland respondents rated sweet potato second or third in preference after other staples such as taro. Several respondents in both the Highlands and the Lowlands said that, if sweet potato roots were substituted by rice or bread, such a substitution would be tolerated for only a short time (a few days) before a person would want to return to eating sweet potato. Older people were said not to feel full even after a large plate of rice unless they had eaten sweet potato. These replies were interesting in that respondents only considered the possibility of other foods substituting for sweet potato and not the reverse, whereas the question was designed to determine the consequences of sweet potato acting as a substitute for other foods.

Highland and Lowland rural dwellers all eat sweet potato with a high frequency of twice a day at main meals and in between as snacks. The urban dwellers use sweet potato once a day at a main meal or less frequently, three times a week. Sweet potato is considered to be a food suitable for small children, including a weaning food. In the Highlands it is given to children either as a staple or snack prepared in the same way as for adults. All Lowland respondents mentioned that sweet potato is eaten mashed, alone or with other foods, or as soup, by children.

With two exceptions of an anecdotal nature, it was generally stated that there are no taboos associated with eating sweet potato. On the other hand, sweet potato was not believed to bestow any special beneficial effects, apart from a feeling of repletion or satisfaction. Only one person mentioned a medicinal use – that of applying sweet potato to a boil to cure it – a practice which she said had been discontinued.

As regards livestock feeding, roots are fed either raw or cooked to pigs. Roots were said to always be cooked before feeding to poultry. Raw leaves are given to both pigs and poultry. Several respondents mentioned feeding cooked roots to cassowarys.

Marketing

The domestic market for the sweet potato crop is limited to local urban markets. Markets are not a traditional feature of New Guinea life having been introduced in the last 30 years, but are increasing in size and importance. However, the amounts of sweet potato on sale, especially in the Highlands, are small, since most consumers have access to their own home-grown supplies (Joughin, J., personal communication). Moreover, prices paid for sweet potato at urban food markets are high (Kanua and Rangat, 1987). Marketing is time consuming and therefore expensive in strictly commercial terms.

Marketing of sweet potato is a major problem for growers. It has been suggested that institutional demand may be the only real avenue for advance (Kanua and Rangat, 1987). However, institutions such as schools, hospitals and jails may require a guaranteed regular or consistent supply of roots which local growers often fail to meet. These heavily programmed institutions cannot accommodate risks and therefore resort to alternative purchases of imported rice. There would appear to be a potential here for increasing sales of roots through increasing yields and linking this to processing for greater stability of supplies.

Pig feeding

The great importance which sweet potato has in Papua New Guinea as a pig food has already been stressed. The same people who utilize sweet potato as a major staple food also use much of their crop as fodder, a portion estimated in some places to be as high as 63% (sources cited by Watson, 1977). However, the role this may seem to suggest of sweet potato as an alternative or indirect human food is only one aspect of the pig–sweet potato interrelationship which forms the basis of Highland society. Indeed, as can be appreciated from Tables 8.19 and 8.20, the contribution of pig meat to dietary and nutrient intakes is extremely small, owing to the custom of very infrequent slaughter and consequent consumption of animals only for special occasions.

In what has been called the 'prestige syndrome', the ownership and exchange of pigs satisfies a series of powerful needs within society in which dietary considerations play a minor part. Pigs are highly valued as exchange or payments for scarce imports of great value, for brides, children, death payments, establishing, cementing and maintaining alliances, for bribes, peace making, medico-magical purposes, and as tokens of personal standing and status (Watson, 1977). Sweet potato has advantages as a fodder over traditional staples such as taro (which cannot be used raw due to high concentrations of oxalate). It provides the means to maintain large pig herds and as such underpins a vital part of the economic and political structure of New Guinean Highland society. In fact, Watson (1977) has put forward a persuasive case that the conversion of Highland societies to sweet potato as their main food, with consequent changes in their ecological circumstances, originated from the principal, initial attraction of the new crop as a fodder source for the required increasing numbers of domestic pigs. Be that as it may, the roles of sweet potato as human food and livestock feed are evidently of equal importance in some societies of Papua New Guinea. This situation is no doubt unique for sweet potato in the world today.

Processing

So far as is known, there is still no sweet potato processing industry. The 1977 attempt to produce a dried product from sweet potato roots, known as *kaukau* rice, has already been described in Chaper 6. Production ceased reportedly due to inadequate supplies of roots, low sales due to high cost, poor consumer acceptance and poor marketing (Thomas, 1982).

In 1975 and 1976 dried sweet potato chips were ground into a flour and mixed with wheat flour to make cakes for students at a high school in Wapenamanda. Promising results were obtained in that the resulting cake was acceptable in taste. Other institutions such as schools, hospitals and jails could explore the economic benefits of such sweet potato/wheat products at the local level (Kanua and Rangat, 1987). This would help to promote a local food at the expense of an imported cereal. Provision of consistent supplies of roots, as in the case of marketing fresh roots, would seem to be the key to future processing possibilities. This is no doubt dependent among other factors on yield improvements.

Conclusion

Sweet potato will obviously remain a highly important food in Papua New Guinea for the foreseeable future. It is still the major staple in the Highlands and its consumption is increasing in the Lowlands. In recent times its status has changed from that of a solely subsistence crop, to that of a cash crop also, no doubt reflecting the urbanization which has taken place in the country. The increasing dependence in both rural and urban areas on imported foods at the expense of sweet potato, which has been noted by many researchers, has resulted in some improvement in energy and protein intakes. However, disquiet about a possible parallel increase in the incidence of degenerative diseases such as heart disease, diabetes mellitus and obesity has been expressed (Harvey and Heywood, 1983). A lessening of the food dependence, while at the same time maintaining an improvement in dietary nutritional value, might be sought by diversifying locally grown foods and improving the yield and nutritional quality of sweet potato.

There must be possibilities of sweet potato product development and commercialization given the increasing urban population. The failure of a previous attempt to introduce a processed product, however, illustrates the forward planning which must be involved to ensure success. The type of product, economics of the process and promotion through consumer studies and appropriate marketing strategies are some important factors to be considered.

Rwanda and Uganda

Rwanda and Uganda are among the few countries which have significantly increased sweet potato production in recent years. In doing so they have exploited characteristics of the crop which enable it to be grown intensively in areas where cereals and cassava are less successful. In its roles as home garden and field crops, it acts as a major staple, securing food needs and nutritional benefits for much of the population.

General background

Although there is little up-to-date information on consumption patterns, Rwanda and Uganda have been included to illustrate how sweet potato may serve as a major staple food and security crop, and to provide a further contrast with the situation in China and Japan, where sweet potato is an important industrial commodity. According to FAO statistics (Horton, 1988), Rwanda and Uganda have both increased production of sweet potato significantly in the last 30 years. Production in Rwanda rose from 452,000 tonnes in 1961 to 900,000 tonnes in 1985, with most of the increase occurring between 1973 and 1981. Production in Uganda rose steadily from 495,000 tonnes in 1961 to 2,002,000 tonnes in 1976; it registered a decrease to 1,200,000 tonnes in 1980 and then rose steadily again to 2,000,000 tonnes in 1985. In both countries the significant increase in production has apparently been largely due to an expansion in cultivated area, with yields remaining about constant (Horton, 1988). There is evidence that production is increasing in areas of highest population density (Ewell, P., personal communication). According to the sweet potato balance sheet, over 80% of total production is used as food in Rwanda and Uganda, with the remainder being registered as waste. This results in very high quantities of sweet potato being available for consumption – 124 and 127 kg per capita per year in Rwanda and Uganda, respectively (Horton, 1988). Although sweet potato roots are a major staple in Rwanda and Uganda, the leaves are very rarely consumed. They are infrequently cut and used for animal feeding in Uganda, but hardly used for this purpose at all in Rwanda (Ewell, P., personal communication). There is apparently no storage apart from leaving roots in the ground and harvesting progressively. No village-scale processing is reported apart from some sun-drying of root slices (McDowell, 1970) in parts of Uganda to enable storage. The dried roots are ground into a flour for use in gruels, porridges and soups. There is no evidence of industrial processing for starch or alcohol etc.

Rwanda

The exact date at which sweet potatoes entered Rwanda is not known, but legend has it that a Rwandan king invaded Uganda in the eighteenth century and his soldiers brought vines back wrapped around their spears. They were apparently impressed with the crop because it gives

Table 8.21. *Production of sweet potato and other principal food crops (in thousands of tonnes) in Rwanda, 1984 and 1986*

	Year	
Crop	1984	1986
Bananas	1982.3	2193.8
Sweet potatoes	*730.9*	*861.9*
Cassava	324.2	361.9
Beans	256.5	278.3
Potatoes	251.4	241.5
Sorghum	171.1	158.9
Maize	111.5	135.8
Peas	16.9	19.0
Peanuts	14.7	16.3
Soybeans	4.5	3.8

Notes:
From: Ewell, 1988.

milk (latex) when cut. Sweet potatoes became the major root/tuber staple in place of yams, which have high fertility requirements, need much labour time for staking and must be stored for 2 months before planting to break dormancy.

Rwanda is a densely populated country of 6.5 million people with an average of 540 people/km^2. Ninety five per cent of the population is engaged in agriculture. It is estimated that 86% of the farmers in the country grow sweet potatoes on an average of 9.2% of their cropped area. In 1984 and 1986, it was shown that sweet potatoes were second in weight of total production only to bananas (see Table 8.21).

The intensive sweet potato production system which pertains in this heavily populated country has been suggested as a useful and encouraging model for other African countries to follow in tackling their current food crisis (Alvarez, 1987). There are three main characteristics to this model. The first is an increase in productivity as a result of more intensive and less extensive cropping systems in terms of resource inputs, land use and management. This intensive crop management fully exploits yield potentials and reduces inputs of fertilizer, pest control and weeding. Secondly, because of the crop's versatility and adaptability to local climatic variations and soil factors several crops per year can be produced. Sweet potatoes are grown throughout the country but most intensively in the central plateau. Cultivation takes place on the hill slopes in the principal rainy season which starts in October, and in the bottoms of valleys in the drier season starting in March or April. The

Figure 8.15. Sweet potato production on raised beds in the lowland *marais* (swamp) of central Rwanda (P. Ewell).

poorly drained bottom lands or *marais* are marshes which have been reclaimed for agriculture by the farmers. The farmers constantly alternate sweet potatoes between seasons and areas, planting vines in the *marais* as they harvest in the hills and vice versa (Ewell, 1988; and see Figure 8.15). In the third instance, the intensive utilization of sweet potato as a home garden plant and as a field crop has contributed significantly to meeting nutritional and year-round food security needs in Rwanda. The government recognizes the vital role of sweet potato as a food bank and farmers are encouraged always to have staggered plots of sweet potato and cassava in their fields and home gardens (Alvarez, 1987).

Consumption patterns

Although average annual consumption has been estimated by the FAO as 124 kg per capita, a recent income and expenditure survey by the Rwandan Ministry of Planning (1988) found that adult consumption was higher than this and also that it varied somewhat from one region of the country to another. Tables 8.22 and 8.23 show the importance of sweet potato in relation to other selected foods, in terms of consumption levels in five major regions, and of nutritional contributions towards energy and protein intakes in rural households.

The type of storage root most widely preferred in both Rwanda and Uganda is white or cream to light yellow, firm, moderately sweet and

Table 8.22. *Average consumption (kg/year) per adult equivalent of sweet potatoes and other selected foods in rural households of five Rwandan regions, 1986*

	North-west	South-west	Centre-north	Centre-south	East
Sweet potato	*137* (17%)	*211* (34%)	*309* (40%)	*213* (37%)	*131* (17%)
Cassava	174	40	43	19	8
Bananas	51	81	112	41	244
Banana beer	59	42	43	32	68
Legumes	102	66	131	88	138
Sorghum beer	79	48	31	61	24
Cereals	56	26	7	6	8
Animal products	8	5	4	6	26
Other	97	42	31	24	32
Total	787	615	756	580	774

Notes:
From: Ewell, 1988.

Table 8.23. *The percentage contribution of sweet potato and other selected foods to energy and protein intakes of Rwandan rural households, 1986*

Food	% energy	% protein
Sweet potato	17.4	7.5
Cassava	12.3	2.6
Potatoes	3.5	2.6
Bananas	8.2	2.5
Banana beer	4.7	0.0
Sub-total	46.1	15.2
Legumes	35.2	71.3
Cereal products	14.3	8.8
Animal products	1.4	2.5
Other	3.0	2.2
Total	100	100

Notes:
From: Ewell, 1988.

'floury' (Ewell, P., personal communication). Soft, orange-fleshed clones introduced from other parts of the world have been universally rejected by consumers even if the roots possess other desirable agronomic characteristics.

Most farmers plant a mixture of clones to provide protection against risk. Plots are harvested section by section thereby obtaining a mixture of underdeveloped and fully developed roots, rather than harvesting individual plants piecemeal. Only about 12% of the harvest is marketed, the rest being used for home consumption. Farmers harvest when they need food and/or money. Increasing population pressure is making farmers more interested in yield improvement.

Marketing

A wholesale assembly market has been described (Ewell, 1988; and see Figure 8.16). Two days per week merchants from Kigali, where there is a major retail market, take the bus 50 km to Gitarama and set up in a regular area near the road. Farmers walk into town with baskets of sweet potatoes on their heads to sell. The merchants purchase the sweet potatoes, pack them into big double bags (because freight rates are by the bag) and hire pick-up trucks to ferry them to their stands in the market in Kigali.

Sweet potato prices have increased over time. Between 1969/70 and 1975/6 the price of sweet potatoes in Kigali market more than doubled (Durr, 1983). This was the greatest increase of the nine selected food items observed apart from fish. A study of sweet potato marketing in Rwanda is in progress.

Given the desire of people in Rwanda to continue eating sweet potato as a major staple food, production will have to be increased further to maintain supplies in the face of population growth. This means breeding high yielding clones with improved disease and insect resistance as well as looking to ways of improving cultivation techniques. The energy density of fresh roots is important to the nutritional status of individual consumers and clones with high dry matter content should therefore be sought. At the same time, more knowledge is needed about the specific organoleptic preferences of consumers, which are known to be pronounced, so that clones with improved agronomic traits are combined with the desired characteristics of taste, colour, texture etc.

Uganda

However sweet potatoes were introduced to Uganda, they already occupied an entrenched position and considerable importance as a crop

Figure 8.16. (a) and (b) Sweet potato wholesale assembly market in Gitarama, Rwanda, where farmers sell sweet potato to middlemen, for transport to Kigali (P. Ewell).

at the time of establishment of the British administration in the 1890s. A contemporary agricultural report describes them as 'next to plantain the great food of the Bantu tribes'. Their spread to northern Uganda seems to be more recent as the crop only arrived in Lango (Northern Province) shortly before 1923 (Jana, 1982). They have not apparently always enjoyed the favour commensurate with their importance. Roscoe in 1911 (McMaster, 1962) wrote of the Ganda tribe that 'sweet potatoes were looked on as a food for peasants and servants and for use in times of drought when plantains were scarce. No chief would consent to have them served to him under ordinary circumstances'. In 1962 McMaster maintained that something of this attitude still pertained, the plantain remaining supreme among the Baganda with the sweet potato being accorded a secondary status as a food. However, he also stated that in northern and eastern parts there had been a spread of sweet potato associated with various pressures which the crop helped to meet. These included unfavourable seasons for growth and yield of cereal crops, declining soil fertility due to heavier cropping and reductions of fallow periods consequent upon population growth, heavy locust and weevil damage to other crops and shortages of agricultural labour due to migrations to towns. Even in areas where cereals remain predominant, sweet potatoes act as a buffer against famine and help to supplement food supplies in the hungry months preceding the grain harvest. In this situation they would be fulfilling the role of seasonal or complementary staple defined in the typology of consumption.

Sweet potatoes are now widely cultivated in all parts of Uganda with their distribution relating quite closely to that of population (Jana, 1982). The increase of production which has taken place in the last 20 years has already been noted above. The most marked concentration of production is in Kigezi (in the southwest). The scale of production there is mainly in response to the pressures given above enhanced by the fact that in much of the district cassava does not thrive. In parts of Uganda where sweet potatoes are less important they are found growing on small plots around homes, being treated as a garden vegetable crop. In most parts with a pronounced dry season they are grown in swampy areas during the dry months to ensure a supply of planting material and to provide food in the hungry season prior to the harvesting of the main rains crop. In some districts of Uganda the planting of these plots is compulsory under legislation providing for famine reserves (Jana, 1982). In Uganda as a whole sweet potatoes occupy about 9% of food crops area and rank fourth in importance after *Eleusine* millet, bananas and cassava. As in Rwanda, sweet potatoes are a women's crop once the heaviest work of preparing the plots for planting has been carried out by both sexes.

Table 8.24. *Estimated daily per capita consumption (in grams) of sweet potato and other selected foods in rural households of four Ugandan districts, 1968*

Food	W. Province Ankole	E. Province Busoga	Buganda Masaka	N. Province W. Nile
Sweet potato	278.0	41.3	183.0	25.5
Matooke (banana)	498.0	348.0	991.0	5.0
Fresh cassava	22.0	74.6	112.6	13.9
Dried cassava	2.5	8.4	5.1	470.0
Millet	196.4	68.0	2.4	61.0
Sorghum	1.2	2.4	—	48.6
Dry beans	42.3	4.8	40.0	79.6
Groundnuts	10.0	21.2	21.3	21.3
Green leaves	19.0	34.2	4.5	3.3
Milk	132.1	38.7	9.4	1.3

Notes:
As calculated by McDowell (1970).

Sweet potatoes may be cooked by steaming, for example wrapped in banana leaves. They may also be boiled, or roasted peeled or unpeeled in the ashes of a fire. Methods of dietary preparation of sweet potato have been described for two tribal groups (McDowell, 1970). Sweet potato is most frequently steamed together with pre-boiled beans (*omugoyo*) or cooked separately from the beans and then mashed with them in the ratio 2:1 sweet potato:beans (*acok*). The sweet potato staple alone or mixed with beans is then eaten with a sauce or relish of some kind.

In 1970 McDowell calculated the daily per capita consumption of foods, including sweet potato, in four rural districts of Uganda, using the average daily household consumption of foods given in a rural food consumption survey conducted by the Ugandan government in 1968, and the average size of households in each area. The estimated per capita daily consumption of sweet potato and some other major foods is shown in Table 8.24. The variation in consumption levels between regions and the relative importance of sweet potato and other staples can be clearly seen. In addition, McDowell (1970) has also reported on a detailed study of food consumption over 1 year of the Adhola tribe in Bukedi, southeastern Uganda. Among these people sweet potato was ranked third in preference for staples after finger millet and plantains, but sweet potato consumption levels were still higher than the preferred staples, as can be seen from Table 8.25.

In 1968 a rural food consumption survey (as cited by McDowell, 1970)

Table 8.25. *Quantities of sweet potato and other staples used in meals of one tribe in Bukedi district of Uganda 1968*

Staple		g per capita
Millet bread (millet flour)		200
(cassava flour)		90
Plantains		450
Sweet potatoes		560
Cassava		560
Sweet potatoes	mixed	230
Cowpeas	mixed	170
Cowpeas		170 or more

Notes:
From: Sharman, 1968, as cited by McDowell (1970).

found that in three out of four districts (in Buganda, Western and Eastern Provinces) the percentage of households using sweet potato was high, ranging from 64% to 87%. The percentage in West Nile (Northern Province) was much lower at 29%. In the first three districts sweet potato represented only 1–3% of purchased food, whereas it represented 13% of food purchased in West Nile. However, sweet potatoes are entering increasingly into the cash economy in response to urbanization and the necessarily greater amounts of purchased food. Although sun-drying of root slices is practised in some areas of Uganda, the potential for this is limited, since many areas have either fairly high humidity or fairly frequent and heavy rainfall. In the subsistence sector fresh roots surplus to family needs are sold to avoid storage and rapid deterioration. I am not aware of any detailed study of sweet potato marketing in Uganda.

Conclusions

By reason of the ease of sweet potato cultivation in regions where cereals and cassava grow poorly, sweet potato has become established as a major staple in many parts of Rwanda and Uganda. Other factors such as shortage of labour on the land due to rural migration and urban population expansion have also played a part in production increase. This increase seems to have taken place through expansion in land area and intensification of crop management, with yields per hectare remaining static. Apparently the agronomic advantages of the crop have overcome the earlier disdain in which sweet potato seems to have been held in parts of Uganda.

There will be continuing use of sweet potato as a major staple food in the two countries, with increasing demand due to population pressures. These two factors coupled with the presently low average yields indicate a great potential for increasing production by increasing yields. A successful system of crop management found in Rwanda should be studied and improved further if possible. Knowledge of factors affecting sweet potato consumption levels, preferences, utilization and marketing practices in Rwanda and Uganda is out-dated, fragmented or non-existent. Socio-economic studies are needed to determine future consumer requirements and to plan crop improvement efforts. There would also seem to be scope for investigating possibilities of simple storage methods and expanded processing.

The important role of sweet potato as a home garden crop suggests possibilities for promotion of yellow to orange-fleshed roots, and the increased use of leaves, in children's diets. However, this would require studies of local attitudes to these two items as well as educational programmes to disseminate information about their nutritional virtues.

There would also appear to be a great potential for use of sweet potato leaves as a cattle forage, something which has been a considerable success in Kenya.

Peru

During the last decade sweet potato has played an increasingly important role in the diets of low income people, especially along Peru's central coast. In doing so it has substituted for more expensive cereal-derived products such as bread. As a food which is well known to all sectors of the population, it could be further exploited as a local alternative to imported basic foodstuffs which require substantial foreign exchange spending and create food dependency.

The Cañete Valley in the central coast of Peru serves as an example of an area in which sweet potato has evolved into a successful, market-oriented crop which not only allows the nourishment of low income families, but also the establishment of a flourishing dairy cattle industry in an environment lacking natural pasture land. In depth studies of the Cañete Valley provide pointers to investigations which may lead to increased future potentials for the crop both in the valley and elsewhere.

Historical background

Sweet potato was most probably cultivated for the first time by people inhabiting the area which now encompasses Peru, Ecuador and Colombia. Certainly the most ancient sweet potato remains discovered so far, which date from the Neolithic period (8000–10,000 B.C. according to Engel, 1970), have been found in caves in the Chilca canyon of Peru. Examples of sweet potato have also been unearthed at numerous Peruvian coastal archaeological sites representing both Inca and pre-Inca cultures (Ugent and Peterson, 1988). In almost all cases the sweet potato has been found along with remains of other food plants indigenous to the Andean culture such as potato, groundnut, bean, quinoa, guava, calabash, pepper and others. The richness of genetic variation existing in Peru is manifested by the diverse names given to sweet potato types in the *quechua* language: the sweet and moist types being known as *apichu* and the drier types as *kumara*. Sweet potatoes of different skin colours were also known by individual names.

During the twentieth century the sweet potato has suffered changing fortunes reflecting agricultural policies towards cash and food crops (Collins, 1989). However at the present time sweet potato is grown throughout Peru and ranks twelfth among annual crops, accounting for about 1.2% of the total value of agricultural output. Shipments of sweet potato are the third largest in volume entering the wholesale market in the capital city, Lima (Collins, 1989). In spite of this importance, production for the whole of Peru over the last 20 years has suffered a reduction from a maximum of 168,000 tonnes in 1971 to 123,000 tonnes in 1987.

Harvested roots are utilized principally for human food in the fresh state. Studies have indicated (Beltran Chavez, 1989) that only about 10% of production is used for processing, mainly into starch and snack chips. In addition the vines are used for animal feed in some production zones, a notable example being their use as cattle forage in the Cañete Valley, 150 km south of Lima. They may also, however, be fed to small livestock such as guinea pigs. In Lima most purchases of sweet potato by middle and high income groups is for dog food.

In recent years regional production of sweet potato has undergone great changes. In 1944, the north coast and sierra (mountain) regions were responsible for 45% of the total production. This is now reduced to only 18%, whereas the central coast, where during the early decades of the twentieth century sweet potato was an unimportant subsistence crop, now produces 74% of the national production. Of the central coast's production, 85% is grown in the Cañete Valley, which supplies some 60% of the sweet potato entering the Lima wholesale market (Achata et al., 1988).

There was a low positive trend of sweet potato production in the world, and negative production trends in Latin America and in Peru as a whole during a 20 year period from the 1960s to the 1980s. In contrast, the Cañete Valley showed significant positive trends in area planted,

Table 8.26. *Trends in area planted, yield and production of sweet potato during a 20 year period*

	% Change		
	Area	Yield	Production
World[a]	−30	+56	+14
South America[a]	−36	−10	−41
Peru[a]	−30	+17	−18
Cañete Valley[b]	+24	+37	+61

Notes:
[a] Horton, 1988; 1963/65 to 1983/85.
[b] Achata et al., 1988.

yields and hence production (Table 8.26). In the valley sweet potato has evolved from an unimportant subsistence crop of landless workers and farm labourers in the early years of this century to a highly successful and market-oriented crop of major importance at the present time. It has done so in spite of extremely limited support from both state and private sectors. The evolution has taken place largely during the last ten years. This is due to a series of circumstances including the valley's proximity to Lima with its burgeoning population. The growth of sweet potato in Cañete and its link to the Lima market has become the subject of present and projected in depth studies. Cañete, while illustrating the characteristics of sweet potato production destined for the large urban market, also provides a contrast with the department of Cajamarca in the northern mountain area of Peru, where sweet potato is still largely a subsistence crop. Although the department of Cajamarca has the greatest concentration of sweet potato production in this mountain area, it produces only 1% of national production (Benavides, 1990).

Great emphasis has been placed by researchers on the potential role of sweet potato in Peru's food policy (Achata et al., 1988; Collins, 1989), given greater state and private support for research. Over the last 70 years the tendency has been to rely on imported foods, especially for Lima. Dependency on imported foods coupled with dwindling foreign exchange reserves have motivated successive governments to promote traditional crops (Collins, 1989). However, the promotion of little known 'native' crops such as the cereal *quinoa* (*Chenopodium quinoa*), the legume *tarwi* (*Lupinus mutabilis*) and the tuber *oca* (*Oxalis tuberosa*) has had limited success. In contrast, sweet potatoes, which are well known to consumers, relatively cheap to grow and high in energy content, could provide a viable alternative to imported foods. Diversifying sweet

Table 8.27. *Per capita consumption of sweet potato (kg/year) and some other basic foods in rural and urban areas of Peru, 1972*

	Peru	Lima	Large cities	Population centres	Rural areas
Sweet potato	4	6	4	4	4
Rice	23	30	30	31	16
Potato	73	42	35	51	100
Cassava	12	3	6	9	17

Notes:
Ministry of Economy and Finance, and General Directorate of Information and Statistics of the Ministry of Food, 1975. Per capita and family consumption of 54 selected food items by area, region and zone of Peru, August 1971–2.

potato utilization could therefore be vital not only for the agricultural economy but also for the country's food policy.

Consumption patterns

The combined influences of the Humboldt ocean current and the Andean mountain chain have given rise to a series of ecological zones within Peru which influence rural food habits. In contrast, those of the large urban centres have tended to an increasing uniformity in recent years, being based more and more on imported wheat-based products and marketed foods. Lima has the greatest consumption per capita of sweet potato which in general lags far behind those of other staples (Table 8.27).

Detailed studies of sweet potato production and use have been conducted for the Cañete Valley in the central coast department of Lima. Initial investigations in the rural area of the northern sierra department of Cajamarca also furnish information. These departments provide contrasting examples of sweet potato consumption in different ecological zones and in urban and rural areas. The reasons for the success of sweet potato as a commercial crop in the Cañete Valley will be discussed briefly. The role played by sweet potato in the diets of low income groups in both the city of Lima and the Cañete Valley is particularly noteworthy.

Table 8.28. *Per capita availability of sweet potatoes in Lima (1971–1988) as estimated from shipments to the wholesale market*

Year	Share of yellow cultivars	Share of purple cultivars	Availability[a] (kg per capita)
1971	87	13	15
1972	84	16	16
1975	87	13	18
1976	82	18	16
1977	83	17	17
1978	78	23	18
1979	80	20	17
1980	75	25	14
1981	75	25	15
1985	72	28	14
1986	69	31	18
1987	62	38	12
1988	67	33	—

Notes:
From: Collins, 1989.
[a] Calculated by dividing total sweet potato shipments by the population in Lima in December of that year.

Central coast

Lima

Insight into sweet potato consumption in the large urban centres of Peru has been gained by a study of consumption trends and patterns in Lima. Two ways of estimating sweet potato consumption here have been used: the volume of sweet potato shipments to the city's vegetable wholesale market, and the perusal of household food consumption surveys (Collins, 1989). The former method, although providing a good approximation of root weight available for consumption is not entirely satisfactory, as a proportion of roots entering the Lima market are re-shipped to other destinations. Household surveys are useful for estimating per capita consumption of sweet potato by income level and over time. The results of the 'Encuesta Nacional de Consumo de Alimentos' (ENCA) or National Food Consumption Survey carried out among 8000 country-wide households in 1971–2, as well as surveys carried out in the late 1970s and 1980, are a valuable source of information. There is some discrepancy between the estimates obtained by the two methods (see Tables 8.28 and 8.29). This is not unexpected as at least 20% of

Table 8.29. *Estimated sweet potato consumption in Lima from household food surveys*

Year	Sweet potatoes: Average annual consumption (kg per capita)	Sweet potatoes: Share of total food consumed (%)
1971–2	6.5[a]	
1976	8.6[b]	2.8
	10.6[c]	2.9
1977	10.3[b]	3.3
	9.6[c]	2.8
1978	12.5[b]	4.2
	10.7[c]	2.9
1979	13.6[b]	4.6
	10.4[c]	2.8
1980	12.9[b]	4.4

Notes:
Compiled by Collins (1989) from local statistics.
[a] Disaggregate figures for 1971–2 were: low income families 6.5; middle income families 6.6; high income families 6.1.
[b] Low income families. Annual estimates based on average daily family consumption multiplied by 360 days divided by average family size.
[c] Middle income families. Calculated as for note *b*.

purple-skinned sweet potatoes are re-shipped out of the wholesale market and are therefore unavailable for consumption in Lima. Moreover wholesale market estimations do not include roots used for processing, animal feed and loss from spoilage (Collins, 1989).

Results from the ENCA survey (Table 8.29) estimated the average annual consumption for Lima to be 6.5 kg per capita. In the late 1970s studies by the Ministry of Agriculture showed increases in consumption for low and middle income groups with respect to 1971–2. In the case of low income groups consumption showed a steady increase and more than doubled between 1971–2 and 1979. Only in 1980 did consumption by low income groups show a decline. The per capita consumption of middle income groups, in contrast, fluctuated around 10 kg. Whereas consumption of sweet potatoes was higher in middle than low income families in 1971–2 and 1976, this situation was reversed in 1977–9. Although statistics are not available, a researcher has recently reported that, in 1989, given reduced consumer purchasing power due to inflation, consumption of sweet potato increased among low income groups due to its low price relative to the potato, one of the two most important staple Limeñan foods (Beltran Chavez, 1989). Other basic foods such as noodles and bread, derived from wheat, have also suffered price increases relative to sweet potato. If this trend continues sweet potato could become an important alternative staple food for low income households.

Human consumption of fresh roots is the most important use of sweet potato in Peru. Sweet potato may be used as the principal component of a dish, a supplement to other dishes or a main ingredient of desserts. In Lima it is usually eaten as a complementary vegetable. Sweet potato is usually served boiled as an accompaniment to *ceviche*, a dish of raw fish marinaded in lime juice. Boiled, baked or fried roots accompany various dishes of meat or fish. Purée and other preparations based on sweet potato are served in typical dishes prepared for the Christmas dinner. According to one researcher (Beltran Chavez, 1989) the Limeñan consumption of sweet potato is basically a weekly affair. However, a common sight in Lima, by bus stops or near large work centres is the street food vendor. Among other foods, they sell sweet potato with bread which is eaten for breakfast by work people (Benavides, 1990). This may be an example of a case of specialized consumption as a breakfast food.

In a 1989 survey undertaken in Lima (Beltran Chavez, 1989) consumers cited agreeable taste, low price and ease of preparation as the most important reasons for eating sweet potato. In addition it has been noted that size, shape and origin of roots may have an impact on consumer purchasing (Collins, 1989). Small roots are preferred by low income consumers for boiling, as they save on energy expended for cooking. They also require a shorter cooling time. In contrast, larger roots are preferred for frying. The types of sweet potato particularly sought after are 'Amarillo verdadero' (yellow type), and 'Morado legitimo' (purple type) which is especially preferred for accompaniment to *ceviche*.

Production origin may influence consumer buying. Wholesalers generally specify the origin of sweet potato roots from the Lima area which are said to be of superior quality (Collins, 1989). Such roots are rounder in shape and easier to peel than others. In addition they are considered cleaner and better graded than those from the Cañete Valley. This may be because the Lima area has traditionally been the supplier of sweet potatoes, resulting in greater farmer experience in growing and grading them. In contrast Cañete's importance as a sweet potato growing area is relatively recent and farmers still focus more on quantity rather than quality.

Table 8.30. *Quantities of energy and protein acquired per monetary unit from sweet potato and some other basic foods in Peru*

Food	Per 100 g of edible portion		Price (August 1988) I/kg	Quantity acquired per I10	
	Energy (kJ)	Protein (g)		Energy (kJ)	Protein (g)
Sweet potato	473	1.3	8	5912	16
Cassava	553	1.0	14	3950	7
Potato	314	1.8	18	1744	10
Wheat bread	1285	9.3	77	1674	12

Notes:
I, Intis.
From: Achata et al., 1988; energy values converted from kcal to kJ.

It has also been noted that sweet potatoes compare favourably with many of Lima's staple foods in terms of nutrients (energy and protein) purchased per monetary unit (see Table 8.30). In the case of another popular food – noodles (spaghetti) – the contents of major vitamins and minerals are higher in boiled sweet potato than in an equal weight of cooked noodles. It has been argued that low income consumers in Peru are aware of the nutritional value of food (Creed, H., as cited by Collins, 1989). Therefore they should be willing to increase their purchases of sweet potatoes at the expense of noodles when noodle prices rise more than those of sweet potato. This could have important implications for estimating the demand for sweet potatoes, especially since low income groups purchase over half the sweet potatoes consumed in Lima (Collins, 1989).

Some of the information obtained in estimating the demand for sweet potatoes in Lima may be obscured by the large purchases for dog food among the middle and upper income groups. This type of purchase is evidenced by the low quality of roots for sale in some middle and high income neighbourhoods. Supermarkets sell sweet potatoes in large units (5 or 10 kg bags) for this purpose. Since a suitable cheap alternative for dog food does not exist, consumers are more concerned about quantity than price in buying sweet potato. Hence purchases are fairly price inelastic (Collins, 1989).

The Cañete Valley

Most sweet potato from Cañete is shipped to Lima, the rest being sold in regional and local markets. Since 1980 sweet potato has taken its place

Table 8.31. *Percentage production cost of four important crops above (+) or below (−) the cost of sweet potato production in the Cañete valley, Peru*

	Per hectare	Per kilo
Maize	+20	+282
Cotton	+32	+750
Potato	+165	+188
Beans	−18	+840
Sweet potato	0	0

Note:
From: Achata et al., 1988.

with cotton, an export crop, maize the major poultry feed, and potato, one of the basic staples of the Limeñan diet, as one of the principal crops of the valley. This situation owes much to changes in land ownership and usage. At the beginning of the 1980s the large agricultural cooperatives formed as a result of the land reform of 1969 were broken up and divided between the members. The new small holders, with little experience of administering the land, sought to secure their incomes by seeking low risk crops needing little investment and producing returns within a short period of time. One such crop was sweet potato, thanks to a series of factors including access to land with good soil, the possibility in the prevailing climatic conditions of planting sweet potato all the year round, access to seed of a significant number of locally adapted improved sweet potato cultivars with high yields and good eating qualities, and the low cost of production in comparison to other important crops of the valley. The existence of new early and high yielding cultivars with characteristics preferred by consumers in Lima, which has contributed greatly to the expansion of the area dedicated to sweet potato planting, is owed to the efforts of a Peruvian geneticist, Romulo del Carpio Burga. Yields per hectare of sweet potato vary with ecological conditions in the production zones and with size of farm. They range from yields of more than 15 tonnes/ha to more than 30 tonnes/ha (Achata et al., 1988), which is usually obtained only on research stations in other developing countries. The relatively low production cost of sweet potato can be appreciated from Table 8.31. This stems from the low investment made by producers in inflationary inputs of seed, fertilizer and pesticides. The cost of the last two does not exceed one fifth of the total cost of production. This fact coupled with the proximity of the Lima market results in profitability for all types of sweet potato producers in the

Table 8.32. *Frequency (%) of sweet potato consumption in various areas of the Cañete valley, Peru*

	Rural	Semi-urban	Urban	Marginal
Daily or almost daily	54	57	61	17
Weekly	18	10	8	0
Every 2 weeks	9	0	0	0
Monthly	9	14	0	0
Occasionally	0	0	8	0
No information	10	19	23	83

Note:
From: Achata et al., 1988.

region and does not incur losses even in times of low market prices. A fuller description of the system can be found elsewhere (Achata et al., 1988).

Only a very minor amount of the valley's sweet potato harvest is retained by small farmers for home consumption. Although no figures are available for levels of consumption among consumers in the valley itself (as opposed to the city of Lima), a high frequency of consumption has been noted (Table 8.32). More than 50% of rural or small town dwellers eat sweet potato daily or almost daily. It was also found (Achata et al., 1988) that sweet potato was eaten most frequently and by the most diverse methods of preparation by low income groups, due to its low cost in relation to wheat-derived products such as bread.

Distribution and marketing of roots

The market-oriented production of roots (and vines, see next section) has given rise to a high level of organization among sweet potato intermediaries and has generated new sources of employment. Roots are transported from the valley on a weekly basis to Lima, or occasionally to other large urban centres, with especially large quantities moving in the months of greatest demand. Roots are bought by middlemen who are usually responsible for transport and who determine prices paid for the roots at the farm level, according to the prices they are likely to receive from the wholesale merchants in Lima, with whom they maintain regular contact by telephone. The same merchants usually deal with potato. Some of the more important middlemen as well as possessing high capacity lorries also own warehouses for temporary storage of sweet potato and potato. Although the middle merchants may not receive high sales margins because of their dependency on the Lima wholesalers, they still make satisfactory profits due to the large volumes of roots sold (Achata et al., 1988).

According to farmers, prices paid for roots at the farm vary considerably depending on the season, the type of root and its size. On the other hand the middlemen maintain that they establish prices according to four criteria: skin colour, size, shape and keeping quality. In other words merchants appear to be somewhat sensitive to quality and the characteristics demanded by sweet potato consumers. Purple-skinned roots may cost 80–100% more than yellow types. This may be due partly to the greater perishability of yellow roots. Purple roots apparently have a lower water content and are more resistant to damage during handling and transport. Some yellow cultivars also have less commercial appeal than purple roots in that they may be elongated and irregular in appearance whereas purple roots are more rounded in shape. Roots of non-commercial size may cost only one third as much as roots of saleable size and appearance. Moreover, in times of abundance all types of roots may fall in price by three or four times that in times of relative scarcity.

The importance of investigating reasons for such price fluctuations which at times since 1983 have been quite dramatic has been emphasized by several researchers (Achata et al., 1988; Collins, 1989). Large fluctuations can cause uncertainty among buyers and producers and can have a detrimental effect on them. Producers may sustain heavy losses with a drop in price. If prices go up, on the other hand, consumers in low income groups may have to spend more on purchases or eat fewer sweet potatoes. Studies directed to the workings of the sweet potato market and consumer interaction with it could contribute to understanding price and quantity fluctuations, hence aiding efforts to implement a coherent future food policy (Collins, 1989).

Although the Lima wholesale market is supplied with roots from the producers' principal harvest, the internal market for roots in Cañete is supplied by means of a phenomenon known as *rastrojeo* (gleaning). A significant quantity of roots, which remain in the ground after the end of the principal harvest, is lifted by casual labourers, small farmers and cattle keepers who supplement their incomes in this manner. This activity is permanent in nature, highly organized and efficient and represents the main supply of roots to local urban markets and marginal communities. Those gleaners whose main objective is sale of the roots take them from the farm to the various main population centres of the valley and sell them at a price which is somewhat below that of the established market. Casual labourers as well as keeping roots for their own family's consumption sell some to an intermediary or sell directly to the consumer. Alternatively they may harvest roots for a transporter who then makes a further selection or grading in order to sell the roots in urban markets at different prices. The small cattle keeper harvests the *rastrojo* roots to feed to his stock.

It is not known at present whether *rastrojeo* is a social phenomenon in that the farmer is somehow obliged to leave a part of his harvest to the gleaners who are basically casual labourers or landless families, or whether it stems from a lack of technical expertise or equipment applied to sweet potato harvesting. It has been suggested that a diminution of the quantity of roots left for *rastrojeo* could significantly increase sweet potato farmers' yields (Achata et al., 1988). However, any research in this direction should take account not only of farmers' profits, but also of the effect on gleaners' livelihoods.

Vines and roots as livestock feed

Although both roots and vines are used as livestock feed in the Cañete valley, by both dairy and meat producers, roots are much less important for cattle than are vines. Both items are more frequently used as feed in dairy than in meat operations. Use of roots is only occasional (Achata et al., 1988), usually when they are in abundance and if a merchant or producer is inclined to transport them to a large dairy establishment. Roots can also be used for feeding smaller animals such as pigs, chickens and ducks (Collins, 1989). Feed roots are usually culls or those obtained from a second harvest or *rastrojo* as these cost less than good quality roots or other feed products such as maize.

The most important sweet potato product for animal feeding is the vine. As in the case of roots for human consumption its distribution is highly commercialized; the level of organization in the production and sale of sweet potato vines as forage for animals is well developed and increasingly efficient. In contrast to roots, however, only 70% of production of vines is commercialized, a further 20% being utilized by the producers themselves and the remaining 10% being employed as planting material for the following crop. The commercialization of vines has developed to such an extent that it has been seen as the main influence behind the growth and increasing strength of the valley's dairy cattle industry (Achata et al., 1988). Unlike Cajamarca, where pastures are abundant, the use of land for crops in Cañete is so intensive that there is no pasture land for cattle and other livestock. In spite of this, the use of sweet potato vine forage as an alternative to pasture has encouraged a significant increase, during the last 5 years, in the number of small and medium dairy cattle farmers situated in all parts of the valley (Achata et al., 1988). Sale of sweet potato vines, in the prevailing circumstances of scarce pasture and lack of access to alternative animal feedstuffs, is highly profitable and, as in the case of roots for human consumption, has provided new sources of employment. These have involved not only farmers but numbers of different commercial agents. Vines for forage supply only the internal Cañete market and are less profitable than roots when considered on an equal unit of comparison basis (e.g. weight). However, the volume of vines transported daily through the valley is very high. It is especially impressive in the summer months when production in the valley is greatest. A variety of transport, including lorries, pick-up trucks, tricycles and mule carts, is used to move the vines. There is some domestic use of vines by sweet potato farmers who also keep small numbers of cattle. They may also sell their vines to larger establishments. Merchants who also own their own transport may sell vines to large dairy establishments by previous contract. Those with smaller forms of transport may specialize in selling to small and medium dairy farmers. These farmers may also, however, buy their own vines directly from sweet potato farmers. For each type of intermediary the quantities sold and the profit margins achieved are different.

So important is the sweet potato vine in relation to the dairy cattle industry that small farmers who also raise dairy cows are willing to accept some loss of root yield as long as the leaves of that particular cultivar are appetizing to their cattle (Achata et al., 1988). There is a clear preference for certain cultivars especially purple types, which may have lower root yields but are actually preferred by farmers because of their appeal to the cattle.

Vines for use in feeding smaller forms of livestock such as pigs and goats, and domestic animals including poultry and guinea pigs, are also commercialized albeit on a smaller scale. However, there is a constant daily demand and small-scale transporters act as intermediaries to supply local markets on a daily basis.

Northern Highlands

The department of Cajamarca situated in the northern sierra region of Peru provides two contrasting examples of sweet potato consumption patterns. One is characteristic of those rural areas of the region where sweet potato is grown, the other of a typical mountain town.

The countryside

Unlike the northern and southwestern parts of the department, which are basically rice growing, the central and southeast zones are less specialized, growing a variety of mainly subsistence crops, among them sweet potato. Most sweet potato is grown for home consumption, but

Table 8.33. *Uses of sweet potato in rural Cajamarca, Peru*

Use	Percentage of farmers employing each use[a]
Home consumption	96
Sale	85
Gift	52
Food or payment to peasant labourers	34
Exchange/barter	32

Notes:
Prain, G. 1989, unpublished results. Analysis of sweet potato in Cajamarca.
[a] Each farmer utilizes sweet potato in several ways.

may also be sold in small quantities when income is required (Table 8.33). As a gift to neighbours or other farmers it plays an important part in the mutual help which aids in binding the society together as in Papua New Guinea (see above). It may also be used for payment in kind to peasant labourers and for barter involving products not produced within the region. It is not utilized as animal feed due to the presence of adequate pasture land within the area.

An important characteristic of sweet potato production in this region is the number of cultivars grown. One study (Prain, G., cited by Benavides, 1990) identified ten sweet types and eight non-sweet types. This access to a variety of distinct types is important to the cultivator not only in the agricultural sense but in the specific role which each has in the diet. Moreover they are prepared in a great variety of ways. Sixteen ways of preparing the sweet type of root and 12 ways of preparing the non-sweet type have been identified (Prain, G., cited by Benavides, 1990). In general the sweet potato is used in the production zones as an important co-staple consumed daily in place of bread, since these zones are situated at a considerable distance from large population centres (Benavides, 1990). It may also play the role of supplementary or seasonal staple, alternating with cassava, maize or potato. As a festival food it is consumed at carnival time with *chicharon* of pork (Prain, G., personal communication). Due to the high temperatures pertaining in these zones potato production is limited. The sweet potato, especially the non-sweet type, takes the place of potato in the diet.

Cajamarca city

The city of Cajamarca, with 60,000 inhabitants, is situated in the mountainous region of the department of Cajamarca. In contrast to the rural sweet potato production zones potato can be grown freely and is the basic staple food. Sweet potato consumed in the city originates mainly from two coastal production zones, with only a minor quantity coming from the immediate surroundings.

It is eaten mainly as a complementary vegetable on a weekly or twice weekly basis, boiled with *ceviche*, or accompanying soups or different meats. It can also be a breakfast food fried in slices with bread. However, Benavides (1990) mentions that, in the case of very low income consumers, sweet potato may be eaten boiled accompanied exclusively by rice. In this case it could truly be said to be fulfilling the role of co-staple as defined by Poats (see Woolfe, 1987). Middle or higher income groups use it in the form of desserts, purée or pie filling. As in Lima, the highest income groups utilize the roots as a dog food.

Consumers in the city have very marked preferences for type, colour and quality of sweet potato. The yellow or purple-skinned types originating from the coast are preferred, since they are said to be sweeter and more floury than those coming from the vicinity of Cajamarca, which are described as having an insipid taste, a watery consistency and fibrous, hard texture. Although until 1986 sweet potatoes from the coastal zones, when sold in Cajamarca, were lower in price than potatoes or cassava, a rise in transport costs since 1987 has raised the price of roots to a level similar to that of potatoes. It was observed that, although local types were sold in the market at Cajamarca at a lower price than the types originating from the coast, quality was variable and they were not necessarily preferred due to their lower cost (Benavides, 1990). Considering the increased costs of transporting roots from the coastal production zones it would seem that locally produced sweet potatoes have a potential market in Cajamarca if quality could be improved through better handling, and if production of those cultivars demanded by consumers could be expanded.

Processing

The volume of sweet potato used as an industrial raw material for food or other products is very small in Peru, accounting for only about 10% of total production. However, researchers have stressed the potential importance of sweet potato processing in future efforts to stimulate agricultural production and diversify demand (Achata et al., 1988; Beltran Chavez, 1989; Bellatin, 1988; Collins, 1989).

Processed products already made include flour used as a thickening agent, starch which is used to prepare a Limeñan dessert called *mazamorra morada*, sweet potato chips (crisps) and sweet potato bread (see Figure 3.6, p. 131). Processors generally purchase supplies of sweet

Table 8.34. *Experimental and commercial utilization of sweet potato for industrial purposes in Peru*

Use	State of development
Bread making	Commercial
Flour as thickening agent	Commercial
Chips	Commercial
Starch	Commercial
Hot-dog	Experimental[a]
Pastes and purées	Experimental[b]
Biscuits	Experimental[b]
Baked goods	Experimental[b]
Pre-gelatinized starch	Experimental[b]
Electrolyte paste for dry batteries	Experimental[b]
Powdered leaves	Experimental[b]

Notes:
From Achata et al., 1988.
[a] National Agricultural University, La Molina.
[b] Private industry.

potato straight from the wholesale market and are not concerned with the need to store roots which are available all year round. Thus one of the constraints which operates against sweet potato processing elsewhere, namely the difficulty of maintaining a stable year-round supply of roots is not present in Peru. Fifteen processing plants utilizing sweet potato as a raw material exist in Lima. Of these one makes sweet potato flour, two starch, six chips (crisps), three bread and one animal feed. In addition some industries are investigating further possibilities of utilizing sweet potato (see Table 8.34). It will be noted that one of the proposed products is not a food, but an electrolyte paste for use in dry batteries.

Sweet potato starch is made by a firm specializing in maize starch production. It manufactures sweet potato starch when the market price of roots is low. Of the six plants making chips (crisps), three use sophisticated, and three simple, technology. The plant with the highest production makes 10 tonnes of chips per month.

The firm with the biggest sweet potato bread production makes popular types of bread using 35% sweet potato on a wet weight basis (and see Appendix 1). The Agricultural University near Lima and a firm in Cañete also make sweet potato bread but on a smaller scale. Sweet potato can be used either as a mash, as finely grated raw pieces of root, or in the form of a flour, as a substitute for wheat. If a mash is used it can be incorporated at the rate of 35–50% on a wet basis (Beltran Chavez,

1989), whereas flour can be used at a maximum rate of only 10% according to Peruvian researchers (Beltran Chavez, 1989; Bellatin, 1988). Sweet potato flour is also being investigated as a substitute (up to 20%) for wheat flour in biscuits and other baked goods (Achata et al., 1988). Such substitutions would create a demand for sweet potato in excess of the current production (Collins, 1989). This would have an impact not only on the sweet potato industry but would decrease the demand for wheat flour, decrease imports of wheat and slightly improve the balance of payments (Collins, 1989). It has been recommended that proposed processed items should be products, such as the most popular form of bread, directed towards the lowest income sectors of the population. Products such as chips are expensive and are used at present only by higher income groups, for example as an item in school lunch boxes. The use of sweet potato roots as an alternative to maize for a raw material ingredient in balanced feeds for chickens is being investigated (Gonzales, J., cited by Beltran Chavez, 1989).

Among the problems cited by industrialists as being responsible for the low level of sweet potato product development in Peru the one most frequently mentioned was lack of state support. Others included the scarcity of research into the technical processes needed, technical problems involved with using sweet potato roots as a raw material and insufficient information about the actual and potential market for the final products. In cases where the use of sweet potato for processing has been unsuccessful it is attributed to the fact that sweet potato does not possess technical or economic advantages over alternative raw materials and that the market for products is very small, selective and demanding (Achata et al., 1988).

Conclusions

While not consumed in large quantities, sweet potato is found with a high frequency in the diets of Peruvian families, especially those on low incomes. Consumption patterns vary, however, in rural and urban areas and in different ecological zones. In the cities sweet potato is a complementary vegetable whereas it acts as a co-staple in the countryside, where it alternates with other basic foods in the meal-time pattern. Due to its relatively low price it has recently become an increasingly popular food with the low income groups in Lima. If this tendency continues sweet potato could become an important alternative staple food for this section of the population.

Although use of sweet potato in industry is small, an increase could help to diversify utilization and increase demand. A particularly promising area might be that of a substitute for wheat in bread and other baked goods. Greater support is needed for research which should include industrialists' experience to date and more careful technical and economic feasibility studies. Researchers have stressed the important role which sweet potato could play in the future food security and self-sufficiency of the country, provided that appropriate research is financed by state and private institutions.

On Peru's central coast, the sweet potato is intensively exploited at a commercial level for both its roots and vines by all, but especially small, farmers. Commercialization of the entire sweet potato plant has provided new forms of employment and sources of income for people from different social strata. This has resulted not only from the specific ecological and organizational circumstances of the Cañete Valley, but also from farmers' ready access to a large number of cultivars with desirable agronomic, culinary and nutritional qualities. Not only has production increased, but prices have remained low relative to many other basic foods, enabling sweet potato to increase its importance as a food of low income families and thus contributing to their nutritional wellbeing. In addition a flourishing dairy cattle industry has been established in an area with scarcity of pastures and other animal feeds. To consolidate and expand this state of affairs, research must be directed towards avoiding the conversion of sweet potato into a crop dependent on outside technology and on high priced inputs. Among other research priorities must be placed (Achata et al., 1988) development of cultivars with distinct characteristics appropriate to the needs of producers, consumers and cattle keepers. In relation to the last mentioned, the quality of leaves is as important as that of roots. Consumer studies should investigate qualities most preferred by consumers in relation to popular dishes containing sweet potato. The determination of factors influencing price fluctuations in roots and vines is another priority for producers and consumers alike.

Although detailed studies have so far concentrated on the Cañete Valley sweet potato system, other less successful areas, especially in the northern parts of the country, should undergo similar analyses. Such areas as well as possessing the greatest genetic diversity of sweet potato also have a greater potential for increasing production. Efforts to bring about expanded production are attractive for two reasons: sweet potato could have a positive influence on the nutritional status of the population and could help towards the development of a dairy and meat cattle industry in those areas with limited pasture.

Survey results on consumption and utilization

Surveys conducted so far on production and utilization patterns and trends have largely been confined to Asia and the Pacific, where more than 90% of the world's sweet potatoes are grown. Surveys have also

Table 8.35. *Research priorities for sweet potato as perceived by researchers from Asia and the Pacific*

Total ($n=50$)	All Asian countries[a] ($n=29$)	Oceania ($n=14$)
1. Cultivar improvt.	1. Cultivar improvt.	1. Cultivar improvt.
2. Storage/processing	2. Crop protection	2. Crop protection
3. Marketing	3. Storage/processing	3. Crop management
4. Crop protection	4. Marketing	4. Marketing
5. Crop management	5. Crop management	5. Crop physiology
6. Crop physiology	6. Crop physiology	6. Storage/processing

Notes:
From: Lin et al., 1983. *n*, number of replies to the questionnaire.
[a] Bangladesh, India, Indonesia, Japan, Korea, Malaysia, Philippines, Sri Lanka, Taiwan and Thailand.

been limited in their usefulness by being directed at root/tuber crop scientists or interested persons closely connected with sweet potato research, rather than at farmers, producers, marketeers or consumers. They do however, provide initial information on which to build future research.

A survey conducted by the AVRDC in 1982 (Lin et al., 1983) collected data on utilization patterns and constraints in Asia and the Pacific. Completed questionnaires were received from 52 respondents, including those in most key countries and territories in Asia and the Pacific, with the exception of China, Nepal, Pakistan, Burma, Vietnam, and the Maldive Islands, three being from Central and South America and four from the United States. A further survey conducted by the International Potato Center at a workshop on improvement of sweet potato in Asia held in India in 1988 (Horton, 1989) also provided information from sweet potato scientists from 39 locations within 12 countries: Bangladesh, China, India, Indonesia, Korea, Laos, Malaysia, Papua New Guinea, Philippines, Sri Lanka, Thailand and Vietnam.

The most interesting finding of both surveys was the high priority given by researchers to postharvest problems of storage and processing, marketing and, in the case of the International Potato Center survey, limited consumer demand (see Tables 8.35 and 8.36 and Figure 8.17), especially in view of the fact that most respondents were agronomists or plant breeders. Postharvest problems on the whole were seen as more serious than pre-planting or field production problems (Figure 8.17), except in the case of Oceania (Table 8.35).

Table 8.36. *The most important production constraints of sweet potatoes as perceived by researchers in Asia and the Pacific region*

Constraints	% people answering affirmatively	
	Total ($n=48$)	Asia and the Pacific ($n=43$)
Storage, processing and marketing problems	60	63
Lack of improved cultivars	63	60
Lack of proper management	33	35
Adverse environments	40	42
Diseases and insects	73	70

Note:
From: Lin et al., 1983.

In the AVRDC survey, clonal improvement was ranked as having the highest priority for research in all countries and territories surveyed (see Table 8.35). The most important characteristics listed for improvement (see Table 8.37) were eating quality and nutritive value (chiefly protein and beta-carotene content). Appearance was of considerable importance. It was also interesting to note that eating quality and nutritive value were ranked above high yield in the survey. This indicated that sweet potato was considered more as a vegetable or supplemental food than as a major staple.

The AVRDC survey indicated some of the sensory characters most preferred by Asian and Pacific consumers. Red was the preferred skin colour and more than 50% of respondents noted a preference for white or yellow flesh. Deep orange flesh was the least favoured, but light orange was somewhat acceptable; 18% responded favourably to this type. There appears to be some discrepancy between the researchers views of preferences for white flesh and their concern for adequate beta-carotene content. It should be noted that in China, which was not included in the AVRDC survey, the maintenance of a minimum beta-carotene content of 5 mg/100 g (fwb) is a nutritional objective in breeding new lines for human consumption. Yellow to light orange-fleshed cultivars containing 1–5 mg/100 g beta-carotene are common in several parts of China particularly in the form of baked roots sold on the streets as a snack.

A survey conducted among associates of the Caribbean Sweet Potato Working Group (CSPWG, 1986) which received only eight completed

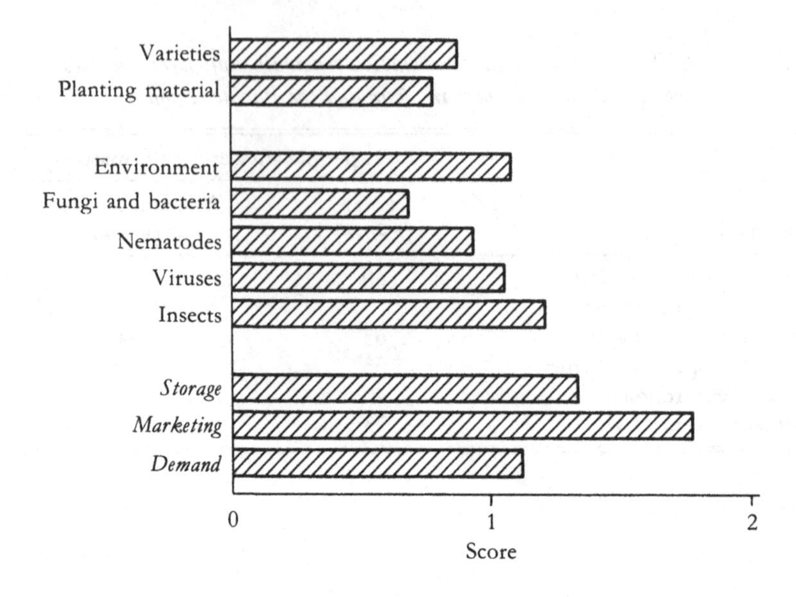

Figure 8.17. Constraints on exploitation of sweet potatoes as perceived by scientists from Asian countries, 1988. Scores: 0, not important; 1, little importance; 2, somewhat important; 3, very important.

questionnaires representing Puerto Rico, Costa Rica, Nevis, Bahamas, Argentina, Philippines and West and East Java also noted the preference for white or yellow flesh among all except East Java, where orange or purple flesh was preferred. It would be valuable to determine whether there is any future in orange-fleshed cultivars as children's food, especially as a backyard crop, particularly if strenuous efforts were made to publicize the nutritional benefits through the media.

Seventy seven per cent of AVRDC survey respondents selected dry or moderately dry mouthfeel as the preferred textural characteristic of roots. Moist mouthfeel was preferred only in the United States. Furthermore, a high percentage (76%) of total respondents felt that sweet or moderately sweet roots were preferable to the low- or non-sweet types. This rose to 82% when only Asian respondents were considered (see Table 8.38). Medium or high sweetness was also felt to be an important characteristic for roots by seven out of eight respondents to the CSPWG survey. Only the respondent from Argentina felt that low sweetness was an important factor. Two out of the three Central or South American AVRDC respondents also considered non-sweetness to be the desired taste characteristic (Table 8.38). The importance of sweetness in Asia was confirmed by answers to questions on the acceptability of sweet potatoes in some areas of Japan and the Philip-

Table 8.37. *The most important characteristics requiring improvement in sweet potatoes for human consumption in Asia and the Pacific*

Characteristics	Frequency of replies	
	Staple	Snack
Eating qualities[a]	17 (1)	17 (1)
Nutritional composition[b]	16 (2)	11 (2)
Insect resistance	13 (3)	6 (5)
High yield	12 (4)	6 (5)
Early maturity	11 (5)	8 (4)
Appearance and uniformity	10 (6)	10 (3)
Disease resistance	7 (7)	3 (6)
High dry matter	5 (8)	3 (6)
High starch	4 (9)	2 (7)
Keeping quality	3 (10)	3 (6)
Plant type	3 (10)	0
Adaptability	1 (11)	1 (8)

Notes:
From: Lin et al., 1983. Numbers in parentheses signify ranking of the importance of characters from the replies.
[a] Flavour, taste, texture.
[b] Mainly protein and beta-carotene content.

Table 8.38. *Preferences for sweetness among researchers from Asia and the Pacific*

Taste classification	Total ($n=52$)	Asia and the Pacific ($n=43$)	Asian countries[a] ($n=24$)	USA ($n=4$)	S and C America ($n=3$)
Sweet (%)	40	39	37	50	33
Moderately sweet (%)	36	37	45	50	0
Not sweet (%)	10	7	0	0	67
Others (%)	14	0	0	17	18

Notes:
From: Lin et al., 1983. *n*, number of responses to questionnaires.
[a] Bangladesh, India, Indonesia, Malaysia, Philippines, Sri Lanka and Thailand. Excludes Japan, Korea and Taiwan.

pines, as noted in the case studies above, where sweet potatoes were liked for their sweet taste. It would be interesting to know whether sweetness is now acceptable in the sweet potato because it has become a complementary vegetable, dessert or snack food, for which sweetness may be an asset. The preferences in countries where sweet potatoes are a major staple food, such as Rwanda, Uganda, Papua New Guinea and the Solomon Islands, should be investigated. It is possible that the cultivars available encompass a wide range of flavours and sweetness levels, which can be exploited for different culinary purposes (see the case study of the northern sierra of Peru, above). Such varied cultivars would provide consumers with a range of taste sensations which do not pall. Taste preferences in Central and South America should be researched further to confirm the liking for non-sweetness and to determine the reason for this. In Asia, however, the low- or non-sweet cultivars may have more future in food processing than as fresh roots for direct consumption.

The highest percentage of AVRDC respondents (43%) preferred uniformly tapered (elliptical) roots, with an elongated shape being least preferred. Improvement of shape was also noted, among respondents to the CSPWG survey, to be of high importance in breeding better clones. The importance of shape to consumers in Japan and Peru has already been described in the corresponding case studies.

Respondents to the AVRDC survey also answered questions about preferences for sweet potato greens. Tender terminal tips of 10–15 cm length were the main portion consumed which indicated a lack of tender cultivars for this purpose. In areas where leaves and petioles were used, cultivars with large leaves and petioles are preferred. Green or purplish-green were the reported preferred stem and leaf colours. Yellow or yellow-green cultivars were not considered acceptable, even though they were usually more tender than the green leafed types. Leaf shape was a minor concern. The most common method of preparation was reported as boiling (47%) followed by frying, steaming and a very small percentage of pickling. Hence the authors considered that the problem of browning of leaves after boiling (mentioned in Chapter 5) should receive consideration.

Producers' and users' perceived attitudes to sweet potato roots and tips were also examined in the AVRDC survey (see Tables 8.39 and 8.40). When only developing Asian countries were considered, it can be seen that a high percentage of respondents considered sweet potato roots to be a low status, poor man's food. A much lower percentage held this opinion when all countries views were included. A lower percentage of respondents in all countries surveyed, and even in Asian developing countries only, thought that sweet potato greens were a low status food. However, a high proportion of countries outside Asian developing

Table 8.39. *Researchers perceptions of how the people who grow and/or utilize sweet potato in Asia and the Pacific feel about the root crop*

Classification	Total (*n*=52)	Asia and the Pacific (*n*=45)	Asian countries[a] (*n*=25)
Low status, poor man's food	25	27	48
Food with no special status	48	47	36
Food with high status	15	16	4
Only suitable for animals	2	2	0
Only grown for industrial use	0	0	0
Other (ceremonial use; export etc.)	10	8	12

Notes:
From: Lin et al., 1983.
[a] Bangladesh, India, Indonesia, Malaysia, Philippines, Sri Lanka and Thailand. Excludes Japan, Korea and Taiwan.

Table 8.40. *Researchers perceptions of how the people who grow or utilize sweet potato in Asia and the Pacific feel about sweet potato tops (tips or leaves)*

Classification	Total (*n*=52)	Asia and the Pacific (*n*=45)	Asian countries[a] (*n*=25)
Low status, poor man's food (%)	12	13	25
Veg. with no special status (%)	19	20	21
Veg. with high social status (%)	4	4	4
Only fed to animals (%)	17	20	28
Not normally consumed (%)	46	40	20
Other (%)	2	3	2

Notes:
From: Lin et al., 1983.
[a] Bangladesh, India, Indonesia, Malaysia, Philippines, Sri Lanka and Thailand. Excludes Japan, Korea and Taiwan.

countries did not use sweet potato greens as human food and a significant proportion fed them only to animals. It should be noted however, that there are some African countries where the use of sweet potato greens for human food is as important as that of roots, for example Liberia.

Factors influencing consumption and utilization

When discussing future trends in consumption and utilization of sweet potato, factors affecting both production and use need to be considered. All too often production problems are addressed to the exclusion of utilization factors. This section attempts to redress the imbalance. Although it is outside the scope of this book to discuss agricultural aspects of sweet potato cultivation, production and utilization are necessarily and closely interlinked. It should not be assumed therefore that in concentrating on postharvest factors affecting consumption and utilization the author was unaware of the preharvest variables associated with these factors.

In summarizing the most important factors which affect sweet potato consumption and utilization patterns and trends, an attempt was made to assess their significance in the light of information provided in the case studies and from further situations described in the literature. For convenience and ease of referral, these factors have been classed separately. However, many are closely related or interconnected. Where possible such a relationship is indicated. Moreover, certain factors will be of greater significance in some countries than in others. The major objective was to indicate promising lines of investigation and positive goals to which researchers should aim in order for sweet potato to realize its potential in future agricultural and food programmes. The main factors considered were:

- Availability
  - Seasonality of supply
  - Handling and transport
  - Storage
  - Industrial use and animal feeding
- Price
- Quality
- Consumer preferences
- Social stigmas
- Promotion

Roots

Availability

A food will not occupy an important position in the diet or be used frequently even in minor quantities if it is not readily available. A low per capita consumption per se does not necessarily imply low or declining demand for a commodity. A declining consumption may also reflect the influence of changes in supply. A number of surveys have indicated the importance of availability in the preference for, or acceptability of, sweet potatoes (Alexander, 1969; Villareal, 1977; and see Japan, above).

Many developing countries are far from self-sufficient in food and consequently have to import to make up the deficit. If government policy towards importation is to be changed and the sweet potato production encouraged, sweet potato must be available to consumers in the same fashion as the imported alternative would have been. An important role of sweet potato is often as a seasonal or supplementary staple (see Table 8.1). It could be used in this role wherever needed if it was available to people at the appropriate times in necessary quantities.

Supplies of sweet potato may be more accessible to consumers in rural rather than urban areas, or within, rather than outside, production zones, due to a poor distribution system. In Trinidad, for example, the relatively high consumption of potatoes was attributed to their availability. In contrast to potatoes, sweet potatoes and other root crops had more limited channels of distribution and more limited distributive outlets (Alexander, 1969). The dietary role of sweet potato often changes as a result of urbanization. In large towns and cities sweet potato is no longer a staple food but a vegetable or snack. This changing role may be accompanied by an alteration of its price relative to other foods which puts it beyond the reach of low income consumers, especially in times of scarcity. In India, for example, sweet potatoes are inexpensive in many rural producing areas but they cost more than potatoes in Delhi. Moreover, due to storage problems they are not available in Delhi during nearly half the year (Jaraith, 1989). The difficulties of distribution and maintenance of supplies are associated not only with the seasonality of production but also the high perishability of sweet potato roots.

Seasonality of supplies

Pronounced production seasons may create alternating glut and dearth situations, associated sharp price fluctuations and uncertainty among producers, consumers and processors. This problem has been emphasized by researchers (Wilson, 1984; Siddique and Rashid, 1989; see also

Japan and China, above). According to a recent international survey of researchers (Horton and Gallegos, 1990), the single most important constraint to expanded utilization of sweet potato is the instability of supplies and prices. The key to increasing availability lies as much with extending the supply season as with increased production. There are several possible ways that this could be achieved. The use of additional land during the dry season by irrigation could extend the growing season in some countries. The provision of storage facilities could help to maintain roots in marketable condition for a longer period of time. The processing of roots would convert them into a dry, storable and more easily transportable form such as chips or flour, which could be used as such or processed further. In many developing country situations where production is highly seasonal, there is little possibility of increasing per capita consumption of sweet potato unless increasing production is accompanied by such measures.

Handling and transport

Another important factor affecting national and regional availability is *postharvest loss* associated with poor harvesting, handling, transport and packaging. The sweet potato is a high moisture perishable commodity susceptible to damage and disease (see Chapter 5). Until recent times most sweet potatoes were produced on small farms for household or local use. Perishability was of little importance as sweet potatoes could be stored in the ground and harvested as required. However, it has increased in significance with the need to transport sweet potato long distances to large urban wholesale markets. Lack of feeder roads and inadequate means of transport result in wastage where roots cannot be moved to market in quantity immediately after harvest. Too often, sweet potatoes arrive at the market in poor condition through rough handling combined with inadequate protection from damage during transport (Figure 8.18). As a result there is a considerable loss of produce between the farm and the consumer, who ultimately pays higher prices for surviving roots. Poor appearance and relatively high prices combine to reduce the image of sweet potato in consumer eyes. These are problems which affect not only regional marketing but also attempts to export produce abroad. There is some evidence that cultivars vary in their susceptibility to deterioration (see Chapter 5). Thus cultivars more resistant to postharvest handling could be identified where there is a need to transport them long distances to regional or overseas markets. Careful handling at all times during the distribution chain could help to minimize losses. Rapid transformation into processed products would also reduce wastage. It has been argued that increasing production of

Figure 8.18. Sweet potatoes near Kuningan, West Java, being loaded for transport to Jakarta. Lack of careful handling is often the cause of damage to roots leading to loss of quality and customer confidence (M. Potts).

tropical root crops such as sweet potato will lead to higher percentage losses unless the postharvest system of handling these roots is considerably improved (Wilson, 1984).

Storage

Storage plays a role in increasing availability by damping fluctuations in supply and prices. It was seen by Asian scientists (see Figure 8.17) as one of the most important constraints to sweet potato utilization. However, bulkiness and perishability of roots makes storage expensive and risky. In many tropical areas where sweet potatoes are produced losses may be high unless roots can be maintained in high cost temperature- and humidity-controlled environments. Investment in this type of store may be applicable to areas with very pronounced seasonal surges in supplies. Producers or market agents will not, however, store sweet potatoes unless the expected price rise will cover storage costs and losses. The erection of large government-built stores may not be the answer to stabilization of supplies if storage costs outweigh the eventual price received for the produce.

Where large investment in storage is not practical, there is still a strong justification for research into simple technical improvements designed for on-farm storage. The identification of cultivars with better keeping qualities, simple methods for more purposeful curing and improvements in existing traditional storage methods can all be investigated (see Chapter 5).

An alternative to consider where technical difficulties of fresh root storage cannot be surmounted is that of processing into less bulky storable dried products. This might be especially applicable to subsequent utilization of sweet potato as a raw material for industry or as an animal feed.

Industrial use and animal feeding

Researchers often stress processing as a key to promoting production and utilization of sweet potato. Processing of sweet potato into a local alternative to, or partial substitute for, imported foods or animal feeds could justify increased interest on the part of planners and policy makers. The significance of processing is three-fold:

Processing can extend the period for which supplies are available (even beyond the inter-season storage period).

Processing can enhance the diversity and convenience of sweet potato-based foods to include, for example flour (bread, baked goods, noodles), instant flakes (breakfast food), chips (snacks) etc.

Processing can lead to better utilization of the whole crop in terms of human food, industrial raw material and animal feed.

The many possibilites which sweet potato presents for food and non-food production and the factors to be considered in future research programmes are fully discussed in Chapter 6.

In addition to the use of processed roots for human food there is a great potential for their use as an animal feed in many tropical developing countries. Moreover, a supply of roots for feed purposes ensures a supply of food which could be diverted for human use in a famine or an emergency. However, as in the case of sweet potato as an industrial raw material, its adoption for feed will depend on how economical is its use compared with that of competing crops. The major need is to keep the price low; production efficiencies must be increased in tandem with the use of specialized cultivars. The quantity of energy supplied by the sweet potato must be maximized if it is to compete with other feed ingredients based on cereals (see Chapter 7). In addition, a country growing sweet potato as an energy source for animal feed must also be able to grow a cheap source of protein to supplement animal diets based on sweet potato. In many cases enhanced utilization of sweet potato which could come to pass through improved production strategies will only succeed where there is a determination on the part of policy makers to promote local, as opposed to imported, feed ingredients.

Price

Studies in several countries have shown that sweet potato consumption is highest in the low-income groups (Alkuino, 1983; Calkins, 1978; Duell, 1989; Mathia, 1975). The importance of sweet potato to such groups has frequently been mentioned and was illustrated by the situation in Peru. Purchases of sweet potatoes do not depend only on their price per se, but on their price in relation to other foods. Again, the Peru case study suggested that low income families in Lima have increased their consumption of sweet potatoes in recent years in response to price reductions for sweet potatoes relative to potatoes.

Urbanization makes it necessary to transport roots from production areas and market them to consumers. Transportation, marketing intermediates, handling and processing increase the cost of sweet potato to consumers. This can be exacerbated by inefficient marketing systems (see Philippines, above), especially where roots are produced by a large number of small farmers who take only small and sporadic quantities to market. Waste due to postharvest losses raises prices of sound roots. Seasonality of supplies can also cause large price fluctuations (see China, above). Such fluctuations are sources of uncertainty for farmers and buyers and can have a detrimental effect on the food intake and nutritional status of low income consumers.

Researchers have also shown that consumers are sensitive to value for money in terms of the edible portion they receive for a unit price. Sweet potatoes are generally peeled before cooking. The peelings represent waste unless they are fed to animals which are later consumed. Wastage is increased when roots are difficult to peel because they are irregular in shape, or when damaged or diseased parts of low quality roots have to be removed as well as peel. It has been estimated that the edible material retained after peeling a unit weight of sweet potatoes was about 78% of that of an equal weight of rice which has an edible yield of 100%. Thus although the actual price of sweet potatoes in a Trinidad market was slightly lower than that of an equal weight of rice, the price adjusted to account for wastage was significantly higher than that of rice (Alexander, 1969). Consumers may only purchase sweet potatoes in preference to other commodities of higher edible yield if the actual price ratio is favourable to sweet potatoes.

The central coast of Peru presents a case where sweet potato prices have been kept low relative to those of potatoes and wheat products.

Figure 8.19. Market conditions in many countries do little to enhance the sweet potato's image in consumer eyes. Sweet potatoes piled up on the ground in a Kampala market, Uganda (M. Iwanaga).

This has been achieved through increased production, low production costs, an efficient marketing system and the large volumes of roots handled, so that producers and merchants can still make a profit in spite of low markups.

In some countries, consumption has been shown to rise with incomes in the lowest income groups and then fall again as incomes increase further (Diaz de Sumaza and Roderiguez de Zapata, 1972; see also Philippines, above). This indicates that increasing production efficiency and lowering the price of sweet potatoes would benefit lower income groups but may have less effect on increasing consumption among higher income groups.

Quality

The rapid deterioration of sweet potato after harvest and its susceptibility to damage during postharvest handling have been mentioned. The sight of dirty, blemished or diseased roots in the market erodes consumers confidence in locally produced food and promotes an inferior image or status in relation to other commodities (Figure 8.19). Mixtures of roots of widely varying shapes and sizes may also be undesirable in consumer eyes, due to preferences associated with ease of peeling and cooking. Moreover the lack of any grading system, associated with different price levels may increase the consumers' doubts over the relative value for money of the roots.

Solving problems of deterioration may be easier in some countries than others. Difficulties arise when roots are produced by small farmers from scattered farms who transport the roots to market as best they can by head, animal or bicycle. Where possible, however, information about the need for careful handling at all stages and the use of appropriate packing for transportation of roots over long distances should be made available to all those who are links in the distribution chain. The advantages of washing, drying and curing of roots, where facilities or climatic conditions are suitable, should also be stressed. However, it should be noted that at least one researcher found that assuming washed roots to be regarded as superior in quality was erroneous. His experience indicated that some dirt on the roots was desirable to hide blemishes and scars which, although not affecting quality in any way, were regarded by the consumers as indicative of damaged and unsound roots. The washed roots in fact had poorer eye appeal and a shorter shelf life than unwashed roots (Alexander, 1969).

A system of grading by soundness and size (Figure 8.20) would raise the appeal of sweet potatoes for fresh consumption in consumer eyes and also provide a source of cull roots selling at lower prices for feed/processing purposes (Figure 8.21).

Consumer preferences

Efforts to introduce new cultivars into a region should initially take strong consumer preferences into account. However productive a cultivar, there is no point in introducing it to an area where no one wants to eat it. Preferences may be based not only on sensory qualities of skin and flesh colour, flavour, sweetness and texture, but also on shape and size (for Asian consumer preference, see Survey results, above). Several examples of the existence and type of consumer preference can also be found in the case studies.

General assumptions thought to fit the preferences of an entire population may be invalid when applied to different ethnic groups within a population or when sweet potatoes are used for varying types of preparation. Thus, a dryish mouthfeel, white-fleshed sweet potato might be preferred for a main savoury dish, but a more moist, yellow or orange-fleshed type could be more acceptable for a baked snack or dessert. In countries with street vendors and many food service outlets quantities of roots used as snacks, or fast foods might be considerable (as in China). Furthermore sweet moist orange cultivars might be highly acceptable to

a

b

Figure 8.21. Good quality sweet potatoes (right) in a Mauritian market fetch twice the price of damaged and shrivelled roots (left) (J.A. Woolfe).

children even in societies where they are unacceptable to adults. It has been suggested that Asian consumers, who prefer dryish white-fleshed cultivars, might find the more nutritious carotene-rich cultivars acceptable if dry mouthfeel could be combined with yellow/orange flesh. An expressed preference for a certain trait which exists mainly in a particular income group could be extremely important if a high percentage of roots are bought by that income group (see Peru, above). Studying the reasons for preferences for roots originating from one production area over another could provide insights into needed improvements in quality as well as characteristics.

Surveys to determine major patterns of consumption, methods of preparation and above all consumer preferences should be conducted among consumers themselves in any given area. This will aid planning by plant breeders, producers and buyers alike for development of appropriate supplies of the most popular cultivars. If customers can obtain the type of root that they want in the market they will be encouraged to consume more.

Figure 8.20. (a) Sweet potato roots being graded by weight in Japan (J.A. Woolfe). (b) Careful hand packing of selected roots into cardboard boxes, as at this packing station in Japan, ensures that roots reach customers in prime condition (J.A. Woolfe).

Consumer preferences are expressed not only for particular cultivars of sweet potato, but for sweet potato above other foods or vice versa. The influence of price, quality and availability on this type of preference have been mentioned in the preceding discussions.

Social stigmas

The inferior status of sweet potato as a 'poor man's food', a 'famine' food or a 'fall-back' food has been cited by researchers as a significant constraint to increased sweet potato consumption (Tsou and Villareal, 1982; Sheng and Wang, 1988). Examples have been given of nationalities, such as the Chinese, Filipinos, Taiwanese and Japanese who resist eating sweet potato because of its associations with war and hardship. The situation is further aggravated in some countries, such as Taiwan, where much sweet potato has traditionally been used as pig feed, or in income groups which regard it mainly as a dog food, as in Peru. Surveys among scientists suggest that the image of sweet potato as an inferior food persists in some developing Asian countries. Others have suggested that sweet potato is in general 'acceptable', but of a lower status than rice and wheat products (Villareal, 1977). The true significance of such attitudes needs to be researched more fully. The importance of social stigmas may well have been overemphasized. People in countries which survived on potato during the Second World War, for example, have not developed an aversion to potatoes. In countries where sweet potatoes have been used in times of hardship and war, for example Japan, a younger generation no longer maintains these associations. More than 100 families questioned in two different areas of Japan found sweet potato highly acceptable (see Japan, above). Contrary to expectations, a high percentage of respondents to a survey in two areas of the Philippines regarded sweet potato as a food for all (Table 8.17). Earlier rejection of sweet potatoes and other tropical root crops on the grounds of inferiority or associations with poverty in the Caribbean Community (CARICOM) region is said to be outdated (Wilson, 1984). Strangely, a dual situation has now arisen in some countries, for example Taiwan, where sweet potatoes are still regarded as an inferior food in rural areas, but are being served as a luxury food in urban restaurants.

Stigma is concerned not only with 'image' or 'status'. The phenomenon of taboo foods is common to cultures in several parts of the world. It is frequently associated with the female reproductive cycle, especially pregnancy and lactation, often to the detriment of female nutritional status. There is a strong taboo against eating sweet potato for the whole of a woman's lactation period in parts of Tamilnadu, India. Among roots and tubers sweet potato is the most dreaded for its believed harmful effects (Ferro-Luzzi, 1974), something which may be associated with Hindu influence or may be attributed to its distinctive colour and sweetness compared to other roots and tubers. However, such a reference to the sweet potato as a taboo dietary item appears only once in the literature as far as could be ascertained and it seems unlikely that this phenomenon would be a significant constraint to consumption in most societies.

The sweetness of sweet potato has been put forward as a reason for its lack of popularity in areas where more bland staples hold sway. Sweetness is obviously not a barrier to high levels of consumption in parts of Africa and Papua New Guinea. Moreover, sweet potato used to be acceptable as a staple food in many countries where its role has changed with changing social circumstances. Surveys suggest that sweet potato is enjoyed or preferred for its sweetness, although it is likely that this could be associated with its role as vegetable or snack, and that among many people it would still not be acceptable as a staple for this reason. Sweetness is often associated in peoples minds with a soggy or moist texture. If sweetness and dryness could be combined, moderate sweetness in a staple might be more acceptable. However, in some areas a sweet taste undoubtedly does limit the inclusion of sweet potato in the diet every day; in the Dominican Republic, for example, everyone interviewed about sweet potato indicated that its taste palls if it is eaten too often (Boy and Horton, 1988).

The discovery of low or non-sweet cultivars enlarges the spectrum of possible preparations and products and increases the versatility of the root further. Even where sweet potato is used on a seasonal basis more out of necessity than desire, such low sweet cultivars might have an increased acceptability. Moreover, the use of non-sweet cultivars in food processing might reduce prejudices against sweet potato products which at present are regarded as inferior to similar products made from, for example, potatoes.

Flatulence, with its associated social unacceptability, has been cited as a possible cause for sweet potato's dwindled popularity in some quarters (Tsou and Villareal, 1982). Though there is little documentary evidence for the significance of flatulence in sweet potato consumption, the presence of flatulence factors has been investigated (see Chapter 4). If they are eventually identified, it may be possible to select cultivars where they are absent or at low levels for fresh root consumption purposes. Prejudices against sweet potato by reason of its poor appearance and presentation in market places have already been mentioned and are another reason to improve handling and selling strategies.

Surveys undertaken to determine consumer attitudes and preferences in a given area should uncover any social stigmas which could prejudice increased utilization and should suggest ways of overcoming them. Developing cultivars specialized for human food, industrial use and animal feed is one approach. Breeding programmes for human food use could then emphasize good eating quality with particular attention to consumer preferences. Up-grading the quality of fresh roots by better postharvest handling as suggested above could improve their image. Social stigmas have been cited as a possible reason for the abandonment of sweet potato in favour of higher status foods, especially those based on cereals, as incomes rise. An important factor might be simply the greater number of food choices available to higher income consumers. Alternatively the shift could be in favour of more convenient foods, more easily prepared and stored. If so, scientists should process sweet potato into forms which can challenge the supremacy of existing convenience foods. Advertising or promoting the nutritional virtues of sweet potato especially in terms of vitamin and dietary fibre content could encourage consumption by all strata of society. It is interesting to note that in Japan, where the image of sweet potato as a 'healthy' food seems to have taken hold, consumption levels were found to be greatest in both the lowest and highest income groups (see Figure 8.11).

Promotion

Several researchers have stressed the importance of promotion in increasing the consumption of sweet potato (Tsou and Villareal, 1982; Law et al., 1981). Some have even indicated a particular group, for example children (Law et al., 1981), young home makers (Huang, Epperson and Law, 1980) or low income families (Mathia, 1975) to which this promotion should be addressed. I believe strongly that a strenuous campaign of promotion should be carried out in conjunction with other strategies in any country where it is desired to increase sweet potato production and use. Such a campaign might be funded with part of the savings in foreign exchange made by substituting sweet potato for imported commodities. In particular the nutritional advantages of sweet potato should be made known widely through the media (press, radio and television), poster campaigns and dissemination of information to clinics, schools, hospitals and other institutions. This nutrition information should include not only nutrient content per se, but also value for money in terms of nutrients purchased per monetary unit compared with other readily available foods. A study in a poor Brazilian squatter community, which consumed a very restricted diet based mainly on five items, found that some foods such as sweet potato, providing moderately good value for money in terms of nutrients, were being ignored by health workers giving dietary advice (Reichenheim and Ebrahim, 1986). Extension workers should be encouraged to explore the nutritional possibilites of sweet potato in their areas and adopt it as an item for promotion in health and diet education plans. This could take place through the medium of short training courses on sweet potato aimed at extension or health promotors.

In addition to nutritional promotion, efforts to expand the range of dishes and preparations which can be made with sweet potato for use not only in the home but also in food service outlets and large institutions could be initiated. Where school meals programmes exist, the use of cultivars rich in beta-carotene could promote good nutritional status and enjoyment of sweet potato among the young. Home economics classes in schools and colleges could feature sweet potato from time to time. Attractive, eye-catching posters and leaflets can combine nutrition information and recipes.

Tops

Little research has so far been carried out on the utilization of sweet potato tops. The reasons for their total rejection as human food in some societies or underexploitation in others have not been explored. Their advantages of dependability in the face of adverse climatic conditions, relative cheapness in comparison to other green vegetables, high nutritional value and utilization in fresh and preserved form as an animal forage have already been discussed, but are infrequently promoted in agricultural or food programmes. With the exception of parts of China, Peru and Indonesia, where vines are commercialized as animal feed, and the Philippines where tips are a commonly marketed vegetable, the green tops hardly enter the market economy.

Removal of some of the constraints affecting sweet potato root consumption could contribute to an increase in the popularity of tops. *Consumer attitudes* to their consumption could be explored. They appear to be regarded as an inferior vegetable, but this could be connected with the failure to identify cultivars with tender leaves which maintain their colour on cooking. The *cooking and eating qualities* of different cultivars must be investigated in a local context to find those with qualities most attractive to consumers. Prejudices might also be overcome (and nutritional value preserved) by *improving market presentation*, including handling, packaging and shading to prevent wilting, drying and darkening. Increased interest might be aroused in tops by finding ways to *preserve* them such as drying, salting or pickling. As in the case of roots, promotion of their *nutritional qualities* allied with efforts to find new and *varied methods of preparation* could help to raise their profile in consumer eyes. They should particularly be included in the home garden programmes of extension and health workers.

If sweet potato tops are to become a marketable commodity, *problems associated with harvesting* must be overcome. These include the development of bush types to increase the possibility of mechanical harvesting in bulk. Breeding tender cultivars would extend harvesting to a greater part of the plant instead of confining it to the tips alone. Where plants are to be used in a dual purpose system for root and top (roots for human consumption and tops for animal feed), breeders need to develop cultivars which not only regenerate top growth rapidly after cutting, but also give an optimum yield of roots and tops.

Perspectives

In the introduction to this chapter, a number of questions were posed about the potential for greater consumption and utilization of sweet potato. The first concerned the type of people consuming and utilizing the crop. Although a definitive answer to this question awaits the completion of in-depth field studies, one or two general conclusions can be offered. There is no doubt that sweet potato is eaten to some extent by a wide range of social groups in every producer country. However, there are significant regional differences in consumption even within producer countries. The sweet potato is still most frequently consumed, and in the greatest quantity, by people living in rural areas, especially those in relatively isolated, mountainous districts. There is a tendency for sweet potato to be eaten more by people on low incomes than those more comfortably off. Some of the problems associated with these inequalities, namely availability, distribution, quality and image, have been discussed above. However, additional information is needed on whether those who consume little sweet potato would like to consume more and in what form.

The only countries who have taken significant steps to utilize sweet potato as an industrial raw material or animal feed are in Asia. They all have breeding programmes that concentrate on yield improvement and the development of cultivars that can compete with other crops as industrial raw materials. Many other countries could utilize sweet potato on a range of scales.

Answers to the second question concerning factors affecting increased utilization have been addressed in the preceding section and in the overview to Chapter 6. Many of the constraints implicit in these factors will readily yield to application of appropriate research efforts.

The immense possibilites which sweet potato holds for improving nutritional status, livestock maintenance and industrial output in many tropical developing countries have all been propounded in depth in preceding chapters and need no further emphasis here. What is now needed is the political and scientific will to recognize these possibilities and overcome the obstacles which at present restrain the realization of sweet potato's full potential. It cannot be overemphasized that much depends on the attitude of government planners and policy makers towards positive discrimination in favour of local crops such as sweet potato at the expense of imported alternatives. Countries must formulate policies, adopt strategies and fully finance and encourage sweet potato research, development and training programmes which at present are all too often lacking or given low priority in agricultural research and development efforts. Evidence suggests that particular emphasis must be placed on the postharvest sector. Effective collaboration at ministerial, institutional, disciplinary, and international levels is indispensible to ensure maximum interchange of ideas and smooth implementation of policies and research findings. Though extensive information already exists, it is not readily available to many scientists or planners, although it is hoped that this book will make much postharvest information more widely accessible. An organized system of data collection for all aspects of sweet potato research, providing bibliographies and abstracts should be internationally funded.

The task is not insuperable. The potential benefits in terms of improved quality of life for human beings make it imperative.

References

Achata, A., Fano, H., Goyas, H., Chiang, O. and Andrade, M. 1988. [*The sweet potato (Ipomoea batatas L. (Lam.)) in the Cañete valley of Peru. Analysis of the present situation and of the research requirements of producers and consumers*] Spanish. International Potato Center/INIAA, Lima.

Alcober, D.I. and Parrilla, L.S. 1987. Gender roles on sweet potato production, processing and utilization in Eastern Visayas, Philippines. Paper presented at an International Sweet Potato Symposium, 20–26 May, ViSCA, Baybay, Leyte.

Alexander, M.N. 1969. Some factors affecting the demand for starchy roots and tubers in Trinidad. In: Tai, E.A., Charles, W.B. and Haynes, P.H. (eds.), *Proceedings of an International Symposium on Tropical Root Crops*, St Augustine, Trinidad, 1967, pp. 45–56.

Alkuino, J.M. 1983. Factors affecting household demand for sweet potato in two regions of the Philippines. *Ann. Trop. Res.* **5** (1): 29–37.

Alvarez, M.N. 1987. Sweet potato and the African food crisis. In: Terry, E.R., Akoroda, M.O. and Arene, O.B. (eds.), *Tropical root crops. Root crops and the African food crisis*, Proceedings of the Third Triennial Symposium of the International Society for Tropical Root Crops – Africa Branch. IDRC, Ottawa, pp. 66–9.

Aviguetero, E.F., San Antonio, F.V., Serrano, I.G., Castillo, H.A. del and Cabilangan, C.K. 1976. *Regional consumption patterns for major foods.* Special Studies Division, Department of Agriculture, Diliman, Quezon City.

1977. *Proportion of families using selected foods by region and income*. Special Studies Division, Department of Agriculture, Diliman, Quezon City.

Bailey, K.V. 1963. Nutrition in New Guinea. *Food Nutr. Notes Rev.* **20** (7–8): 1–26.

Bellatin, A. 1988. [The hour of sweet potato. A feasible proposal to lower the dependence on external supplies of wheat] Spanish. *Agronoticias* **108** (October).

Beltran Chavez, N. 1989. [Study of the commercialization, consumption and processing of sweet potato in the metropolitan area of Lima] Spanish. Draft, available from the Social Science Department, International Potato Center, Lima.

Benavides, M. 1990. [Notes on the production and consumption of sweet potato in Peru] Spanish. Social Science Department, International Potato Center, Lima. [Mimeo]

Bourke, R.M. 1985. Sweet potato (*Ipomoea batatas*) production and research in Papua New Guinea. *Papua New Guinea J. Agric. Forest. Fish.* **33** (3–4): 89–108.

Boy, A. and Horton, D. 1988. [Sweet potato cultivation in the Dominican Republic: potential, technical problems and socio-economic factors] Spanish. Report, International Potato Center, Lima. [Mimeo]

Brookfield, H.C. and White, J.P. 1968. Revolution or evolution in the prehistory of the New Guinea highlands: a seminar report. *Ethnology* **7** (1): 43–52.

Calkins, P.H. 1978. *Vegetable consumption patterns in five cities of Taiwan*. AVRDC Tech. Bull. No. 5 (78–94), Shanhua, T'ainan.

Chandra, S. and de Boer, A.J. 1975. Root crop diets in two Sigatoka valley villages. *Trop. Root Tuber Crops Newsl.* **8**: 19–22.

Clarke, W.C. 1977. A change of subsistence staple in prehistoric New Guinea. In: Leakey, C.L.A. (ed.), *Proceedings of the Third Symposium of the International Society for Tropical Root Crops*, IITA, Ibadan, 1973, pp. 159–63.

Collins, M.I. 1989. Economic analysis of wholesale demand for sweet potatoes in Lima, Peru. M.S. thesis, University of Florida.

CSPWG 1986. Report No. 3. Tropical Agriculture Research Station, Mayaguez, Puerto Rico.

Diaz de Sumaza, Z. and Roderiguez de Zapata, L. 1972. Estimation on the income elasticity of houshold demand for sweet potatoes, *Ipomoea batatas* and yams, *Dioscorea* spp. *J. Agric. Univ. Puerto Rico* **56** (4): 432–8.

Duell, B. 1983. Anthropological problems connected with the introduction and diffusion of the sweet potato into Japan (I). *J. Int. Coll. Comm. Econ.* **28** (9): 47–62.

1984. Anthropological problems connected with the introduction and diffusion of the sweet potato into Japan (II). *J. Int. Coll. Comm. Econ.* **29** (3): 51–73.

1985. Post-World War II change in sweet potato production and use in Japan. *J. Int. Coll. Comm. Econ.* **32** (9): 43–64.

1989. Variations in sweet potato consumption in Japan. *J. Tokyo Int. Univ.* **39**: 55–67.

1990. Ways of eating sweet potatoes in Japan. In: *Tropical root and tuber crops changing role in a modern world*, Proceedings of the Eighth Symposium of the International Society for Tropical Root Crops, 30 October–5 November, 1988, Bangkok (in press).

Duke, J.A. and Ayensu, E.S. 1985. *Medicinal plants of China*. Vol.1. Reference Publications Inc., Michigan.

Durr, G. 1983. *Potato production and utilization in Rwanda*. International Potato Center Social Science Department Working Paper 1983–1. International Potato Center, Lima.

Dwyer, P.D. 1985. Choice and constraint in a Papua New Guinea food quest. *Hum. Ecol.* **13** (1): 49–70.

Engel, F. 1970. Exploration of the Chilca Canyon, Peru. *Curr. Anthropol.* **11**: 55–8.

Ewell, P.T. 1988. Trip report, Rwanda. International Potato Center, Lima. [Mimeo]

Ferro-Luzzi, G.E. 1974. Food avoidance during the puerperium and lactation in Tamilnad. *Ecol. Food Nutr.* **3** (1): 7–15.

Food and Nutrition Research Institute. 1981. *First nationwide nutrition survey, Philippines, 1978*. National Science Development Board and Food and Nutrition Research Institute Publication GP-11, 2nd revision.

Gitomer, C.S. 1987. *Sweet potato and white potato development in China*. International Food Policy Research Institute, Washington, DC.

Golson, J. 1982. The Ipomoean revolution revisited: society and sweet potato in the Upper Wahgi Valley. In: Strathern, A. (ed.), *Inequality in New Guinea highland societies*, Cambridge Papers in Social Anthropology II. Cambridge University Press, Cambridge, pp. 109–36.

Harvey, P.W. and Heywood, P.F. 1983. Twenty-five years of dietary change in Simbu Province, Papua New Guinea. *Ecol. Food Nutr.* **13**: 27–35.

Heywood, P. and Nakikus, M. 1982. Protein, energy and nutrition in Papua New Guinea. In: Bourke, R.M. and Kesavan, V. (eds.), *Proceedings of the Second Papua New Guinea Food Crops Conference*. Department of Primary Industry, Port Moresby, pp. 303–24.

Horton, D. 1988. *Underground crops. Long-term trends in production of roots and tubers*. Winrock International, Morrilton, AR.

1989. Constraints to sweet potato production and use. International Potato Center, Lima. [Mimeo]

Horton, D. and Gallegos, C. 1990. Constraints to potato and sweet potato production and use in developing countries: results of an international survey. Social Science Department Working Paper, International Potato Center, Lima. (Forthcoming)

Horton, D., Prain, G. and Gregory, P. 1989. *Economic rationale for international sweet potato research and development*. International Potato Center, Lima.

Huang, C.L., Epperson, J.E. and Law, J.M. 1980. Sweet potato purchase behaviour: an analysis of the household decision-making process. *J. Agric. Univ. Puerto Rico* **64** (4): 418–23.

International Potato Center 1987. *Sweet potato research in the People's Republic of China*, a CIP/AVRDC/IFPRI study. International Potato Center, Lima.

International Potato Center/ViSCA. 1988. Report of a survey on food needs and preferences of consumers. Phase 1 of a consumer-oriented approach for the development of processed sweet potato food products for low- and middle-income urban groups. Unpublished report, International Potato Center, Lima and ViSCA, Baybay, Leyte. [Mimeo]

Jaraith, M.S. 1989. Progress report on sweet potato in food systems of India. International Potato Center, Lima. [Mimeo]

Jana, R.K. 1982. Status of sweet potato cultivation in East Africa and its future. In: Villareal, R.L. and Griggs, T.D. (eds.), *Sweet potato*, Proceedings of the First International Symposium. AVRDC, Shanhua, T'ainan, pp. 63–72.

Kainuma, K. 1984. Uses of sweet potato starch. *Farming Japan* **18** (5): 36–40.

Kanua, M.B. and Rangat, S.S. 1987. Indigenous technologies and recent advances in sweet potato production, processing, utilization and marketing in Papua New Guinea. Paper presented at an International Sweet Potato Symposium, 20–26 May, ViSCA, Baybay, Leyte.

Kawabata, A., Garcia, V.V. and Rosario, R.R. del. 1984. Processing and utilization of root crops in the tropics. In: Uritani, I. and Reyes, E.D. (eds.), *Tropical root crops: postharvest physiology and processing*. Japan Scientific, Tokyo, pp. 183–203.

Kimber, A.J. 1972. The sweet potato in subsistence agriculture. *Papua New Guinea Agric. J.* **23** (3–4): 80–95.

Law, J.M., Fielder, L., Huang, C.L. and Epperson, J.E. 1981. Sweet potato purchases in relation to demographic characteristics of consumer households. *J. Food Distrib. Res.* **12** (1): 143–8.

Lin, S.M., Peet, C.C., Chen, D-M. and Lo, H-F. 1983. Breeding goals for sweet potato in Asia and the Pacific. A survey on sweet potato production and utilization. *Proc. Am. Soc. Hort. Sci. Trop. Reg.* **27B**: 42–60.

Mathia, G.A. 1975. An economic evaluation of consumer characteristics affecting sweet potato consumption. *J. Am. Soc. Hort. Sci.* **100** (5): 529–31.

McDowell, J. 1970. Food utilization in Uganda. Protein project interim report No. 3, prepared for the Nestlé Foundation, Makerere University, Kampala. [Mimeo]

McMaster, D.N. 1962. A subsistence geography of Uganda. *The World Land Use Survey Occasional Papers* No.2, Geographical Publications Ltd, Bude, Cornwall.

Medina, J.P., Mendoza, C.F., Sungcaya, N.G., Horigue, C.T. and Atega, P.C. 1979. *Sweet potato socio-economic and marketing study, Philippines*. Special Studies Division, Planning Service, Ministry of Agriculture, Diliman, Quezon City.

O'Brien, P.J. 1972. The sweet potato: its origin and dispersal. *Am. Anthropol.* **74** (3): 342–65.

Oomen, H.A.P.C., Spoon, W., Heesterman, J.E., Ruinard, J., Luyken, R. and Slump, P. 1961. The sweet potato as the staff of life of the highland Papuan. *Trop. Geogr. Med.* **13**: 55–66.

Orr-Ewing, A. 1983. Papua New Guinea: what people eat. *Nutr. Health* **2** (1): 26–32.

Pascual, N.P., Abamo, A.P. and Binongo, M.S.G. 1987. Economic tests for profitability, marketability, and alternative uses of sweet potato in the Philippines. Paper presented at an International Sweet Potato Symposium, 20–26 May, ViSCA, Baybay, Leyte.

Philippine Root Crops Information Service and PCARRD. 1986. *State of the art abstract bibliography: sweet potato research*. Crops Bibliography Series No. 9, PCARRD, Los Baños, Laguna.

Pospisil, L. 1963. *The Kapauku Papuans of West New Guinea*. Holt Rinehart and Winston, New York.

Reichenheim, M. and Ebrahim, G.J. 1986. Obtaining best value for money in nutrition: an emerging new priority for the urban poor. *J. Trop. Ped.* **32** (2): 93–6.

Sakamoto, S. 1984. Dissemination to Japan and breeding of sweet potatoes. *Farming Japan* **18** (5): 14–20.

Santos, C.L.G. 1977. *Sweet potato marketing in the Philippines*. Department of Agriculture, Diliman, Quezon City.

Sheng, J. and Wang, S. 1988. The status of sweet potato in China from 1949 to the present. Xuzhou Institute of Sweet Potato, Jiangsu. [Mimeo]

Siddique, M.A. and Rashid, M.M. 1989. Present status and future prospects of sweet potatoes in Bangladesh. In: *Improvement of sweet potato (*Ipomoea batatas*) in Asia*. Report of the Workshop on sweet potato improvement in Asia, held at ICAR, India, 24–28 October, 1988. International Potato Center, Lima, pp. 7–20.

Sinnett, P.F. 1975. *The people of Murapin*. Institute of Medical Research Monograph Series No. 4. E.W. Classey Ltd., Faringdon, Oxon.

Strahan, M., Wilson, M. and Lindbeck, K. 1989. Solomon Island foods: identification and composition. *Food Austr.* **41** (7): 842–5.

Tavioni, M. 1982. Kumara (sweet potato) in the Cook Islands. In: *Proceedings of the Fifth International Symposium on Tropical Root and Tuber Crops*, Los Baños, Philippines, pp. 67–79.

Thomas, G.S. 1982. Review of the prospects for food processing in Papua New Guinea. In: Bourke, R.M. and Kesavan, V. (eds.), *Proceedings of the Second Papaua New Guinea Food Crops Conference*, Department of Primary Industry, Port Moresby, pp. 408–20.

Thomason, J.A., Jenkins, C.L. and Heywood, P. 1986. Child feeding patterns amongst the Au of the West Sepik, Papua New Guinea. *J. Trop. Ped.* **32** (2): 90–2.

Tsou, S.C.S. and Villareal, R.L. 1982. Resistance to eating sweet potato. In: Villareal, R.L. and Griggs, T.D. (eds.), *Sweet potato*, Proceedings of the First International Symposium. AVRDC, Shanhua, T'ainan, pp. 37–44.

Ugent, D. and Peterson, L. 1988. Archeological remains of potato and sweet potato in Peru. *CIP Circular* **16** (3): 1–10.

Urbino, M.C., Torres, E.B. and Darrah, L.B. 1972. *Consumption patterns for selected vegetables*. Department of Agricultural Economics, University of the Philippines College of Agriculture Staff Paper Ser. No. 124.

Villamayor, F.G. 1989. Sweet potato varieties in the Philippines. *Int. Sweet Pot. Newsl.* **2** (1): 1, 4–5.

Villareal, R.L. 1977. Sweet potato: its present and potential role in the food production of developing countries. *South Pacific Commission Technical Paper* No. 174, pp. 170–80.

Wang, J. 1984. [Development of starch resources from sweet potato] Chinese. *Hunan Agric. Sci.* **5**: 44–6.

Watson, J.B. 1965. The significance of a recent ecological change in the central highlands of New Guinea. *J. Polynesian Soc.* **74** (4): 438–50.

1977. Pigs, fodder and the Jones effect in post-ipomoean New Guinea. *Ethnology* **16** (1): 57–70.

Wiersema, S.G., Hesen, J.C. and Song B.F. 1989. Report on a sweet potato postharvest advisory visit to the People's Republic of China, 12–27 January, 1989. International Potato Center, Lima. [Mimeo]

Wilson, L.A. 1984. Problems of utilization of tropical root crops for food in the CARICOM region. In: Dolly, D. (ed.), *Root crops in the Caribbean*, Proceedings of a Caribbean Regional Workshop on Tropical Root Crops, Jamaica, 1983, pp. 189–97.

Woolfe, J.A. 1987. *The potato in the human diet.* Cambridge University Press, Cambridge.

Yen, D.E. 1974. *The sweet potato and Oceania.* B.P. Bishop Museum Bull. No. 236. Bishop Museum Press, Honolulu.

1982. Sweet potato in historical perspective. In: Villareal, R.L. and Griggs, T.D. (eds.), *Sweet potato*, Proceedings of the First International Symposium. AVRDC, Shanhua, T'ainan, pp. 17–30.

APPENDIX

A selection of sweet potato dishes from around the world

(1 cup = 225 ml)

Dishes especially suited to the needs of young children

The following two recipes are suitable for preparation in a family cooking pot, after which the portions shown below are removed and served separately to the child. Each recipe was devised to provide about 300–350 kcal of energy (approximately one-third of a 2-year-old child's daily need) and the equivalent of 5–6 g of good quality protein.

Papua New Guinea

Ingredient	Quantity
Sweet potato, peeled	1 small 150 g
Coconut milk	2 tbsp 20 g
Groundnuts (peanuts)	10 peanuts 20 g
Sweet potato leaves or other dark green leaves	15 g
Tomato	1 slice 10 g
Spring onion	1 small 5 g
Pineapple	$\frac{1}{4}$ small 100 g

(Cooked volume about 1 cup)

Preparation

1. Cut the peeled sweet potato root into small pieces
2. Wash and chop onion, tomato and green leaves
3. Place a leaf inside a half coconut shell
4. Put the sweet potato and vegetables inside the shell with the groundnuts
5. Squeeze coconut milk over the vegetables

6. Cover with a leaf. Put the other half shell on top and tie tightly into place
7. Boil in a pan of water for nearly 1 hour
8. Open shell and give a mashed portion to the child
9. Serve the pineapple after the meal

From: Cameron and Hofvander, 1983.

Tonga

Ingredient	Quantity
Sweet potato, peeled	100 g
Fresh fish	15 g
Sweet potato leaves or other dark green leaves	40 g
Coconut, shredded	25 g
Banana, ripe	60 g
Salt	To taste

(Cooked volume about $\frac{3}{4}$–1 cup)

Preparation

1. Prepare coconut milk
2. Cut the sweet potato root into pieces and put in a pot with half the coconut milk, a small amount of water and a little salt
3. Cook gently for about 20 min
4. Wash the leaves, cut into pieces and put into a pot with the rest of the coconut milk, a little water, salt and any onion to flavour
5. Cook gently for about 20 min
6. Wash the fish, remove any visible bones, put on top of the vegetables and continue cooking for another 10 min
7. Serve with the sweet potato, which can be mashed with any remaining cooking liquid
8. Serve banana after the meal

From: Cameron and Hofvander, 1983.

Recipes for general use

Sweet potato flour breakfast cereal (Serves 2–3)

Ingredient	Quantity
Sweet potato flour	1 cup
Milk	4 cups
Salt	1 tsp

Preparation

1. Heat the milk over boiling water until it simmers
2. Gradually add the flour, mixing to avoid lumps
3. Stir in the salt
4. Cook for 20 min or until thickened

From: Ruberte et al., 1987.

Iddiapam (steamed breakfast item, South India) *(Serves 2–3)*

Ingredient	Quantity
Sweet potato flour	200 g
Oil	2 tsp
Mustard	½ tsp
Black gram dhal	¼ tsp
Onions, chopped	2 tsp
Green chilis, chopped	1 tsp
Salt	To taste

Preparation

1. Add sufficient water to flour to make a stiff dough
2. Pass through a mould with holes in to obtain strings (like spaghetti)
3. Steam strings for 5 min
4. Season with oil, mustard, black gram dhal, onion, chilis and salt
5. Serve hot

From: Mrs Aruna Seralathan and Mrs Susheela Thirumaran, Tamil Nadu Agricultural University, Coimbatore.

Sweet potato, meat and vegetable soup *(Serves 8)*

Ingredient	Quantity
Sweet potato (cubed)	900 g
Meat (cubed, pork or beef)	450 g
Onion (diced)	1
Tomatoes (diced)	2
Green pepper (diced)	1
Plantains (cubed)	2
Cabbage (shredded)	2 cups
Hot pepper	1
Garlic clove (crushed)	1

Oregano (shredded)	1 tsp
Water	6 cups

Preparation

1. Cook meat in water with herbs and spices until tender
2. Add the vegetables
3. Boil, and then reduce heat to simmer for about 1 hour

From: Ruberte et al., 1987.

Sweet potato cream soup
(Serves 6)

Ingredient	**Quantity**
Sweet potato vine tips (chopped)	4 cups
Milk	2 cups
Flour	2 tbsp
Soup stock	1 cup
Onion (minced)	$\frac{1}{4}$ cup
Salt	To taste

Preparation

1. Boil the chopped leaves until soft
2. Coarsely grind with blender or pestle and mortar
3. Make a white sauce with flour, milk, and soup stock (about 4 cups of liquid)
4. Add the ground leaves, onion and salt to the white sauce

From: Ruberte et al., 1987.

Kumala salad (Western Samoa)
(Serves 2)

Ingredient	**Quantity**
Cooked sweet potato root (kumala), cubed	$1\frac{1}{2}$ cups
Lemon juice	1 tbsp
Oil	1 tbsp
Mint, finely chopped	1 tbsp
Onion, grated	1 small
Orange rind, shredded	1 tbsp
Vinegar	2 tbsp

Preparation

1. Mix all ingredients except mint and orange rind and toss gently
2. Add a little salt and pepper to taste if desired
3. Garnish with mint and orange rind

From: Broderick, 1984.

Sweet potato tops salad (Western Samoa)
(Serves 6)

Ingredient	Quantity
Sweet potato tops	6 cups
Lemon juice	$\frac{1}{2}$ cup
Salt	2 tsp
Soy sauce	$\frac{1}{4}$ cup
Mayonnaise (optional)	1 tbsp

Preparation

1. Blanch tops by dipping in boiling water
2. Chop, then mix with rest of ingredients
3. Chill and serve

From: Broderick, 1984.

Stir-fried sweet potato leaves (South Pacific)

Preparation

1. Cut up leaves
2. Heat a little oil or fat in a pot (1 tsp for every cup of leaves)
3. Add chopped garlic, ginger or other flavouring, and fry for 1 min
4. Stir in cut up leaves and fry, stirring constantly, for about 5 min
5. Add a few tsp water if necessary, cover the pot and steam for 2–5 min. Serve hot

Coconut cream can be added to the leaves for extra flavour and food value. Eating green leaves with coconut cream, margarine, meat drippings or other fat aids bodily absorption of provitamin A from the leaves. Green leaves can be added to any kind of soup or stew, and can be mixed into main dishes to make meat or fish go further.

From: South Pacific Commission Community Education Training Centre, 1983.

Sweet potatoes and rice (Asia)
(Serves 4)

Ingredient	Quantity
Sweet potato, peeled and cubed	2 cups
Rice	2 cups
Salt	1 tsp
Butter or margarine (optional)	3 tbsp

Preparation

1. Boil the sweet potato cubes until just tender
2. Cook the rice until tender
3. Mix the sweet potato cubes and rice and add ½ cup of water and the salt
4. Cook for another 5 min
5. Place in serving dish and top with butter or margarine if desired

Sweet potato croquettes
(Serves 6)

Ingredient	Quantity
Cooked, mashed sweet potatoes	700 g
Milk	½ cup
Crumbs (bread or cracker)	
Cooking oil	

Preparation

1. Combine mashed sweet potato with crumbs and milk. Make balls and flatten them
2. Fry in greased pan until brown on both sides

From: Ruberte et al., 1987.

Tuna-sweet potato cakes (Australia)
(Serves 2)

Ingredient	Quantity
Canned tuna	1 can (200 g)
Cooked, mashed sweet potato	1 cup
Onion, grated	1
Breadcrumbs	½ cup

Preparation

1. Mix all ingredients together
2. Roll into balls and then press flat; cover with breadcrumbs

3. Shallow fry in hot oil, browning on both sides

From Anon., 1979.

Tempura (Japan)
(Serves 2)

Ingredient	Quantity
Sweet potato	100 g
Flour	20 g
Egg	10 g
Water	240 ml
Soy sauce	30 ml
Soup stock	40 ml
Salt	2 g
Radish	20 g

Preparation

1. Peel and cut sweet potatoes into round slices
2. Stir flour, egg and water together into a batter and coat sweet potato slices in this batter
3. Deep fry the coated sweet potato slices in hot oil
4. Mix soy sauce and stock to give *tentsuyu*. Eat sweet potato with grated radish dipped in *tentsuyu*

From: Professor Machiko Ono, Nagoya Women's University.

Sweet potatoes with beans or peas (Uganda)
(Serves 1–2)

Ingredient	Quantity
Dry sweet potato pieces or	1 cup
Fresh sweet potato	1 medium
Dry cow peas/pigeon peas or beans	1 cup
Water	0.5 litre
Salt	To taste

Preparation

1. Boil peas/beans until half cooked
2. Add dry sweet potato/fresh sweet potato cut into small pieces
3. Add salt to taste
4. Cook together until done and fairly dry
5. Mash and serve warm

From: Mwanamugimu Nutrition Services, undated.

Sweet potato pork bun (Taiwan)
(Serves 10)

Ingredient	Quantity
Mashed sweet potato	1.2 kg
Sweet potato starch	2 cups
Salt	½ tsp
Ground pork	0.5 kg
Mixed vegetables	0.9 kg
Dried shrimp	10 g
Onions	To taste
Sesame oil	1 tsp
Soy sauce	1 soup spoon
Wine	½ soup spoon
Salad oil	3 soup spoons

Preparation

1. Mix sweet potato mash, starch and salt and roll into balls
2. Mix ground pork, soy sauce, wine and sesame oil
3. Cook and mash mixed vegetables, add salt and heat to dry a little
4. Mince onion and wash dried shrimp
5. Sauté ground pork, onion and dried shrimp in salad oil
6. Mix sautéed ground pork, onion, shrimp and vegetables and stuff a portion inside each sweet potato ball
7. Steam for 8 min

From: Hualian District Agricultural Improvement Station (recipe leaflet).

San Martin Pie (Argentina)
(Serves 8)

Ingredient	Quantity
Fresh sweet potatoes	1 kg
Onion	0.5 kg
Margarine or oil	100 g
Salt, pepper, and minced chili	To taste
Meat, coarsely ground	0.5 kg
Hard-boiled eggs	3
Seedless raisins	50 g
Green olives, chopped	50 g
Butter	50 g
Powdered sugar (optional)	2 or 3 tbsp

Preparation

1. Wash the sweet potatoes well and boil them until they are just tender
2. Meanwhile, chop the onion and transfer to a casserole with the margarine or oil and salt. Replace the lid and cook until the onion is transparent
3. Add the meat, brown it and remove from the heat
4. Season with salt, pepper and chili
5. Add the raisins and olives, mix and allow to cool
6. Purée the peeled sweet potatoes, add butter and season with salt and pepper
7. Put half the purée in an oven-proof dish, then spread over the meat mixture, layer on top the hard-boiled eggs cut into slices and finally cover with the remaining sweet potato purée
8. Dust the surface with the powdered sugar (this step may be omitted if not desired)
9. Bake the pie in a hot oven until the surface is golden brown

From: Asociación Argentina de Economas y Gastronomas in a recipe leaflet issued by INTA, Estación Experimental Agropecuaria, San Pedro, Buenos Aires.

Sweet potato groundnut mpotonpoto *(Ghana)* *(Serves* 1*)*

Ingredient	Quantity
Sweet potatoes	250 g
Water	$\frac{2}{3}$ cup
Groundnut paste (peanut butter)	1 tbsp
Chopped onion	$\frac{1}{2}$ tsp
Powdered pepper	$\frac{1}{4}$ tsp
Palm oil	1 tbsp

Preparation

1. Wash, peel and cook sweet potatoes until tender
2. Mash into a fine meal and mix well with groundnut paste
3. Put mixture back on the heat and add onion and all seasonings
4. Simmer for 15 min
5. Garnish with palm oil and serve

From: Osei-Opare, 1984.

Sweet potato pancakes
(Serves 2)

Ingredient	Quantity
Sweet potato flour (made with a low-sweet cultivar)	½ cup
Wheat flour	½ cup
Milk	½ cup
Egg	1
Sugar	1 tbsp
Salt	½ tsp
Baking powder	2 tsp
Cooking oil	As necessary

Preparation

1. Mix flours with salt, baking powder, sugar
2. Add the beaten egg and milk
3. Preheat the oil in a frying pan until just smoking
4. Pour mixture into the frying pan until desired pancake size is reached
5. When bubbles form and break on the upper side of the pancake, turn it over and brown the other side

From: Ruberte et al., 1987.

Spiced glazed Louisiana yams (United States of America)
(Serves 6)

Ingredient	Quantity
Fresh moist, orange-fleshed, sweet potatoes	1 kg
or	
Canned, drained orange sweet potatoes	700 g
Brown sugar	70 g
Water, hot	140 ml
Butter/margarine	17 g
Cinnamon	½ tsp
Nutmeg	¼ tsp
Salt	½ tsp

Preparation

1. If fresh sweet potatoes are used, wash them and then steam or cook in boiling water until tender. Cool, peel and slice

2. For canned sweet potatoes, drain syrup off, add butter or margarine to syrup and boil to reduce volume by one third. Slice sweet potatoes
3. For fresh roots, mix sugar, water, butter or margarine and seasonings; let syrup simmer for about 10 min
4. Add seasonings to syrup if canned sweet potatoes used
5. Arrange sliced sweet potatoes in greased baking pans. Pour syrup over them
6. Bake in oven at 177°C (350°F) for 30–35 min

From: Harris, 1963.

Bou-Loi-Muntesh (Thailand)

Ingredient	Quantity
Steamed, mashed sweet potato	1½ cups
Glutinous rice flour	1½ cups
Water	10 tbsp
Coconut milk	3½ cups
Coconut sugar	1 cup
Sucrose	3 tbsp
Salt	1 tsp

Preparation

1. Mix sweet potato, flour and water, knead together and roll into small balls
2. Boil together coconut milk, coconut sugar, sucrose and salt to make a syrup
3. Boil balls of dough in this syrup

From: Sweet potato day, Bangpahan, Thailand, held 26 July 1988.

Hawaiian style sweet potato (Hawaii)
(Serves 5–6)

Ingredient	Quantity
Cooked, peeled, mashed sweet potato roots	5–6 cups
Cream	½ cup
Salt	½ tsp
Butter	2 tbsp
Crushed pineapple	1½ cups
Sugar	1 cup
Fresh grated coconut	To garnish

Preparation

1. Add the cream, salt and butter to the mashed sweet potato
2. Cook the crushed pineapple and sugar in a saucepan for about 10–15 min
3. Put about one-third of the sweet potato mixture into a casserole, top with pineapple and follow with more sweet potato and pineapple. There should be three layers of each
4. Top with the grated coconut
5. Bake in oven at 177°C (350°F) until brown, 20–40 min

From: Na Lima Kokua, 1983.

Sweet potatoes with orange (Israel)
(Serves 4)

Ingredient	Quantity
Sweet potato (orange-fleshed)	900 g
Salt	To taste
Oranges	2 large
Black peppercorns	1 tbsp
Oil	2 tbsp
Butter/margarine	25 g
Cinnamon (ground)	$\frac{1}{2}$ tsp
Fresh parsley, chopped	2 tbsp

Preparation

1. Boil sweet potatoes for about 20 min until almost tender
2. Peel and segment one orange and squeeze the juice from the other
3. Crush the peppercorns
4. Drain sweet potatoes, peel and cut into large chunks
5. Heat oil and butter/margarine in a large frying pan and when hot tip in sweet potatoes
6. Fry, turning occasionally until golden brown and beginning to flake
7. Remove pan from heat, stir in cinnamon, peppercorns, parsley, orange juice and orange segments and salt
8. Mix well, transfer to a warmed serving dish and serve immediately

From: Carmel Information Bureau, London (recipe leaflet).

Dulce de batata *(Argentina)*

Ingredient	Quantity
Sweet potatoes	3 kg
Sugar	750 g per kg of cooked sweet potato flesh
Cloves	4 or 5

Preparation

1. Choose small and sound sweet potatoes, wash them well and cook them in boiling water
2. Drain, cool, and take off the skins
3. Weigh out sugar as given above
4. Place sugar in a pan, cover with water and boil for 5 min
5. Add the sweet potatoes and the cloves
6. Allow to cook on low heat until the sweet potatoes have become a rich golden colour and the mixture has thickened

From: INTA, Buenos Aires (recipe leaflet).

Three sweet potato dessert pastes

Sweet potato paste

Ingredient	Quantity
Mashed yellow/orange fleshed sweet potato	900 g
Sugar	4 cups
Vanilla	½ tsp

Sweet potato pineapple paste

Ingredient	Quantity
Mashed sweet potato (as above)	450 g
Pineapple (crushed)	2 cups
Sugar	4 cups
Vanilla	½ tsp

Sweet potato coconut paste

Ingredient	Quantity
Mashed sweet potato (as above)	4 cups
Coconut milk	1 cup
Sugar	5 cups
Almond extract	To taste

Preparation

1. Mix the basic ingredients except for the vanilla or almond extract
2. Simmer over low heat until sugar is dissolved
3. Add the extracts and mix in
4. Pour into a baking dish and allow the mixture to cook and harden
5. Use when cool or store refrigerated and use when desired as a stuffing in puddings and pies

From: Rubarte et al., 1987.

Imo-Youkan (Japan)

Ingredient	Quantity
Sweet potato	1 kg
Sugar	300–500 g
Agar-agar	16 g
Water	1 cup

Preparation

1. Peel and cut sweet potato into pieces
2. Boil pieces in a large quantity of water until soft
3. Strain, add sugar and maintain over a low heat while kneading into a paste
4. Dissolve agar-agar in 1 cup of water, add to sweet potato paste, mix well
5. Press into a mould, cool
6. When cold and solid, turn out of mould and cut into squares. Serve with tea

From: Professor Machiko Ono, Nagoya Women's University.

Sweet potato flapjacks (Israel)

Ingredient	Quantity
Sweet potatoes (orange-fleshed)	450 g
Butter/margarine	175 g
Salt	½ tsp
Light brown sugar	225 g
Eggs	2
Grated orange rind	2 tbsp
Powdered ginger	½ tsp

Nutmeg	½ tsp
Whipped cream (optional)	
Rum (optional)	2 tbsp

Preparation

1. Peel and finely grate sweet potato
2. Cream butter/margarine and salt and gradually add sugar. Beat until light and fluffy
3. Add the eggs and then beat in remaining ingredients until well mixed
4. Turn into a buttered baking dish 20 cm × 30 cm
5. Cook in pre-heated oven at 180°C (350°F) for 50–60 min
6. Allow to cool, cut into slices and serve topped with rum-flavoured whipped cream if desired

From: Carmel Information Bureau, London (recipe leaflet).

Picarones (Peru)

Ingredient	**Quantity**
Sweet potato	250 g
Yellow squash	250 g
All-purpose flour	500 g
Cinnamon	1 stick
Cloves	2
Anise granules	½ tsp
Yeast	150 g
Water	1 litre

Preparation

1. Into the water put cut and peeled pieces of sweet potato and squash, cinnamon stick and cloves. Boil until sweet potato is tender
2. Remove squash and sweet potato from water and mash. When cold, mix them with the flour and pour sufficient water into the mixture with stirring to form a dough
3. In a separate bowl, boil ½ cup of water with the anise granules for 10 min. Drain and pour into the dough
4. Dissolve the yeast in a little water and add to the dough while kneading
5. After kneading dough well, transfer to a large bowl. Cover bowl with a wet cloth and allow dough to rise until it doubles in size (approx. 2 hours)

6. Knead dough again to remove air and cover it for another 3 hours
7. In a deep frying pan, heat the oil to fry the *picarones*
8. Wet one hand with salted water, take a small piece of dough with the finger tips, forming a hole in the middle (as for doughnuts). Drop into hot oil and use a stick to form the *picaron* while it is cooking in the oil. When it turns light brown, turn it over and brown the other side
9. Wet the fingers each time before forming a *picaron*
10. Serve hot with syrup

Sweet potato bread (Peru) *(Commercial preparation)*

Ingredient	Quantity
Wheat flour	15 kg
Raw, washed, unpeeled, grated sweet potato	5–7 kg
Sugar	1.5 kg
Salt	200 g
Fresh yeast	250 g
Butter	1.2 kg
Dough improver (preparatory mix of dough improvers and raising agents)	150 g
Warm water (maximum 50°C)	7 litres

Preparation

1. Introduce the ingredients into the mixer in the following order: water (6 litres), sugar, salt and wheat flour
2. Switch mixer on to first speed
3. Add the yeast, dough improver and the grated sweet potato, adding the latter in three portions; a little later add the butter
4. Mix for 5–8 min until a homogeneous dough has been achieved and all the water is absorbed
5. Stop the mixer, remove half the dough and switch to a higher speed
6. Add 500 ml of water slowly and mix for a further 8–10 min until an elastic dough is formed
7. Divide the dough into equal portions of desired size
8. Place in proving room at 30°C and 100% relative humidity for 2 hours or until doubled in size
9. Bake at 240°C for 15 min

10. Repeat steps 6–9 inclusive with the other half of the dough

From: SR Industrias S.A., Lima, Peru (as used in Novapan, Lima).

Chaphatis (South India)
(Serves 2)

Ingredient	Quantity
Sweet potato flour	50 g
Refined wheat flour	50 g
Salt	1 g
Water	60 ml

Preparation

1. Make dough with flours, salt and water
2. Knead well and set aside for 10 min
3. Divide into lime-sized balls
4. Roll out into flat rounds
5. Shallow fry and serve hot

From: Mrs Aruna Seralathan and Mrs Susheela Thirumaran, Tamil Nadu Agricultural University, Coimbatore.

Wheatless bread
Paste

Ingredient	Quantity
Sweet potato starch	400 g
Water	2200 ml

Preparation

1. Mix ingredients and boil until translucent
2. Remove from heat and cool slightly

Batter

Ingredient	Quantity
Sweet potato flour	2 kg
Sugar	100 g
Salt	40 g
Cooking oil	

Preparation

3. Add these ingredients to the paste

4. Mix and allow to cool to touch
5. Add 25 g of fresh yeast in 160 ml of water containing 5 g of sugar
6. Mix for 5 min and pour batter into baking tin. Allow to rise
7. Bake at 210°C (410°F) for 35–40 min
8. Allow to cool for 12 hours
9. Slice and serve

From: Mr Morton Satin, FAO, Rome.

Sweet potato spread (Australia)

Ingredient	Quantity
Sweet potato (orange-fleshed)	Unspecified quantity
Tomato	1
Egg	1
Mixed herbs	¼ tsp
Onion	1 medium
Margarine	1 tsp
Breadcrumbs	1 tbsp
Salt and pepper	To taste

Preparation

1. Peel, cook and mash sweet potato
2. Peel onion and tomato, cut up finely and simmer in melted margarine for 10 min
3. Beat egg, add breadcrumbs and herbs
4. Add to onion and tomato mixture and boil for 1 min
5. Remove from stove, add salt and pepper to taste and enough mashed sweet potato to form a desirable consistency for a spread or dip

From: Anon., 1979.

References

Anon. 1979. Sweet potato recipes. *Qld. Agric. J.*, January-February: 94–5.

Broderick, B. 1984. Cooking with coconuts and other Samoan foods. Peace Corps, Western Samoa. [Mimeo]

Cameron, M. and Hofvander, Y. 1983. *Manual on feeding infants and young children*, 3rd edn. Oxford University Press, Oxford.

Harris, M. 1963. Gold in school lunch menus. *School Lunch J.* **17** (10): 22.

Mwanamugimu Nutrition Services. *Good food recipes in child nutrition.* Ministry of Health and Department of Home Economics, Ministry of Agriculture and Forestry, Kampala.

Na Lima Kokua 1983. *Sweet potato,* ('Uala) uses and recipes. Pacific Tropical Botanical Garden, Lawai, Hawaii.

Osei-Opare, F. 1984. The varied uses of sweet potatoes. Home Science Department, University of Ghana, Legon. [Mimeo]

Ruberte, R.M., Martin, F.W. and Herrera, J.L. 1987. The sweet potato cookbook. [Mimeo] (Available from F.W. Martin – see Appendix 2)

South Pacific Commission Community Education Training Centre 1983. *Green leaves. Nutritious Pacific plants,* South Pacific Foods Leaflet No. 6. South Pacific Commission, Noumea Cedex, New Caledonia.

INDEX